Student's Solutions Manual

Jeffery A. Cole
Anoka-Ramsey Community College

Basic College Mathematics

Seventh Edition

Margaret L. Lial
American River College

Stanley A. Salzman
American River College

Diana L. Hestwood
Minneapolis Community and Technical College

Boston San Francisco New York
London Toronto Sydney Tokyo Singapore Madrid
Mexico City Munich Paris Cape Town Hong Kong Montreal

Reproduced by Pearson Addison-Wesley from electronic files supplied by the author.

Publishing as Pearson Addison-Wesley, 75 Arlington Street, Boston, MA 02116.

ISBN 0-321-27938-7

8 9 10 11 12 13 BB 10 09 08 07

Preface

This *Student's Solutions Manual* contains solutions to selected exercises in the text *Basic College Mathematics, Seventh Edition* by Margaret L. Lial, Stanley A. Salzman, and Diana L. Hestwood. It contains solutions to all margin exercises, the odd-numbered exercises in each section, all Relating Concepts exercises, as well as solutions to all the exercises in the review sections, the chapter tests, and the cumulative review sections.

This manual is a text supplement and should be read along *with* the text. You should read all exercise solutions in this manual because many concept explanations are given and then used in subsequent solutions. All concepts necessary to solve a particular problem are not reviewed for every exercise. If you are having difficulty with a previously covered concept, refer back to the section where it was covered for more complete help.

A significant number of today's students are involved in various outside activities, and find it difficult, if not impossible, to attend all class sessions; this manual should help meet the needs of these students. In addition, it is my hope that this manual's solutions will enhance the understanding of all readers of the material and provide insights to solving other exercises.

I appreciate feedback concerning errors, solution correctness or style, and manual style. Any comments may be sent directly to me at the address below, at jeff.cole@anokaramsey.edu, or in care of the publisher, Pearson Addison-Wesley.

I would like to thank Ken Grace, of Anoka-Ramsey Community College, for typesetting the manuscript and providing assistance with many features of the manual; Marv Riedesel and Mary Johnson, of Inver Hills Community College, for their careful accuracy checking and valuable suggestions; and Maureen O'Connor and Lauren Morse, of Pearson Addison-Wesley, for entrusting me with this project.

Jeffery A. Cole
Anoka-Ramsey Community College
11200 Mississippi Blvd. NW
Coon Rapids, MN 55433

Table of Contents

CHAPTER 1 WHOLE NUMBERS

1.1 Reading and Writing Whole Numbers

1.1 Margin Exercises

1. **(a)** 341

   4 is in the *tens* place.

   (b) 714

   4 is in the *ones* place.

   (c) 479

   4 is in the *hundreds* place.

2. The place value of each digit:

   (a) 14,218; ones: 8; tens: 1; hundreds: 2;

   thousands: 4; ten thousands: 1

   (b) 460,329; ones: 9; tens: 2; hundreds: 3;

   thousands: 0; ten thousands: 6;

   hundred thousands: 4

3. The digits in each period (group) of 3,251,609,328:

   (a) 3 is in the billions period,

   (b) 251 is in the millions period,

   (c) 609 is in the thousands period,

   (d) 328 is in the ones period.

4. **(a)** 18 is eighteen.

   (b) 36 is thirty-six.

   (c) 418 is four hundred eighteen.

   (d) 902 is nine hundred two.

5. **(a)** 3104 is three thousand, one hundred four.

   (b) 95,372 is ninety-five thousand, three hundred seventy-two.

   (c) 100,075,002 is one hundred million, seventy-five thousand, two.

   (d) 11,022,040,000 is eleven billion, twenty-two million, forty thousand.

6. **(a)** One thousand, four hundred thirty-seven is 1437.

   (b) Nine hundred seventy-one thousand, six is 971,006.

   (c) Eighty-two million, three hundred twenty-five is 82,000,325.

7. **(a)** 14 million in digits is 14,000,000.

   (b) 13 million in digits is 13,000,000.

   (c) $42, 936 is forty-two thousand, nine hundred thirty-six dollars.

   (d) $28,050 is twenty-eight thousand, fifty dollars.

1.1 Section Exercises

1. 3065; thousands: 3; tens: 6

3. 18,015; ten thousands: 1; hundreds: 0

5. 7,628,592,183; millions: 8; thousands: 2

7. 3,561,435; millions: 3; thousands: 561; ones: 435

9. 60,000,502,109; billions: 60; millions: 0; thousands: 502; ones: 109

11. Evidence suggests that this is true. It is common to count using fingers.

13. 23,115 is twenty-three thousand, one hundred fifteen.

15. 346,009 is three hundred forty-six thousand, nine.

17. 25,756,665 is twenty-five million, seven hundred fifty-six thousand, six hundred sixty-five.

19. Sixty-three thousand, one hundred sixty-three is 63,163.

21. Ten million, two hundred twenty-three is 10,000,223.

23. Three million, two hundred thousand in digits is 3,200,000.

25. Two billion in digits is 2,000,000,000.

27. Eight hundred fifty-four thousand, seven hundred ninety-five in digits is 854,795.

29. Eight hundred trillion, six hundred twenty-one million, twenty thousand, two hundred fifteen in digits is 800,000,621,020,215.

31. The least-used method of transportation is public transportation. 6,069,589 in words is six million, sixty-nine thousand, five hundred eighty-nine.

33. The number of people who walk to work or work at home is 7,894,911. In words, 7,894,911 is written as seven million, eight hundred ninety-four thousand, nine hundred eleven.

1.2 Adding Whole Numbers

1.2 Margin Exercises

1. **(a)** $2 + 6 = 8$; $6 + 2 = 8$

   (b) $9 + 5 = 14$; $5 + 9 = 14$

(c) $7+8=15;\ 8+7=15$

(d) $6+9=15;\ 9+6=15$

2. (a)
$$\begin{array}{rl} 3 & \\ 8 & 3+8=11 \\ 5 & 11+5=16 \\ 4 & 16+4=20 \\ +6 & 20+6=26 \\ \hline 26 & \end{array}$$

(b)
$$\begin{array}{rl} 5 & \\ 6 & 5+6=11 \\ 3 & 11+3=14 \\ 2 & 14+2=16 \\ +4 & 16+4=20 \\ \hline 20 & \end{array}$$

(c)
$$\begin{array}{rl} 9 & \\ 6 & 9+6=15 \\ 8 & 15+8=23 \\ 7 & 23+7=30 \\ +3 & 30+3=33 \\ \hline 33 & \end{array}$$

(d)
$$\begin{array}{rl} 3 & \\ 8 & 3+8=11 \\ 6 & 11+6=17 \\ 4 & 17+4=21 \\ +8 & 21+8=29 \\ \hline 29 & \end{array}$$

3. (a)
$$\begin{array}{r} 26 \\ +73 \\ \hline 99 \end{array}$$

(b)
$$\begin{array}{r} 534 \\ +265 \\ \hline 799 \end{array}$$

(c)
$$\begin{array}{r} 42{,}305 \\ +11{,}563 \\ \hline 53{,}868 \end{array}$$

4. (a)
$$\begin{array}{rl} \overset{1}{6}6 & \\ +27 & 6+7=13 \\ \hline 93 & \end{array}$$

(b)
$$\begin{array}{rl} \overset{1}{5}8 & \\ +33 & 8+3=11 \\ \hline 91 & \end{array}$$

(c)
$$\begin{array}{rl} \overset{1}{5}6 & \\ +37 & 6+7=13 \\ \hline 93 & \end{array}$$

(d)
$$\begin{array}{rl} \overset{1}{3}4 & \\ +49 & 4+9=13 \\ \hline 83 & \end{array}$$

5. (a)
$$\begin{array}{r} \scriptstyle 21 \\ 42 \\ 651 \\ 396 \\ +\ 87 \\ \hline 1176 \end{array}$$

(b)
$$\begin{array}{r} \scriptstyle 1\ 21 \\ 162 \\ 4\ 271 \\ 372 \\ +8\ 976 \\ \hline 13{,}781 \end{array}$$

(c)
$$\begin{array}{r} \scriptstyle 22 \\ 57 \\ 4 \\ 392 \\ 804 \\ 51 \\ +\ 27 \\ \hline 1335 \end{array}$$

(d)
$$\begin{array}{r} \scriptstyle 2\ 21 \\ 7\ 821 \\ 435 \\ 72 \\ 305 \\ +1\ 693 \\ \hline 10{,}326 \end{array}$$

6. (a)
$$\begin{array}{r} 816 \\ 363 \\ 17 \\ 2 \\ 5 \\ +7654 \\ \hline 8857 \end{array}$$

(b)
$$\begin{array}{r} 3305 \\ 650 \\ 708 \\ 29 \\ 40 \\ 6 \\ +\ 3 \\ \hline 4741 \end{array}$$

(c)
```
   15,829
      765
       78
       15
        9
        7
 + 13,179
   29,882
```

(d)
```
   21,892
   11,746
 + 43,925
   77,563 (79,563 is incorrect.)
```

7. The shortest route from Lake Buena Vista to Conway is as follows:

```
  4   Lake Buena Vista to Resort Area
  6   Resort Area to Pine Castle
  3   Pine Castle to Belle Isle
+ 6   Belle Isle to Conway
 19   miles
```

8. The next shortest route from Orlando to Clear Lake is as follows:

```
  5   Orlando to Pine Hills
  8   Pine Hills to Altamonte Springs
  5   Altamonte Springs to Castleberry
  6   Castleberry to Bertha
  7   Bertha to Winter Park
+ 7   Winter Park to Clear Lake
 38   miles
```

9.
```
   13
   818
   348
   818
 + 348
  2332 feet
```

The amount of fencing needed is 2332 feet which is the perimeter of (distance around) the park.

10. (a)
```
    63
     4
     9
 +  28
   104 correct
```

(b)
```
   927
   395
    64
 + 251
  1637 correct
```

(c)
```
    79
   218
     7
 + 639
   943 (953 is incorrect.)
```

1.2 Section Exercises

1.
```
   43
 + 54
   97
```

3.
```
   56
 + 33
   89
```

5.
```
   317
 + 572
   889
```

7.
```
   318
   151
 + 420
   889
```

9.
```
   6310
    252
 + 1223
   7785
```

11. Line up:
```
   932
    44
 + 613
  1589
```

13. Line up:
```
   1251
   4311
 + 2114
   7676
```

15. Line up:
```
   12,142
   43,201
 + 23,103
   78,446
```

17. Line up:
```
   3213
 + 5715
   8928
```

19. Line up:
```
   38,204
 + 21,020
   59,224
```

21.
```
   1
   87
 + 63
  150
```

23. $\begin{array}{r} {}^{1} \\ 86 \\ +\,69 \\ \hline 155 \end{array}$

25. $\begin{array}{r} {}^{1} \\ 47 \\ +\,74 \\ \hline 121 \end{array}$

27. $\begin{array}{r} {}^{1} \\ 67 \\ +\,78 \\ \hline 145 \end{array}$

29. $\begin{array}{r} {}^{1} \\ 73 \\ +\,29 \\ \hline 102 \end{array}$

31. $\begin{array}{r} {}^{1} \\ 746 \\ +\,905 \\ \hline 1651 \end{array}$

33. $\begin{array}{r} {}^{1} \\ 306 \\ +\,848 \\ \hline 1154 \end{array}$

35. $\begin{array}{r} {}^{11} \\ 278 \\ +\,135 \\ \hline 413 \end{array}$

37. $\begin{array}{r} {}^{1} \\ 928 \\ +\,843 \\ \hline 1771 \end{array}$

39. $\begin{array}{r} {}^{11} \\ 526 \\ +\,884 \\ \hline 1410 \end{array}$

41. $\begin{array}{r} {}^{1\ 1} \\ 3574 \\ +\,2817 \\ \hline 6391 \end{array}$

43. $\begin{array}{r} {}^{111} \\ 7\,896 \\ +\ \ 3\,728 \\ \hline 11{,}624 \end{array}$

45. $\begin{array}{r} {}^{111} \\ 9\,625 \\ +\ \ 7\,986 \\ \hline 17{,}611 \end{array}$

47. $\begin{array}{r} {}^{22} \\ 9\,056 \\ 78 \\ 6\,089 \\ +\ \ \ 731 \\ \hline 15{,}954 \end{array}$

49. $\begin{array}{r} {}^{112} \\ 18 \\ 708 \\ 9\,286 \\ +\ \ \ 636 \\ \hline 10{,}648 \end{array}$

51. $\begin{array}{r} {}^{111} \\ 422 \\ 6\,074 \\ 435 \\ +\ \ 8\,663 \\ \hline 15{,}594 \end{array}$

53. $\begin{array}{r} {}^{1\ 1} \\ 321 \\ 9\,603 \\ 8 \\ 21 \\ +\ 1\,604 \\ \hline 11{,}557 \end{array}$

55. $\begin{array}{r} {}^{112} \\ 2\,109 \\ 63 \\ 16 \\ 3 \\ +\ \ 9\,887 \\ \hline 12{,}078 \end{array}$

57. $\begin{array}{r} {}^{233} \\ 553 \\ 97 \\ 2772 \\ 437 \\ 63 \\ +\ \ 328 \\ \hline 4250 \end{array}$

59. $\begin{array}{r} {}^{212} \\ 413 \\ 85 \\ 9\,919 \\ 602 \\ 31 \\ +\ \ 1\,218 \\ \hline 12{,}268 \end{array}$

61. Add up to check addition.

$\begin{array}{r} 2091 \\ \hline 832 \\ 468 \\ +\,791 \\ \hline 2091 \end{array}$ correct

63. Add up to check addition.

```
   769
  ----
   179
   214
 + 376
  ----
   759 incorrect; should be 769
```

65. Add up to check addition.

```
  5420
  ----
  4713
    28
   615
  + 64
  ----
  5420 correct
```

67. Add up to check addition.

```
  11,577
  ------
     678
    7952
      56
     718
  + 2173
  ------
  11,377 incorrect; should be 11,577
```

69. Add up to check addition.

```
  14,332
  ------
    4714
      27
      77
    8878
   + 636
  ------
  14,332 correct
```

71. Changing the order in which numbers are added does not change the sum. You can add from bottom to top when checking addition.

73. The shortest route between Southtown and Rena is through Thomasville.

```
   21  Southtown to Thomasville
 + 12  Thomasville to Rena
 ----
   33  miles
```

75. The shortest route between Thomasville and Murphy is through Rena and Austin.

```
   12  Thomasville to Rena
   15  Rena to Austin
 + 11  Austin to Murphy
 ----
   38  miles
```

77.

```
    $3482  flea market
 + 12,860  annual auction
  -------
  $16,342  total amount raised
```

79.

```
   413  women
 + 286  men
  ----
   699  total people
```

81.

```
  $114  leaf blower
  + 98  weed trimmer
  ----
  $212  total cost
```

83. To find the perimeter, add the 4 sides in any order.

```
   220
    55
   220
  + 55
  ----
   550  feet
```

550 feet of curbing will be needed.

85. To find the perimeter, add the 3 sides in any order.

```
   30
   24
 + 18
 ----
   72  feet
```

Martin will need 72 feet of lumber.

87. The largest four-digit number possible, using the digits 4, 1, 9, and 2 each once, will begin with the largest digit in the thousands place and the remaining digits will descend in size until the smallest digit is in the ones place. Therefore, the largest four-digit number possible is 9421.

88. The smallest four-digit number possible, using the digits 4, 1, 9, and 2 each once, will begin with the smallest digit in the thousands place and the remaining digits will ascend in size until the largest digit is in the ones place. Therefore, the smallest four-digit number possible is 1249.

89. The largest five-digit number possible, using the digits 6, 2, and 7 at least once, will use the largest digit in the ten thousands, thousands, and hundreds place and the remaining digits will descend in size until the smallest digit is in the ones place. Therefore, the largest five-digit number possible is 77,762.

90. The smallest five-digit number possible, using the digits 6, 2, and 7 at least once, will use the smallest digit in the ten thousands, thousands, and hundreds place and the remaining digits will ascend in size until the largest digit is in the ones place. Therefore, the smallest five-digit number possible is 22,267.

91. The largest seven-digit number possible is 9,994,433.

92. The smallest seven-digit number possible is 3,334,499.

93. To write the largest seven-digit number possible, using the digits 4, 3, and 9, and using each digit at least twice, write the largest digits on the left, and use the smaller digits as you move right.

94. To write the smallest seven-digit number possible, using the digits 4, 3, and 9, and using each digit at least twice, write the smallest digits on the left, and use the larger digits as you move right.

1.3 Subtracting Whole Numbers

1.3 Margin Exercises

1. **(a)** $5 + 3 = 8$:

$8 - 3 = 5$ or $8 - 5 = 3$

(b) $7 + 4 = 11$:

$11 - 7 = 4$ or $11 - 4 = 7$

(c) $15 + 22 = 37$:

$37 - 15 = 22$ or $37 - 22 = 15$

(d) $23 + 55 = 78$:

$78 - 55 = 23$ or $78 - 23 = 55$

2. **(a)** $7 - 5 = 2$:

$7 = 5 + 2$ or $7 = 2 + 5$

(b) $9 - 4 = 5$:

$9 = 4 + 5$ or $9 = 5 + 4$

(c) $21 - 15 = 6$:

$21 = 15 + 6$ or $21 = 6 + 15$

(d) $58 - 42 = 16$:

$58 = 42 + 16$ or $58 = 16 + 42$

3. **(a)**

$$\begin{array}{r} 74 \\ -\ 43 \\ \hline 31 \end{array} \quad \begin{array}{l} 4 - 3 = 1 \\ 7 - 4 = 3 \end{array}$$

(b)

$$\begin{array}{r} 68 \\ -\ 24 \\ \hline 44 \end{array} \quad \begin{array}{l} 8 - 4 = 4 \\ 6 - 2 = 4 \end{array}$$

(c)

$$\begin{array}{r} 429 \\ -\ 318 \\ \hline 111 \end{array}$$

(d)

$$\begin{array}{r} 3927 \\ -\ 2614 \\ \hline 1313 \end{array}$$

(e)

$$\begin{array}{r} 5464 \\ -\ 324 \\ \hline 5140 \end{array}$$

4. **(a)**

$$\begin{array}{r} 76 \\ -\ 45 \\ \hline 31 \end{array} \begin{array}{l} \textit{Subtraction} \\ \textit{problem} \end{array} \qquad \begin{array}{r} 45 \\ +\ 31 \\ \hline 76 \end{array} \begin{array}{l} \textit{Addition} \\ \textit{problem} \end{array}$$

31 is correct.

(b)

$$\begin{array}{r} 53 \\ -\ 22 \\ \hline 21 \end{array} \begin{array}{l} \textit{Subtraction} \\ \textit{problem} \end{array} \qquad \begin{array}{r} 22 \\ +\ 21 \\ \hline 43 \end{array} \begin{array}{l} \textit{Addition} \\ \textit{problem} \end{array}$$

21 is incorrect.

Rework.

$$\begin{array}{r} 53 \\ -\ 22 \\ \hline 31 \end{array}$$ is correct.

(c)

$$\begin{array}{r} 374 \\ -\ 251 \\ \hline 113 \end{array} \begin{array}{l} \textit{Subtraction} \\ \textit{problem} \end{array} \qquad \begin{array}{r} 251 \\ +\ 113 \\ \hline 364 \end{array} \begin{array}{l} \textit{Addition} \\ \textit{problem} \end{array}$$

113 is incorrect.

Rework.

$$\begin{array}{r} 374 \\ -\ 251 \\ \hline 123 \end{array}$$ is correct.

(d)

$$\begin{array}{r} 7531 \\ -\ 4301 \\ \hline 3230 \end{array} \begin{array}{l} \textit{Subtraction} \\ \textit{problem} \end{array} \qquad \begin{array}{r} 4301 \\ +\ 3230 \\ \hline 7531 \end{array} \begin{array}{l} \textit{Addition} \\ \textit{problem} \end{array}$$

3230 is correct.

5. **(a)**

$$\begin{array}{r} {\scriptstyle 4\,18} \\ \not{5}\,\not{8} \\ -\ 1\,9 \\ \hline 3\,9 \end{array}$$

(b)

$$\begin{array}{r} {\scriptstyle 7\,16} \\ \not{8}\,\not{6} \\ -\ 3\,8 \\ \hline 4\,8 \end{array}$$

(c)

$$\begin{array}{r} {\scriptstyle 3\,11} \\ \not{4}\,\not{1} \\ -\ 2\,7 \\ \hline 1\,4 \end{array}$$

(d)

$$\begin{array}{r} {\scriptstyle 5\,13} \\ 8\,\not{6}\,\not{3} \\ -\ 4\,7 \\ \hline 81\,6 \end{array}$$

(e)

$$\begin{array}{r} {\scriptstyle 5\,12} \\ 7\,\not{6}\,\not{2} \\ -\ 15\,7 \\ \hline 60\,5 \end{array}$$

6. **(a)**

```
  812
  927
-  43
  884
```

(b)

```
  51615
  675
-  86
  589
```

(c)

```
  31617
  477
- 389
   88
```

(d)

```
  0131217
  1437
-  988
   449
```

(e)

```
  71613
  8739
- 3892
  4847
```

7. **(a)**

```
    9
  11016
  206
- 177
   29
```

(b)

```
    9
  61013
  703
- 415
  288
```

(c)

```
    9
  61012
  7024
- 2632
  4392
```

8. **(a)**

```
    9
  21018
  308
- 159
  149
```

(b)

```
   610
  570
- 368
  202
```

(c)

```
  0141610
  1570
-  983
   587
```

(d)

```
   9 9
  6101011
  7001
- 5193
  1808
```

(e)

```
   9 9
  3101010
  4000
- 1782
  2218
```

9. **(a)**

```
  357      168
- 168    + 189
  189      357
```

Match: 189 is correct.

(b)

```
  570      328
- 328    + 252
  252      580
```

Not a match: 252 is incorrect.

Rework.

```
  570      328
- 328    + 242
  242      570
```

Match: 242 is correct.

(c)

```
  14,726      8839
-  8 839    + 5887
   5 887    14,726
```

Match: 5887 is correct.

10. **(a)**

$52,458	*Bachelor's degree*
− 39,902	*Associate of Arts degree*
$12,556	*More earnings*

(b)

$39,902	*Associate of Arts degree*
− 25,371	*Not a high-school graduate*
$14,531	

1.3 Section Exercises

1.

```
  48    Check:   32
- 32           + 16
  16             48
```

3.

```
  86    Check:   53
- 53           + 33
  33             86
```

5.

```
  77    Check:   60
- 60           + 17
  17             77
```

7.

```
  335   Check:   213
- 122          + 122
  213            335
```

9.

```
  552   Check:   451
- 451          + 101
  101            552
```

11. 7352 − 241 = 7111 Check: 241 + 7111 = 7352

13. 5546 − 2134 = 3412 Check: 3412 + 2134 = 5546

15. 6259 − 4148 = 2111 Check: 4148 + 2111 = 6259

17. 24,392 − 11,232 = 13,160 Check: 13,160 + 11,232 = 24,392

19. 46,253 − 5143 = 41,110 Check: 41,110 + 5143 = 46,253

21. 54 − 42 = 12 Check: 42 + 12 = 54 correct

23. 89 − 27 = 63 Check: 63 + 27 = 90 incorrect

Rework.

89 − 27 = 62 is the correct answer.

25. 382 − 261 = 131 Check: 261 + 131 = 392 incorrect

Rework.

382 − 261 = 121 is the correct answer.

27. 4683 − 3542 = 1141 Check: 3542 + 1141 = 4683 correct

29. 8643 − 1421 = 7212 Check: 7212 + 1421 = 8633 incorrect

Rework.

8643 − 1421 = 7222 is the correct answer.

31. 615
75
− 37
38

33. 814
94
− 49
45

35. 417
57
− 38
19

37. 712
828
− 547
281

39. 611
771
− 252
519

41. 41218
7538
− 479
7059

43. 81718
9988
− 2399
7589

45. 217 121215
38,335
− 29,476
8 859

47. 310
40
− 37
3

49. 510
60
− 37
23

51. 9
21018
308
− 289
19

53. 310 311
4041
− 1208
2833

55. 81210
9305
− 1530
7775

57.

```
   710
  1580
− 1077
   503
```

59.

```
   9
  11010
  2006
− 1850
   156
```

61.

```
   11310
  8240
− 6056
  2184
```

63.

```
     9
  7141013
  8503
− 2816
  5687
```

65.

```
         9
  710 61015
  80,705
− 61,667
  19,038
```

67.

```
    9 9
  5 101010
  66,000
− 34,444
  31,556
```

69.

```
   9  9
  110 101710
  20,080
− 13,496
   6 584
```

71.

```
  9428      4509
− 4509    + 4919
  4919      9428   correct
```

73.

```
  2548      2278
− 2278    + 270
  270       2548   correct
```

75.

```
  93,758      52,869
− 52,869    + 40,889
  40,889      93,758   correct
```

77.

```
  36,778      19,373
− 17,405    + 17,405
  19,373      36,778   correct
```

79. Possible answers are:

$3 + 2 = 5$ could be changed to $5 - 2 = 3$ or $5 - 3 = 2$.

$6 - 4 = 2$ could be changed to $2 + 4 = 6$ or $4 + 2 = 6$.

81.

```
  187   calories man burns
− 140   calories woman burns
   47   fewer calories woman burned
```

The woman burned 47 fewer calories.

83.

```
  1821   height of CN Tower
− 1454   height of Sears Tower
   367   difference in heights
```

There is a difference in height of 367 feet.

85.

```
  1815   passengers on the ship
− 1348   passengers that go ashore
   467   passengers that remain on the ship
```

467 passengers are left on the ship.

87. There were more people at the game on Friday.

```
  7994   Friday's attendance
− 5822   Tuesday's attendance
  2172
```

There were 2172 more people on Friday.

89.

```
  14,608   flags manufactured
− 5 069    flags sold
  9 539    flags remaining
```

There are 9539 flags that remain.

91.

```
  $913   monthly housing expense
− 650    monthly rent
  $263   increase in monthly housing expense
```

93. More people visited the park on Tuesday.

```
  12,352   visitors to the park on Tuesday
− 11,594   visitors to the park on Monday
     758   more visitors on Tuesday
```

There were 758 more visitors on Tuesday.

95. Rewrite one hundred one thousand, five hundred dollars using digits.

```
  $101,500   salary of a general manager
−  44,000    salary of a director
  $57,500    difference in salaries
```

A general manager's average salary is $57,500 more than a director's.

97.

```
  137   deliveries made on Monday
− 89    deliveries made on Thursday
   48   more deliveries made on Monday
```

Diana made 48 more deliveries on Monday than on Thursday.

99.

$$\begin{array}{rl} 147 & \textit{deliveries made on Friday} \\ +\,137 & \textit{deliveries made on Monday} \\ \hline 284 & \textit{deliveries made on the two busiest days} \end{array}$$

There were 284 deliveries made on the two busiest days.

1.4 Multiplying Whole Numbers

1.4 Margin Exercises

1. **(a)**

$$\begin{array}{rl} 8 & \textit{factor} \\ \times\,5 & \textit{factor} \\ \hline 40 & \textit{product} \end{array}$$

(b)

$$\begin{array}{rl} 6 & \textit{factor} \\ \times\,4 & \textit{factor} \\ \hline 24 & \textit{product} \end{array}$$

(c)

$$\begin{array}{rl} 7 & \textit{factor} \\ \times\,6 & \textit{factor} \\ \hline 42 & \textit{product} \end{array}$$

(d)

$$\begin{array}{rl} 3 & \textit{factor} \\ \times\,9 & \textit{factor} \\ \hline 27 & \textit{product} \end{array}$$

2. **(a)** $7 \times 4 = 28$

(b) $0 \times 9 = 0$

(c) $8(5) = 40$

(d) $6 \cdot 5 = 30$

(e) $(3)(8) = 24$

3. **(a)** $(3 \times 2) \times 5 = 6 \times 5 = 30$

(b) $(4 \cdot 7) \cdot 1 = 28 \cdot 1 = 28$

(c) $(8)(3)(0) = 24(0) = 0$

4. **(a)**

$$\begin{array}{r} {}^{1} \\ 53 \\ \times\,5 \\ \hline 265 \end{array}$$

(b)

$$\begin{array}{r} 79 \\ \times\,0 \\ \hline 0 \end{array}$$

(c)

$$\begin{array}{r} {}^{46} \\ 758 \\ \times\,8 \\ \hline 6064 \end{array}$$

(d)

$$\begin{array}{r} {}^{52} \\ 2831 \\ \times\,7 \\ \hline 19{,}817 \end{array}$$

(e)

$$\begin{array}{r} {}^{513} \\ 4714 \\ \times\,8 \\ \hline 37{,}712 \end{array}$$

5. **(a)** $63 \times 10 = 630$ *Attach* 0.

(b) $305 \times 100 = 30{,}500$ *Attach* 00.

(c) $714 \times 1000 = 714{,}000$ *Attach* 000.

6. **(a)** 16×50

$$\begin{array}{r} 16 \\ \times\,5 \\ \hline 80 \end{array}$$

$16 \times 50 = 800$ *Attach* 0.

(b) 73×400

$$\begin{array}{r} 73 \\ \times\,4 \\ \hline 292 \end{array}$$

$73 \times 400 = 29{,}200$ *Attach* 00.

(c) 180×30

$$\begin{array}{r} 18 \\ \times\,3 \\ \hline 54 \end{array}$$

$180 \times 30 = 5400$ *Attach* 00.

(d) 4200×80

$$\begin{array}{r} 42 \\ \times\,8 \\ \hline 336 \end{array}$$

$4200 \times 80 = 336{,}000$ *Attach* 000.

(e) 800×600

$$\begin{array}{r} 8 \\ \times\,6 \\ \hline 48 \end{array}$$

$800 \times 600 = 480{,}000$ *Attach* 0000.

7. **(a)**

$$\begin{array}{rl} 35 & \\ \times\,54 & \\ \hline 140 & \\ 175 & \textit{Add} \\ \hline 1890 & \end{array}$$

(b)

$$\begin{array}{rl} 76 & \\ \times\,49 & \\ \hline 684 & \\ 304 & \textit{Add} \\ \hline 3724 & \end{array}$$

8. **(a)**

$$\begin{array}{rl} 52 & \\ \times\,16 & \\ \hline 312 & \leftarrow 6 \times 52 \\ 52 & \leftarrow 1 \times 52 \\ \hline 832 & \end{array}$$

(b)

$$\begin{array}{rl} 81 & \\ \times\ 49 & \\ \hline 729 & \leftarrow 9 \times 81 \\ 324 & \leftarrow 4 \times 81 \\ \hline 3969 & \end{array}$$

(c)

$$\begin{array}{rl} 75 & \\ \times\ 63 & \\ \hline 225 & \leftarrow 3 \times 75 \\ 450 & \leftarrow 6 \times 75 \\ \hline 4725 & \end{array}$$

(d)

$$\begin{array}{rl} 234 & \\ \times\ 73 & \\ \hline 702 & \leftarrow 3 \times 234 \\ 1638 & \leftarrow 7 \times 234 \\ \hline 17{,}082 & \end{array}$$

(e)

$$\begin{array}{rl} 835 & \\ \times\ 189 & \\ \hline 7515 & \leftarrow 9 \times 835 \\ 6680 & \leftarrow 8 \times 835 \\ 835 & \leftarrow 1 \times 835 \\ \hline 157{,}815 & \end{array}$$

9. **(a)**

$$\begin{array}{r} 28 \\ \times\ 60 \\ \hline 1680 \end{array}$$

(b)

$$\begin{array}{r} 728 \\ \times\ 50 \\ \hline 36{,}400 \end{array}$$

(c)

$$\begin{array}{rl} 562 & \\ \times\ 109 & \\ \hline 5058 & \\ 5620 & 1 \times 562 = 562 \quad \textit{Insert } 0 \\ \hline 61{,}258 & \end{array}$$

(d)

$$\begin{array}{rl} 3526 & \\ \times\ 6002 & \\ \hline 7052 & \\ 2115600 & 6 \times 3526 = 21{,}156 \quad \textit{Insert } 00 \\ \hline 21{,}163{,}052 & \end{array}$$

10. **(a)**

$$\begin{array}{r} 289 \\ \times\ 12 \\ \hline 578 \\ 289 \\ \hline 3468 \end{array}$$

The total cost of 289 redwood planters is $3468.

(b)

$$\begin{array}{r} 205 \\ \times\ 89 \\ \hline 1845 \\ 1640 \\ \hline 18{,}245 \end{array}$$

The total cost of 205 DVD/CD players is $18, 245.

(c)

$$\begin{array}{r} 24{,}300 \\ \times\ 12 \\ \hline 48600 \\ 24300 \\ \hline 291{,}600 \end{array}$$

The total cost of 12 delivery vans is $291, 600.

1.4 Section Exercises

1. $2 \times 6 \times 2 = (2 \times 6) \times 2 = 12 \times 2 = 24$
or $2 \times (6 \times 2) = 2 \times 12 = 24$.

3. $8 \times 6 \times 1 = (8 \times 6) \times 1 = 48 \times 1 = 48$
or $8 \times (6 \times 1) = 8 \times 6 = 48$.

5. $7 \cdot 8 \cdot 0 = (7 \cdot 8) \cdot 0 = 56 \cdot 0 = 0$
or $7 \cdot (8 \cdot 0) = 7 \cdot 0 = 0$.

7. $4 \cdot 1 \cdot 6 = (4 \cdot 1) \cdot 6 = 4 \cdot 6 = 24$
or $4 \cdot (1 \cdot 6) = 4 \cdot 6 = 24$.

9. $(4)(5)(2) = [(4)(5)](2) = 20(2) = 40$
or $(4)[(5)(2)] = (4)(10) = 40$.

11. $(3)(0)(7) = [(3)(0)](7) = 0(7) = 0$
or $(3)[(0)(7)] = (3)(0) = 0$.

The product of any number and 0 is 0.

13. Factors may be multiplied in any order to get the same answer. The commutative properties of addition and multiplication are similar since with either you may add or multiply numbers in any order.

15.

$$\begin{array}{r} \scriptstyle 3 \\ 35 \\ \times\ 6 \\ \hline 210 \end{array}$$

$6 \cdot 5 = 30$ Write 0, carry 3 tens.
$6 \cdot 3 = 18$ Add 3 to get 21. Write 21.

17.

$$\begin{array}{r} \scriptstyle 2 \\ 34 \\ \times\ 7 \\ \hline 238 \end{array}$$

$7 \cdot 4 = 28$ Write 8, carry 2 tens.
$7 \cdot 3 = 21$ Add 2 to get 23. Write 23.

19.
$$\begin{array}{r} {}^{21} \\ 642 \\ \times\ 5 \\ \hline 3210 \end{array}$$

$5 \cdot 2 = 10$ Write 0, carry 1 ten.
$5 \cdot 4 = 20$ Add 1 to get 21. Write 1, carry 2.
$5 \cdot 6 = 30$ Add 2 to get 32. Write 32.

21.
$$\begin{array}{r} {}^{1} \\ 624 \\ \times\ 3 \\ \hline 1872 \end{array}$$

$3 \cdot 4 = 12$ Write 2, carry 1 ten.
$3 \cdot 2 = 6$ Add 1 to get 7. Write 7.
$3 \cdot 6 = 18$ Write 18.

23.
$$\begin{array}{r} {}^{21} \\ 2153 \\ \times\ 4 \\ \hline 8612 \end{array}$$

$4 \cdot 3 = 12$ Write 2, carry 1 ten.
$4 \cdot 5 = 20$ Add 1 to get 21. Write 1, carry 2.
$4 \cdot 1 = 4$ Add 2 to get 6. Write 6.
$4 \cdot 2 = 8$ Write 8.

25.
$$\begin{array}{r} {}^{2} \\ 2521 \\ \times\ 4 \\ \hline 10{,}084 \end{array}$$

$4 \cdot 1 = 4$ Write 4.
$4 \cdot 2 = 8$ Write 8.
$4 \cdot 5 = 20$ Write 0, carry 2.
$4 \cdot 2 = 8$ Add 2 to get 10. Write 10.

27.
$$\begin{array}{r} {}^{44} \\ 2561 \\ \times\ 8 \\ \hline 20{,}488 \end{array}$$

$8 \cdot 1 = 8$ Write 8.
$8 \cdot 6 = 48$ Write 8, carry 4.
$8 \cdot 5 = 40$ Add 4 to get 44. Write 4, carry 4.
$8 \cdot 2 = 16$ Add 4 to get 20. Write 20.

29.
$$\begin{array}{r} {}^{46\ 1} \\ 36{,}921 \\ \times\ 7 \\ \hline 258{,}447 \end{array}$$

$7 \cdot 1 = 7$ Write 7.
$7 \cdot 2 = 14$ Write 4, carry 1.
$7 \cdot 9 = 63$ Add 1 to get 64. Write 4, carry 6.
$7 \cdot 6 = 42$ Add 6 to get 48. Write 8, carry 4.
$7 \cdot 3 = 21$ Add 4 to get 25. Write 25.

31.
$$\begin{array}{r} 40 \\ \times\ 7 \\ \hline \end{array} \quad \begin{array}{r} 4 \\ \times\ 7 \\ \hline 28 \end{array} \quad \begin{array}{r} 40 \\ \times\ 7 \\ \hline 280 \end{array} \quad \textit{Attach } 0.$$

33.
$$\begin{array}{r} 80 \\ \times\ 6 \\ \hline \end{array} \quad \begin{array}{r} 8 \\ \times\ 6 \\ \hline 48 \end{array} \quad \begin{array}{r} 80 \\ \times\ 6 \\ \hline 480 \end{array} \quad \textit{Attach } 0.$$

35.
$$\begin{array}{r} 740 \\ \times\ 3 \\ \hline \end{array} \quad \begin{array}{r} 74 \\ \times\ 3 \\ \hline 222 \end{array} \quad \begin{array}{r} 740 \\ \times\ 3 \\ \hline 2220 \end{array} \quad \textit{Attach } 0.$$

37.
$$\begin{array}{r} 600 \\ \times\ 6 \\ \hline \end{array} \quad \begin{array}{r} 6 \\ \times\ 6 \\ \hline 36 \end{array} \quad \begin{array}{r} 600 \\ \times\ 6 \\ \hline 3600 \end{array} \quad \textit{Attach } 00.$$

39.
$$\begin{array}{r} 125 \\ \times\ 30 \\ \hline \end{array} \quad \begin{array}{r} 125 \\ \times\ 3 \\ \hline 375 \end{array} \quad \begin{array}{r} 125 \\ \times\ 30 \\ \hline 3750 \end{array} \quad \textit{Attach } 0.$$

41.
$$\begin{array}{r} 1635 \\ \times\ 40 \\ \hline \end{array} \quad \begin{array}{r} 1635 \\ \times\ 4 \\ \hline 6540 \end{array} \quad \begin{array}{r} 1635 \\ \times\ 40 \\ \hline 65{,}400 \end{array} \quad \textit{Attach } 0.$$

43.
$$\begin{array}{r} 900 \\ \times\ 300 \\ \hline \end{array} \quad \begin{array}{r} 9 \\ \times\ 3 \\ \hline 27 \end{array} \quad \begin{array}{r} 900 \\ \times\ 300 \\ \hline 270{,}000 \end{array} \quad \textit{Attach } 0000.$$

45.
$$\begin{array}{r} 43{,}000 \\ \times\ 2000 \\ \hline \end{array} \quad \begin{array}{r} 43 \\ \times\ 2 \\ \hline 86 \end{array} \quad \begin{array}{r} 43{,}000 \\ \times\ 2000 \\ \hline 86{,}000{,}000 \end{array} \quad \textit{Attach } 000000.$$

47. $970 \cdot 50$
$$\begin{array}{r} 97 \\ \times\ 5 \\ \hline 485 \end{array}$$

$970 \cdot 50 = 48{,}500$ *Attach* 00.

49. $800 \cdot 900$
$$\begin{array}{r} 8 \\ \times\ 9 \\ \hline 72 \end{array}$$

$800 \cdot 900 = 720{,}000$ *Attach* 0000.

51. $9700 \cdot 200$
$$\begin{array}{r} 97 \\ \times\ 2 \\ \hline 194 \end{array}$$

$9700 \cdot 200 = 1{,}940{,}000$ *Attach* 0000.

53.
$$\begin{array}{rl} 28 & \\ \times\ 17 & \\ \hline 196 & \leftarrow 7 \times 28 \\ 28 & \leftarrow 1 \times 28 \\ \hline 476 & \end{array}$$

55.
$$\begin{array}{rl} 75 & \\ \times\ 32 & \\ \hline 150 & \leftarrow 2 \times 75 \\ 225 & \leftarrow 3 \times 75 \\ \hline 2400 & \end{array}$$

57.
$$\begin{array}{rl} 83 & \\ \times\ 45 & \\ \hline 415 & \leftarrow 5\times 83 \\ 332 & \leftarrow 4\times 83 \\ \hline 3735 & \end{array}$$

59. (58)(41)
$$\begin{array}{rl} 58 & \\ \times\ 41 & \\ \hline 58 & \leftarrow 1\times 58 \\ 232 & \leftarrow 4\times 58 \\ \hline 2378 & \end{array}$$

61. (67)(92)
$$\begin{array}{rl} 67 & \\ \times\ 92 & \\ \hline 134 & \leftarrow 2\times 67 \\ 603 & \leftarrow 9\times 67 \\ \hline 6164 & \end{array}$$

63. (28)(564)
$$\begin{array}{rl} 564 & \\ \times\ 28 & \\ \hline 4512 & \leftarrow 8\times 564 \\ 1128 & \leftarrow 2\times 564 \\ \hline 15{,}792 & \end{array}$$

65. (619)(35)
$$\begin{array}{rl} 619 & \\ \times\ 35 & \\ \hline 3095 & \leftarrow 5\times 619 \\ 1857 & \leftarrow 3\times 619 \\ \hline 21{,}665 & \end{array}$$

67. (55)(286)
$$\begin{array}{rl} 286 & \\ \times\ 55 & \\ \hline 1430 & \leftarrow 5\times 286 \\ 1430 & \leftarrow 5\times 286 \\ \hline 15{,}730 & \end{array}$$

69.
$$\begin{array}{rl} 735 & \\ \times\ 112 & \\ \hline 1470 & \leftarrow 2\times 735 \\ 735 & \leftarrow 1\times 735 \\ 735 & \leftarrow 1\times 735 \\ \hline 82{,}320 & \end{array}$$

71.
$$\begin{array}{rl} 538 & \\ \times\ 342 & \\ \hline 1076 & \leftarrow 2\times 538 \\ 2152 & \leftarrow 4\times 538 \\ 1614 & \leftarrow 3\times 538 \\ \hline 183{,}996 & \end{array}$$

73.
$$\begin{array}{rl} 9352 & \\ \times\ 264 & \\ \hline 37408 & \leftarrow 4\times 9352 \\ 56112 & \leftarrow 6\times 9352 \\ 18704 & \leftarrow 2\times 9352 \\ \hline 2{,}468{,}928 & \end{array}$$

75.
$$\begin{array}{rl} 215 & \\ \times\ 307 & \\ \hline 1505 & \leftarrow 7\times 215 \\ 6450 & \leftarrow 30\times 215 \\ \hline 66{,}005 & \end{array}$$

77.
$$\begin{array}{rl} 428 & \\ \times\ 201 & \\ \hline 428 & \leftarrow 1\times 428 \\ 8560 & \leftarrow 20\times 428 \\ \hline 86{,}028 & \end{array}$$

79.
$$\begin{array}{rl} 6310 & \\ \times\ 3078 & \\ \hline 50480 & \leftarrow 8\times 6310 \\ 44170 & \leftarrow 7\times 6310 \\ 189300 & \leftarrow 30\times 6310 \\ \hline 19{,}422{,}180 & \end{array}$$

81.
$$\begin{array}{rl} 2195 & \\ \times\ 1038 & \\ \hline 17560 & \leftarrow 8\times 2195 \\ 6585 & \leftarrow 3\times 2195 \\ 21950 & \leftarrow 10\times 2195 \\ \hline 2{,}278{,}410 & \end{array}$$

83. To multiply by 10, 100, or 1000, just add one, two, or three zeros to the number you are multiplying and that's your answer.

85.
$$\begin{array}{rl} 300 & \textit{cartons} \\ \times\ 10 & \textit{balls per carton} \\ \hline 3000 & \textit{balls} \end{array}$$

3000 balls were purchased.

87.
$$\begin{array}{rl} 12 & \textit{tomato plants per flat} \\ \times\ 18 & \textit{flats} \\ \hline 96 & \\ 12 & \\ \hline 216 & \textit{plants} \end{array}$$

He bought 216 plants.

89.
$$\begin{array}{rl} 22 & \textit{miles per gallon} \\ \times\ 16 & \textit{gallons} \\ \hline 132 & \\ 22 & \\ \hline 352 & \textit{miles} \end{array}$$

The Toyota Highlander can go 352 miles on 16 gallons of gas.

91. 75 first-aid kits at \$8 per kit

$$\begin{array}{r} 75 \\ \times\ 8 \\ \hline 600 \end{array}$$

The total cost is \$600.

93. 65 rebuilt alternators at \$24 per alternator

$$\begin{array}{r} 65 \\ \times\ 24 \\ \hline 260 \\ 130 \\ \hline 1560 \end{array}$$

The total cost is \$1560.

95. 206 desktop computers at \$548 per computer

$$\begin{array}{r} 548 \\ \times\ 206 \\ \hline 3288 \\ 10960 \\ \hline 112{,}888 \end{array}$$

The total cost is \$112, 888.

97. $21 \cdot 43 \cdot 56 = (21 \cdot 43) \cdot 56$

$$\begin{array}{r} 21 \\ \times\ 43 \\ \hline 63 \\ 84 \\ \hline 903 \end{array} \qquad \begin{array}{r} 903 \\ \times\ 56 \\ \hline 5418 \\ 4515 \\ \hline 50{,}568 \end{array}$$

99.
$$\begin{array}{rl} 450 & \textit{trees per acre} \\ \times\ 85 & \textit{acres} \\ \hline 2250 & \\ 3600 & \\ \hline 38{,}250 & \textit{trees} \end{array}$$

38,250 trees are needed to plant 85 acres.

101.
$$\begin{array}{rl} 11{,}988 & \textit{miles driven in 2000} \\ -\ 8813 & \textit{miles driven in 1980} \\ \hline 3175 & \textit{more miles} \end{array}$$

The average American drove 3175 more miles in 2000 than in 1980.

103.
$$\begin{array}{r} 880 \\ \times\ 6 \\ \hline 5280 \end{array} \qquad \begin{array}{r} 235 \\ \times\ 6 \\ \hline 1410 \end{array} \qquad \begin{array}{r} 140 \\ \times\ 5 \\ \hline 700 \end{array}$$

The total cost is the sum of these values:

$$\$5280 + \$1410 + \$700 = \$7390$$

105. **(a)** $189 + 263 = 452$

(b) $263 + 189 = 452$

106. commutative

107. **(a)** $(65 + 81) + 135 = 146 + 135 = 281$

(b) $65 + (81 + 135) = 65 + 216 = 281$

108. associative

109. **(a)** $220 \times 72 = 15{,}840$

(b) $72 \times 220 = 15{,}840$

110. commutative

111. **(a)** $(26 \times 18) \times 14 = 468 \times 14 = 6552$

(b) $26 \times (18 \times 14) = 26 \times 252 = 6552$

112. associative

113. No. Some examples are

1. $7 - 5 = 2$, but $5 - 7$ does not equal 2.
2. $12 - 6 = 6$, but $6 - 12$ does not equal 6.
3. $(8 - 2) - 5 = 1$, but $8 - (2 - 5)$ does not equal 1.

114. No. Some examples are

1. $10 \div 2 = 5$, but $2 \div 10$ does not equal 5.
2. $(16 \div 8) \div 2 = 1$, but $16 \div (8 \div 2)$ does not equal 1.

1.5 Dividing Whole Numbers

1.5 Margin Exercises

1. **(a)** $24 \div 6 = 4$: $\begin{array}{r} 4 \\ 6\overline{)24} \end{array}$ or $\dfrac{24}{6} = 4$

(b) $\begin{array}{r} 4 \\ 9\overline{)36} \end{array}$: $36 \div 9 = 4$ or $\dfrac{36}{9} = 4$

(c) $48 \div 6 = 8$: $\begin{array}{r} 8 \\ 6\overline{)48} \end{array}$ or $\dfrac{48}{6} = 8$

(d) $\dfrac{42}{6} = 7$: $42 \div 6 = 7$ or $\begin{array}{r} 7 \\ 6\overline{)42} \end{array}$

2. **(a)** $15 \div 3 = 5$
dividend: 15; divisor: 3; quotient: 5

(b) $18 \div 6 = 3$
dividend: 18; divisor: 6; quotient: 3

(c) $\dfrac{28}{7} = 4$
dividend: 28; divisor: 7; quotient: 4

(d) $\begin{array}{r} 3 \\ 9\overline{)27} \end{array}$
dividend: 27; divisor: 9; quotient: 3

3. **(a)** $0 \div 5 = 0$

(b) $\dfrac{0}{9} = 0$

(c) $\dfrac{0}{24} = 0$

(d) $\begin{array}{r} 0 \\ 37\overline{)0} \end{array}$

4. **(a)** $\begin{array}{r} 3 \\ 5\overline{)15} \end{array}$; $5 \cdot 3 = 15$ or $3 \cdot 5 = 15$

(b) $\dfrac{32}{4} = 8$; $4 \cdot 8 = 32$ or $8 \cdot 4 = 32$

(c) $48 \div 8 = 6$; $8 \cdot 6 = 48$ or $6 \cdot 8 = 48$

5. **(a)** $\dfrac{4}{0}$; undefined

(b) $\dfrac{0}{4} = 0$

(c) $0\overline{)36}$; undefined

(d) $\begin{array}{r} 0 \\ 36\overline{)0} \end{array}$

(e) $100 \div 0$; undefined

(f) $0 \div 100 = 0$

6. **(a)** $8 \div 8 = 1$

(b) $\begin{array}{r} 1 \\ 15\overline{)15} \end{array}$

(c) $\dfrac{37}{37} = 1$

7. **(a)** $9 \div 1 = 9$

(b) $\begin{array}{r} 18 \\ 1\overline{)18} \end{array}$

(c) $\dfrac{43}{1} = 43$

8. **(a)** $\begin{array}{r} 12 \\ 2\overline{)24} \end{array}$ $\dfrac{2}{2} = 1$, $\dfrac{4}{2} = 2$

(b) $\begin{array}{r} 31 \\ 3\overline{)93} \end{array}$ $\dfrac{9}{3} = 3$, $\dfrac{3}{3} = 1$

(c) $\begin{array}{r} 22 \\ 4\overline{)88} \end{array}$ $\dfrac{8}{4} = 2$, $\dfrac{8}{4} = 2$

(d) $\begin{array}{r} 312 \\ 2\overline{)624} \end{array}$ $\dfrac{6}{2} = 3$, $\dfrac{2}{2} = 1$, $\dfrac{4}{2} = 2$

9. **(a)** $\begin{array}{r} 62\ \mathbf{R}1 \\ 2\overline{)125} \end{array}$ $\dfrac{12}{2} = 6$, $\dfrac{5}{2} = 2\,\mathbf{R}1$

(b) $\begin{array}{r} 71\ \mathbf{R}2 \\ 3\overline{)215} \end{array}$ $\dfrac{21}{3} = 7$, $\dfrac{5}{3} = 1\,\mathbf{R}2$

(c) $\begin{array}{r} 1\ 3\ 4\ \mathbf{R}2 \\ 4\overline{)5\,{}^{1}3\,{}^{1}8} \end{array}$ $\dfrac{5}{4} = 1\,\mathbf{R}1$, $\dfrac{13}{4} = 3\,\mathbf{R}1$,
$\dfrac{18}{4} = 4\,\mathbf{R}2$

(d) $\begin{array}{r} 1\ 6\ 3\ \mathbf{R}4 \\ 5\overline{)8\,{}^{3}1\,{}^{1}9} \end{array}$ $\dfrac{8}{5} = 1\,\mathbf{R}3$, $\dfrac{31}{5} = 6\,\mathbf{R}1$,
$\dfrac{19}{5} = 3\,\mathbf{R}4$

10. **(a)** $\begin{array}{r} 1\ 30\ \mathbf{R}3 \\ 4\overline{)5\,{}^{1}2\,3} \end{array}$

(b) $\begin{array}{r} 7\ 3\ \mathbf{R}4 \\ 7\overline{)5\,1\,{}^{2}5} \end{array}$

(c) $\begin{array}{r} 62\ 8\ \mathbf{R}1 \\ 3\overline{)1\,8\,8\,{}^{2}5} \end{array}$

(d) $\begin{array}{r} 2\ 3\ 5\ \mathbf{R}5 \\ 6\overline{)1\,4\,{}^{2}1\,{}^{3}5} \end{array}$

11. **(a)** $\begin{array}{r} 32\ \mathbf{R}1 \\ 2\overline{)65} \end{array}$

divisor × *quotient* + *remainder* = *dividend*

$$\begin{array}{ccccccc} \downarrow & & \downarrow & & \downarrow & & \downarrow \\ 2 & \times & 32 & + & 1 & & \\ & & 64 & + & 1 & = & 65 \end{array}$$

The answer is correct.

(b) $\begin{array}{r} 83\ \mathbf{R}4 \\ 7\overline{)586} \end{array}$

divisor × *quotient* + *remainder* = *dividend*

$$\begin{array}{ccccccc} \downarrow & & \downarrow & & \downarrow & & \downarrow \\ 7 & \times & 83 & + & 4 & & \\ & & 581 & + & 4 & = & 585 \\ & & & & & & \uparrow \\ & & & & & & \textit{incorrect} \end{array}$$

continued

$$\begin{array}{r} 8\ 3\ \mathbf{R}5 \\ 7\overline{)5\,8\,{}^{2}6} \end{array}$$

The correct answer is 83 **R**5.

(c) $\begin{array}{r} 407\ \mathbf{R}2 \\ 3\overline{)1223} \end{array}$

$$\begin{array}{ccccccc} \textit{divisor} \times & \textit{quotient} & + & \textit{remainder} & = & \textit{dividend} \\ \downarrow & \downarrow & & \downarrow & & \downarrow \\ 3 \times & 407 & + & 2 & & \\ & 1221 & + & 2 & = & 1223 \end{array}$$

The answer is correct.

(d) $\begin{array}{r} 476\ \mathbf{R}3 \\ 5\overline{)2383} \end{array}$

$$\begin{array}{ccccccc} \textit{divisor} \times & \textit{quotient} & + & \textit{remainder} & = & \textit{dividend} \\ \downarrow & \downarrow & & \downarrow & & \downarrow \\ 5 \times & 476 & + & 3 & & \\ & 2380 & + & 3 & = & 2383 \end{array}$$

The answer is correct.

12. **(a)** 258: ends in 8, divisible by 2

(b) 307: ends in 7, not divisible by 2

(c) 4216: ends in 6, divisible by 2

(d) 73,000: ends in 0, divisible by 2

13. **(a)** 743: $7 + 4 + 3 = 14$, not divisible by 3

(b) 5325: $5 + 3 + 2 + 5 = 15$, divisible by 3

(c) 374,214: $3 + 7 + 4 + 2 + 1 + 4 = 21$, divisible by 3

(d) 205,633: $2 + 0 + 5 + 6 + 3 + 3 = 19$, not divisible by 3

14. **(a)** 180: ends in 0, divisible by 5

(b) 635: ends in 5, divisible by 5

(c) 8364: does not end in 0 or 5, not divisible by 5

(d) 206,105: ends in 5, divisible by 5

15. **(a)** 270: ends in 0, divisible by 10

(b) 495: does not end in 0, not divisible by 10

(c) 5030: ends in 0, divisible by 10

(d) 14,380: ends in 0, divisible by 10

1.5 Section Exercises

1. $24 \div 4 = 6$: $\begin{array}{r} 6 \\ 4\overline{)24} \end{array}$ or $\frac{24}{4} = 6$

3. $\frac{45}{9} = 5$: $\begin{array}{r} 5 \\ 9\overline{)45} \end{array}$ or $45 \div 9 = 5$

5. $\begin{array}{r} 8 \\ 2\overline{)16} \end{array}$: $16 \div 2 = 8$ or $\frac{16}{2} = 8$

7. $9 \div 9 = 1$

Any nonzero number divided by itself is 1.

9. $\frac{14}{2} = 7$

11. $22 \div 0$ is undefined

13. $\frac{24}{1} = 24$

15. $\begin{array}{r} 0 \\ 15\overline{)0} \end{array}$

Zero divided by any nonzero number is zero.

17. $0\overline{)43}$ is undefined

19. $\frac{15}{1} = 15$

21. $\frac{8}{1} = 8$

23. $\begin{array}{r} 2\ 5 \\ 3\overline{)7\,{}^{1}5} \end{array}$ Check: $\begin{array}{r} 25 \\ \times\ 3 \\ \hline 75 \end{array}$

25. $\begin{array}{r} 1\ 8 \\ 7\overline{)1\,2\,{}^{5}6} \end{array}$ Check: $\begin{array}{r} 18 \\ \times\ 7 \\ \hline 126 \end{array}$

27. $\begin{array}{r} 3\,0\,4 \\ 4\overline{)1\,2\,1\,6} \end{array}$ Check: $\begin{array}{r} 304 \\ \times\ 4 \\ \hline 1216 \end{array}$

29. $\begin{array}{r} 6\ 2\ 7\ \mathbf{R}1 \\ 4\overline{)2\,5\,{}^{1}0\,{}^{2}9} \end{array}$

Check: $4 \times 627 + 1 = 2508 + 1 = 2509$

31. $\begin{array}{r} 1\ 5\ 2\ 2\ \mathbf{R}5 \\ 6\overline{)9\,{}^{3}1\,{}^{1}3\,{}^{1}7} \end{array}$

Check: $6 \times 1522 + 5 = 9132 + 5 = 9137$

33. $\begin{array}{r} 3\,0\,9 \\ 6\overline{)1\,8\,5\,4} \end{array}$

Check: $6 \times 309 = 1854$

35. $\begin{array}{r} 3\,005 \\ 4\overline{)12{,}020} \end{array}$

Check: $4 \times 3005 = 12{,}020$

37. $\begin{array}{r} 5\,006 \\ 6\overline{)30{,}036} \end{array}$ Check: $6 \times 5006 = 30{,}036$

39. $\begin{array}{r} 811\ \textbf{R1} \\ 3\overline{)2434} \end{array}$

Check: $3 \times 811 + 1 = 2433 + 1 = 2434$

41. $\begin{array}{r} 2\ 5\ 8\ 9\ \textbf{R2} \\ 5\overline{)12,{}^{2}9{}^{4}4{}^{4}7} \end{array}$

Check: $5 \times 2589 + 2 = 12{,}945 + 2 = 12{,}947$

43. $\begin{array}{r} 7\ \ 32\ 4\ \textbf{R2} \\ 4\overline{)29,{}^{1}29{}^{1}8} \end{array}$

Check: $4 \times 7324 + 2 = 29{,}296 + 2 = 29{,}298$

45. $\begin{array}{r} 3\ 1\ 5\ 7\ \textbf{R2} \\ 4\overline{)12,6{}^{2}3{}^{3}0} \end{array}$

Check: $4 \times 3157 + 2 = 12{,}628 + 2 = 12{,}630$

47. $\begin{array}{r} 2\ \ 6\ 30 \\ 8\overline{)21,{}^{5}0{}^{2}40} \end{array}$ Check: $8 \times 2630 = 21{,}040$

49. $\begin{array}{r} 1\ 2,\ 4\ 5\ 8\ \textbf{R3} \\ 6\overline{)7{}^{1}4,{}^{2}7{}^{3}5{}^{5}1} \end{array}$

Check: $6 \times 12{,}458 + 3 = 74{,}748 + 3 = 74{,}751$

51. $\begin{array}{r} 10,\ 2\ 5\ 3\ \textbf{R5} \\ 7\overline{)71,{}^{1}7{}^{3}7{}^{2}6} \end{array}$

Check: $7 \times 10{,}253 + 5 = 71{,}771 + 5 = 71{,}776$

53. $\begin{array}{r} 1\ 8,\ 3\ 7\ 7\ \textbf{R6} \\ 7\overline{)12{}^{5}8,{}^{2}6{}^{5}4{}^{5}5} \end{array}$

Check: $7 \times 18{,}377 + 6 = 128{,}639 + 6 = 128{,}645$

55. $\begin{array}{r} 375\ \textbf{R2} \\ 5\overline{)1877} \end{array}$

Check: $5 \times 375 + 2 = 1875 + 2 = 1877$ *correct*

57. $\begin{array}{r} 1908\ \textbf{R2} \\ 3\overline{)5725} \end{array}$

Check: $3 \times 1908 + 2 = 5724 + 2 = 5726$
incorrect

Rework:

$\begin{array}{r} 1\ 90\ 8\ \textbf{R1} \\ 3\overline{)5{}^{2}72{}^{2}5} \end{array}$

59. $\begin{array}{r} 650\ \textbf{R2} \\ 7\overline{)4692} \end{array}$

Check: $7 \times 650 + 2 = 4550 + 2 = 4552$
incorrect

Rework:

$\begin{array}{r} 6\ 70\ \textbf{R2} \\ 7\overline{)46{}^{4}92} \end{array}$

Check: $7 \times 670 + 2 = 4690 + 2 = 4692$ *correct*

61. $\begin{array}{r} 3\ 568\ \textbf{R2} \\ 6\overline{)21,409} \end{array}$

Check: $6 \times 3568 + 2 = 21{,}408 + 2 = 21{,}410$
incorrect

Rework:

$\begin{array}{r} 3\ \ 5\ 6\ 8\ \textbf{R1} \\ 6\overline{)21,{}^{3}4{}^{4}0{}^{4}9} \end{array}$

63. $\begin{array}{r} 2\ 002\ \textbf{R3} \\ 8\overline{)16,019} \end{array}$

Check: $8 \times 2002 + 3 = 16{,}016 + 3 = 16{,}019$
correct

65. $\begin{array}{r} 11{,}523\ \textbf{R2} \\ 6\overline{)69,140} \end{array}$

Check: $6 \times 11{,}523 + 2 = 69{,}138 + 2 = 69{,}140$
correct

67. $\begin{array}{r} 9\ 628\ \textbf{R7} \\ 9\overline{)86,655} \end{array}$

Check: $9 \times 9628 + 7 = 86{,}652 + 7 = 86{,}659$
incorrect

Rework:

$\begin{array}{r} 9\ \ 6\ 2\ 8\ \textbf{R3} \\ 9\overline{)86,{}^{5}6{}^{2}5{}^{7}5} \end{array}$

Check: $9 \times 9628 + 3 = 86{,}652 + 3 = 86{,}655$
correct

69. $\begin{array}{r} 27{,}822 \\ 8\overline{)222,576} \end{array}$

Check: $8 \times 27{,}822 = 222{,}576$ *correct*

71. Multiply the quotient by the divisor and add any remainder. The result should be the dividend.

73. $\begin{array}{r} 3\ 2\ 8 \\ 8\overline{)26{}^{2}2{}^{6}4} \end{array}$

328 tables can be set.

75. $\begin{array}{r} 11,\ 200 \\ 5\overline{)56,{}^{1}000} \end{array}$

The daily attendance was 11,200 people.

77. $\begin{array}{r} 4\ 8,\ 500 \\ 9\overline{)43{}^{7}6,{}^{4}500} \end{array}$

Each employee received $48,500.

79. $\begin{array}{r} 1\ 3\ 5 \\ 8\overline{)10{}^{2}8{}^{4}0} \end{array}$

135 acres can be fertilized.

81. $\begin{array}{r} 1,1\ 3\ 7,\ 5\ 0\ 0 \\ 6\overline{)6,8\,{}^{2}2\,{}^{4}5,{}^{3}0\,0\,0} \end{array}$

Each person received $1,137,500.

83. Kaci Salmon earns $36,540 in one year. Since there are 12 months in one year, $36,540 ÷ 12 = $3045 are the earnings for each month. For three months the earnings would be 3 × $3045 = $9135.

Alternate solution: Since there are four 3-month periods in a year, we can divide $36,540 by 4 to find her earnings. $36,540 ÷ 4 = $9135

85. 60 ends in 0, so it is divisible by 2, 5, and 10. The sum of its digits, 6, is divisible by 3, so 60 is divisible by 3.

87. 92 ends in 2, so it is divisible by 2, but not divisible by 5 or 10. The sum of its digits, 11, is not divisible by 3, so 92 is not divisible by 3.

89. 445 ends in 5, so it is divisible by 5, but not divisible by 2 or 10. The sum of its digits, 13, is not divisible by 3, so 445 is not divisible by 3.

91. 903 ends in 3, so it is not divisible by 2, 5, or 10. The sum of its digits, 12, is divisible by 3, so 903 is divisible by 3.

93. 5166 ends in 6, so it is divisible by 2, but not divisible by 5 or 10. The sum of its digits, 18, is divisible by 3, so 5166 is divisible by 3.

95. 21,763 ends in 3, so it is not divisible by 2, 5, or 10. The sum of its digits, 19, is not divisible by 3, so 21,763 is not divisible by 3.

1.6 Long Division

1.6 Margin Exercises

1. **(a)**
$$\begin{array}{rl} 82 & \\ 28\overline{)2296} & \\ 224 & \leftarrow 8\times 28 \\ \hline 56 & \\ 56 & \leftarrow 2\times 28 \\ \hline 0 & \end{array}$$

(b)
$$\begin{array}{rl} 64 & \\ 16\overline{)1024} & \\ 96 & \leftarrow 6\times 16 \\ \hline 64 & \\ 64 & \leftarrow 4\times 16 \\ \hline 0 & \end{array}$$

(c)
$$\begin{array}{rl} 144 & \\ 61\overline{)8784} & \\ 61 & \leftarrow 1\times 61 \\ \hline 268 & \\ 244 & \leftarrow 4\times 61 \\ \hline 244 & \\ 244 & \leftarrow 4\times 61 \\ \hline 0 & \end{array}$$

(d)
$$\begin{array}{rl} 29 & \\ 93\overline{)2697} & \\ 186 & \leftarrow 2\times 93 \\ \hline 837 & \\ 837 & \leftarrow 9\times 93 \\ \hline 0 & \end{array}$$

2. **(a)**
$$\begin{array}{rl} 56 & \\ 24\overline{)1344} & \\ 120 & \leftarrow 5\times 24 \\ \hline 144 & \\ 144 & \leftarrow 6\times 24 \\ \hline 0 & \end{array}$$

(b)
$$\begin{array}{rl} 62\ \mathbf{R8} & \\ 72\overline{)4472} & \\ 432 & \leftarrow 6\times 72 \\ \hline 152 & \\ 144 & \leftarrow 2\times 72 \\ \hline 8 & \end{array}$$

(c)
$$\begin{array}{rl} 83\ \mathbf{R21} & \\ 65\overline{)5416} & \\ 520 & \leftarrow 8\times 65 \\ \hline 216 & \\ 195 & \leftarrow 3\times 65 \\ \hline 21 & \end{array}$$

(d)
$$\begin{array}{rl} 74\ \mathbf{R63} & \\ 89\overline{)6649} & \\ 623 & \leftarrow 7\times 89 \\ \hline 419 & \\ 356 & \leftarrow 4\times 89 \\ \hline 63 & \end{array}$$

3. **(a)**
$$\begin{array}{rl} 107\ \mathbf{R4} & \\ 17\overline{)1823} & \\ 17 & \leftarrow 1\times 17 \\ \hline 123 & \\ 119 & \leftarrow 7\times 17 \\ \hline 4 & \end{array}$$

(b)
$$\begin{array}{rl} 208\ \mathbf{R7} & \\ 23\overline{)4791} & \\ 46 & \leftarrow 2\times 23 \\ \hline 191 & \\ 184 & \leftarrow 8\times 23 \\ \hline 7 & \end{array}$$

(c)

```
      4 0 8  R21
39 ) 1 5, 9 3 3
     1 5 6          ← 4 × 39
       3 3 3
       3 1 2        ← 8 × 39
           2 1
```

(d)

```
      3 0 0  R62
78 ) 2 3, 4 6 2
     2 3 4          ← 3 × 78
           6 2
```

4. **(a)** $70 \div 10 = 7$

One zero is dropped.

(b) $2600 \div 100 = 26$

Two zeros are dropped.

(c) $505{,}000 \div 1000 = 505$

Three zeros are dropped.

5. **(a)** 50) 6250

Drop 1 zero from the divisor and the dividend.

```
    1 2 5
5 ) 6 2 5
    5
    1 2
    1 0
      2 5
      2 5
        0
```

The quotient is 125.

(b) 130) 131,040

Drop 1 zero from the divisor and the dividend.

```
      1 0 0 8
13 ) 1 3, 1 0 4
     1 3
         1 0 4
         1 0 4
             0
```

The quotient is 1008.

(c) 3400) 190,400

Drop 2 zeros from the divisor and the dividend.

```
        5 6
34 ) 1 9 0 4
     1 7 0
       2 0 4
       2 0 4
           0
```

The quotient is 56.

6. **(a)**

```
      3 8           38   Multiply the
16 ) 6 0 8        × 16   quotient and
     4 8           228   the divisor.
     1 2 8         38
     1 2 8         608   ← correct
         0
```

(b)

```
        4 2  R178            426
426 ) 1 9, 1 7 0          ×   42
      1 7 0 4                852
        1 1 3 0            17 04
          9 5 2            17,892
          1 7 8          +    178
                           18,070
```

The result does not match the dividend. Rework.

```
        4 5
426 ) 1 9, 1 7 0
      1 7 0 4
        2 1 3 0
        2 1 3 0
              0
```

(c)

```
        5 7  R18
514 ) 2 9, 3 1 6
      2 5 7 0
        3 6 1 6
        3 5 9 8
            1 8
```

```
      514   Multiply the
  ×    57   quotient and
    3 598   the divisor.
   25 70
   29,298
  +    18   Add the remainder
   29,316   ← correct
```

1.6 Section Exercises

1.

```
      5
50 ) 2650      5; 53; 530
```

5 goes over the 5, because $\frac{265}{50}$ is about 5. The answer must then be a two-digit number or 53.

3.

```
      2
18 ) 4500      2; 25; 250
```

2 goes over the 5, because $\frac{45}{18}$ is about 2. The answer must then be a three-digit number or 250.

5.

```
       1
86 ) 10,327
```

12; 120 **R**7; 1200

1 goes over the 3, because $\frac{103}{86}$ is about 1. The answer must then be a three-digit number or 120 **R**7.

7.

```
       1
26 ) 28,735
```

11; 110; 1105 **R**5

1 goes over the 8, because $\frac{28}{26}$ is about 1. The answer must then be a four-digit number or 1105 **R**5.

9.

```
        7
21 ) 149,826
```

71; 713; 7134 **R**12

7 goes over the 9, because $\frac{149}{21}$ is about 7. The answer must then be a four-digit number or 7134 **R**12.

11.

```
         9
523 ) 470,800
```

9 **R**100; 90 **R**100; 900 **R**100

9 goes over the 8, because $\frac{4708}{523}$ is about 9. The answer must then be a three-digit number or 900 **R**100.

13.

```
        7 3 R5     Check:    73
18 ) 1 3 1 9                × 18
     1 2 6                   584
       5 9                   73
       5 4                  1314
         5                 +   5
                            1319
```

15.

```
        4 7 6 R15    Check:    476
23 ) 1 0, 9 6 3               × 23
       9 2                    1428
       1 7 6                  952
       1 6 1                 10,948
         1 5 3              +   15
         1 3 8               10,963
           1 5
```

17.

```
       2 4 0 7 R1   Check:   2407
26 ) 6 2, 5 8 3              × 26
     5 2                     14442
     1 0 5                   4814
     1 0 4                   62,582
         1 8 3              +    1
         1 8 2               62,583
             1
```

19.

```
       1 1 4 6 R15   Check:    1146
74 ) 8 4, 8 1 9                × 74
     7 4                       4584
     1 0 8                     8022
       7 4                     84,804
       3 4 1                  +    15
       2 9 6                   84,819
         4 5 9
         4 4 4
           1 5
```

21.

```
         3 3 3 1 R82   Check:    3331
153 ) 5 0 9, 7 2 5              × 153
      4 5 9                      9993
        5 0 7                   16655
        4 5 9                   3331
          4 8 2                509,643
          4 5 9               +     82
            2 3 5              509,725
            1 5 3
              8 2
```

23.

```
           8 5 0   Check:     850
420 ) 3 5 7, 0 0 0          × 420
      3 3 6 0                17000
        2 1 0 0              3400
        2 1 0 0             357,000
              0
```

25.

```
       1 0 1 R4   Check:     101
35 ) 3 5 4 9               ×  35
                             505
                             303
                            3535
                          +    4
                            3539 incorrect
```

Rework.

```
       1 0 1 R14
35 ) 3 5 4 9
     3 5
       0 4 9
         3 5
         1 4
```

The correct answer is 101 **R**14.

27.

```
        6 5 8 R9   Check:      658
28 ) 1 8, 4 2 4              ×  28
                              5 264
                              13 16
                             18,424
                            +     9
                             18,433 incorrect
```

Rework.

```
      6 5 8
28 ) 1 8, 4 2 4
     1 6 8
       1 6 2
       1 4 0
         2 2 4
         2 2 4
             0
```

The correct answer is 658.

29.

```
       6 2 R3   Check:     614
614 ) 3 8, 0 6 8        ×   62
                          1 228
                         36 84
                         38,068
                       +      3
                         38,071 incorrect
```

Rework.

```
          6 2
614 ) 3 8, 0 6 8
      3 6  8 4
         1 2 2 8
         1 2 2 8
               0
```

The correct answer is 62.

31. When dividing by 10, 100, or 1000, drop the same number of zeros from the dividend to get the quotient. One example is $2500 \div 100 = 25 \div 1 = 25$.

33.

```
        3 5
90 ) 3 1 5 0
     2 7 0
       4 5 0
       4 5 0
           0
```

It will take 35 days.

35. First, add the number of wall and table clocks.

$$\begin{array}{rl} 272 & \textit{wall clocks} \\ +\ 308 & \textit{table clocks} \\ \hline 580 & \textit{wall and table sum} \end{array}$$

Subtract the sum from the total number of clocks.

$$\begin{array}{rl} 636 & \textit{total number of clocks} \\ -\ 580 & \textit{wall and table sum} \\ \hline 56 & \textit{standing floor clocks} \end{array}$$

He worked on 56 floor clocks this year.

37. Divide.

```
         3 5 5
96 ) 3 4, 0 8 0
     2 8  8
       5  2 8
       4  8 0
          4 8 0
          4 8 0
              0
```

Judy's monthly payment is \$355.

39.

$$\begin{array}{rl} 1000 & \textit{wagons per hour} \\ \times\ \ 8 & \textit{hours per day} \\ \hline 8000 & \textit{wagons per day} \\ \times\ \ 22 & \textit{days} \\ \hline 176{,}000 & \textit{wagons} \end{array}$$

41. Divide the yearly amount spent for eating away from home by 52 (the number of weeks in one year).

```
         3 9
52 ) 2 0 2 8
     1 5 6
       4 6 8
       4 6 8
           0
```

The average weekly amount spent for eating away from home is \$39 per household.

43. $\frac{\$0}{3} = \$0$

44. When 0 is divided by any nonzero number, the result is 0.

45. $8 \div 0$ is undefined.

46. It is *impossible*. If you have 6 cookies, it is not possible to divide them among zero people.

47. **(a)** $14 \div 1 = 14$

(b)

```
    1 7
1 ) 1 7
```

(c) $\frac{38}{1} = 38$

48. Yes. Some examples are $18 \cdot 1 = 18$; $26 \cdot 1 = 26$; and $43 \cdot 1 = 43$.

49. **(a)** $32{,}000 \div 10 = 3200$

Drop 1 zero from the dividend and 1 zero from the divisor.

(b) $32{,}000 \div 100 = 320$

Drop 2 zeros from the dividend and 2 zeros from the divisor.

(c) $32{,}000 \div 1000 = 32$

Drop 3 zeros from the dividend and 3 zeros from the divisor.

50. Drop the same number of zeros that appear in the divisor. The result is the quotient. With the divisor 10, drop one zero; with 100, drop two zeros; with 1000, drop three zeros.

Summary Exercises on Whole Numbers

1. 631,548; ten thousands: 3; tens: 4

3. 9,181,576,423; hundred millions: 1; thousands: 6

5. 425,208,733 is four hundred twenty-five million, two hundred eight thousand, seven hundred thirty-three.

7.
$$\begin{array}{r} {\scriptstyle 11} \\ 166 \\ +\ 739 \\ \hline 905 \end{array}$$

9.
$$\begin{array}{r} {\scriptstyle 8\,18} \\ 798 \\ -\ 389 \\ \hline 409 \end{array}$$

11.
$$\begin{array}{r} {\scriptstyle 1\ 12} \\ 75 \\ 81{,}579 \\ 506 \\ +\quad 4 \\ \hline 82{,}164 \end{array}$$

13.
$$\begin{array}{r} {\scriptstyle 4\,14\ 9\,9\,10} \\ 55{,}000 \\ -\ 17{,}326 \\ \hline 37{,}674 \end{array}$$

15. $56 \times 10 = 560$ *Attach* 0.

17.
$$\begin{array}{r} 500 \\ \times\ 700 \\ \hline \end{array} \quad \begin{array}{r} 5 \\ \times\ 7 \\ \hline 35 \end{array} \quad \begin{array}{r} 500 \\ \times\ 700 \\ \hline 350{,}000 \end{array}$$
Attach 0000.

19. Eight million, six hundred forty-two thousand, one hundred eighty is written as 8,642,180.

21. $8 \div 8 = 1$

23. $\frac{12}{0}$ is undefined.

25. $7 \times 8 = 56$

27. $8 \cdot 4 \cdot 3 = (8 \cdot 4) \cdot 3 = 32 \cdot 3 = 96$

29.
$$\begin{array}{r} 2\ 7\ 5\ 0\ \textbf{R2} \\ 3\overline{)8\ {}^{2}2\ {}^{1}5\ 2} \end{array}$$

Check: $3 \times 2750 + 2 = 8250 + 2 = 8252$

31.
$$\begin{array}{r} 65 \\ \times\ 52 \\ \hline 130 \\ 325 \\ \hline 3380 \end{array}$$

33. $(28)(72)$
$$\begin{array}{rl} 72 & \\ \times\ 28 & \\ \hline 576 & \leftarrow 8 \times 72 \\ 144 & \leftarrow 2 \times 72 \\ \hline 2016 & \end{array}$$

35.
$$\begin{array}{rl} 78 & \\ 25\overline{)1950} & \\ 175 & \leftarrow 7 \times 25 \\ \hline 200 & \\ 200 & \leftarrow 8 \times 25 \\ \hline 0 & \end{array}$$

37.
$$\begin{array}{r} 3602 \\ \times\ 5008 \\ \hline 28816 \\ 1801000 \\ \hline 18{,}038{,}816 \end{array}$$

39. $630\overline{)32{,}760}$ Drop 1 zero.
$$\begin{array}{rl} 52 & \\ 63\overline{)3276} & \\ 315 & \leftarrow 5 \times 63 \\ \hline 126 & \\ 126 & \leftarrow 2 \times 63 \\ \hline 0 & \end{array}$$

1.7 Rounding Whole Numbers

1.7 Margin Exercises

1. (a) 373 (nearest ten)

Underline the tens place. 373

373 is closer to 370.

(b) 1482 (nearest thousand)

Underline the thousands place. 1482

1482 is closer to 1000.

(c) 89,512 (nearest hundred)

Underline the hundreds place. 89,512

89,512 is closer to 89,500.

(d) 546,325 (nearest ten thousand)

Underline the ten thousands place. 546,325

546,325 is closer to 550,000.

2. **(a)** 62 Next digit is 4 or less.

Underline the tens place. Change the digit to the right of the underlined place to zero.

62 rounded to the nearest ten is 60.

(b) 94 Next digit is 4 or less.

Underline the tens place. Change the digit to the right of the underlined place to zero.

94 rounded to the nearest ten is 90.

(c) 134 Next digit is 4 or less.

Underline the tens place. Change the digit to the right of the underlined place to zero.

134 rounded to the nearest ten is 130.

(d) 7543 Next digit is 4 or less.

Underline the tens place. Change the digit to the right of the underlined place to zero.

7543 rounded to the nearest ten is 7540.

3. **(a)** 3683 Next digit is 5 or more.

Change 3 to 4. All digits to the right of the underlined place are changed to zero.

3683 rounded to the nearest thousand is 4000.

(b) 6502 Next digit is 5 or more.

Change 6 to 7. All digits to the right of the underlined place are changed to zero.

6502 rounded to the nearest thousand is 7000.

(c) 84,621 Next digit is 5 or more.

Change 4 to 5. All digits to the right of the underlined place are changed to zero.

84,621 rounded to the nearest thousand is 85,000.

(d) 55,960 Next digit is 5 or more.

Change 5 to 6. All digits to the right of the underlined place are changed to zero.

55,960 rounded to the nearest thousand is 56,000.

4. **(a)** 3458 to the nearest ten

3458 Next digit is 5 or more.

Change 5 to 6. All digits to the right of the underlined place are changed to zeros.

3458 rounded to the nearest ten is 3460.

(b) 6448 to the nearest hundred

6448 Next digit is 4 or less.

Leave 4 as 4.

All digits to the right of the underlined place are changed to zeros.

6448 rounded to the nearest hundred is 6400.

(c) 73,077 to the nearest hundred

73,077 Next digit is 5 or more.

Change 0 to 1. All digits to the right of the underlined place are changed to zeros.

73,077 rounded to the nearest hundred is 73,100.

(d) 85,972 to the nearest hundred

85,972 Next digit is 5 or more.

Change 9 to 10. Write zero and carry the 1. All digits to the right of the underlined place are changed to zeros.

85,972 rounded to the nearest hundred is 86,000.

5. **(a)** 14,598 to the nearest ten thousand

14,598

The ten thousands place does not change because the digit to the right is 4 or less. All digits to the right of the underlined place are changed to zeros.

14,598 rounded to the nearest ten thousand is 10,000.

(b) 724,518,715 to the nearest million

724,518,715

Change 4 to 5 because the digit to the right is 5 or more.

All digits to the right of the underlined place are changed to zeros.

724,518,715 rounded to the nearest million is 725,000,000.

6. **(a) To the nearest ten:** 458

Next digit is 5 or more.

Tens place $(5 + 1 = 6)$ changes.

458 rounded to the nearest ten is 460.

To the nearest hundred: 458

Next digit is 5 or more.

Hundreds place $(4 + 1 = 5)$ changes. All digits to the right of the underlined place are changed to zeros.

458 rounded to the nearest hundred is 500.

(b) To the nearest ten: 549

Next digit is 5 or more.

Tens place $(4 + 1 = 5)$ changes.

549 rounded to the nearest ten is 550.

To the nearest hundred: 549

Next digit is 4 or less.

Hundreds place stays the same.

549 rounded to the nearest hundred is 500.

(c) To the nearest ten: 9308

Next digit is 5 or more.

Tens place $(0 + 1 = 1)$ changes.

9308 rounded to the nearest ten is 9310.

To the nearest hundred: 9308

Next digit is 4 or less.

Hundreds place stays the same.

9308 rounded to the nearest hundred is 9300.

7. **(a) To the nearest ten:** 4078

Next digit is 5 or more.

Tens place $(7 + 1 = 8)$ changes.

4078 rounded to the nearest ten is 4080.

To the nearest hundred: 4078

Next digit is 5 or more.

Hundreds place $(0 + 1 = 1)$ changes.

4078 rounded to the nearest hundred is 4100.

To the nearest thousand: 4078

Next digit is 4 or less.

Thousands place stays the same.

4078 rounded to the nearest thousand is 4000.

(b) To the nearest ten: 46,364

Next digit is 4 or less.

Tens place stays the same.

46,364 rounded to the nearest ten is 46,360.

To the nearest hundred: 46,364

Next digit is 5 or more.

Hundreds place $(3 + 1 = 4)$ changes.

46,364 rounded to the nearest hundred is 46,400.

To the nearest thousand: 46,364

Next digit is 4 or less.

Thousands place stays the same.

46,364 rounded to the nearest thousand is 46,000.

(c) To the nearest ten: 268,328

Next digit is 5 or more.

Tens place $(2 + 1 = 3)$ changes.

268,328 rounded to the nearest ten is 268,330.

To the nearest hundred: 268,328

Next digit is 4 or less.

Hundreds place stays the same.

268,328 rounded to the nearest hundred is 268,300.

To the nearest thousand: 268,328

Next digit is 4 or less.

Thousands place stays the same.

268,328 rounded to the nearest thousand is 268,000.

8. **(a)**

16	20	rounded to the nearest ten
74	70	
58	60	
+ 31	+ 30	
	180	estimated answer

(b)

53	50	rounded to the nearest ten
− 19	− 20	
	30	estimated answer

(c)

46	50	rounded to the nearest ten
× 74	× 70	
	3500	estimated answer

9. **(a)**

358	400	rounded to the nearest
743	700	hundred
822	800	
+ 978	+ 1000	
	2900	estimated answer

(b)

842	800	rounded to the nearest
− 475	− 500	hundred
	300	estimated answer

(c)

723	700	rounded to the nearest
× 478	× 500	hundred
	350,000	estimated answer

10. **(a)**

36	40	first digit rounded; all
3852	4000	others changed to zero
749	700	
+ 5474	+ 5000	
	9740	estimated answer

(b)

2583	3000	first digit rounded; all
− 765	− 800	others changed to zero
	2200	estimated answer

(c)

648	600	only one nonzero digit
× 67	× 70	remains
	42,000	estimated answer

1.7 Section Exercises

1. 624 rounded to the nearest ten: 620

6$\underline{2}$4 Next digit is 4 or less. Tens place does not change. All digits to the right of the underlined place change to zero.

3. 855 rounded to the nearest ten: 860

8$\underline{5}$5 Next digit is 5 or more. Tens place changes $(5 + 1 = 6)$. All digits to the right of the underlined place change to zero.

5. 6771 rounded to the nearest hundred: 6800

6$\underline{7}$71 Next digit is 5 or more. Hundreds place changes $(7 + 1 = 8)$. All digits to the right of the underlined place change to zero.

7. 86,813 rounded to the nearest hundred: 86,800

86,$\underline{8}$13 Next digit is 4 or less. Hundreds place does not change. All digits to the right of the underlined place change to zero.

9. 28,472 rounded to the nearest hundred: 28,500

28,$\underline{4}$72 Next digit is 5 or more. Hundreds place changes $(4 + 1 = 5)$. All digits to the right of the underlined place change to zero.

11. 5996 rounded to the nearest hundred: 6000

5$\underline{9}$96 Next digit is 5 or more. Hundreds place changes $(9 + 1 = 10)$. Write 0 and carry 1. All digits to the right of the underlined place change to zero.

13. 15,758 rounded to the nearest thousand: 16,000

1$\underline{5}$,758 Next digit is 5 or more. Thousands place changes $(5 + 1 = 6)$. All digits to the right of the underlined place change to zero.

15. 78,499 rounded to the nearest thousand: 78,000

7$\underline{8}$,499 Next digit is 4 or less. Thousands place does not change. All digits to the right of the underlined place change to zero.

17. 7760 rounded to the nearest thousand: 8000

$\underline{7}$760 Next digit is 5 or more. Thousands place changes $(7 + 1 = 8)$. All digits to the right of the underlined place change to zero.

19. 12,987 rounded to the nearest ten thousand: 10,000

$\underline{1}$2,987 Next digit is 4 or less. Ten thousands place does not change. All digits to the right of the underlined place change to zero.

21. 595,008 rounded to the nearest ten thousand: 600,000

5$\underline{9}$5,008 Next digit is 5 or more. Ten thousands place changes $(9 + 1 = 10)$. Write 0 and carry 1. All digits to the right of the underlined place change to zero.

23. 4,860,220 rounded to the nearest million: 5,000,000

$\underline{4}$,860,220 Next digit is 5 or more. Millions place changes $(4 + 1 = 5)$. All digits to the right of the underlined place change to zero.

25. **To the nearest ten:** 44$\underline{7}$6

Next digit is 5 or more. All digits to the right of the underlined place are changed to zero. Add 1 to 7. **4480**

To the nearest hundred: 4$\underline{4}$76

Next digit is 5 or more. All digits to the right of the underlined place are changed to zero. Add 1 to 4. **4500**

To the nearest thousand: $\underline{4}$476

Next digit is 4 or less. All digits to the right of the underlined place are changed to zero. Leave 4 as 4. **4000**

27. **To the nearest ten:** 33$\underline{7}$4

Next digit is 4 or less. All digits to the right of the underlined place are changed to zero. Leave 7 as 7. **3370**

To the nearest hundred: 3$\underline{3}$74

Next digit is 5 or more. All digits to the right of the underlined place are changed to zero. Add 1 to 3. **3400**

To the nearest thousand: $\underline{3}$374

Next digit is 4 or less. All digits to the right of the underlined place are changed to zero. Leave 3 as 3. **3000**

29. **To the nearest ten:** $60\underline{4}8$

Next digit is 5 or more. All digits to the right of the underlined place are changed to zero. Add 1 to 4. **6050**

To the nearest hundred: $6\underline{0}48$

Next digit is 4 or less. All digits to the right of the underlined place are changed to zero. Leave 0 as 0. **6000**

To the nearest thousand: $\underline{6}048$

Next digit is 4 or less. All digits to the right of the underlined place are changed to zero. Leave 6 as 6. **6000**

31. **To the nearest ten:** $53\underline{4}3$

Next digit is 4 or less. All digits to the right of the underlined place are changed to zero. Leave 4 as 4. **5340**

To the nearest hundred: $5\underline{3}43$

Next digit is 4 or less. All digits to the right of the underlined place are changed to zero. Leave 3 as 3. **5300**

To the nearest thousand: $\underline{5}343$

Next digit is 4 or less. All digits to the right of the underlined place are changed to zero. Leave 5 as 5. **5000**

33. **To the nearest ten:** $19{,}5\underline{3}9$

Next digit is 5 or more. All digits to the right of the underlined place are changed to zero. Add 1 to 3. **19,540**

To the nearest hundred: $19{,}\underline{5}39$

Next digit is 4 or less. All digits to the right of the underlined place are changed to zero. Leave 5 as 5. **19,500**

To the nearest thousand: $1\underline{9}{,}539$

Next digit is 5 or more. All digits to the right of the underlined place are changed to zero. Add 1 to 9. Write 0 and carry 1. **20,000**

35. **To the nearest ten:** $26{,}2\underline{9}2$

Next digit is 4 or less. All digits to the right of the underlined place are changed to zero. Leave 9 as 9. **26,290**

To the nearest hundred: $26{,}\underline{2}92$

Next digit is 5 or more. All digits to the right of the underlined place are changed to zero. Add 1 to 2. **26,300**

To the nearest thousand: $2\underline{6}{,}292$

Next digit is 4 or less. All digits to the right of the underlined place are changed to zero. Leave 6 as 6. **26,000**

37. **To the nearest ten:** $93{,}7\underline{0}6$

Next digit is 5 or more. All digits to the right of the underlined place are changed to zero. Add 1 to 0. **93,710**

To the nearest hundred: $93{,}\underline{7}06$

Next digit is 4 or less. All digits to the right of the underlined place are changed to zero. Leave 7 as 7. **93,700**

To the nearest thousand: $9\underline{3}{,}706$

Next digit is 5 or more. All digits to the right of the underlined place are changed to zero. Add 1 to 3. **94,000**

39. *Step 1* Locate the place to be rounded and underline it.

Step 2 Look only at the next digit to the right. If this digit is 5 or more, increase the underlined digit by 1.

Step 3 Change all digits to the right of the underlined place to zeros.

41. *Estimate:* *Exact:*

Estimate	Exact
30	25
60	63
50	47
+ 80	+ 84
220	219

43. *Estimate:* *Exact:*

Estimate	Exact
80	78
− 40	− 43
40	35

45. *Estimate:* *Exact:*

Estimate	Exact
70	67
× 30	× 34
2100	268
	201
	2278

47. *Estimate:* *Exact:*

$$\begin{array}{r} 900 \\ 700 \\ 400 \\ +\,800 \\ \hline 2800 \end{array} \qquad \begin{array}{r} 863 \\ 735 \\ 438 \\ +\,792 \\ \hline 2828 \end{array}$$

49. *Estimate:* *Exact:*

$$\begin{array}{r} 900 \\ -\,400 \\ \hline 500 \end{array} \qquad \begin{array}{r} 883 \\ -\,448 \\ \hline 435 \end{array}$$

51. *Estimate:* *Exact:*

$$\begin{array}{r} 800 \\ \times\,400 \\ \hline 320{,}000 \end{array} \qquad \begin{array}{r} 752 \\ \times\,375 \\ \hline 3760 \\ 5264 \\ 2256 \\ \hline 282{,}000 \end{array}$$

53. *Estimate:* *Exact:*

$$\begin{array}{r} 8000 \\ 60 \\ 700 \\ +\,4000 \\ \hline 12{,}760 \end{array} \qquad \begin{array}{r} 8215 \\ 56 \\ 729 \\ +\,3605 \\ \hline 12{,}605 \end{array}$$

55. *Estimate:* *Exact:*

$$\begin{array}{r} 700 \\ -\,500 \\ \hline 200 \end{array} \qquad \begin{array}{r} 687 \\ -\,529 \\ \hline 158 \end{array}$$

57. *Estimate:* *Exact:*

$$\begin{array}{r} 900 \\ \times\,30 \\ \hline 27{,}000 \end{array} \qquad \begin{array}{r} 939 \\ \times\,29 \\ \hline 8451 \\ 1878 \\ \hline 27{,}231 \end{array}$$

59. Perhaps the best explanation is that 3492 is closer to 3500 than 3400, but 3492 is closer to 3000 than to 4000.

61. $\underline{7}6{,}000{,}000$ Next digit is 5 or more. Change digit to the right of the underlined place to zero. Add 1 to 7. **80 million people**

$2\underline{9}6{,}000{,}000$ Next digit is 5 or more. Change digit to the right of the underlined place to zero. Add 1 to 9. Write 0 and carry 1.
300 million people

63. $\underline{4}7$ Next digit is 5 or more. Change digit to the right of the underlined place to zero. Add 1 to 4.
50 years

$\underline{7}6$ Next digit is 5 or more. Change digit to the right of the underlined place to zero. Add 1 to 7.
80 years

65. **To the nearest ten thousand:** $1{,}6\underline{6}7{,}300$

Next digit is 5 or more, so add 1 to 6. Change digits to the right of the underlined place to zeros.

1,670,000 pounds

To the nearest hundred thousand: $1{,}\underline{6}67{,}300$

Next digit is 5 or more, so add 1 to 6. Change digits to the right of the underlined place to zeros.

1,700,000 pounds

To the nearest million: $\underline{1}{,}667{,}300$

Next digit is 5 or more, so add 1 to 1. Change digits to the right of the underlined place to zeros.

2,000,000 pounds

67. **To the nearest hundred thousand:** $\$25{,}765{,}\underline{4}75{,}000$

Next digit is 5 or more. All digits to the right of the underlined place are changed to zero. Add 1 to 4. **$25,765,500,000**

To the nearest hundred million: $\$25{,}\underline{7}65{,}475{,}000$

Next digit is 5 or more. All digits to the right of the underlined place are changed to zero. Add 1 to 7. **$25,800,000,000**

To the nearest billion: $\$2\underline{5}{,}765{,}475{,}000$

Next digit is 5 or more. All digits to the right of the underlined place are changed to zero. Add 1 to 5. **$26,000,000,000**

69. 71,499 would round to 71,000, so the smallest whole number it could have been is 71,500.

70. 72,500 would round to 73,000, so the largest whole number it could have been is 72,499.

71. 7500 is the smallest possible whole number that will round to 8000 using front end rounding.

72. 8499 is the largest possible whole number that will round to 8000 using front end rounding.

73. **Exact:** **Rounding to the nearest ten:**

Exact	Rounding to the nearest ten
1155	1160
998	1000
979	980
937	940
794	790

74. **Exact:** **Front end rounded:**

Exact	Front end rounded
1155	1000
998	1000
979	1000
937	900
794	800

75. **(a)** When using front end rounding, all digits are 0 except the first digit. These numbers are easier to work with when estimating answers.

(b) Sometimes when using front end rounding, the estimated answer can vary greatly from the exact answer.

1.8 Exponents, Roots, and Order of Operations

1.8 Margin Exercises

1. **(a)** 4^2: exponent is 2; base is 4.

$4^2 = 4 \times 4 = 16$

(b) 5^3: exponent is 3; base is 5.

$5^3 = 5 \times 5 \times 5 = 125$

(c) 3^4: exponent is 4; base is 3.

$3^4 = 3 \times 3 \times 3 \times 3 = 81$

(d) 2^6: exponent is 6; base is 2.

$2^6 = 2 \times 2 \times 2 \times 2 \times 2 \times 2 = 64$

2. **(a)** Because $2^2 = 4$, $\sqrt{4} = 2$.

(b) Because $5^2 = 25$, $\sqrt{25} = 5$.

(c) Because $6^2 = 36$, $\sqrt{36} = 6$.

(d) Because $15^2 = 225$, $\sqrt{225} = 15$.

(e) Because $1^2 = 1$, $\sqrt{1} = 1$.

3. **(a)** $4 + 5 + 2^2$ *Evaluate exponent*
$4 + 5 + 2 \cdot 2$ *Multiply*
$4 + 5 + 4$ *Add from left to right*
$9 + 4 = 13$

(b) $3^2 + 2^3$ *Evaluate exponent*
$3 \cdot 3 + 2 \cdot 2 \cdot 2$ *Multiply from left to right*
$9 + 8 = 17$ *Add*

(c) $4 \cdot 6 \div 12 - 2$ *Multiply*
$24 \div 12 - 2$ *Divide*
$2 - 2 = 0$ *Subtract*

(d) $60 \div 6 \div 2$ *Divide*
$10 \div 2 = 5$ *Divide*

(e) $8 + 6(14 \div 2)$ *Work inside parentheses*
$8 + 6(7)$ *Multiply*
$8 + 42 = 50$ *Add*

4. **(a)** $12 - 6 + 4^2$ *Evaluate exponent*
$12 - 6 + 16$ *Subtract*
$6 + 16 = 22$ *Add*

(b) $2^3 + 3^2 - (5 \cdot 3)$ *Work inside parentheses*
$2^3 + 3^2 - 15$ *Evaluate exponents*
$8 + 9 - 15$ *Add*
$17 - 15 = 2$ *Subtract*

(c) $2 \cdot \sqrt{64} - 5 \cdot 3$ *Square root*
$2 \cdot 8 - 5 \cdot 3$ *Multiply*
$16 - 15 = 1$ *Subtract*

(d) $20 \div 2 + (7 - 5)$ *Parentheses*
$20 \div 2 + 2$ *Divide*
$10 + 2 = 12$ *Add*

(e) $15 \cdot \sqrt{9} - 8 \cdot \sqrt{4}$ *Square roots*
$15 \cdot 3 - 8 \cdot 2$ *Multiply*
$45 - 16 = 29$ *Subtract*

1.8 Section Exercises

1. 3^2: exponent is 2, base is 3.

$3^2 = 3 \cdot 3 = 9$

3. 5^2: exponent is 2, base is 5.

$5^2 = 5 \cdot 5 = 25$

5. 8^2: exponent is 2, base is 8.

$8^2 = 8 \cdot 8 = 64$

7. 15^2: exponent is 2, base is 15.

$15^2 = 15 \cdot 15 = 225$

9. From the table, $4^2 = 16$, so $\sqrt{16} = 4$.

11. From the table, $8^2 = 64$, so $\sqrt{64} = 8$.

13. From the table, $10^2 = 100$, so $\sqrt{100} = 10$.

15. From the table, $12^2 = 144$, so $\sqrt{144} = 12$.

17. $6^2 = \underline{36}$, so $\sqrt{36} = 6$.

19. $20^2 = \underline{400}$, so $\sqrt{400} = 20$.

21. $35^2 = \underline{1225}$, so $\sqrt{1225} = 35$.

23. $30^2 = \underline{900}$, so $\sqrt{900} = 30$.

25. $100^2 = \underline{10{,}000}$, so $\sqrt{10{,}000} = 100$.

27. A perfect square is the square of a whole number. The number 25 is the square of 5 because $5 \cdot 5 = 25$.

The number 50 is not a perfect square. There is no whole number that can be squared to get 50.

29. $3^2 + 8 - 5$ *Exponent*
$9 + 8 - 5$ *Add*
$17 - 5 = 12$ *Subtract*

31. $3 \cdot 7 - 6$ *Multiply*
$21 - 6 = 15$ *Subtract*

33. $8 \cdot 5 \div 10$ *Multiply*
$40 \div 10 = 4$ *Divide*

35. $25 \div 5(8 - 4)$ *Parentheses*
$25 \div 5(4)$ *Divide*
$5(4) = 20$ *Multiply*

37. $5 \cdot 3^2 + \frac{0}{8}$ *Exponent*
$5 \cdot 9 + \frac{0}{8}$ *Multiply*
$45 + \frac{0}{8}$ *Divide*
$45 + 0 = 45$ *Add*

39. $4 \cdot 1 + 8(9 - 2) + 3$ *Parentheses*
$4 \cdot 1 + 8 \cdot 7 + 3$ *Multiply*
$4 + 56 + 3$ *Add*
$60 + 3 = 63$ *Add*

41. $2^2 \cdot 3^3 + (20 - 15) \cdot 2$ *Parentheses*
$2^2 \cdot 3^3 + 5 \cdot 2$ *Exponents*
$4 \cdot 27 + 5 \cdot 2$ *Multiply*
$108 + 10 = 118$ *Add*

43. $5 \cdot \sqrt{36} - 2(4)$ *Square root*
$5 \cdot 6 - 2(4)$ *Multiply*
$30 - 8 = 22$ *Subtract*

45. $8(2) + 3 \cdot 7 - 7$ *Multiply*
$16 + 21 - 7$ *Add*
$37 - 7 = 30$ *Subtract*

47. $2^3 \cdot 3^2 + 3(14 - 4)$ *Parentheses*
$2^3 \cdot 3^2 + 3(10)$ *Exponents*
$8 \cdot 9 + 3(10)$ *Multiply*
$72 + 30 = 102$ *Add*

49. $7 + 8 \div 4 + \frac{0}{7}$ *Divide*
$7 + 2 + \frac{0}{7}$ *Divide*
$7 + 2 + 0$ *Add*
$9 + 0 = 9$ *Add*

51. $3^2 + 6^2 + (30 - 21) \cdot 2$ *Parentheses*
$3^2 + 6^2 + 9 \cdot 2$ *Exponents*
$9 + 36 + 9 \cdot 2$ *Multiply*
$9 + 36 + 18$ *Add*
$45 + 18 = 63$ *Add*

53. $7 \cdot \sqrt{81} - 5 \cdot 6$ *Square root*
$7 \cdot 9 - 5 \cdot 6$ *Multiply*
$63 - 30 = 33$ *Subtract*

55. $8 \cdot 2 + 5(3 \cdot 4) - 6$ *Parentheses*
$8 \cdot 2 + 5(12) - 6$ *Multiply*
$16 + 60 - 6$ *Add*
$76 - 6 = 70$ *Subtract*

57. $4 \cdot \sqrt{49} - 7(5 - 2)$ *Parentheses*
$4 \cdot \sqrt{49} - 7 \cdot 3$ *Square root*
$4 \cdot 7 - 7 \cdot 3$ *Multiply*
$28 - 21 = 7$ *Subtract*

59. $7(4 - 2) + \sqrt{9}$ *Parentheses*
$7(2) + \sqrt{9}$ *Square root*
$7(2) + 3$ *Multiply*
$14 + 3 = 17$ *Add*

61. $7^2 + 3^2 - 8 + 5$ *Exponents*
$49 + 9 - 8 + 5$ *Add*
$58 - 8 + 5$ *Subtract*
$50 + 5 = 55$ *Add*

63. $5^2 \cdot 2^2 + (8 - 4) \cdot 2$ *Parentheses*
$5^2 \cdot 2^2 + 4 \cdot 2$ *Exponents*
$25 \cdot 4 + 4 \cdot 2$ *Multiply*
$100 + 8 = 108$ *Add*

65. $5 + 9 \div 3 + 6 \cdot 3$ *Divide*
$5 + 3 + 6 \cdot 3$ *Multiply*
$5 + 3 + 18$ *Add*
$8 + 18 = 26$ *Add*

67. $8 \cdot \sqrt{49} - 6(9 - 4)$ *Parentheses*
$8 \cdot \sqrt{49} - 6(5)$ *Square root*
$8 \cdot 7 - 6(5)$ *Multiply*
$56 - 30 = 26$ *Subtract*

69. $5^2 - 4^2 + 3 \cdot 6$ *Exponent*
$25 - 16 + 3 \cdot 6$ *Multiply*
$25 - 16 + 18$ *Subtract*
$9 + 18 = 27$ *Add*

71. $8+8\div 8+6+\frac{5}{5}$ *Divide*
$8+1+6+1$ *Add*
$9+6+1$ *Add*
$15+1=16$ *Add*

73. $6\cdot\sqrt{25}-7(2)$ *Square root*
$6\cdot 5-7\cdot 2$ *Multiply*
$30-14=16$ *Subtract*

75. $9\cdot\sqrt{16}-3\cdot\sqrt{25}$ *Square roots*
$9\cdot 4-3\cdot 5$ *Multiply*
$36-15=21$ *Subtract*

77. $7\div 1\cdot 8\cdot 2\div(21-5)$ *Parentheses*
$7\div 1\cdot 8\cdot 2\div 16$ *Divide*
$7\cdot 8\cdot 2\div 16$ *Multiply*
$56\cdot 2\div 16$ *Multiply*
$112\div 16=7$ *Divide*

79. $15\div 3\cdot 2\cdot 6\div(14-11)$ *Parentheses*
$15\div 3\cdot 2\cdot 6\div 3$ *Divide*
$5\cdot 2\cdot 6\div 3$ *Multiply*
$10\cdot 6\div 3$ *Multiply*
$60\div 3=20$ *Divide*

81. $6\cdot\sqrt{25}-4\cdot\sqrt{16}$ *Square roots*
$6\cdot 5-4\cdot 4$ *Multiply*
$30-16=14$ *Subtract*

83. $5\div 1\cdot 10\cdot 4\div(17-9)$ *Parentheses*
$5\div 1\cdot 10\cdot 4\div 8$ *Divide*
$5\cdot 10\cdot 4\div 8$ *Multiply*
$50\cdot 4\div 8$ *Multiply*
$200\div 8=25$ *Divide*

85. $8\cdot 9\div\sqrt{36}-4\div 2+(14-8)$ *Parentheses*
$8\cdot 9\div\sqrt{36}-4\div 2+6$ *Square root*
$8\cdot 9\div 6-4\div 2+6$ *Multiply*
$72\div 6-4\div 2+6$ *Divide*
$12-2+6$ *Subtract*
$10+6=16$ *Add*

87. $2+1-2\cdot\sqrt{1}+4\cdot\sqrt{81}-7\cdot 2$ *Square roots*
$2+1-2\cdot 1+4\cdot 9-7\cdot 2$ *Multiply*
$2+1-2+36-14$ *Add*
$3-2+36-14$ *Subtract*
$1+36-14$ *Add*
$37-14=23$ *Subtract*

89. $5\cdot\sqrt{36}\cdot\sqrt{100}\div 4\cdot\sqrt{9}+8$ *Square roots*
$5\cdot 6\cdot 10\div 4\cdot 3+8$ *Multiply*
$30\cdot 10\div 4\cdot 3+8$ *Multiply*
$300\div 4\cdot 3+8$ *Divide*
$75\cdot 3+8$ *Multiply*
$225+8=233$ *Add*

1.9 Reading Pictographs, Bar Graphs, and Line Graphs

1.9 Margin Exercises

1. **(a)** The United Kingdom had the greatest number of U.S. students attending their colleges and universities since that row has the most pencil symbols.

(b) There is about one more pencil in France's row than Mexico's row. Since one pencil represents 5000 students, there are about 5000 more students studying in France than in Mexico.

2. **(a)** Pro football: 28 out of 100

(b) Pro baseball: 17 out of 100

(c) Pro basketball: 13 out of 100

(d) College basketball: 6 out of 100

(e) Golf: 6 out of 100

3. **(a)** For the year 2050, the dot is very close to the horizontal line marked 4, so the predicted population is about 400,000,000 people in 2050.

(b) About 475,000,000 people in 2075

(c) About 575,000,000 people in 2100

1.9 Section Exercises

1. $9\cdot 500=4500$ stores

3. From the pictograph, Dollar General has the greatest number of retail stores.

5. $5750-4500=1250$ stores

7. From the bar graph, 9 people out of 100 found their career as a result of training for a job.

9. **(a)** According to the bar graph, the greatest number of people found their job by seeing an ad.

(b) From the bar graph, 25 people out of 100 found their career as a result of seeing an ad.

11. According to the bar graph, $18-9=9$ more people out of 100 found their career as a result of "Studied in school" than "Luck or chance."

13. According to the line graph, the year with the greatest number of home sales is 2006, with 7000 home sales.

15. According to the line graph the increase in the number of home sales from 2005 to 2006 is $7000-2500=4500$.

17. Possible answers are

1. shortage of homes for sale
2. lack of qualified buyers
3. poor economy
4. high interest rates on home loans.

19. We want to insert parentheses in $7 - 2 \cdot 3 - 6$ so that it evaluates to 9. There isn't a method to use, so just use trial and error.

$$(7-2) \cdot 3 - 6 = 5 \cdot 3 - 6 = 15 - 6 = 9$$

20. $(4+2) \cdot (5+1) = 6 \cdot 6 = 36$

21. $36 \div (3 \cdot 3) \cdot 4 = 36 \div 9 \cdot 4 = 4 \cdot 4 = 16$

22. $56 \div (2 \cdot 2 \cdot 2) + \frac{0}{6} = 56 \div 8 + 0 = 7 + 0 = 7$

23. **(a)** $100 + 50 + 65 + (50 - 15) + (100 - 65) + 15$

(b) $100 + 50 + 65 + (50 - 15) + (100 - 65) + 15$

$100 + 50 + 65 + 35 + 35 + 15$

$150 + 65 + 35 + 35 + 15$

$215 + 35 + 35 + 15$

$250 + 35 + 15$

$285 + 15$

300 *amount of fencing for one corral*

$12 \times 300 = 3600$ feet for 12 corrals

1.10 Solving Application Problems

1.10 Margin Exercises

1. **(a)** A grocery clerk's hourly wage: \$1.40, [\$14], \$140

(b) The total length of five sports utility vehicles: 8 feet; 18 feet; [80 feet]; 800 feet

(c) The cost of heart bypass surgery: \$1000; [\$100,000]; \$10,000,000

2. **(a)** *Step 1*
The total number of fossils is given, and the number each person receives must be found.

Step 2
"Divided equally" indicates division should be used.

Step 3
80 fossils divided equally among 4 people gives an estimate of 20 fossils per person.

Step 4

$$4\overline{)84}\quad = 21$$

Step 5
Each person receives 21 fossils.

Step 6

21 *amount received by each person*
× 4 *number of people*
84 *total fossils; matches*

(b) *Step 1*
The total number of children and the number of children assigned to each counselor is given. The number of counselors needed must be found.

Step 2
"12 children are assigned to *each* camp counselor" indicates division should be used.

Step 3
If we replace 408 and 12 with 400 and 10, our estimate is $400 \div 10 = 40$.

Step 4

$$12\overline{)408} = 34$$
$$36$$
$$48$$
$$48$$
$$0$$

Step 5
There are 34 counselors needed.

Step 6

34 *counselors*
× 12 *children per counselor*
408 *total children; matches*

3. **(a)** *Step 1*
The number of points on examinations and quizzes is given. The total number of points must be found.

Step 2
"Her total points" indicates addition should be used.

Step 3
Rounding each score to the nearest ten gives 90, 80, 80, 100, 20, 10, 20, and 10. The sum of these scores gives an estimate of 410 points.

Step 4

$$\begin{array}{r} \overset{3}{92} \\ 81 \\ 83 \\ 98 \\ 15 \\ 14 \\ 15 \\ +\ 12 \\ \hline 410 \end{array}$$ *matches estimate*

Step 5
Her total is 410 points.

Step 6
The answer is the same as the estimate, so it is reasonable. Add the numbers again to check.

(b) *Step 1*
The customer contacts for each day are given. The total number of contacts for the week must be found.

Step 2
Adding the daily numbers to find the weekly total seems reasonable.

Step 3
Rounding each day's contacts to the nearest ten gives 80, 60, 120, 100, and 200. The sum gives an estimate of 560 customer contacts.

Step 4

$$\begin{array}{r} \overset{22}{78} \\ 64 \\ 118 \\ 102 \\ +196 \\ \hline 558 \end{array}$$

Step 5
Stephanie had 558 customer contacts for the week.

Step 6
The answer is reasonable. A check shows that the answer is correct.

4. **(a)** *Step 1*
The number of square feet for the home and the apartment is given. The difference must be found.

Step 2
"Difference" indicates subtraction should be used.

Step 3
Rounding each number of square feet to the nearest hundred gives $1500 - 1000 = 500$ square feet.

Step 4

$$\begin{array}{r} 1450 \\ -\ 980 \\ \hline 470 \end{array}$$

Step 5
The difference in the number of square feet is 470 square feet.

Step 6
The answer is reasonable.

Check: $\begin{array}{r} 980 \\ +\ 470 \\ \hline 1450 \end{array}$ *matches*

(b) *Step 1*
We know the beginning balance and the check amount, and need to find the new balance.

Step 2
Subtracting the numbers seems reasonable.

Step 3
Using the values $14,900 and $1200 in place of $14,863 and $1180 gives us
$\$14,900 - \$1200 = \$13,700$ as an estimate.

Step 4

$$\begin{array}{r} 14,863 \\ -\ 1180 \\ \hline 13,683 \end{array}$$

Step 5
The amount remaining in the club account is $13,683.

Step 6
The answer is reasonable.

Check: $\begin{array}{r} 13,683 \\ +\ 1180 \\ \hline 14,863 \end{array}$ *matches*

5. **(a)** *Step 1*
The amount per job, number of jobs, and expenses are given. The amount made is to be found.

Step 2
Use multiplication to find the amount received for all jobs, and then use subtraction to find the amount made.

Step 3
An estimate would be
$6 \cdot \$700 - \$300 = \$4200 - \$300 = \$3900.$

Step 4

$$\begin{array}{r} \overset{53}{\$685} \\ \times\ \ 6 \\ \hline \$4110 \end{array} \qquad \begin{array}{r} \$4110 \\ -\ 320 \\ \hline \$3790 \end{array}$$

Step 5
Brenda will make $3790 after deducting her expenses.

Step 6
The answer is close to the estimate, so it is reasonable. Check by adding the expenses to the amount made and then dividing by 6.
$3790 + $320 = $4110

$$\begin{array}{r} 685 \\ 6\overline{)4110} \\ \underline{36} \\ 51 \\ \underline{48} \\ 30 \end{array}$$

$685 matches the amount per job given in the problem.

(b) *Step 1*
The number of books sold and the number of those returned is given. The profit for each book sold is also given. The total profit on the books sold and not returned must be found.

Step 2
"Remaining" indicates subtraction. The number of books returned should be subtracted from the number sold and that number multiplied by the profit per book.

Step 3
Estimate that 13,000 books are sold, and 1000 of those are returned. A reasonable estimate is 13,000 − 1000 = 12,000. Each book has a profit of $6, so 12,000 × $6 = $72,000.

Step 4

$$\begin{array}{r} 12{,}628 \\ -\ 863 \\ \hline 11{,}765 \end{array} \qquad \begin{array}{r} 11{,}765 \\ \times\ \ 6 \\ \hline \$70{,}590 \end{array}$$

Step 5
The total profit for the books remaining after the returns is $70,590.

Step 6
The answer is reasonable since it is close to the estimate.

Check: $\begin{array}{r} 11{,}765 \\ 6\overline{)70{,}590} \end{array} \qquad \begin{array}{r} 11{,}765 \\ +\ 863 \\ \hline 12{,}628 \end{array}$

1.10 Section Exercises

1. *Step 1*
Given the number of invoices sent out for five days, find the total number of invoices sent out in a five-day period.

Step 2
Addition seems reasonable since we want a total.

Step 3
The total number of invoices should be about 80 + 80 + 100 + 40 + 50 = 350 invoices.

Step 4

$$\begin{array}{r} {\scriptstyle 21} \\ 80 \\ 75 \\ 135 \\ 40 \\ +\ 52 \\ \hline 382 \end{array}$$

Step 5
The total number of invoices in a five-day period is 382.

Step 6
The answer is reasonably close to the estimate. Check by adding the values again.

3. *Step 1*
Given the average credit card debt for a Junior and a Sophomore, find the difference.

Step 2
"How much more" indicates subtraction could be used.

Step 3
An estimate is $3000 − $2000 = $1000.

Step 4

$$\begin{array}{r} \$2705 \\ -\ 1825 \\ \hline \$880 \end{array}$$

Step 5
On average, a Junior owes $880 more than a Sophomore on credit card debt.

Step 6
The answer is reasonably close to the estimate.
Check: $880 + $1825 = $2705

5. *Step 1*
The number of kits packaged in one hour is given and the total number of kits packaged in 24 hours must be found.

Step 2
Multiply the number of kits packaged in 1 hour by 24.

Step 3
A reasonable answer would be 200 × 20 = 4000.

Step 4

$$\begin{array}{r} 236 \\ \times\ 24 \\ \hline 944 \\ 472 \\ \hline 5664 \end{array}$$

Step 5
5664 kits are packaged in 24 hours.

Step 6
The answer is reasonably close to the estimate, which is clearly low.

Check: $24\overline{)5664}$ = 236

7. *Step 1*
Find the number of toys each child will receive.

Step 2
"Same number of toys to each" indicates division should be used.

Step 3
3000 divided by 700 gives an estimate of about 4 toys.

Step 4

$$657\overline{)2628} = 4$$
$$\begin{array}{r} 2628 \\ \hline 0 \end{array}$$

Step 5
Each child will receive 4 toys.

Step 6
$4 \times 657 = 2628$

9. *Step 1*
The number of people at the lake on Friday is higher than that on Wednesday. The number on Friday and the number higher is given. The number of people on Wednesday must be found.

Step 2
Subtract the number higher from the number on Friday.

Step 3
A reasonable answer would be
$8000 - 4000 = 4000$ people.

Step 4

$$\begin{array}{r} 8392 \\ -\ 4218 \\ \hline 4174 \end{array}$$

Step 5
4174 people were at the lake on Wednesday.

Step 6
The answer is reasonably close to the estimate.

Check: $\begin{array}{r} 4218 \\ +\ 4174 \\ \hline 8392 \end{array}$ *matches*

11. *Step 1*
Find the amount saved for five months.

Step 2
Use multiplication to find the amount.

Step 3
The amount saved is about \$30 per month, making the total $\$30 \times 5 = \$150$.

Step 4

$$\begin{array}{r} \$34 \\ \times\ 5 \\ \hline \$170 \end{array}$$

Step 5
The amount saved in 5 months is \$170.

Step 6
The answer is reasonably close to the estimate.
Check: $5\overline{)170}$ = 34

13. *Step 1*
Find how much more an executive chef earns than a jet mechanic.

Step 2
"More than" indicates subtraction.

Step 3
An estimate is $\$50,000 - \$50,000 = \$0$.

Step 4

$$\begin{array}{rl} \$54,300 & \textit{executive chef} \\ -\ 45,000 & \textit{jet mechanic} \\ \hline \$9\,300 & \textit{difference} \end{array}$$

Step 5
An executive chef earns \$9300 more than a jet mechanic.

Step 6
The estimate isn't close to the exact value, but the check verifies that we are correct.

Check: $\begin{array}{r} \$9\,300 \\ +\ 45,000 \\ \hline \$54,300 \end{array}$ *matches*

15. *Step 1*
Find which couple has higher earnings and the difference in their earnings.

Step 2
Add the salaries of each couple to find the total for each. Then subtract to find the "difference."

Step 3
Estimate:

$30,000 + $40,000 = $70,000 Garrett
$30,000 + $50,000 = $80,000 Harcos

```
  $80,000  Harcos
– 70,000   Garrett
  $10,000  difference
```

Step 4

```
Garrett:   $32,800     Harcos:   $33,000
         + 35,200              + 49,900
           $68,000               $82,900
```

```
  $82,900  Harcos
– 68,000   Garrett
  $14,900  difference
```

Step 5
(a) Mr. and Mrs. Harcos have higher earnings.

(b) The difference between the couples' earnings is $14,900.

Step 6
The answers are reasonably close to the estimates. Check the couples' earnings by adding again. Check the difference by adding $14,900 to $68,000 to get $82,900.

17. *Step 1*
Find her monthly savings.

Step 2
Her monthly take home pay and expenses are given. "Remainder" indicates subtraction may be used.

Step 3
Estimate:
$2000 – $700 – $300 – $400 – $200 – $200 = $200

Step 4

```
  $695       $2240
   340      – 1890
   435        $350
   240
 + 180
  1890
```

Step 5
Her monthly savings are $350.

Step 6
The answer seems reasonable for the given estimate.

Check: $350 + $1890 = $2240. Re-add the expenses to check the sum of $1890.

19. *Step 1*
The square feet in one acre is given and the square feet in the given number of acres must be found.

Step 2
Multiply the square feet in one acre by the total acres.

Step 3
An estimate is $40{,}000 \times 100 = 4{,}000{,}000$ ft².

Step 4

```
    43,560
   ×   138
   348 480
 1 306 80
 4 356 0
 6,011,280
```

Step 5
There are 6,011,280 square feet in 138 acres. The answer is reasonable considering the rounding.

Step 6
Check: $138\overline{)6{,}011{,}280}$ = 43,560

21. *Step 1*
Find the total cost of all Safety and Security Items.

Step 2
Add each cost to find the total cost.

Step 3
Estimate:
$200 + $400 + $600 + $200 + $200 + $200 = $1800

Step 4

```
    42
   225
   390
   565
   195
   185
 + 160
  1720
```

Step 5
The total cost is $1720.

Step 6
The answer is reasonably close to the estimate. Check by re-adding the costs.

23. *Step 1*
An option package is offered. We must find how much can be saved if the customer buys the option package instead of paying for each option separately.

Step 2
First, add the cost of the options. Then subtract the cost of the option package from this total.

Step 3
Estimate:
$\$200 + \$600 + \$200 + \$400 = \$1400$
$\$1400 - \$1200 = \$200$

Step 4

Add.	$\overset{11}{\$225}$	*Child seats*
	565	*Antilock brakes*
	150	*Keyless entry*
	+ 400	*Power sliding door*
	\$1340	*Total cost*

Subtract.	\$1340	*Total cost*
	− 1220	*Option package*
	\$120	*Amount saved*

Step 5
Jill can save \$120 by buying the option package.

Step 6
The answer is reasonably close to the estimate.
Check: $\$1220 + 120 = \$1340$

25. *Step 1*
The number and cost of wheelchairs and recorder-players are given and the total cost of all items must be found.

Step 2
Find the cost of all wheelchairs and the cost of all recorder-players. Then add these costs to get the total cost.

Step 3
The cost of the wheelchairs is about $\$1000 \times 6 = \$6000$.

The cost of the recorder-players is about $\$900 \times 20 = \$18,000$.

Estimate: $\$6000 + \$18,000 = \$24,000$

Step 4
cost of wheelchairs: $\$1256 \times 6 = \$7536$

cost of recorder-players:

$$\begin{array}{r} \$895 \\ \times\ 15 \\ \hline 4475 \\ 895 \\ \hline \$13{,}425 \end{array}$$

Step 5
The total cost is $\$13,425 + \$7536 = \$20,961$.

Step 6
The answer is reasonably close to the estimate. Check by repeating Step 4.

27. Possible answers are

Addition: more; total; gain of
Subtraction: less; loss of; decreased by
Multiplication: twice; of; product
Division: divided by; goes into; per
Equals: is; are

29. Estimating the answer can help you avoid careless mistakes like decimal or calculation errors. Examples of reasonable answers in daily life might be a \$25 bag of groceries, \$20 to fill the gas tank, or \$45 for a phone bill.

31. *Step 1*
The cost of packages of undershirts and socks is given and the total cost of a different number of these items must be found.

Step 2
First, find the number of packages of undershirts and socks required. Then find the cost of each item. Finally, add these costs to get the total cost.

Step 3
Packages of undershirts is $30 \div 3 = 10$
Packages of socks is $18 \div 6 = 3$
Cost of undershirts is about $\$10 \times 10 = \$100$
Cost of socks is about $\$20 \times 3 = \$60$
An estimate is $\$100 + \$60 = \$160$.

Step 4
cost of undershirts:

10	*Packages*
× \$12	*Cost per package*
\$120	*Total*

cost of socks:

3	*Packages*
× \$15	*Cost per package*
\$45	*Total*

Step 5
The total cost is $\$120 + \$45 = \$165$.

Step 6
The answer is reasonably close to the estimate. Check by repeating Step 4.

33. *Step 1*
Find the final weight of the car.

Step 2
The second engine weighs more than the first engine. Find the difference and add it to the weight of the car.

Step 3
582 and 634 both round to 600, so the difference is 0 and an estimate of the car is just its original weight, 2425 pounds.

Step 4

$$\begin{array}{rl} 634 & \textit{new engine} \\ -\,582 & \textit{old engine} \\ \hline 52 & \textit{difference} \end{array} \qquad \begin{array}{rl} 2425 & \textit{original weight} \\ +\,52 & \textit{additional weight} \\ \hline 2477 & \textit{new weight} \end{array}$$

Step 5
The car will weigh 2477 pounds.

Step 6
The answer is reasonable since it's just slightly more than the original weight.

Check:
$$\begin{array}{r} 2425 \\ -\,582 \\ \hline 1843 \end{array} \qquad \begin{array}{r} 1843 \\ +\,634 \\ \hline 2477 \end{array}$$

35. *Step 1*
The costs of two hotels are given. Find the amount saved by staying at the less expensive hotel.

Step 2
First subtract the cost of the least expensive hotel and then multiply this savings by five nights.

Step 3
Estimate: $\$160 - \$60 = \$100$
$\$100 \times 5 = \$500$

Step 4
Subtract.
$$\begin{array}{rl} \$159 & \textit{Harrah's Lake Tahoe} \\ -\,59 & \textit{Harrah's Reno} \\ \hline \$100 & \textit{Savings for 1 night} \end{array}$$

Multiply.
$$\begin{array}{r} \$100 \\ \times\,5 \\ \hline \$500 \end{array}$$

Step 5
The savings for 5 nights is $500.

Step 6
The answer matches the estimate.

Check: Lake Tahoe cost: $5 \times \$159 = \$795$
Reno cost: $5 \times \$59 = \$295$
Difference: $\$795 - \$295 = \$500$

37. *Step 1*
Find out how much money each team received given the amount of money raised, the expenses, and the number of teams.

Step 2
Subtract the expenses from the amount raised, then divide the result by the number of teams.

Step 3
Estimate: $\$8000 - \$800 = \$7200$
$\$7200 \div 20 = \$360$

Step 4
Subtract.
$$\begin{array}{r} \$7588 \\ -\,838 \\ \hline \$6750 \end{array}$$

Divide.
$$\begin{array}{r} \$\,3\,7\,5 \\ 18\overline{)\$6\,7\,5\,0} \\ 5\,4 \\ \hline 1\,3\,5 \\ 1\,2\,6 \\ \hline 9\,0 \\ 9\,0 \\ \hline 0 \end{array}$$

Step 5
Each team will receive $375.

Step 6
The answer is reasonably close to the estimate.
Check: $18 \times \$375 = \$6750$
$\$6750 + \$838 = \$7588$

39. *Step 1*
The total seating is given along with the information to find the number of seats on the main floor. The number of rows of seats in the balcony is given and the number of seats in each row of the balcony must be found.

Step 2
First, the number of seats on the main floor must be found. Next, the number of seats on the main floor must be subtracted from the total number of seats to find the number of seats in the balcony. Finally, the number of seats in the balcony must be divided by the number of rows of seats in the balcony.

Step 3
Seats on main floor: $30 \times 25 = 750$
Seats in balcony: $1250 - 750 = 500$
Seats in each row in balcony: $500 \div 25 = 20$
Since we didn't round, our estimate matches our exact answer.

Step 4
See the calculations in Step 3.

Step 5
The number of seats in each row of the balcony is 20.

Step 6
The answer matches the estimate, as expected.
Check: Seats in balcony: $20 \times 25 = 500$
Seats on main floor: $30 \times 25 = 750$
Total seats: $500 + 750 = 1250$

Chapter 1 Review Exercises

1. 6,573; thousands: 6; ones: 573
2. 36,215; thousands: 36; ones: 215
3. 105,724; thousands: 105; ones: 724
4. 1,768,710,618; billions: 1; millions: 768; thousands: 710; ones: 618
5. 728 is seven hundred twenty-eight.
6. 15,310 is fifteen thousand, three hundred ten.
7. 319,215 is three hundred nineteen thousand, two hundred fifteen.
8. 62,500,005 is sixty-two million, five hundred thousand, five.
9. Ten thousand, eight is 10,008.
10. Two hundred million, four hundred fifty-five is 200,000,455.

11. $\begin{array}{r} {}^{1}\\ 72 \\ +38 \\ \hline 110 \end{array}$

12. $\begin{array}{r} {}^{1}\\ 54 \\ +67 \\ \hline 121 \end{array}$

13. $\begin{array}{r} {}^{1\ 1}\\ 807 \\ 4606 \\ +\ \ 51 \\ \hline 5464 \end{array}$

14. $\begin{array}{r} {}^{1}\\ 8\,215 \\ 9 \\ +7\,433 \\ \hline 15{,}657 \end{array}$

15. $\begin{array}{r} {}^{11}\\ 2\,130 \\ 453 \\ 8\,107 \\ +\ \ 296 \\ \hline 10{,}986 \end{array}$

16. $\begin{array}{r} {}^{221}\\ 5684 \\ 218 \\ 2960 \\ +\ \ 983 \\ \hline 9845 \end{array}$

17. $\begin{array}{r} {}^{11\ 33}\\ 5\,732 \\ 11{,}069 \\ 37 \\ 1\,595 \\ +22{,}169 \\ \hline 40{,}602 \end{array}$

18. $\begin{array}{r} {}^{1\ 31}\\ 3\,451 \\ 12{,}286 \\ 43 \\ 1\,291 \\ +32{,}784 \\ \hline 49{,}855 \end{array}$

19. $\begin{array}{r} {}^{5\,14}\\ 6\,4 \\ -2\,8 \\ \hline 3\,6 \end{array}$ Check: $\begin{array}{r} {}^{1}\\ 28 \\ +36 \\ \hline 64 \end{array}$

20. $\begin{array}{r} {}^{3\,16}\\ 4\,6 \\ -1\,9 \\ \hline 2\,7 \end{array}$ Check: $\begin{array}{r} {}^{1}\\ 19 \\ +27 \\ \hline 46 \end{array}$

21. $\begin{array}{r} {}^{2\,16\,15}\\ 3\,7\,5 \\ -1\,8\,6 \\ \hline 1\,8\,9 \end{array}$ Check: $\begin{array}{r} {}^{11}\\ 186 \\ +189 \\ \hline 375 \end{array}$

22. $\begin{array}{r} {}^{4\,16\,13}\\ 5\,7\,3 \\ -3\,8\,9 \\ \hline 1\,8\,4 \end{array}$ Check: $\begin{array}{r} {}^{11}\\ 389 \\ +184 \\ \hline 573 \end{array}$

23. $\begin{array}{r} {}^{6\,13\,10\,16}\\ 7\,4\,1\,6 \\ -\ \ 5\,6\,7 \\ \hline 6\,8\,4\,9 \end{array}$ Check: $\begin{array}{r} {}^{111}\\ 567 \\ +6849 \\ \hline 7416 \end{array}$

24. $\begin{array}{r} {}^{4\,11\,10\,10}\\ 5\,2\,1\,0 \\ -\ \ 8\,8\,3 \\ \hline 4\,3\,2\,7 \end{array}$ Check: $\begin{array}{r} {}^{111}\\ 4327 \\ +883 \\ \hline 5210 \end{array}$

25. $\begin{array}{r} {}^{1\,11\,10\,10}\\ 2\,2\,1\,0 \\ -1\,9\,8\,6 \\ \hline 2\,2\,4 \end{array}$ Check: $\begin{array}{r} {}^{111}\\ 1986 \\ +224 \\ \hline 2210 \end{array}$

26. $\begin{array}{r} {}^{9}\\ {}^{8\ 16\,10\,14}\\ 99{,}7\,0\,4 \\ -73{,}8\,3\,8 \\ \hline 25{,}8\,6\,6 \end{array}$ Check: $\begin{array}{r} {}^{1\ 1\ 1}\\ 25{,}866 \\ +73{,}838 \\ \hline 99{,}704 \end{array}$

27. $$\begin{array}{r} 7 \\ \times 7 \\ \hline 49 \end{array}$$

28. $$\begin{array}{r} 8 \\ \times 0 \\ \hline 0 \end{array}$$

29. $8(4) = 32$

30. $8(8) = 64$

31. $(5)(9) = 45$

32. $(6)(7) = 42$

33. $7 \cdot 8 = 56$

34. $9 \cdot 9 = 81$

35. $5 \times 4 \times 2$
$(5 \times 4) \times 2$
$20 \times 2 = 40$

36. $9 \times 1 \times 5$
$(9 \times 1) \times 5$
$9 \times 5 = 45$

37. $4 \times 4 \times 3$
$(4 \times 4) \times 3$
$16 \times 3 = 48$

38. $2 \times 2 \times 2$
$(2 \times 2) \times 2$
$4 \times 2 = 8$

39. $(6)(0)(8) = 0$ Any number times 0 equals 0.

40. $(7)(1)(6)$
$(7 \cdot 1) \cdot 6$
$7 \cdot 6 = 42$

41. $6 \cdot 1 \cdot 8$
$(6 \cdot 1) \cdot 8$
$6 \cdot 8 = 48$

42. $7 \cdot 7 \cdot 0 = 0$ Any number times 0 equals 0.

43. $$\begin{array}{r} {\scriptstyle 2\,\,} \\ 28 \\ \times 3 \\ \hline 84 \end{array}$$

44. $$\begin{array}{r} {\scriptstyle 4\,\,} \\ 46 \\ \times 8 \\ \hline 368 \end{array}$$

45. $$\begin{array}{r} {\scriptstyle 7\,\,} \\ 58 \\ \times 9 \\ \hline 522 \end{array}$$

46. $$\begin{array}{r} 98 \\ \times 1 \\ \hline 98 \end{array}$$

47. $$\begin{array}{r} {\scriptstyle 2\,4\,\,} \\ 625 \\ \times 8 \\ \hline 5000 \end{array}$$

48. $$\begin{array}{r} {\scriptstyle 5\,3\,\,} \\ 374 \\ \times 8 \\ \hline 2992 \end{array}$$

49. $$\begin{array}{r} {\scriptstyle 1\,1\,3\,\,} \\ 1349 \\ \times 4 \\ \hline 5396 \end{array}$$

50. $$\begin{array}{r} {\scriptstyle 3\,1\,\,\,\,} \\ 9163 \\ \times 5 \\ \hline 45{,}815 \end{array}$$

51. $$\begin{array}{r} {\scriptstyle 1\,1\,\,\,\,} \\ 7456 \\ \times 2 \\ \hline 14{,}912 \end{array}$$

52. $$\begin{array}{r} {\scriptstyle 6\,5\,\,\,\,} \\ 2880 \\ \times 7 \\ \hline 20{,}160 \end{array}$$

53. $$\begin{array}{r} {\scriptstyle 1\,\,\,\,\,\,\,2\,\,\,\,} \\ 93{,}105 \\ \times 5 \\ \hline 465{,}525 \end{array}$$

54. $$\begin{array}{r} {\scriptstyle 1\,6\,\,\,5\,2\,\,} \\ 21{,}873 \\ \times 8 \\ \hline 174{,}984 \end{array}$$

55. $$\begin{array}{r} 35 \\ \times 25 \\ \hline 175 \\ 70 \\ \hline 875 \end{array}$$

56. $$\begin{array}{r} 74 \\ \times 32 \\ \hline 148 \\ 222 \\ \hline 2368 \end{array}$$

57. $$\begin{array}{rl} 98 & \\ \times 12 & \\ \hline 196 & \leftarrow 2 \times 98 \\ 98 & \leftarrow 1 \times 98 \\ \hline 1176 & \end{array}$$

58.
$$\begin{array}{r l} 68 & \\ \times\ 75 & \\ \hline 340 & \leftarrow 5 \times 68 \\ 476 & \leftarrow 7 \times 68 \\ \hline 5100 & \end{array}$$

59.
$$\begin{array}{r l} 472 & \\ \times\ 33 & \\ \hline 1416 & \leftarrow 3 \times 472 \\ 1416 & \\ \hline 15{,}576 & \end{array}$$

60.
$$\begin{array}{r l} 392 & \\ \times\ 77 & \\ \hline 2744 & \leftarrow 7 \times 392 \\ 2744 & \\ \hline 30{,}184 & \end{array}$$

61.
$$\begin{array}{r} 4051 \\ \times\ 219 \\ \hline 887{,}169 \end{array}$$

62.
$$\begin{array}{r} 1527 \\ \times\ 328 \\ \hline 500{,}856 \end{array}$$

63.
$$\begin{array}{r l} \$12 & \textit{cost per calculator} \\ \times\ 30 & \textit{calculators} \\ \hline \$360 & \textit{total cost} \end{array}$$

64.
$$\begin{array}{r l} \$14 & \textit{cost of subscription} \\ \times\ 76 & \textit{subscribers} \\ \hline 84 & \\ 98 & \\ \hline \$1064 & \textit{total cost} \end{array}$$

65.
$$\begin{array}{r l} \$64 & \textit{cost per set} \\ \times\ 318 & \textit{sets} \\ \hline 512 & \\ 64 & \\ 192 & \\ \hline \$20{,}352 & \textit{total cost} \end{array}$$

66.
$$\begin{array}{r l} 114 & \textit{ear plugs} \\ \times\ \$6 & \textit{cost per plug} \\ \hline \$684 & \textit{total cost} \end{array}$$

67.
$$\begin{array}{r} 280 \\ \times\ 50 \\ \hline \end{array} \quad \begin{array}{r} 28 \\ \times\ 5 \\ \hline 140 \end{array} \quad \begin{array}{r} 280 \\ \times\ 50 \\ \hline 14{,}000 \end{array} \quad \textit{Attach } 00.$$

68.
$$\begin{array}{r} 340 \\ \times\ 70 \\ \hline \end{array} \quad \begin{array}{r} 34 \\ \times\ 7 \\ \hline 238 \end{array} \quad \begin{array}{r} 340 \\ \times\ 70 \\ \hline 23{,}800 \end{array} \quad \textit{Attach } 00.$$

69.
$$\begin{array}{r} 517 \\ \times\ 400 \\ \hline \end{array} \quad \begin{array}{r} 517 \\ \times\ 4 \\ \hline 2068 \end{array} \quad \begin{array}{r} 517 \\ \times\ 400 \\ \hline 206{,}800 \end{array} \quad \textit{Attach } 00.$$

70.
$$\begin{array}{r} 637 \\ \times\ 500 \\ \hline \end{array} \quad \begin{array}{r} 637 \\ \times\ 5 \\ \hline 3185 \end{array} \quad \begin{array}{r} 637 \\ \times\ 500 \\ \hline 318{,}500 \end{array} \quad \textit{Attach } 00.$$

71.
$$\begin{array}{r} 16{,}000 \\ \times\ 8000 \\ \hline \end{array} \quad \begin{array}{r} 16 \\ \times\ 8 \\ \hline 128 \end{array} \quad \begin{array}{r} 16{,}000 \\ \times\ 8000 \\ \hline 128{,}000{,}000 \end{array} \quad \textit{Attach } 000000.$$

72.
$$\begin{array}{r} 43{,}000 \\ \times\ 2100 \\ \hline \end{array} \quad \begin{array}{r} 43 \\ \times\ 21 \\ \hline 903 \end{array} \quad \begin{array}{r} 43{,}000 \\ \times\ 2100 \\ \hline 90{,}300{,}000 \end{array} \quad \textit{Attach } 00000.$$

73. $20 \div 4 = 5$

74. $35 \div 5 = 7$

75. $42 \div 7 = 6$

76. $18 \div 9 = 2$

77. $\frac{54}{9} = 6$

78. $\frac{36}{9} = 4$

79. $\frac{49}{7} = 7$

80. $\frac{0}{6} = 0$

81. $\frac{148}{0}$ is undefined.

82. $\frac{0}{23} = 0$

83. $\frac{64}{8} = 8$

84. $\frac{81}{9} = 9$

85.
$$\begin{array}{r} 82 \\ 4\overline{)328} \end{array} \quad \text{Check: } \begin{array}{r} 82 \\ \times\ 4 \\ \hline 328 \end{array}$$

86.
$$\begin{array}{r} 9\ 8 \\ 3\overline{)29\,{}^{2}4} \end{array} \quad \text{Check: } \begin{array}{r} 98 \\ \times\ 3 \\ \hline 294 \end{array}$$

87.
```
   4 4 2 2        Check:   211
6|26,²5¹3¹2                4422
                          ×   6
                         26,532
```

88.
```
     3 5 2        Check:    352
76|2 6, 7 5 2              × 76
   2 2 8                   2112
     3 9 5                2464
     3 8 0               26,752
       1 5 2
       1 5 2
           0
```

89. $2704 \div 18$
```
      1 5 0 R4
18|2 7 0 4
   1 8
     9 0
     9 0
       0 4
       0 0
         4
```
```
Check:    150
        ×  18
         1200
         150
         2700
        +   4
         2704
```

90. $15{,}525 \div 125$
```
         1 2 4 R25
125|1 5, 5 2 5
    1 2 5
      3 0 2
      2 5 0
        5 2 5
        5 0 0
          2 5
```
```
Check:    124
        × 125
          620
         2 48
         12 4
       15,500
      +    25
       15,525
```

91. 817 rounded to the nearest ten: 820

8$\underline{1}$7 Next digit is 5 or more. Tens place changes $(1+1=2)$. The digit to the right of the underlined place changes to zero.

92. 15,208 rounded to the nearest hundred: 15,200

15,$\underline{2}$08 Next digit is 4 or less. Hundreds place does not change. All digits to the right of the underlined place change to zero.

93. 20,643 rounded to the nearest thousand: 21,000

2$\underline{0}$,643 Next digit is 5 or more. Thousands place changes $(0+1=1)$. All digits to the right of the underlined place change to zero.

94. 67,485 rounded to the nearest ten thousand: 70,000

$\underline{6}$7,485 Next digit is 5 or more. Ten thousands place changes $(6+1=7)$. All digits to the right of the underlined place change to zero.

95. **To the nearest ten:** 34$\underline{8}$7

Next digit is 5 or more. Tens place changes $(8+1=9)$. The digit to the right of the underlined place changes to zero. **3490**

To the nearest hundred: 3$\underline{4}$87

Next digit is 5 or more. Hundreds place changes $(4+1=5)$. All digits to the right of the underlined place are changed to zero. **3500**

To the nearest thousand: $\underline{3}$487

Next digit is 4 or less. Thousands place does not change. All digits to the right of the underlined place are changed to zero. **3000**

96. **To the nearest ten:** 20,0$\underline{6}$5

Next digit is 5 or more. Tens place changes $(6+1=7)$. The digit to the right of the underlined place changes to zero. **20,070**

To the nearest hundred: 20,$\underline{0}$65

Next digit is 5 or more. Hundreds place changes $(0+1=1)$. All digits to the right of the underlined place are changed to zero. **20,100**

To the nearest thousand: 2$\underline{0}$,065

Next digit is 4 or less. Thousands place does not change. All digits to the right of the underlined place are changed to zero. **20,000**

97. **To the nearest ten:** 98,2$\underline{0}$1

Next digit is 4 or less. Tens place does not change. The digit to the right of the underlined place changes to zero. **98,200**

To the nearest hundred: 98,$\underline{2}$01

Next digit is 4 or less. Hundreds place does not change. All digits to the right of the underlined place are changed to zero. **98,200**

continued

To the nearest thousand: 98,201

Next digit is 4 or less. Thousands place does not change. All digits to the right of the underlined place are changed to zero. **98,000**

98. **To the nearest ten:** 352,118

Next digit is 5 or more. Tens place changes $(1 + 1 = 2)$. The digit to the right of the underlined place changes to zero. **352,120**

To the nearest hundred: 352,118

Next digit is 4 or less. Hundreds place does not change. All digits to the right of the underlined place are changed to zero. **352,100**

To the nearest thousand: 352,118

Next digit is 4 or less. Thousands place does not change. All digits to the right of the underlined place are changed to zero. **352,000**

99. From the table, $4^2 = 16$, so $\sqrt{16} = 4$.

100. From the table, $7^2 = 49$, so $\sqrt{49} = 7$.

101. From the table, $12^2 = 144$, so $\sqrt{144} = 12$.

102. From the table, $14^2 = 196$, so $\sqrt{196} = 14$.

103. 7^3: exponent is 3; base is 7.
$7^3 = 7 \cdot 7 \cdot 7 = 343$

104. 3^6: exponent is 6; base is 3.
$3^6 = 3 \cdot 3 \cdot 3 \cdot 3 \cdot 3 \cdot 3 = 729$

105. 5^3: exponent is 3; base is 5.
$5^3 = 5 \cdot 5 \cdot 5 = 125$

106. 4^5: exponent is 5; base is 4.
$4^5 = 4 \cdot 4 \cdot 4 \cdot 4 \cdot 4 = 1024$

107. $7^2 - 15$ *Exponent*
$49 - 15 = 34$ *Subtract*

108. $6^2 - 10$ *Exponent*
$36 - 10 = 26$ *Subtract*

109. $2 \cdot 3^2 \div 2$ *Exponent*
$2 \cdot 9 \div 2$ *Multiply*
$18 \div 2 = 9$ *Divide*

110. $9 \div 1 \cdot 2 \cdot 2 \div (11 - 2)$ *Parentheses*
$9 \div 1 \cdot 2 \cdot 2 \div 9$ *Divide*
$9 \cdot 2 \cdot 2 \div 9$ *Multiply*
$18 \cdot 2 \div 9$ *Multiply*
$36 \div 9 = 4$ *Divide*

111. $\sqrt{9} + 2(3)$ *Square root*
$3 + 2 \cdot 3$ *Multiply*
$3 + 6 = 9$ *Add*

112. $6 \cdot \sqrt{16} - 6 \cdot \sqrt{9}$ *Square root*
$6 \cdot 4 - 6 \cdot 3$ *Multiply*
$24 - 18 = 6$ *Subtract*

113. From the bar graph, 8 parents out of 100 nagged their children about washing hands after using the bathroom.

114. From the bar graph, 5 parents out of 100 nagged their children about taking shoes off when coming inside.

115. From the bar graph, the greatest number of parents nagged their children about keeping bedroom clean.

116. From the bar graph, the least number of parents nagged their children about hanging up wet bath towels.

117. *Estimate:* *Exact:*

$$\begin{array}{r} \$20 \\ \times\ 70 \\ \hline \$1400 \end{array} \qquad \begin{array}{r} \$18 \\ \times\ 65 \\ \hline \$1170 \end{array}$$

118. *Step 1*
Find the total revolutions.

Step 2
We know the revolutions per minute and the number of minutes.

Number of revolutions $\times$ minutes = total revolutions.

Step 3
An estimate is $1000 \times 60 = 60{,}000$ revolutions.

Step 4
$1400 \times 60 = 84{,}000$ revolutions

Step 5
There were 84,000 revolutions.

Step 6
The answer is reasonably close to the estimate considering the rounding.

Check: $84{,}000 \div 60 = 1400$

119. *Estimate:* *Exact:*

$$\begin{array}{r} 100 \\ \times\ 20 \\ \hline 2000 \end{array}\ \textit{forks} \qquad \begin{array}{r} 144 \\ \times\ 15 \\ \hline 2160 \end{array}\ \begin{array}{l} \textit{forks per box} \\ \textit{boxes} \\ \textit{forks} \end{array}$$

120. *Estimate and Exact:*

$$\begin{array}{r} 6000 \\ \times\ 30 \\ \hline 180{,}000 \end{array}\ \begin{array}{l} \textit{brackets per drum} \\ \textit{drums} \\ \textit{brackets} \end{array}$$

121. *Estimate:*

$$\begin{array}{r} 2000 \\ \times\ 10 \\ \hline 20{,}000 \end{array} \text{ hours}$$

Exact:

$$\begin{array}{rl} 2000 & \textit{hours per home} \\ \times\ 12 & \textit{homes} \\ \hline 24{,}000 & \textit{hours} \end{array}$$

122. *Estimate and Exact:*

$$\begin{array}{rl} 80 & \textit{miles per hour} \\ \times\ 5 & \textit{hours} \\ \hline 400 & \textit{miles} \end{array}$$

123. *Step 1*
Find the total cost to admit adults and children.

Step 2
Multiply the number of adults times the adult admission fee and the number of children by the admission fee for children. Add to find the total fee.

Step 3
Estimate:

Adults amount: $30 × 20 = $600
Children amount: $20 × 30 = $600
Total: $600 + $600 = $1200

Step 4
Adults amount: $33 × 22 = $726
Children amount: $23 × 25 = $575
Total: $726 + $575 = $1301

Step 5
The total cost is $1301.

Step 6
The answer is reasonably close to the estimate. Check by repeating Step 4.

124. *Step 1*
Find the total monthly collections.

Step 2
We know the number of daily customers and the daily rate. We know the number of weekend-only customers and the rate. Number of customers × daily rate + number of customers × weekend rate = total collections.

Step 3
Estimate:
(60 × $20) + (20 × $7) = $1200 + $140 = $1340

Step 4
(62 × $16) + (21 × $7) = $992 + $147 = $1139

Step 5
The total monthly collections are $1139.

Step 6
The answer is reasonably close to the estimate. Check by repeating Step 4.

125. *Step 1*
Find the difference in price between the most and least expensive sinks.

Step 2
Difference indicates subtraction.

Step 3
An estimate is $400 − $40 = $360.

Step 4

$$\begin{array}{r} \$388 \\ -\ \$39 \\ \hline \$349 \end{array}$$

Step 5
The difference in price is $349.

Step 6
The answer is reasonably close to the estimate. Check: $349 + $39 = $388

126. *Step 1*
Find how many hours it takes to produce all the plates.

Step 2
We know the total number of plates and we know how many are produced each hour.

$$\frac{\textit{total number}}{\textit{number per hour}} = \textit{total hours}$$

Step 3
An estimate: 30,000 ÷ 1000 = 30 hours

Step 4

$$\frac{32{,}538}{986} = 33 \text{ hours}$$

Step 5
It will take 33 hours.

Step 6
The answer is reasonably close to the estimate. Check: $33 \cdot 986 = 32{,}538$

127. *Step 1*
Find out how many pounds of pork are needed.

Step 2
We know the total number of cans and we must divide that total by 175 since each group of 175 cans requires 1 pound of pork. The number of groups × 1 pound = total pounds.

Step 3
Estimate: 9000 ÷ 200 = 45 pounds

Step 4

$$\frac{8750}{175} = 50 \text{ pounds}$$

Step 5
50 pounds of pork are needed.

Step 6
The answer is reasonably close to the estimate.

Check:
(175 cans per pound) • (50 pounds) = 8750 cans

128. *Step 1*
Find the new account balance.

Step 2
We know the amount of the checks and the old balance.

Old balance − check amounts = new balance.

Step 3
Estimate: \$2000 − \$500 − \$400 = \$1100

Step 4
\$1924 − \$520 − \$385 = \$1019

Step 5
She has \$1019 in her account.

Step 6
The answer is reasonably close to the estimate.
Check: \$1019 + \$520 + \$385 = \$1924

129. *Step 1*
Find the total acres fertilized.

Step 2
We know the total amount of fertilizer and how much each acre needs. Total pounds ÷ pounds needed per acre = total acres.

Step 3
Estimate: $30{,}000 \div 600 = 50$

Step 4
$\frac{32{,}500}{625} = 52$ acres

Step 5
52 acres can be spread with 32,500 pounds of nitrogen sulfate.

Step 6
The answer is reasonably close to the estimate.
Check: $52 \times 625 = 32{,}500$

130. *Step 1*
Find the number of homes that can be fenced.

Step 2
Divide the number of feet of fencing available by the number of feet needed for each home.

Step 3
Estimate: $6000 \div 200 = 30$ homes

Step 4
$\frac{5760}{180} = 32$ homes

Step 5
32 homes can be fenced.

Step 6
The answer is reasonably close to the estimate.
Check: (180 feet per home) • (32 homes) = 5760 feet

131. [1.4] 4(83)
$$\begin{array}{r} \scriptstyle 1 \\ 83 \\ \times\ 4 \\ \hline 332 \end{array}$$

132. [1.4] 7(64)
$$\begin{array}{r} \scriptstyle 2 \\ 64 \\ \times\ 7 \\ \hline 448 \end{array}$$

133. [1.3]
$$\begin{array}{r} \scriptstyle 2\,10 \\ \not{3}\,\not{0}\,9 \\ -\ \ 5\ 6 \\ \hline 2\ 5\ 3 \end{array}$$

134. [1.3]
$$\begin{array}{r} \scriptstyle 12 \\ \scriptstyle 7\ \not{2}\ 15 \\ \not{8}\ \not{3}\ \not{5} \\ -\ 2\ 4\ 7 \\ \hline 5\ 8\ 8 \end{array}$$

135. [1.2]
$$\begin{array}{r} \scriptstyle 1\,1 \\ 662 \\ +\,379 \\ \hline 1041 \end{array}$$

136. [1.2]
$$\begin{array}{r} \scriptstyle 1\,1 \\ 789 \\ +\,872 \\ \hline 1661 \end{array}$$

137. [1.3]
$$\begin{array}{r} \scriptstyle 13 \\ \scriptstyle 0\ \not{3}\ 10 \\ 38,\not{1}\ 4\ \not{0} \\ -\ \ 6\ 0\ 7\ 8 \\ \hline 32{,}0\ 6\ 2 \end{array}$$
Check:
$$\begin{array}{r} \scriptstyle 1\,1 \\ 32{,}062 \\ +\,6078 \\ \hline 38{,}140 \end{array}$$

138. [1.3]
$$\begin{array}{r} \scriptstyle 8\ 11\,4\ 16 \\ 2\not{9},\not{1}\ \not{5}\ \not{6} \\ -\ \ 4\ 2\ 0\ 9 \\ \hline 24{,}9\ 4\ 7 \end{array}$$
Check:
$$\begin{array}{r} \scriptstyle 1\ \ 1 \\ 24{,}947 \\ +\,4209 \\ \hline 29{,}156 \end{array}$$

139. [1.5] $21 \div 7 = 3$

140. [1.5] $\frac{42}{6} = 7 \quad (7 \cdot 6 = 42)$

141. [1.2]
$$\begin{array}{r} {}^{1\,2\,2\,2} \\ 7\,218 \\ 3 \\ 18 \\ 1\,791 \\ 82{,}623 \\ +\ \ 1\,982 \\ \hline 93{,}635 \end{array}$$

142. [1.2]
$$\begin{array}{r} {}^{1\,2\,1\,2} \\ 3\,812 \\ 5 \\ 22 \\ 1\,836 \\ 75{,}134 \\ +\ \ 2\,369 \\ \hline 83{,}178 \end{array}$$

143. [1.5] $\frac{9}{0}$ is undefined.

144. [1.5] $\frac{7}{1} = 7 \quad (7 \cdot 1 = 7)$

145. [1.5] $27{,}600 \div 4 = 6900$

$$\begin{array}{r} 6\ 9 \\ 4\overline{)27^{3}6} \end{array}$$ *Attach* 00.

146. [1.5] $18{,}480 \div 8$

$$\begin{array}{r} 2\ 3\ 10 \\ 8\overline{)18,{}^{2}480} \end{array}$$

147. [1.4]
$$\begin{array}{rl} 8\ 430 & \\ \times\ \ \ 128 & \\ \hline 67\ 440 & \leftarrow 8 \times 8430 \\ 168\ 60\ \ & \leftarrow 2 \times 8430 \\ 843\ 0\ \ \ \ & \\ \hline 1{,}079{,}040 & \end{array}$$

148. [1.4]
$$\begin{array}{r} {}^{14\ \ 1} \\ 21{,}702 \\ \times\ \ \ \ \ 6 \\ \hline 130{,}212 \end{array}$$

149. [1.6]
$$\begin{array}{r} 1\,0\,8 \\ 34\overline{)3\,6\,7\,2} \\ \underline{3\,4}\ \ \ \ \ \\ 2\,7\ \ \ \\ \underline{0}\ \ \ \\ 2\,7\,2 \\ \underline{2\,7\,2} \\ 0 \end{array}$$

150. [1.6]
$$\begin{array}{r} 2\,0\,7 \\ 68\overline{)1\,4,0\,7\,6} \\ \underline{1\,3\,6}\ \ \ \ \ \\ 4\,7\ \ \ \\ \underline{0}\ \ \ \\ 4\,7\,6 \\ \underline{4\,7\,6} \\ 0 \end{array}$$

151. [1.1] 376,853 is three hundred seventy-six thousand, eight hundred fifty-three in words.

152. [1.1] 408,610 is four hundred eight thousand, six hundred ten in words.

153. [1.7] 8749 rounded to the nearest hundred: $8\underline{7}49$
Next digit is 4 or less. Hundreds place doesn't change. All digits to the right of the underlined place are changed to zero. **8700**

154. [1.7] 400,503 rounded to the nearest thousand: $40\underline{0},503$
Next digit is 5 or more. Thousands place changes $(0 + 1 = 1)$. All digits to the right of the underlined place are changed to zero. **401,000**

155. [1.8] From the table, $8^2 = 64$, so $\sqrt{64} = 8$.

156. [1.8] From the table, $9^2 = 81$, so $\sqrt{81} = 9$.

157. [1.4]
$$\begin{array}{rl} \$3\,0\,8 & \textit{cost per pair} \\ \times\ \ 1\,8 & \textit{pairs} \\ \hline 2\,4\,6\,4 & \\ 3\,0\,8\ \ & \\ \hline \$5\,5\,4\,4 & \textit{total cost} \end{array}$$

158. [1.4]
$$\begin{array}{rl} \$3\,7\,0 & \textit{cost per dishwasher} \\ \times\ \ 8\,4 & \textit{dishwashers} \\ \hline 1\,4\,8\,0 & \\ 2\,9\,6\,0\ \ & \\ \hline \$31{,}0\,8\,0 & \textit{total cost} \end{array}$$

159. [1.4]
$$\begin{array}{rl} 2\,0\,8 & \textit{baseball hats} \\ \times\ \ \$1\,1 & \textit{cost of hat} \\ \hline 2\,0\,8 & \\ 2\,0\,8\ \ & \\ \hline \$2\,2\,8\,8 & \textit{total cost} \end{array}$$

160. [1.4]
$$\begin{array}{rl} 6\,0\,7 & \textit{boxes of avocados} \\ \times\ \ \$2\,6 & \textit{cost per box} \\ \hline 3\,6\,4\,2 & \\ 1\,2\,1\,4\ \ & \\ \hline \$15{,}7\,8\,2 & \textit{total cost} \end{array}$$

161. [1.4]
$$\begin{array}{rl} \overset{1}{5}\,2 & \textit{cards per deck} \\ \times\ \ \ 9 & \textit{decks} \\ \hline 4\,6\,8 & \textit{total cards} \end{array}$$

162. [1.4]

180 *cartons*
× 20 *books per carton*
3600 *textbooks*

163. [1.3]

$380
− $100
$280

A "push-type" mower costs $280.

164. [1.3]

$218,450 *Amount needed*
− $103,815 *Amount raised*
$114,635 *Amount left to be raised*

165. [1.10] Multiply the number of rentals times the sum of the rental fee and the launch fee.

4-person	6 × ($28 + $2) = 6($30)	$180
6-person	15 × ($38 + $2) = 15($40)	600
10-person	10 × ($70 + $2) = 10($72)	720
12-person	3 × ($75 + $2) = 3($77)	231
16-person	2 × ($85 + $2) = 2($87)	+ 174
		$1905

Total receipts were $1905.

166. [1.10] Multiply the number of rentals times the sum of the rental fee and the launch fee.

4-person	38 × ($28 + $2) = 38($30)	$1 140
6-person	73 × ($38 + $2) = 73($40)	2 920
10-person	58 × ($70 + $2) = 58($72)	4 176
12-person	34 × ($75 + $2) = 34($77)	2 618
16-person	18 × ($85 + $2) = 18($87)	+ 1 566
		$12,420

Total receipts were $12,420.

167. [1.9]

1317
0 8 7 13
1 4 8 3 *Petronas Towers*
− 9 8 6 *Eiffel Tower*
4 9 7 *difference*

The Petronas Towers in Malaysia are 497 feet taller than the Eiffel Tower in Paris.

168. [1.9]

14
3 4 14
14 5 4 *Empire State Building*
− 11 8 5 *Canadian National Tower*
2 6 9 *difference*

The Empire State Building in New York is 269 feet taller than the Canadian National Tower in Toronto.

169. [1.9] (a)

231
1483 *Petronas Tower*
1454 *Empire State Building*
1185 *Canadian National Tower*
+ 986 *Eiffel Tower*
5108 *combined height in feet*

(b)

7 10
52 8 0 *feet in one mile*
− 51 0 8 *combined height*
17 2 *difference*

The combined height of the four buildings (counting the Petronas Towers as one building) is 172 feet less than a mile.

170. [1.9]

$$\frac{\text{estimated cost}}{\text{height}} = \text{cost per foot}$$

$$\frac{\$200{,}000{,}000}{2000} = \$100{,}000 \text{ per foot}$$

Chapter 1 Test

1. 9205 is nine thousand, two hundred five.

2. 25,065 is twenty-five thousand, sixty-five.

3. Four hundred twenty-six thousand, five is 426,005.

4.

122
853
66
4022
+ 3589
8530

5.

11 12
17,063
7
12
1 505
93,710
+ 333
112,630

6.

9
8 10 10
9 0 0 9
− 7 9 6 4
1 0 4 5

7.

8 10 6 15
9 0 7 5
− 2 8 6 9
6 2 0 6

8. $7 \times 6 \times 4 = (7 \times 6) \times 4 = 42 \times 4 = 168$

9. $57 \cdot 3000$

```
  57
×  3
 171
```

$57 \cdot 3000 = 171{,}000$ *Attach* 000.

10. 85(19)

```
  85      Check:      85
× 19             19|1615
 765               152
 85                 95
1615                95
                     0
```

11.

```
    7381    Check:        7381
  ×  603          603|4,450,743
   22143               4221
  442860                2297
4,450,743               1809
                         4884
                         4824
                           603
                           603
                             0
```

12.

```
     7047      Check:   7047
16|112,752            ×   16
   112                 42282
    075                7047
     64               112,752
     112
     112
       0
```

13. $\frac{835}{0}$ is undefined

14.

```
     458 R5
42|19,241
   168
    244
    210
     341
     336
       5
```

15.

```
      160
280|44,800
    280
    1680
    1680
      00
```

16. 6347 rounded to the nearest ten: 63<u>4</u>7

Next digit is 5 or more. Tens place changes $(4 + 1 = 5)$. The digit to the right of the underlined place changes to zero. **6350**

17. 76,502 rounded to the nearest thousand: 7<u>6</u>,502

Next digit is 5 or more. Thousands place changes $(6 + 1 = 7)$. Digits to the right of the underlined place change to zero. **77,000**

18. $5^2 + 8(2)$ *Exponent*
$25 + 8(2)$ *Multiply*
$25 + 16 = 41$ *Add*

19. $7 \cdot \sqrt{64} - 14 \cdot 2$ *Square root*
$7 \cdot 8 - 14 \cdot 2$ *Multiply*
$56 - 28 = 28$ *Subtract*

20. Estimate:
$\$500 + \$500 + \$500 + \$400 - \$800 = \$1100$

Exact: Add the rent collected.

```
 $485
  500
  515
+ 425
$1925
```

Subtract expenses.

```
$1925
− 785
$1140  amount left
```

21. Estimate: $70{,}000 \div 500 = 140$ days

Exact: Divide the total number of printers by the number of printers assembled each day.

```
      125
542|67,750
    542
    1355
    1084
     2710
     2710
        0
```

It would take 125 work days.

22. Estimate: $\$2000 - \$500 - \$200 - \$200 = \$1100$
Exact: $\$1906 - \$528 - \$195 - \$235 = \$948$

23. Estimate: $(200 \times 4) + (200 \times 4) = 1600$ cameras

Exact:

$$\begin{array}{r} 208 \text{ digital cameras per hour} \\ \times \quad 4 \text{ hours} \\ \hline 832 \text{ digital cameras} \end{array}$$

$$\begin{array}{r} 238 \text{ camcorders per hour} \\ \times \quad 4 \text{ hours} \\ \hline 952 \text{ camcorders} \end{array}$$

$$\begin{array}{r} 832 \text{ digital cameras} \\ +952 \text{ camcorders} \\ \hline 1784 \text{ total cameras} \end{array}$$

24. (1) Locate the place to which you are rounding and underline it.

(2) Look only at the next digit to the right. If this digit is a 4 or less, do not change the underlined digit. If the digit is a 5 or more, increase the underlined digit by 1.

(3) Change all digits to the right of the underlined place to zeros.

Each person's example will vary, but the following two examples illustrate the two general cases.

(A) 1$\underline{2}$,499 rounds to 12,000

(B) 1$\underline{2}$,500 rounds to 13,000

25. (1) Read the problem carefully.

(2) Work out a plan.

(3) Estimate a reasonable answer.

(4) Solve the problem.

(5) State the answer.

(6) Check your work.

CHAPTER 2 MULTIPLYING AND DIVIDING FRACTIONS

2.1 Basics of Fractions

2.1 Margin Exercises

1. **(a)** The figure has 4 equal parts.

   Three parts are shaded: $\frac{3}{4}$

   One part is unshaded: $\frac{1}{4}$

   (b) The figure has 6 equal parts.

   One part is shaded: $\frac{1}{6}$

   Five parts are unshaded: $\frac{5}{6}$

   (c) The figure has 8 equal parts.

   Seven parts are shaded: $\frac{7}{8}$

   One part is unshaded: $\frac{1}{8}$

2. **(a)** An area equal to 8 of the $\frac{1}{7}$ parts is shaded.

   $$\frac{8}{7}$$

   (b) An area equal to 7 of the $\frac{1}{4}$ parts is shaded.

   $$\frac{7}{4}$$

3. **(a)** $\frac{2}{3}$ ← Numerator / ← Denominator

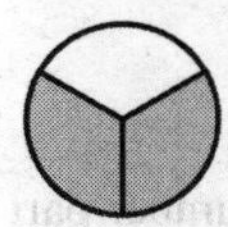

   (b) $\frac{1}{4}$ ← Numerator / ← Denominator

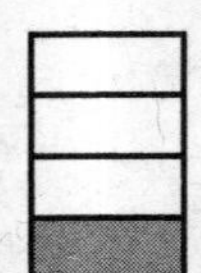

   (c) $\frac{8}{5}$ ← Numerator / ← Denominator

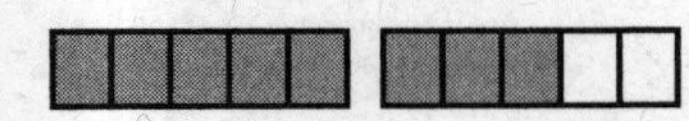

   (d) $\frac{5}{2}$ ← Numerator / ← Denominator

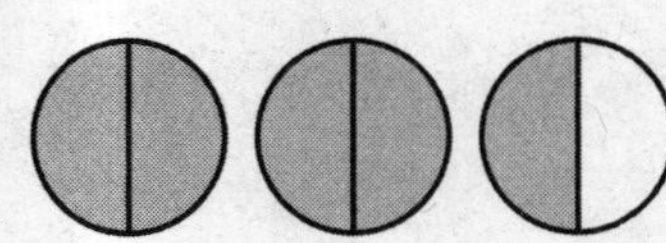

4. **(a)** Proper fractions: numerator *smaller* than denominator.

   $$\frac{2}{3}, \frac{3}{4}, \frac{1}{3}$$

   (b) Improper fractions: numerator *greater than or equal to* denominator.

   $$\frac{4}{3}, \frac{8}{8}, \frac{3}{1}$$

2.1 Section Exercises

1. The figure has 4 equal parts.
   Three parts are shaded: $\frac{3}{4}$
   One part is unshaded: $\frac{1}{4}$

3. The figure has 3 equal parts.
   One part is shaded: $\frac{1}{3}$

   Two parts are unshaded: $\frac{2}{3}$

5. Each of the two figures is divided into 5 parts and 7 are shaded: $\frac{7}{5}$

   Three are unshaded: $\frac{3}{5}$

7. Five of the 6 bills have a lifespan of 2 years or greater:

   $$\frac{5}{6}$$

9. There are 25 students, and 8 are hearing impaired.

   $$\frac{8}{25}$$

11. There are 218 trees. 63 are oak trees, and $218 - 63 = 155$ are not oak trees.

   $$\frac{155}{218}$$

13. $\frac{4}{5}$ ← Numerator / ← Denominator

15. $\frac{9}{8}$ ← Numerator / ← Denominator

17. Proper fractions: numerator *smaller* than denominator.

   $$\frac{1}{3}, \frac{5}{8}, \frac{7}{16}$$

   Improper fractions: numerator *greater than or equal to* denominator.

   $$\frac{8}{5}, \frac{6}{6}, \frac{12}{2}$$

19. Proper fractions: numerator *smaller* than denominator.

$$\frac{3}{4}, \frac{9}{11}, \frac{7}{15}$$

Improper fractions: numerator *greater than or equal to* denominator.

$$\frac{3}{2}, \frac{5}{5}, \frac{19}{18}$$

21. Answers will vary. One possibility is

$3 \leftarrow$ Numerator
$4 \leftarrow$ Denominator

The denominator shows the number of equal parts in the whole and the numerator shows how many of the parts are being considered.

23. The fraction $\frac{3}{8}$ represents 3 of the 8 equal parts into which a whole is divided.

25. The fraction $\frac{5}{24}$ represents 5 of the 24 equal parts into which a whole is divided.

2.2 Mixed Numbers

2.2 Margin Exercises

1. **(a)** The figure shows 1 whole object with 3 equal parts, all shaded, and a second whole with 2 parts shaded, so 5 parts are shaded in all.

$$1\frac{2}{3} = \frac{5}{3}$$

(b) The figure shows 2 wholes with 4 equal parts, all shaded, and a third whole with 1 part shaded, so 9 parts are shaded in all.

$$2\frac{1}{4} = \frac{9}{4}$$

2. **(a)** $6\frac{1}{2}$ $6 \cdot 2 = 12$ Multiply 6 and 2.

$12 + 1 = 13$ Add 1.

$$6\frac{1}{2} = \frac{13}{2}$$

(b) $7\frac{3}{4}$ $7 \cdot 4 = 28$ Multiply 7 and 4.

$28 + 3 = 31$ Add 3.

$$7\frac{3}{4} = \frac{31}{4}$$

(c) $4\frac{7}{8}$ $4 \cdot 8 = 32$ Multiply 4 and 8.

$32 + 7 = 39$ Add 7.

$$4\frac{7}{8} = \frac{39}{8}$$

(d) $8\frac{5}{6}$ $8 \cdot 6 = 48$ Multiply 8 and 6.

$48 + 5 = 53$ Add 5.

$$8\frac{5}{6} = \frac{53}{6}$$

3. **(a)** $\frac{6}{5}$

$$\begin{array}{r} 1 \\ 5\overline{)6} \\ \underline{5} \\ 1 \end{array}$$

1 ← Whole number part; 1 ← Remainder

$$\frac{6}{5} = 1\frac{1}{5}$$

(b) $\frac{9}{4}$

$$\begin{array}{r} 2 \\ 4\overline{)9} \\ \underline{8} \\ 1 \end{array}$$

2 ← Whole number part; 1 ← Remainder

$$\frac{9}{4} = 2\frac{1}{4}$$

(c) $\frac{35}{5}$

$$\begin{array}{r} 7 \\ 5\overline{)35} \\ \underline{35} \\ 0 \end{array}$$

7 ← Whole number part; 0 ← Remainder

$$\frac{35}{5} = 7$$

(d) $\frac{78}{7}$

$$\begin{array}{r} 11 \\ 7\overline{)78} \\ \underline{7} \\ 8 \\ \underline{7} \\ 1 \end{array}$$

11 ← Whole number part; 1 ← Remainder

$$\frac{78}{7} = 11\frac{1}{7}$$

2.2 Section Exercises

1. $1\frac{1}{4}$ $1 \cdot 4 = 4$ Multiply 1 and 4.

$4 + 1 = 5$ Add 1.

$$1\frac{1}{4} = \frac{5}{4}$$

3. $4\frac{3}{5}$ $4 \cdot 5 = 20$ Multiply 4 and 5.
$20 + 3 = 23$ Add 3.
$4\frac{3}{5} = \frac{23}{5}$

5. $6\frac{1}{2}$ $6 \cdot 2 = 12$ Multiply 6 and 2.
$12 + 1 = 13$ Add 1.
$6\frac{1}{2} = \frac{13}{2}$

7. $8\frac{1}{4}$ $8 \cdot 4 = 32$ Multiply 8 and 4.
$32 + 1 = 33$ Add 1.
$8\frac{1}{4} = \frac{33}{4}$

9. $1\frac{7}{11}$ $1 \cdot 11 = 11$ Multiply 1 and 11.
$11 + 7 = 18$ Add 7.
$1\frac{7}{11} = \frac{18}{11}$

11. $6\frac{1}{3}$ $6 \cdot 3 = 18$ Multiply 6 and 3.
$18 + 1 = 19$ Add 1.
$6\frac{1}{3} = \frac{19}{3}$

13. $10\frac{1}{8}$ $10 \cdot 8 = 80$ Multiply 10 and 8.
$80 + 1 = 81$ Add 1.
$10\frac{1}{8} = \frac{81}{8}$

15. $10\frac{3}{4}$ $10 \cdot 4 = 40$ Multiply 10 and 4.
$40 + 3 = 43$ Add 3.
$10\frac{3}{4} = \frac{43}{4}$

17. $3\frac{3}{8}$ $3 \cdot 8 = 24$ Multiply 3 and 8.
$24 + 3 = 27$ Add 3.
$3\frac{3}{8} = \frac{27}{8}$

19. $8\frac{3}{5}$ $8 \cdot 5 = 40$ Multiply 8 and 5.
$40 + 3 = 43$ Add 3.
$8\frac{3}{5} = \frac{43}{5}$

21. $4\frac{10}{11}$ $4 \cdot 11 = 44$ Multiply 4 and 11.
$44 + 10 = 54$ Add 10.
$4\frac{10}{11} = \frac{54}{11}$

23. $32\frac{3}{4}$ $32 \cdot 4 = 128$ Multiply 32 and 4.
$128 + 3 = 131$ Add 3.
$32\frac{3}{4} = \frac{131}{4}$

25. $18\frac{5}{12}$ $18 \cdot 12 = 216$ Multiply 18 and 12.
$216 + 5 = 221$ Add 5.
$18\frac{5}{12} = \frac{221}{12}$

27. $17\frac{14}{15}$ $17 \cdot 15 = 255$ Multiply 17 and 15.
$255 + 14 = 269$ Add 14.
$17\frac{14}{15} = \frac{269}{15}$

29. $7\frac{19}{24}$ $7 \cdot 24 = 168$ Multiply 7 and 24.
$168 + 19 = 187$ Add 19.
$7\frac{19}{24} = \frac{187}{24}$

31. $\frac{4}{3}$

$$\begin{array}{r} 1 \\ 3\overline{)4} \\ \underline{3} \\ 1 \end{array}$$
1 ← Whole number part
1 ← Remainder

$\frac{4}{3} = 1\frac{1}{3}$

33. $\frac{9}{4}$

$$\begin{array}{r} 2 \\ 4\overline{)9} \\ \underline{8} \\ 1 \end{array}$$
2 ← Whole number part
1 ← Remainder

$\frac{9}{4} = 2\frac{1}{4}$

35. $\frac{48}{6}$

$$\begin{array}{r} 8 \\ 6\overline{)48} \\ \underline{48} \\ 0 \end{array}$$
8 ← Whole number part
0 ← Remainder

$\frac{48}{6} = 8$

37. $\frac{38}{5}$

$$\begin{array}{r} 7 \\ 5\overline{)38} \\ \underline{35} \\ 3 \end{array}$$
7 ← Whole number part
3 ← Remainder

$\frac{38}{5} = 7\frac{3}{5}$

39. $\frac{39}{8}$

$$\begin{array}{r} 4 \leftarrow \text{Whole number part} \\ 8\overline{)39} \\ \underline{32} \\ 7 \leftarrow \text{Remainder} \end{array}$$

$\frac{39}{8} = 4\frac{7}{8}$

41. $\frac{27}{3}$

$$\begin{array}{r} 9 \leftarrow \text{Whole number part} \\ 3\overline{)27} \\ \underline{27} \\ 0 \leftarrow \text{Remainder} \end{array}$$

$\frac{27}{3} = 9$

43. $\frac{63}{4}$

$$\begin{array}{r} 15 \leftarrow \text{Whole number part} \\ 4\overline{)63} \\ \underline{4} \\ 23 \\ \underline{20} \\ 3 \leftarrow \text{Remainder} \end{array}$$

$\frac{63}{4} = 15\frac{3}{4}$

45. $\frac{47}{9}$

$$\begin{array}{r} 5 \leftarrow \text{Whole number part} \\ 9\overline{)47} \\ \underline{45} \\ 2 \leftarrow \text{Remainder} \end{array}$$

$\frac{47}{9} = 5\frac{2}{9}$

47. $\frac{65}{8}$

$$\begin{array}{r} 8 \leftarrow \text{Whole number part} \\ 8\overline{)65} \\ \underline{64} \\ 1 \leftarrow \text{Remainder} \end{array}$$

$\frac{65}{8} = 8\frac{1}{8}$

49. $\frac{84}{5}$

$$\begin{array}{r} 16 \leftarrow \text{Whole number part} \\ 5\overline{)84} \\ \underline{5} \\ 34 \\ \underline{30} \\ 4 \leftarrow \text{Remainder} \end{array}$$

$\frac{84}{5} = 16\frac{4}{5}$

51. $\frac{112}{4}$

$$\begin{array}{r} 28 \leftarrow \text{Whole number part} \\ 4\overline{)112} \\ \underline{8} \\ 32 \\ \underline{32} \\ 0 \leftarrow \text{Remainder} \end{array}$$

$\frac{112}{4} = 28$

53. $\frac{183}{7}$

$$\begin{array}{r} 26 \leftarrow \text{Whole number part} \\ 7\overline{)183} \\ \underline{14} \\ 43 \\ \underline{42} \\ 1 \leftarrow \text{Remainder} \end{array}$$

$\frac{183}{7} = 26\frac{1}{7}$

55. Multiply the denominator by the whole number and add the numerator. The result becomes the new numerator, which is placed over the original denominator.

$$2\frac{1}{2} = \frac{2 \times 2 + 1}{2} = \frac{5}{2}$$

57. $250\frac{1}{2}$ $\quad 250 \cdot 2 = 500$

$500 + 1 = 501$

$250\frac{1}{2} = \frac{501}{2}$

59. $333\frac{1}{3}$ $\quad 333 \cdot 3 = 999$

$999 + 1 = 1000$

$333\frac{1}{3} = \frac{1000}{3}$

61. $522\frac{3}{8}$ $\quad 522 \cdot 8 = 4176$

$4176 + 3 = 4179$

$522\frac{3}{8} = \frac{4179}{8}$

63. $\frac{617}{4}$

```
    1 5 4  ← Whole number part
 4|6 1 7
   4
   2 1
   2 0
     1 7
     1 6
       1  ← Remainder
```

$\frac{617}{4} = 154\frac{1}{4}$

65. The commands used will vary. The following is from a TI-83 Plus:

$\frac{2565}{15} = 171$

67. The commands used will vary. The following is from a TI-83 Plus:

```
int(3917/32)
                122
fPart(3917/32)▸F
rac
              13/32
```

$\frac{3917}{32} = 122\frac{13}{32}$

Note: You can use the following procedure on any calculator. Divide 3917 by 32 to get 122.40625. Subtract 122. Multiply by 32 to get 13. The mixed number is $122\frac{13}{32}$.

69. The following fractions are proper fractions.

$\frac{2}{3}, \frac{4}{5}, \frac{3}{4}, \frac{7}{10}$

70. **(a)** The proper fractions in Exercise 69 are the ones where the *numerator* is smaller than the *denominator*.

(b) $\frac{2}{3}$;

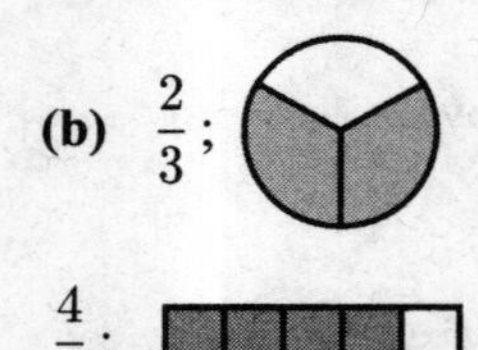

$\frac{4}{5}$;

$\frac{3}{4}$;

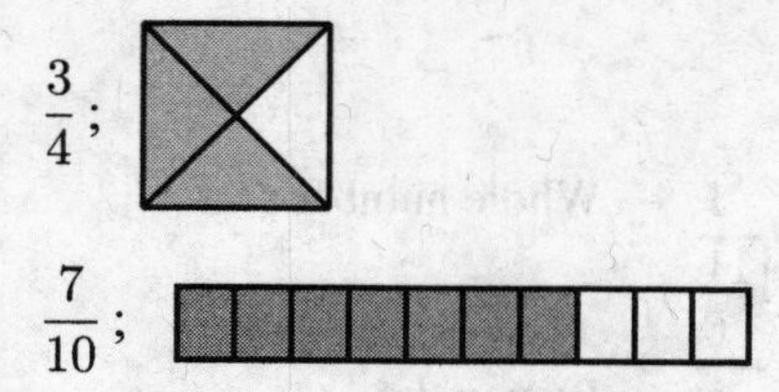

$\frac{7}{10}$;

(c) The proper fractions in Exercise 69 are all *less* than 1.

71. The following fractions are improper fractions.

$\frac{5}{5}, \frac{10}{3}, \frac{6}{5}$

72. **(a)** The improper fractions in Exercise 71 are the ones where the *numerator* is greater than or equal to the *denominator*.

(b) $\frac{5}{5}$;

$\frac{10}{3}$;

$\frac{6}{5}$;

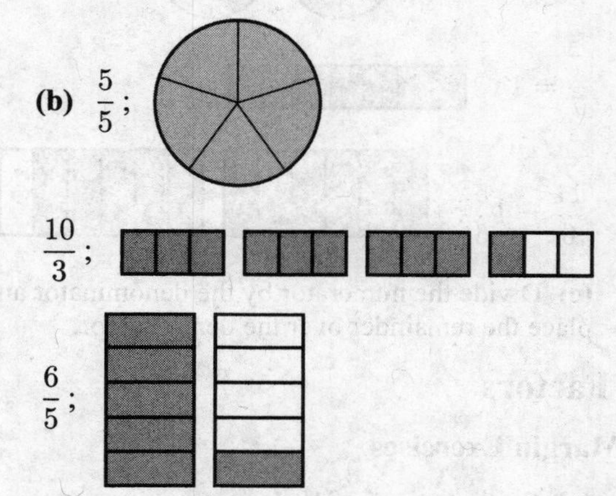

(c) The improper fractions in Exercise 71 are all equal to or *greater* than 1.

73. The following fractions can be written as whole or mixed numbers.

$\frac{5}{3}$

```
    1  ← Whole number part
 3|5
   3
   2  ← Remainder
```

$\frac{5}{3} = 1\frac{2}{3}$

$\frac{7}{7}$

```
    1  ← Whole number part
 7|7
   7
   0  ← Remainder
```

$\frac{7}{7} = 1$

continued

$\frac{11}{6}$

$$\begin{array}{r} 1 \leftarrow \text{Whole number part} \\ 6\overline{)11} \\ \underline{6} \\ 5 \leftarrow \text{Remainder} \end{array}$$

$\frac{11}{6} = 1\frac{5}{6}$

74. **(a)** The fractions that can be written as whole or mixed numbers in Exercise 73 are *improper* fractions, and their value is always *greater than or equal to* 1.

(b) $\frac{5}{3} = 1\frac{2}{3}$;

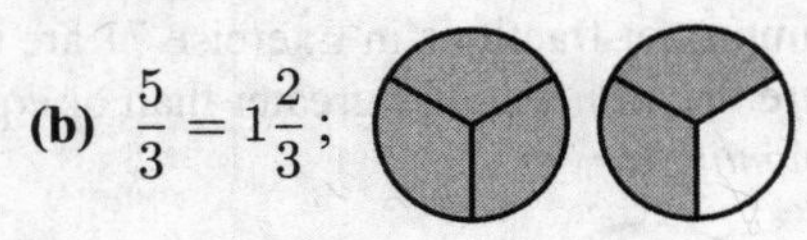

$\frac{7}{7} = 1$;

$\frac{11}{6} = 1\frac{5}{6}$;

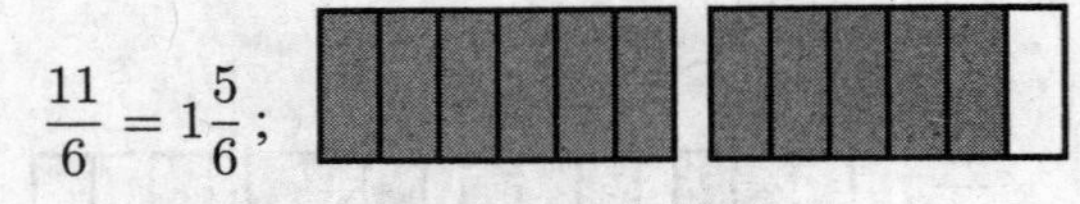

(c) Divide the numerator by the denominator and place the remainder over the denominator.

2.3 Factors

2.3 Margin Exercises

1. **(a)** Factorizations of 10:
$1 \cdot 10 = 10 \quad 2 \cdot 5 = 10$

The factors of 10 are 1, 2, 5, and 10.

(b) Factorizations of 16:

$1 \cdot 16 = 16 \quad 2 \cdot 8 = 16 \quad 4 \cdot 4 = 16$

The factors of 16 are 1, 2, 4, 8, and 16.

(c) Factorizations of 36:

$1 \cdot 36 = 36 \quad 2 \cdot 18 = 36 \quad 3 \cdot 12 = 36 \quad 4 \cdot 9 = 36$
$6 \cdot 6 = 36$

The factors of 36 are 1, 2, 3, 4, 6, 9, 12, 18, and 36.

(d) Factorizations of 80:

$1 \cdot 80 = 80 \quad 2 \cdot 40 = 80 \quad 4 \cdot 20 = 80 \quad 5 \cdot 16 = 80$
$8 \cdot 10 = 80$

The factors of 80 are 1, 2, 4, 5, 8, 10, 16, 20, 40, and 80.

2. 2, 5, 11, 13, and 19 each have no factor other than themselves or 1; 4, 6, 8, 10, 28, 36, and 42 each have a factor of 2; 21, 27, and 33 have a factor of 3. So 4, 6, 8, 10, 21, 27, 28, 33, 36, and 42 are composite.

3. 7, 13, 17, 19, and 29 are prime because they are divisible only by themselves and 1.

4. **(a)** $8 \div 2 = 4$
$4 \div 2 = 2$ *prime*
$8 = 2 \cdot 2 \cdot 2$

(b) $28 \div 2 = 14$
$14 \div 2 = 7$ *prime*
$28 = 2 \cdot 2 \cdot 7$

(c) $18 \div 2 = 9$
$9 \div 3 = 3$ *prime*
$18 = 2 \cdot 3 \cdot 3$

(d) $40 \div 2 = 20$
$20 \div 2 = 10$
$10 \div 2 = 5$ *prime*
$40 = 2 \cdot 2 \cdot 2 \cdot 5$

5. **(a)**
$2\overline{)36}$ = 18 Divide 36 by 2.
$2\overline{)18}$ = 9 Divide 18 by 2.
$3\overline{)9}$ = 3 Divide 9 by 3.
$3\overline{)3}$ = 1 Divide 3 by 3.
Quotient is 1.
$36 = 2 \cdot 2 \cdot 3 \cdot 3 = 2^2 \cdot 3^2$

(b)
$2\overline{)54}$ = 27 Divide 54 by 2.
$3\overline{)27}$ = 9 Divide 27 by 3.
$3\overline{)9}$ = 3 Divide 9 by 3.
$3\overline{)3}$ = 1 Divide 3 by 3.
Quotient is 1.
$54 = 2 \cdot 3 \cdot 3 \cdot 3 = 2 \cdot 3^3$

(c)
$2\overline{)60}$ = 30 Divide 60 by 2.
$2\overline{)30}$ = 15 Divide 30 by 2.
$3\overline{)15}$ = 5 Divide 15 by 3.
$5\overline{)5}$ = 1 Divide 5 by 5.
Quotient is 1.
$60 = 2 \cdot 2 \cdot 3 \cdot 5 = 2^2 \cdot 3 \cdot 5$

(d)

$$\begin{array}{r} 27 \\ 3\overline{)81} \end{array}$$ Divide 81 by 3.

$$\begin{array}{r} 9 \\ 3\overline{)27} \end{array}$$ Divide 27 by 3.

$$\begin{array}{r} 3 \\ 3\overline{)9} \end{array}$$ Divide 9 by 3.

$$\begin{array}{r} 1 \\ 3\overline{)3} \end{array}$$ Divide 3 by 3.

Quotient is 1.

$81 = 3 \cdot 3 \cdot 3 \cdot 3 = 3^4$

6. (a)

$$\begin{array}{r} 24 \\ 2\overline{)48} \end{array}$$ Divide 48 by 2.

$$\begin{array}{r} 12 \\ 2\overline{)24} \end{array}$$ Divide 24 by 2.

$$\begin{array}{r} 6 \\ 2\overline{)12} \end{array}$$ Divide 12 by 2.

$$\begin{array}{r} 3 \\ 2\overline{)6} \end{array}$$ Divide 6 by 2.

$$\begin{array}{r} 1 \\ 3\overline{)3} \end{array}$$ Divide 3 by 3.

Quotient is 1.

$48 = 2 \cdot 2 \cdot 2 \cdot 2 \cdot 3 = 2^4 \cdot 3$

(b)

$$\begin{array}{r} 22 \\ 2\overline{)44} \end{array}$$ Divide 44 by 2.

$$\begin{array}{r} 11 \\ 2\overline{)22} \end{array}$$ Divide 22 by 2.

$$\begin{array}{r} 1 \\ 11\overline{)11} \end{array}$$ Divide 11 by 11.

Quotient is 1.

$44 = 2 \cdot 2 \cdot 11 = 2^2 \cdot 11$

(c)

$$\begin{array}{r} 45 \\ 2\overline{)90} \end{array}$$ Divide 90 by 2.

$$\begin{array}{r} 15 \\ 3\overline{)45} \end{array}$$ Divide 45 by 3.

$$\begin{array}{r} 5 \\ 3\overline{)15} \end{array}$$ Divide 15 by 3.

$$\begin{array}{r} 1 \\ 5\overline{)5} \end{array}$$ Divide 5 by 5.

Quotient is 1.

$90 = 2 \cdot 3 \cdot 3 \cdot 5 = 2 \cdot 3^2 \cdot 5$

(d)

$$\begin{array}{r} 60 \\ 2\overline{)120} \end{array}$$ Divide 120 by 2.

$$\begin{array}{r} 30 \\ 2\overline{)60} \end{array}$$ Divide 60 by 2.

$$\begin{array}{r} 15 \\ 2\overline{)30} \end{array}$$ Divide 30 by 2.

$$\begin{array}{r} 5 \\ 3\overline{)15} \end{array}$$ Divide 15 by 3.

$$\begin{array}{r} 1 \\ 5\overline{)5} \end{array}$$ Divide 5 by 5.

Quotient is 1.

$120 = 2 \cdot 2 \cdot 2 \cdot 3 \cdot 5 = 2^3 \cdot 3 \cdot 5$

(e)

$$\begin{array}{r} 90 \\ 2\overline{)180} \end{array}$$ Divide 180 by 2.

$$\begin{array}{r} 45 \\ 2\overline{)90} \end{array}$$ Divide 90 by 2.

$$\begin{array}{r} 15 \\ 3\overline{)45} \end{array}$$ Divide 45 by 3.

$$\begin{array}{r} 5 \\ 3\overline{)15} \end{array}$$ Divide 15 by 3.

$$\begin{array}{r} 1 \\ 5\overline{)5} \end{array}$$ Divide 5 by 5.

Quotient is 1.

$180 = 2 \cdot 2 \cdot 3 \cdot 3 \cdot 5 = 2^2 \cdot 3^2 \cdot 5$

7. (a)

28 ↙ ↘ [2] 14; 14 ↙ ↘ [2] [7]

$28 = 2 \cdot 2 \cdot 7 = 2^2 \cdot 7$

(b)

35 ↙ ↘ [5] [7]

$35 = 5 \cdot 7$

(c)

78 ↙ ↘ [2] 39; 39 ↙ ↘ [3] [13]

$78 = 2 \cdot 3 \cdot 13$

2.3 Section Exercises

1. Factorizations of 8:

$1 \cdot 8 = 8 \quad 2 \cdot 4 = 8$

The factors of 8 are 1, 2, 4, and 8.

3. Factorizations of 15:

$1 \cdot 15 = 15 \quad 3 \cdot 5 = 15$

The factors of 15 are 1, 3, 5, and 15.

5. Factorizations of 48:

$1 \cdot 48 = 48 \quad 2 \cdot 24 = 48 \quad 3 \cdot 16 = 48 \quad 4 \cdot 12 = 48$
$6 \cdot 8 = 48$

The factors of 48 are 1, 2, 3, 4, 6, 8, 12, 16, 24, and 48.

7. Factorizations of 36:

$1 \cdot 36 = 36 \quad 2 \cdot 18 = 36 \quad 3 \cdot 12 = 36 \quad 4 \cdot 9 = 36$
$6 \cdot 6 = 36$

The factors of 36 are 1, 2, 3, 4, 6, 9, 12, 18, and 36.

9. Factorizations of 40:

$1 \cdot 40 = 40 \quad 2 \cdot 20 = 40 \quad 4 \cdot 10 = 40 \quad 5 \cdot 8 = 40$

The factors of 40 are 1, 2, 4, 5, 8, 10, 20, and 40.

11. Factorizations of 64:

$1 \cdot 64 = 64 \quad 2 \cdot 32 = 64 \quad 4 \cdot 16 = 64 \quad 8 \cdot 8 = 64$

The factors of 64 are 1, 2, 4, 8, 16, 32, and 64.

13. 6 is divisible by 2 and 3, so 6 is composite.

15. 5 is only divisible by itself and 1, so it is prime.

17. 16 is divisible by 2, so 16 is composite.

19. 13 is only divisible by itself and 1, so it is prime.

21. 13 is only divisible by itself and 1, so it is prime.

23. 25 is divisible by 5, so 25 is composite.

25. 48 is divisible by 2 and 3, so 48 is composite.

27. 45 is divisible by 3 and 5, so 45 is composite.

29. 8

$\begin{array}{r} 4 \\ 2\overline{)8} \end{array}$ Divide 8 by 2.

$\begin{array}{r} 2 \\ 2\overline{)4} \end{array}$ Divide 4 by 2.

$\begin{array}{r} 1 \\ 2\overline{)2} \end{array}$ Divide 2 by 2.

Quotient is 1.

$8 = 2 \cdot 2 \cdot 2 = 2^3$

31. We can also use a factor tree.

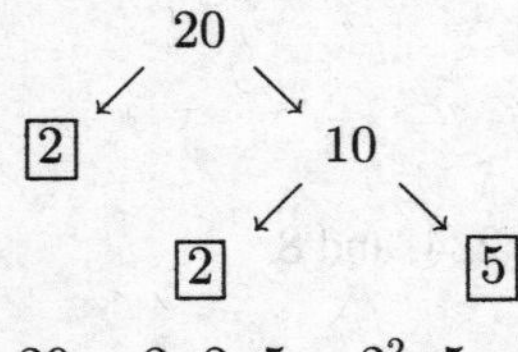

$20 = 2 \cdot 2 \cdot 5 = 2^2 \cdot 5$

33.

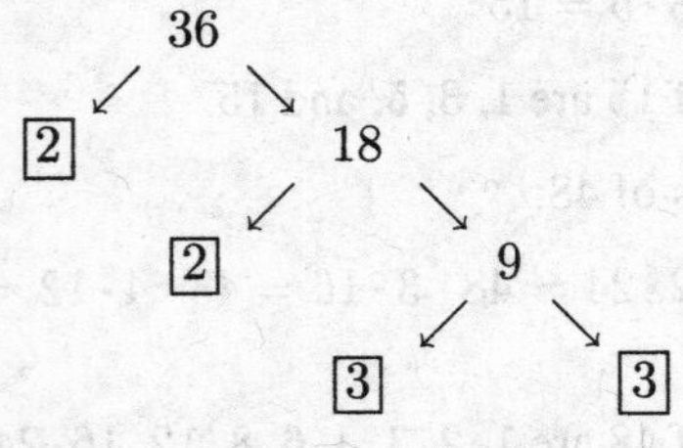

$36 = 2 \cdot 2 \cdot 3 \cdot 3 = 2^2 \cdot 3^2$

35.

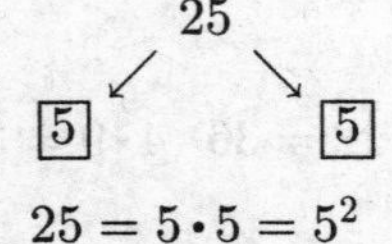

$25 = 5 \cdot 5 = 5^2$

37.

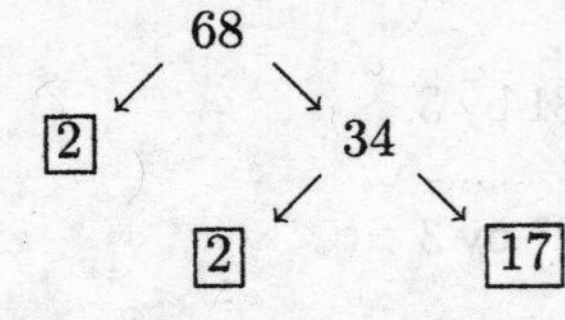

$68 = 2 \cdot 2 \cdot 17 = 2^2 \cdot 17$

39. 72

$\begin{array}{r} 36 \\ 2\overline{)72} \end{array}$ Divide 72 by 2.

$\begin{array}{r} 18 \\ 2\overline{)36} \end{array}$ Divide 36 by 2.

$\begin{array}{r} 9 \\ 2\overline{)18} \end{array}$ Divide 18 by 2.

$\begin{array}{r} 3 \\ 3\overline{)9} \end{array}$ Divide 9 by 3.

$\begin{array}{r} 1 \\ 3\overline{)3} \end{array}$ Divide 3 by 3.

Quotient is 1.

$72 = 2 \cdot 2 \cdot 2 \cdot 3 \cdot 3 = 2^3 \cdot 3^2$

41. 44

$\begin{array}{r} 22 \\ 2\overline{)44} \end{array}$ Divide 44 by 2.

$\begin{array}{r} 11 \\ 2\overline{)22} \end{array}$ Divide 22 by 2.

$\begin{array}{r} 1 \\ 11\overline{)11} \end{array}$ Divide 11 by 11.

Quotient is 1.

$44 = 2 \cdot 2 \cdot 11 = 2^2 \cdot 11$

43. 100

$\begin{array}{r} 50 \\ 2\overline{)100} \end{array}$ Divide 100 by 2.

$\begin{array}{r} 25 \\ 2\overline{)50} \end{array}$ Divide 50 by 2.

$\begin{array}{r} 5 \\ 5\overline{)25} \end{array}$ Divide 25 by 5.

$\begin{array}{r} 1 \\ 5\overline{)5} \end{array}$ Divide 5 by 5.

Quotient is 1.

$100 = 2 \cdot 2 \cdot 5 \cdot 5 = 2^2 \cdot 5^2$

45. 125

$\begin{array}{r} 25 \\ 5\overline{)125} \end{array}$ Divide 125 by 5.

$\begin{array}{r} 5 \\ 5\overline{)25} \end{array}$ Divide 25 by 5.

$\begin{array}{r} 1 \\ 5\overline{)5} \end{array}$ Divide 5 by 5.

Quotient is 1.

$125 = 5 \cdot 5 \cdot 5 = 5^3$

47. 180

$$\begin{array}{rl} & 90 \\ 2 & \overline{)180} \end{array}$$ Divide 180 by 2.

$$\begin{array}{rl} & 45 \\ 2 & \overline{)90} \end{array}$$ Divide 90 by 2.

$$\begin{array}{rl} & 15 \\ 3 & \overline{)45} \end{array}$$ Divide 45 by 3.

$$\begin{array}{rl} & 5 \\ 3 & \overline{)15} \end{array}$$ Divide 15 by 3.

$$\begin{array}{rl} & 1 \\ 5 & \overline{)5} \end{array}$$ Divide 5 by 5.

Quotient is 1.

$180 = 2 \cdot 2 \cdot 3 \cdot 3 \cdot 5 = 2^2 \cdot 3^2 \cdot 5$

49. 320

$$\begin{array}{rl} & 160 \\ 2 & \overline{)320} \end{array}$$ Divide 320 by 2.

$$\begin{array}{rl} & 80 \\ 2 & \overline{)160} \end{array}$$ Divide 160 by 2.

$$\begin{array}{rl} & 40 \\ 2 & \overline{)80} \end{array}$$ Divide 80 by 2.

$$\begin{array}{rl} & 20 \\ 2 & \overline{)40} \end{array}$$ Divide 40 by 2.

$$\begin{array}{rl} & 10 \\ 2 & \overline{)20} \end{array}$$ Divide 20 by 2.

$$\begin{array}{rl} & 5 \\ 2 & \overline{)10} \end{array}$$ Divide 10 by 2.

$$\begin{array}{rl} & 1 \\ 5 & \overline{)5} \end{array}$$ Divide 5 by 5.

Quotient is 1.

$320 = 2 \cdot 2 \cdot 2 \cdot 2 \cdot 2 \cdot 2 \cdot 5 = 2^6 \cdot 5$

51. 360

$$\begin{array}{rl} & 180 \\ 2 & \overline{)360} \end{array}$$ Divide 360 by 2.

$$\begin{array}{rl} & 90 \\ 2 & \overline{)180} \end{array}$$ Divide 180 by 2.

$$\begin{array}{rl} & 45 \\ 2 & \overline{)90} \end{array}$$ Divide 90 by 2.

$$\begin{array}{rl} & 15 \\ 3 & \overline{)45} \end{array}$$ Divide 45 by 3.

$$\begin{array}{rl} & 5 \\ 3 & \overline{)15} \end{array}$$ Divide 15 by 3.

$$\begin{array}{rl} & 1 \\ 5 & \overline{)5} \end{array}$$ Divide 5 by 5.

Quotient is 1.

$360 = 2 \cdot 2 \cdot 2 \cdot 3 \cdot 3 \cdot 5 = 2^3 \cdot 3^2 \cdot 5$

53. Answers will vary. A sample answer follows. A composite number has a factor(s) other than itself or 1. Examples include 4, 6, 8, 9, and 10. A prime number is a whole number that has exactly two *different* factors, itself and 1. Examples include 2, 3, 5, 7, and 11. The numbers 0 and 1 are neither prime nor composite.

55. All the possible factors of 24 are 1, 2, 3, 4, 6, 8, 12, and 24. This list includes both prime numbers and composite numbers. The prime factors of 24 include only prime numbers. The prime factorization of 24 is

$$2 \cdot 2 \cdot 2 \cdot 3 = 2^3 \cdot 3.$$

57. 350

$$\begin{array}{rl} & 175 \\ 2 & \overline{)350} \end{array}$$ Divide 350 by 2.

$$\begin{array}{rl} & 35 \\ 5 & \overline{)175} \end{array}$$ Divide 175 by 5.

$$\begin{array}{rl} & 7 \\ 5 & \overline{)35} \end{array}$$ Divide 35 by 5.

$$\begin{array}{rl} & 1 \\ 7 & \overline{)7} \end{array}$$ Divide 7 by 7.

Quotient is 1.

$350 = 2 \cdot 5 \cdot 5 \cdot 7 = 2 \cdot 5^2 \cdot 7$

59. 960

$$\begin{array}{rl} & 480 \\ 2 & \overline{)960} \end{array}$$ Divide 960 by 2.

$$\begin{array}{rl} & 240 \\ 2 & \overline{)480} \end{array}$$ Divide 480 by 2.

$$\begin{array}{rl} & 120 \\ 2 & \overline{)240} \end{array}$$ Divide 240 by 2.

$$\begin{array}{rl} & 60 \\ 2 & \overline{)120} \end{array}$$ Divide 120 by 2.

$$\begin{array}{rl} & 30 \\ 2 & \overline{)60} \end{array}$$ Divide 60 by 2.

$$\begin{array}{rl} & 15 \\ 2 & \overline{)30} \end{array}$$ Divide 30 by 2.

$$\begin{array}{rl} & 5 \\ 3 & \overline{)15} \end{array}$$ Divide 15 by 3.

$$\begin{array}{rl} & 1 \\ 5 & \overline{)5} \end{array}$$ Divide 5 by 5.

Quotient is 1.

$960 = 2 \cdot 2 \cdot 2 \cdot 2 \cdot 2 \cdot 2 \cdot 3 \cdot 5 = 2^6 \cdot 3 \cdot 5$

61. 1560

$$\begin{array}{rl} & 780 \\ 2 & \overline{)1560} \end{array}$$ Divide 1560 by 2.

$$\begin{array}{rl} & 390 \\ 2 & \overline{)780} \end{array}$$ Divide 780 by 2.

$$\begin{array}{rl} & 195 \\ 2 & \overline{)390} \end{array}$$ Divide 390 by 2.

continued

$3\overline{)195}$ with quotient 65 — Divide 195 by 3.

$5\overline{)65}$ with quotient 13 — Divide 65 by 5.

$13\overline{)13}$ with quotient 1 — Divide 13 by 13.

Quotient is 1.

$1560 = 2 \cdot 2 \cdot 2 \cdot 3 \cdot 5 \cdot 13 = 2^3 \cdot 3 \cdot 5 \cdot 13$

63.

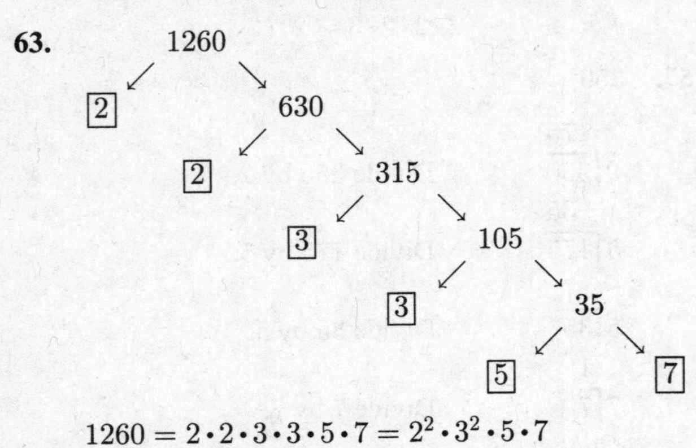

$1260 = 2 \cdot 2 \cdot 3 \cdot 3 \cdot 5 \cdot 7 = 2^2 \cdot 3^2 \cdot 5 \cdot 7$

65. The prime numbers smaller than 50 are 2, 3, 5, 7, 11, 13, 17, 19, 23, 29, 31, 37, 41, 43, and 47.

66. A prime number is a whole number that is evenly divisible only by itself and 1.

67. It is true because any even number is divisible by the number 2 in addition to itself and 1.

68. No. A multiple of a prime number can never be prime because it will always be evenly divisible by the prime number.

69. 2100

$2\overline{)2100}$ with quotient 1050 — Divide 2100 by 2.

$2\overline{)1050}$ with quotient 525 — Divide 1050 by 2.

$3\overline{)525}$ with quotient 175 — Divide 525 by 3.

$5\overline{)175}$ with quotient 35 — Divide 175 by 5.

$5\overline{)35}$ with quotient 7 — Divide 35 by 5.

$7\overline{)7}$ with quotient 1 — Divide 7 by 7.

Quotient is 1.

$2100 = 2 \cdot 2 \cdot 3 \cdot 5 \cdot 5 \cdot 7$

70. $2100 = 2 \cdot 2 \cdot 3 \cdot 5 \cdot 5 \cdot 7 = 2^2 \cdot 3 \cdot 5^2 \cdot 7$

2.4 Writing a Fraction in Lowest Terms

2.4 Margin Exercises

1. **(a)** 6, 12; 2

$6 = 2 \cdot 3 \quad 12 = 2 \cdot 6$

Yes, 2 is a common factor of 6 and 12.

(b) 32, 64; 8

$32 = 8 \cdot 4 \quad 64 = 8 \cdot 8$

Yes.

(c) 32, 56; 16

$32 = 16 \cdot 2$, but 16 is not a factor of 56.

No.

(d) 75, 81; 1

$75 = 75 \cdot 1 \quad 81 = 81 \cdot 1$

Yes.

2. **(a)** $\frac{4}{5}$

4 and 5 have no common factor other than 1. Yes, it is in lowest terms.

(b) $\frac{6}{18}$

6 and 18 have a common factor of 6. No, it is not in lowest terms.

(c) $\frac{9}{15}$

9 and 15 have a common factor of 3. No, it is not in lowest terms.

(d) $\frac{17}{46}$

17 and 46 have no common factor other than 1. Yes, it is in lowest terms.

3. **(a)** $\frac{8}{16} = \frac{8 \div 8}{16 \div 8} = \frac{1}{2}$

(b) $\frac{9}{12} = \frac{9 \div 3}{12 \div 3} = \frac{3}{4}$

(c) $\frac{28}{42} = \frac{28 \div 14}{42 \div 14} = \frac{2}{3}$

(d) $\frac{30}{80} = \frac{30 \div 10}{80 \div 10} = \frac{3}{8}$

(e) $\frac{16}{40} = \frac{16 \div 8}{40 \div 8} = \frac{2}{5}$

4. **(a)** $\frac{12}{36} = \frac{\overset{1}{\cancel{2}} \cdot \overset{1}{\cancel{2}} \cdot \overset{1}{\cancel{3}}}{\underset{1}{\cancel{2}} \cdot \underset{1}{\cancel{2}} \cdot \underset{1}{\cancel{3}} \cdot 3} = \frac{1 \cdot 1 \cdot 1}{1 \cdot 1 \cdot 1 \cdot 3} = \frac{1}{3}$

(b) $\frac{32}{56} = \frac{\overset{1}{\cancel{2}}\cdot\overset{1}{\cancel{2}}\cdot\overset{1}{\cancel{2}}\cdot 2\cdot 2}{\underset{1}{\cancel{2}}\cdot\underset{1}{\cancel{2}}\cdot\underset{1}{\cancel{2}}\cdot 7} = \frac{1\cdot 1\cdot 1\cdot 2\cdot 2}{1\cdot 1\cdot 1\cdot 7} = \frac{4}{7}$

(c) $\frac{74}{111} = \frac{2\cdot\overset{1}{\cancel{37}}}{3\cdot\underset{1}{\cancel{37}}} = \frac{2\cdot 1}{3\cdot 1} = \frac{2}{3}$

(d) $\frac{124}{340} = \frac{\overset{1}{\cancel{2}}\cdot\overset{1}{\cancel{2}}\cdot 31}{\underset{1}{\cancel{2}}\cdot\underset{1}{\cancel{2}}\cdot 5\cdot 17} = \frac{1\cdot 1\cdot 31}{1\cdot 1\cdot 5\cdot 17} = \frac{31}{85}$

5. **(a)** $\frac{24}{48}$ and $\frac{36}{72}$

$\frac{24}{48} = \frac{\overset{1}{\cancel{2}}\cdot\overset{1}{\cancel{2}}\cdot\overset{1}{\cancel{2}}\cdot\overset{1}{\cancel{3}}}{\underset{1}{\cancel{2}}\cdot\underset{1}{\cancel{2}}\cdot\underset{1}{\cancel{2}}\cdot 2\cdot\underset{1}{\cancel{3}}} = \frac{1\cdot 1\cdot 1\cdot 1}{1\cdot 1\cdot 1\cdot 2\cdot 1} = \frac{1}{2}$

$\frac{36}{72} = \frac{\overset{1}{\cancel{2}}\cdot\overset{1}{\cancel{2}}\cdot\overset{1}{\cancel{3}}\cdot\overset{1}{\cancel{3}}}{\underset{1}{\cancel{2}}\cdot\underset{1}{\cancel{2}}\cdot 2\cdot\underset{1}{\cancel{3}}\cdot\underset{1}{\cancel{3}}} = \frac{1\cdot 1\cdot 1\cdot 1}{1\cdot 1\cdot 2\cdot 1\cdot 1} = \frac{1}{2}$

The fractions are *equivalent* $\left(\frac{1}{2} = \frac{1}{2}\right)$.

(b) $\frac{45}{60}$ and $\frac{50}{75}$

$\frac{45}{60} = \frac{3\cdot\overset{1}{\cancel{3}}\cdot\overset{1}{\cancel{5}}}{2\cdot 2\cdot\underset{1}{\cancel{3}}\cdot\underset{1}{\cancel{5}}} = \frac{3\cdot 1\cdot 1}{2\cdot 2\cdot 1\cdot 1} = \frac{3}{4}$

$\frac{50}{75} = \frac{2\cdot\overset{1}{\cancel{5}}\cdot\overset{1}{\cancel{5}}}{3\cdot\underset{1}{\cancel{5}}\cdot\underset{1}{\cancel{5}}} = \frac{2\cdot 1\cdot 1}{3\cdot 1\cdot 1} = \frac{2}{3}$

The fractions are *not equivalent* $\left(\frac{3}{4} \neq \frac{2}{3}\right)$.

(c) $\frac{20}{4}$ and $\frac{110}{22}$

$\frac{20}{4} = \frac{\overset{1}{\cancel{2}}\cdot\overset{1}{\cancel{2}}\cdot 5}{\underset{1}{\cancel{2}}\cdot\underset{1}{\cancel{2}}} = \frac{1\cdot 1\cdot 5}{1\cdot 1} = 5$

$\frac{110}{22} = \frac{\overset{1}{\cancel{2}}\cdot 5\cdot\overset{1}{\cancel{11}}}{\underset{1}{\cancel{2}}\cdot\underset{1}{\cancel{11}}} = \frac{1\cdot 5\cdot 1}{1\cdot 1} = 5$

The fractions are *equivalent* $(5 = 5)$.

(d) $\frac{120}{220}$ and $\frac{180}{320}$

$\frac{120}{220} = \frac{\overset{1}{\cancel{2}}\cdot\overset{1}{\cancel{2}}\cdot 2\cdot 3\cdot\overset{1}{\cancel{5}}}{\underset{1}{\cancel{2}}\cdot\underset{1}{\cancel{2}}\cdot\underset{1}{\cancel{5}}\cdot 11} = \frac{1\cdot 1\cdot 2\cdot 3\cdot 1}{1\cdot 1\cdot 1\cdot 11} = \frac{6}{11}$

$\frac{180}{320} = \frac{\overset{1}{\cancel{2}}\cdot\overset{1}{\cancel{2}}\cdot 3\cdot 3\cdot\overset{1}{\cancel{5}}}{\underset{1}{\cancel{2}}\cdot\underset{1}{\cancel{2}}\cdot 2\cdot 2\cdot 2\cdot 2\cdot\underset{1}{\cancel{5}}}$

$= \frac{1\cdot 1\cdot 3\cdot 3\cdot 1}{1\cdot 1\cdot 2\cdot 2\cdot 2\cdot 2\cdot 1} = \frac{9}{16}$

The fractions are *not equivalent* $\left(\frac{6}{11} \neq \frac{9}{16}\right)$.

2.4 Section Exercises

		2	3	5	10
1.	60	✓	✓	✓	✓
3.	48	✓	✓	X	X
5.	160	✓	X	✓	✓
7.	138	✓	✓	X	X

9. $\frac{6}{8} = \frac{6 \div 2}{8 \div 2} = \frac{3}{4}$

11. $\frac{3}{12} = \frac{3 \div 3}{12 \div 3} = \frac{1}{4}$

13. $\frac{15}{25} = \frac{15 \div 5}{25 \div 5} = \frac{3}{5}$

15. $\frac{36}{42} = \frac{36 \div 6}{42 \div 6} = \frac{6}{7}$

17. $\frac{56}{64} = \frac{56 \div 8}{64 \div 8} = \frac{7}{8}$

19. $\frac{180}{210} = \frac{180 \div 30}{210 \div 30} = \frac{6}{7}$

21. $\frac{72}{126} = \frac{72 \div 18}{126 \div 18} = \frac{4}{7}$

23. $\frac{12}{600} = \frac{12 \div 12}{600 \div 12} = \frac{1}{50}$

25. $\frac{96}{132} = \frac{96 \div 12}{132 \div 12} = \frac{8}{11}$

27. $\frac{60}{108} = \frac{60 \div 12}{108 \div 12} = \frac{5}{9}$

29. $\frac{18}{24} = \frac{\overset{1}{\cancel{2}}\cdot\overset{1}{\cancel{3}}\cdot 3}{\underset{1}{\cancel{2}}\cdot 2\cdot 2\cdot\underset{1}{\cancel{3}}} = \frac{1\cdot 1\cdot 3}{1\cdot 2\cdot 2\cdot 1} = \frac{3}{4}$

31. $\frac{35}{40} = \frac{\overset{1}{\cancel{5}}\cdot 7}{2\cdot 2\cdot 2\cdot\underset{1}{\cancel{5}}} = \frac{1\cdot 7}{2\cdot 2\cdot 2\cdot 1} = \frac{7}{8}$

33. $\frac{90}{180} = \frac{\overset{1}{\cancel{2}}\cdot\overset{1}{\cancel{3}}\cdot\overset{1}{\cancel{3}}\cdot\overset{1}{\cancel{5}}}{\underset{1}{\cancel{2}}\cdot 2\cdot\underset{1}{\cancel{3}}\cdot\underset{1}{\cancel{3}}\cdot\underset{1}{\cancel{5}}} = \frac{1\cdot 1\cdot 1\cdot 1}{1\cdot 2\cdot 1\cdot 1\cdot 1} = \frac{1}{2}$

35. $\frac{36}{12} = \frac{\overset{1}{\cancel{2}}\cdot\overset{1}{\cancel{2}}\cdot\overset{1}{\cancel{3}}\cdot 3}{\underset{1}{\cancel{2}}\cdot\underset{1}{\cancel{2}}\cdot\underset{1}{\cancel{3}}} = \frac{1\cdot 1\cdot 1\cdot 3}{1\cdot 1\cdot 1} = 3$

37. $\frac{72}{225} = \frac{2\cdot 2\cdot 2\cdot\overset{1}{\cancel{3}}\cdot\overset{1}{\cancel{3}}}{\underset{1}{\cancel{3}}\cdot\underset{1}{\cancel{3}}\cdot 5\cdot 5} = \frac{2\cdot 2\cdot 2\cdot 1\cdot 1}{1\cdot 1\cdot 5\cdot 5} = \frac{8}{25}$

39. $\frac{3}{6}$ and $\frac{18}{36}$

$$\frac{3}{6} = \frac{1 \cdot \overset{1}{\cancel{3}}}{2 \cdot \underset{1}{\cancel{3}}} = \frac{1}{2}$$

$$\frac{18}{36} = \frac{\overset{1}{\cancel{2}} \cdot \overset{1}{\cancel{3}} \cdot \overset{1}{\cancel{3}}}{\underset{1}{\cancel{2}} \cdot 2 \cdot \underset{1}{\cancel{3}} \cdot \underset{1}{\cancel{3}}} = \frac{1}{2}$$

The fractions are *equivalent* $\left(\frac{1}{2} = \frac{1}{2}\right)$.

41. $\frac{10}{24}$ and $\frac{12}{30}$

$$\frac{10}{24} = \frac{\overset{1}{\cancel{2}} \cdot 5}{\underset{1}{\cancel{2}} \cdot 2 \cdot 2 \cdot 3} = \frac{5}{12}$$

$$\frac{12}{30} = \frac{\overset{1}{\cancel{2}} \cdot 2 \cdot \overset{1}{\cancel{3}}}{\underset{1}{\cancel{2}} \cdot \underset{1}{\cancel{3}} \cdot 5} = \frac{2}{5}$$

The fractions are *not equivalent* $\left(\frac{5}{12} \neq \frac{2}{5}\right)$.

43. $\frac{15}{24}$ and $\frac{35}{52}$

$$\frac{15}{24} = \frac{\overset{1}{\cancel{3}} \cdot 5}{2 \cdot 2 \cdot 2 \cdot \underset{1}{\cancel{3}}} = \frac{5}{8}$$

$$\frac{35}{52} = \frac{5 \cdot 7}{2 \cdot 2 \cdot 13} = \frac{35}{52}$$

The fractions are *not equivalent* $\left(\frac{5}{8} \neq \frac{35}{52}\right)$.

45. $\frac{14}{16}$ and $\frac{35}{40}$

$$\frac{14}{16} = \frac{\overset{1}{\cancel{2}} \cdot 7}{\underset{1}{\cancel{2}} \cdot 2 \cdot 2 \cdot 2} = \frac{7}{8}$$

$$\frac{35}{40} = \frac{\overset{1}{\cancel{5}} \cdot 7}{2 \cdot 2 \cdot 2 \cdot \underset{1}{\cancel{5}}} = \frac{7}{8}$$

The fractions are *equivalent* $\left(\frac{7}{8} = \frac{7}{8}\right)$.

47. $\frac{48}{6}$ and $\frac{72}{8}$

$$\frac{48}{6} = \frac{2 \cdot 2 \cdot 2 \cdot \overset{1}{\cancel{2}} \cdot \overset{1}{\cancel{3}}}{\underset{1}{\cancel{2}} \cdot \underset{1}{\cancel{3}}} = 8$$

$$\frac{72}{8} = \frac{\overset{1}{\cancel{2}} \cdot \overset{1}{\cancel{2}} \cdot \overset{1}{\cancel{2}} \cdot 3 \cdot 3}{\underset{1}{\cancel{2}} \cdot \underset{1}{\cancel{2}} \cdot \underset{1}{\cancel{2}}} = 9$$

The fractions are *not equivalent* $(8 \neq 9)$.

49. $\frac{25}{30}$ and $\frac{65}{78}$

$$\frac{25}{30} = \frac{5 \cdot \overset{1}{\cancel{5}}}{2 \cdot 3 \cdot \underset{1}{\cancel{5}}} = \frac{5}{6}$$

$$\frac{65}{78} = \frac{5 \cdot \overset{1}{\cancel{13}}}{2 \cdot 3 \cdot \underset{1}{\cancel{13}}} = \frac{5}{6}$$

The fractions are *equivalent* $\left(\frac{5}{6} = \frac{5}{6}\right)$.

51. A fraction is in lowest terms when the numerator and the denominator have no common factors other than 1. Some examples are $\frac{1}{2}$, $\frac{3}{8}$, and $\frac{2}{3}$.

53. $$\frac{160}{256} = \frac{\overset{1}{\cancel{2}} \cdot \overset{1}{\cancel{2}} \cdot \overset{1}{\cancel{2}} \cdot \overset{1}{\cancel{2}} \cdot \overset{1}{\cancel{2}} \cdot 5}{\underset{1}{\cancel{2}} \cdot \underset{1}{\cancel{2}} \cdot \underset{1}{\cancel{2}} \cdot \underset{1}{\cancel{2}} \cdot \underset{1}{\cancel{2}} \cdot 2 \cdot 2 \cdot 2} = \frac{5}{8}$$

55. $$\frac{238}{119} = \frac{2 \cdot \overset{1}{\cancel{7}} \cdot \overset{1}{\cancel{17}}}{\underset{1}{\cancel{7}} \cdot \underset{1}{\cancel{17}}} = \frac{2}{1} = 2$$

Summary Exercises on Fraction Basics

1. The figure has 6 equal parts.
Five parts are shaded: $\frac{5}{6}$

One part is unshaded: $\frac{1}{6}$

3. The figure has 8 equal parts.
Five parts are shaded: $\frac{5}{8}$

Three parts are unshaded: $\frac{3}{8}$

5. 8 ← Numerator
5 ← Denominator

7. Six of the 31 stores are cards, gifts, accessories stores. $\frac{6}{31}$

9. $\frac{5}{2}$

$$2\overline{)5}$$
2 ← Whole number part
4
1 ← Remainder

$$\frac{5}{2} = 2\frac{1}{2}$$

11. $\frac{9}{7}$

$$7\overline{)9}$$
1 ← Whole number part
7
2 ← Remainder

$$\frac{9}{7} = 1\frac{2}{7}$$

13. $\frac{64}{8}$

$$\begin{array}{r} 8 \leftarrow \text{Whole number part} \\ 8\overline{)64} \\ \underline{64} \\ 0 \leftarrow \text{Remainder} \end{array}$$

$\frac{64}{8} = 8$

15. $\frac{36}{5}$

$$\begin{array}{r} 7 \leftarrow \text{Whole number part} \\ 5\overline{)36} \\ \underline{35} \\ 1 \leftarrow \text{Remainder} \end{array}$$

$\frac{36}{5} = 7\frac{1}{5}$

17. $3\frac{1}{3}$ $3 \cdot 3 = 9$ Multiply 3 and 3.

$9 + 1 = 10$ Add 1.

$3\frac{1}{3} = \frac{10}{3}$

19. $6\frac{4}{5}$ $6 \cdot 5 = 30$ Multiply 6 and 5.

$30 + 4 = 34$ Add 4.

$6\frac{4}{5} = \frac{34}{5}$

21. $12\frac{3}{4}$ $12 \cdot 4 = 48$ Multiply 12 and 4.

$48 + 3 = 51$ Add 3.

$12\frac{3}{4} = \frac{51}{4}$

23. $11\frac{5}{6}$ $11 \cdot 6 = 66$ Multiply 11 and 6.

$66 + 5 = 71$ Add 5.

$11\frac{5}{6} = \frac{71}{6}$

25. 10
↙ ↘
[2] [5]

$10 = 2 \cdot 5$

27. 36
↙ ↘
[2] 18
↙ ↘
[2] 9
↙ ↘
[3] [3]

$36 = 2^2 \cdot 3^2$

29.

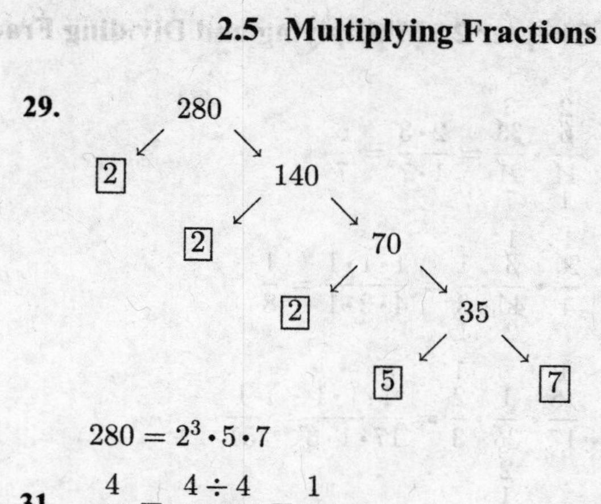

$280 = 2^3 \cdot 5 \cdot 7$

31. $\frac{4}{12} = \frac{4 \div 4}{12 \div 4} = \frac{1}{3}$

33. $\frac{4}{16} = \frac{4 \div 4}{16 \div 4} = \frac{1}{4}$

35. $\frac{18}{24} = \frac{18 \div 6}{24 \div 6} = \frac{3}{4}$

37. $\frac{56}{64} = \frac{56 \div 8}{64 \div 8} = \frac{7}{8}$

39. $\frac{25}{200} = \frac{25 \div 25}{200 \div 25} = \frac{1}{8}$

41. $\frac{88}{154} = \frac{88 \div 22}{154 \div 22} = \frac{4}{7}$

43. $\frac{24}{36} = \frac{\overset{1}{\cancel{2}} \cdot \overset{1}{\cancel{2}} \cdot 2 \cdot \overset{1}{\cancel{3}}}{\underset{1}{\cancel{2}} \cdot \underset{1}{\cancel{2}} \cdot \underset{1}{\cancel{3}} \cdot 3} = \frac{2}{3}$

45. $\frac{126}{42} = \frac{\overset{1}{\cancel{2}} \cdot \overset{1}{\cancel{3}} \cdot 3 \cdot \overset{1}{\cancel{7}}}{\underset{1}{\cancel{2}} \cdot \underset{1}{\cancel{3}} \cdot \underset{1}{\cancel{7}}} = \frac{3}{1} = 3$

2.5 Multiplying Fractions

2.5 Margin Exercises

1. $\frac{1}{4}$ of $\frac{1}{2}$ as read from the figure is the shaded part of the second figure. One of eight equal parts is shaded, or $\frac{1}{8}$.

$\frac{1}{4} \cdot \frac{1}{2} = \frac{1}{8}$

2. **(a)** $\frac{1}{2} \cdot \frac{3}{4} = \frac{1 \cdot 3}{2 \cdot 4} = \frac{3}{8}$

(b) $\frac{3}{5} \cdot \frac{1}{3} = \frac{\overset{1}{\cancel{3}} \cdot 1}{5 \cdot \underset{1}{\cancel{3}}} = \frac{1}{5}$

(c) $\frac{5}{6} \cdot \frac{1}{2} \cdot \frac{1}{8} = \frac{5 \cdot 1 \cdot 1}{6 \cdot 2 \cdot 8} = \frac{5}{96}$

(d) $\frac{1}{2} \cdot \frac{3}{4} \cdot \frac{3}{8} = \frac{1 \cdot 3 \cdot 3}{2 \cdot 4 \cdot 8} = \frac{9}{64}$

3. **(a)** $\frac{\overset{1}{\cancel{3}}}{\underset{2}{\cancel{4}}} \cdot \frac{\overset{1}{\cancel{2}}}{\underset{1}{\cancel{3}}} = \frac{1 \cdot 1}{2 \cdot 1} = \frac{1}{2}$

(b) $\frac{\overset{2}{\cancel{6}}}{\underset{1}{\cancel{11}}}\cdot\frac{\overset{3}{\cancel{33}}}{\underset{7}{\cancel{21}}}=\frac{2\cdot3}{1\cdot7}=\frac{6}{7}$

(c) $\frac{\overset{1}{\cancel{20}}}{4}\cdot\frac{\overset{1}{\cancel{3}}}{\underset{2}{\cancel{40}}}\cdot\frac{1}{\underset{1}{\cancel{3}}}=\frac{1\cdot1\cdot1}{4\cdot2\cdot1}=\frac{1}{8}$

(d) $\frac{\overset{1}{\cancel{18}}}{17}\cdot\frac{1}{\underset{\underset{1}{\cancel{2}}}{\cancel{36}}}\cdot\frac{\overset{1}{\cancel{2}}}{3}=\frac{1\cdot1\cdot1}{17\cdot1\cdot3}=\frac{1}{51}$

4. **(a)** $8\cdot\frac{1}{8}=\frac{\overset{1}{\cancel{8}}}{1}\cdot\frac{1}{\underset{1}{\cancel{8}}}=\frac{1\cdot1}{1\cdot1}=\frac{1}{1}=1$

(b) $\frac{\overset{1}{\cancel{3}}}{\underset{1}{\cancel{4}}}\cdot5\cdot\frac{\overset{1}{\cancel{4}}}{\underset{1}{\cancel{3}}}=\frac{1\cdot5\cdot1}{1\cdot1}=\frac{5}{1}=5$

(c) $\frac{3}{5}\cdot40=\frac{3}{\underset{1}{\cancel{5}}}\cdot\frac{\overset{8}{\cancel{40}}}{1}=\frac{3\cdot8}{1\cdot1}=\frac{24}{1}=24$

(d) $\frac{3}{25}\cdot\frac{5}{11}\cdot99=\frac{3}{\underset{5}{\cancel{25}}}\cdot\frac{\overset{1}{\cancel{5}}}{\underset{1}{\cancel{11}}}\cdot\frac{\overset{9}{\cancel{99}}}{1}$

$=\frac{3\cdot1\cdot9}{5\cdot1\cdot1}$

$=\frac{27}{5}$ or $5\frac{2}{5}$

5. **(a)** Area = length • width

$=\frac{\overset{1}{\cancel{3}}}{4}\cdot\frac{1}{\underset{1}{\cancel{3}}}$

$=\frac{1}{4}\text{yd}^2$

(b) Area = length • width

$=\frac{\overset{1}{\cancel{3}}}{8}\cdot\frac{1}{\underset{1}{\cancel{3}}}$

$=\frac{1}{8}\text{mi}^2$

(c) Area = length • width

$=\frac{7}{\underset{1}{\cancel{5}}}\cdot\frac{\overset{1}{\cancel{5}}}{8}$

$=\frac{7}{8}\text{mi}^2$

2.5 Section Exercises

1. $\frac{1}{3}\cdot\frac{3}{4}=\frac{1}{\underset{1}{\cancel{3}}}\cdot\frac{\overset{1}{\cancel{3}}}{4}=\frac{1\cdot1}{1\cdot4}=\frac{1}{4}$

3. $\frac{2}{7}\cdot\frac{1}{5}=\frac{2\cdot1}{7\cdot5}=\frac{2}{35}$

5. $\frac{8}{5}\cdot\frac{15}{32}=\frac{\overset{1}{\cancel{8}}}{\underset{1}{\cancel{5}}}\cdot\frac{\overset{3}{\cancel{15}}}{\underset{4}{\cancel{32}}}=\frac{1\cdot3}{1\cdot4}=\frac{3}{4}$

7. $\frac{2}{3}\cdot\frac{7}{12}\cdot\frac{9}{14}=\frac{\overset{1}{\cancel{2}}}{\underset{1}{\cancel{3}}}\cdot\frac{\overset{1}{\cancel{7}}}{\underset{\underset{2}{\cancel{6}}}{\cancel{12}}}\cdot\frac{\overset{\overset{1}{\cancel{3}}}{\cancel{9}}}{\underset{2}{\cancel{14}}}=\frac{1\cdot1\cdot1}{1\cdot2\cdot2}=\frac{1}{4}$

9. $\frac{3}{4}\cdot\frac{5}{6}\cdot\frac{2}{3}=\frac{\overset{1}{\cancel{3}}}{\underset{2}{\cancel{4}}}\cdot\frac{5}{6}\cdot\frac{\overset{1}{\cancel{2}}}{\underset{1}{\cancel{3}}}=\frac{1\cdot5\cdot1}{2\cdot6\cdot1}=\frac{5}{12}$

11. $\frac{9}{22}\cdot\frac{11}{16}=\frac{9}{\underset{2}{\cancel{22}}}\cdot\frac{\overset{1}{\cancel{11}}}{16}=\frac{9}{32}$

13. $\frac{5}{8}\cdot\frac{16}{25}=\frac{\overset{1}{\cancel{5}}}{\underset{1}{\cancel{8}}}\cdot\frac{\overset{2}{\cancel{16}}}{\underset{5}{\cancel{25}}}=\frac{1\cdot2}{1\cdot5}=\frac{2}{5}$

15. $\frac{14}{25}\cdot\frac{65}{48}\cdot\frac{15}{28}=\frac{\overset{1}{\cancel{14}}}{\underset{\underset{1}{\cancel{5}}}{\cancel{25}}}\cdot\frac{\overset{13}{\cancel{65}}}{\underset{16}{\cancel{48}}}\cdot\frac{\overset{\overset{1}{\cancel{5}}}{\cancel{15}}}{\underset{2}{\cancel{28}}}=\frac{1\cdot13\cdot1}{1\cdot16\cdot2}=\frac{13}{32}$

17. $\frac{16}{25}\cdot\frac{35}{32}\cdot\frac{15}{64}=\frac{\overset{1}{\cancel{16}}}{\underset{\underset{1}{\cancel{5}}}{\cancel{25}}}\cdot\frac{\overset{7}{\cancel{35}}}{\underset{2}{\cancel{32}}}\cdot\frac{\overset{3}{\cancel{15}}}{64}=\frac{1\cdot7\cdot3}{1\cdot2\cdot64}=\frac{21}{128}$

19. $5\cdot\frac{4}{5}=\frac{\overset{1}{\cancel{5}}}{1}\cdot\frac{4}{\underset{1}{\cancel{5}}}=\frac{4}{1}=4$

21. $\frac{5}{8}\cdot64=\frac{5}{\underset{1}{\cancel{8}}}\cdot\frac{\overset{8}{\cancel{64}}}{1}=\frac{40}{1}=40$

23. $36\cdot\frac{2}{3}=\frac{\overset{12}{\cancel{36}}}{1}\cdot\frac{2}{\underset{1}{\cancel{3}}}=\frac{24}{1}=24$

25. $36\cdot\frac{5}{8}\cdot\frac{9}{15}=\frac{\overset{9}{\cancel{36}}}{1}\cdot\frac{\overset{1}{\cancel{5}}}{\underset{2}{\cancel{8}}}\cdot\frac{\overset{3}{\cancel{9}}}{\underset{\underset{1}{\cancel{3}}}{\cancel{15}}}=\frac{27}{2}=13\frac{1}{2}$

27. $100\cdot\frac{21}{50}\cdot\frac{3}{4}=\frac{\overset{\overset{1}{\cancel{2}}}{\cancel{100}}}{1}\cdot\frac{21}{\underset{1}{\cancel{50}}}\cdot\frac{3}{\underset{2}{\cancel{4}}}=\frac{63}{2}=31\frac{1}{2}$

29. $\frac{2}{5}\cdot200=\frac{2}{\underset{1}{\cancel{5}}}\cdot\frac{\overset{40}{\cancel{200}}}{1}=\frac{80}{1}=80$

31. $142 \cdot \frac{2}{3} = \frac{142}{1} \cdot \frac{2}{3} = \frac{284}{3} = 94\frac{2}{3}$

33. $\frac{28}{21} \cdot 640 \cdot \frac{15}{32} = \frac{\overset{4}{\cancel{28}}}{\underset{\underset{1}{\cancel{3}}}{\cancel{21}}} \cdot \frac{\overset{20}{\cancel{640}}}{1} \cdot \frac{\overset{5}{\cancel{15}}}{\underset{1}{\cancel{32}}}$

$= \frac{4 \cdot 20 \cdot 5}{1 \cdot 1 \cdot 1}$

$= \frac{400}{1} = 400$

35. $\frac{54}{38} \cdot 684 \cdot \frac{5}{6} = \frac{\overset{9}{\cancel{54}}}{\underset{1}{\cancel{38}}} \cdot \frac{\overset{18}{\cancel{684}}}{1} \cdot \frac{5}{\underset{1}{\cancel{6}}}$

$= \frac{9 \cdot 18 \cdot 5}{1 \cdot 1 \cdot 1}$

$= \frac{810}{1} = 810$

37. Area = length • width

$= \frac{3}{4} \cdot \frac{1}{3} = \frac{\overset{1}{\cancel{3}}}{4} \cdot \frac{1}{\underset{1}{\cancel{3}}} = \frac{1}{4}\text{ mi}^2$

39. Area = length • width

$= 12 \cdot \frac{3}{4} = \frac{\overset{3}{\cancel{12}}}{1} \cdot \frac{3}{\underset{1}{\cancel{4}}} = \frac{9}{1} = 9\text{ m}^2$

41. Area = length • width

$= \frac{7}{5} \cdot \frac{3}{14} = \frac{\overset{1}{\cancel{7}}}{5} \cdot \frac{3}{\underset{2}{\cancel{14}}} = \frac{3}{10}\text{ in.}^2$

43. Multiply the numerators and multiply the denominators. An example is

$$\frac{3}{4} \cdot \frac{1}{2} = \frac{3 \cdot 1}{4 \cdot 2} = \frac{3}{8}.$$

45. Area = length • width

$= 2 \cdot \frac{3}{4} = \frac{\overset{1}{\cancel{2}}}{1} \cdot \frac{3}{\underset{2}{\cancel{4}}} = \frac{3}{2} = 1\frac{1}{2}\text{ yd}^2$

47. Area = length • width

$= 2 \cdot \frac{3}{4} = \frac{\overset{1}{\cancel{2}}}{1} \cdot \frac{3}{\underset{2}{\cancel{4}}} = \frac{3}{2} = 1\frac{1}{2}\text{ mi}^2$

49.

Bel Air Grocery	**Raley's Market**
Area = length • width	Area = length • width
$= \frac{1}{4} \cdot \frac{3}{16}$	$= \frac{3}{8} \cdot \frac{1}{8}$
$= \frac{3}{64}\text{ mi}^2$	$= \frac{3}{64}\text{ mi}^2$

They are both the same size.

51. $700{,}000 + 500{,}000 + 400{,}000 + 400{,}000 + 400{,}000 = 2{,}400{,}000$
The estimate of the total customers for the five largest systems is 2,400,000.

52. $698{,}420 + 469{,}035 + 409{,}640 + 400{,}000 + 360{,}822 = 2{,}337{,}917$, which is the exact number of customers for the five largest systems.

53. An estimate of the number of subscribers to the premium channels is

$\frac{3}{5} \cdot 500{,}000 = \frac{3}{\underset{1}{\cancel{5}}} \cdot \frac{\overset{100{,}000}{\cancel{500{,}000}}}{1} = 300{,}000.$

The exact value is

$\frac{3}{5} \cdot 469{,}035 = \frac{3}{\underset{1}{\cancel{5}}} \cdot \frac{\overset{93{,}807}{\cancel{469{,}035}}}{1} = 281{,}421.$

54. An estimate of the number of subscribers to the movie channels is

$\frac{3}{7} \cdot 400{,}000 = \frac{1{,}200{,}000}{7} \approx 171{,}429.$

The exact value is

$\frac{3}{7} \cdot 360{,}822 = \frac{3}{\underset{1}{\cancel{7}}} \cdot \frac{\overset{51{,}546}{\cancel{360{,}822}}}{1} = 154{,}638.$

55. We need a multiple of 5 that is close to 469,035. A reasonable choice is 450,000 and an estimate is

$\frac{3}{5} \cdot 450{,}000 = \frac{3}{\underset{1}{\cancel{5}}} \cdot \frac{\overset{90{,}000}{\cancel{450{,}000}}}{1} = 270{,}000$ customers.

This value is closer to the exact value because using 450,000 as a rounded guess is closer to 469,035 than using 500,000 as a rounded guess.

56. We need a multiple of 7 that is close to 360,822. A reasonable choice is 350,000 and an estimate is

$\frac{3}{7} \cdot 350{,}000 = \frac{3}{\underset{1}{\cancel{7}}} \cdot \frac{\overset{50{,}000}{\cancel{350{,}000}}}{1} = 150{,}000$ customers.

This value is closer to the exact value because using 350,000 as a rounded guess is closer to 360,822 than using 400,000 as a rounded guess.

2.6 Applications of Multiplication

2.6 Margin Exercises

1. **(a)** Find her monthly house payment by multiplying $\frac{1}{4}$ and 4320.

$\frac{1}{4} \cdot 4320 = \frac{1}{\underset{1}{\cancel{4}}} \cdot \frac{\overset{1080}{\cancel{4320}}}{1} = 1080$

Her monthly house payment is $1080.

(b) To find her retirement income, multiply $\frac{5}{8}$ and \$62,504.

$$\frac{5}{8} \cdot 62{,}504 = \frac{5}{\underset{1}{\cancel{8}}} \cdot \frac{\overset{7813}{\cancel{62{,}504}}}{1} = 39{,}065$$

She will receive \$39,065 as retirement income.

2. *Step 1*
The problem asks for the number of prescriptions paid for by a third party.

Step 2
A third party pays for $\frac{5}{16}$ of the total number of prescriptions, 3696.

Step 3

An estimate is $\frac{5}{16} \cdot 4000 = \frac{5}{\underset{1}{\cancel{16}}} \cdot \frac{\overset{250}{\cancel{4000}}}{1} = 1250.$

Step 4
The exact value is

$$\frac{5}{16} \cdot 3696 = \frac{5}{\underset{1}{\cancel{16}}} \cdot \frac{\overset{231}{\cancel{3696}}}{1} = 1155.$$

Step 5
A third party pays for 1155 prescriptions.

Step 6
The answer is reasonably close to our estimate.

3. *Step 1*
The problem asks for the fraction of residents who speak Spanish.

Step 2
$\frac{3}{4}$ of the $\frac{1}{3}$ of the residents who speak a foreign language, speak Spanish.

Step 3

An estimate is $\frac{3}{4} \cdot \frac{1}{3} = \frac{\overset{1}{\cancel{3}}}{4} \cdot \frac{1}{\underset{1}{\cancel{3}}} = \frac{1}{4}.$

Step 4
The exact value is $\frac{1}{4}$; which is the same as the estimate since we didn't round.

Step 5
The fraction of residents who speak Spanish is $\frac{1}{4}$.

Step 6
The answer, $\frac{1}{4}$, matches our estimate.

4. **(a)** The fraction is $\frac{1}{5}$.

(b) Multiply $\frac{1}{5}$ by the number of people in the survey, 2500.

$$\frac{1}{5} \cdot 2500 = \frac{1}{\underset{1}{\cancel{5}}} \cdot \frac{\overset{500}{\cancel{2500}}}{1} = \frac{500}{1} = 500$$

500 children buy food from vending machines.

(c) The fraction is $\frac{1}{10}$.

(d) Multiply $\frac{1}{10}$ by 2500.

$$\frac{1}{10} \cdot 2500 = \frac{1}{\underset{1}{\cancel{10}}} \cdot \frac{\overset{250}{\cancel{2500}}}{1} = \frac{250}{1} = 250$$

250 children buy food from a convenience store or street vendor.

2.6 Section Exercises

1. Multiply the length (depth) and the width.

$$\frac{3}{4} \cdot \frac{2}{3} = \frac{\overset{1}{\cancel{3}} \cdot \overset{1}{\cancel{2}}}{\underset{2}{\cancel{4}} \cdot \underset{1}{\cancel{3}}} = \frac{1}{2}$$

The area of the file cabinet top is $\frac{1}{2}$ yd^2.

3. Multiply the length and the width.

$$\frac{4}{3} \cdot \frac{2}{3} = \frac{8}{9}$$

The area of the cookie sheet is $\frac{8}{9}$ ft^2.

5. Multiply the length and the width.

$$\frac{4}{5} \cdot \frac{3}{8} = \frac{\overset{1}{\cancel{4}} \cdot 3}{5 \cdot \underset{2}{\cancel{8}}} = \frac{3}{10}$$

The area of the top of the table is $\frac{3}{10}$ yd^2.

7. Multiply $\frac{3}{8}$ by 6848.

$$\frac{3}{8} \cdot 6848 = \frac{3}{\underset{1}{\cancel{8}}} \cdot \frac{\overset{856}{\cancel{6848}}}{1} = 2568$$

He earned \$2568 on his job.

9. Multiply the number of freshmen by the fraction receiving scholarships.

$$1800 \cdot \frac{5}{24} = \frac{\overset{75}{\cancel{1800}}}{1} \cdot \frac{5}{\underset{1}{\cancel{24}}} = 375$$

375 students received scholarships.

11. $\frac{7}{12}$ of the 1560 runners are women.

$$\frac{7}{12} \cdot 1560 = \frac{7 \cdot \overset{130}{\cancel{1560}}}{\underset{1}{\cancel{12}} \cdot 1} = 910$$

910 runners are women.

13. The smallest sector of the circle graph is the age group 0-24 years. To find how many books this age group purchased, multiply $\frac{1}{25}$ by the total number of books, 1 billion.

$$\frac{1}{25} \cdot 1{,}000{,}000{,}000 = \frac{1}{25} \cdot \frac{1{,}000{,}000{,}000}{1} = 40{,}000{,}000$$

They purchased 40 million or 40,000,000 books.

15. Multiply the fraction labeled in the 35-44 year age group, $\frac{1}{4}$, by the total number of books, 1 billion.

$$\frac{1}{4} \cdot 1{,}000{,}000{,}000 = \frac{1}{4} \cdot \frac{1{,}000{,}000{,}000}{1} = 250{,}000{,}000$$

They purchased 250 million or 250,000,000 books.

17. Because everyone is included and fractions are given for *all* age groups, the sum of the fractions must be *1*, or *all* of the people.

19. Add the income for all twelve months to find the income for the year.

$$4575 + 4312 + 4988 + 4530 + 4320 + 4898 + 5540 + 3732 + 4170 + 5512 + 4965 + 6458 = 58{,}000$$

The Gomes family had income of $58,000 for the year.

21. From Exercise 19 the total income is $58,000. The circle graph shows that $\frac{1}{5}$ of the income is for rent.

$$\frac{1}{5} \cdot 58{,}000 = \frac{1}{\underset{1}{\cancel{5}}} \cdot \frac{\overset{11{,}600}{\cancel{58{,}000}}}{1} = 11{,}600$$

The amount of their rent is $11,600.

23. Multiply the total income by the fraction saved.

$$\frac{1}{16} \cdot 58{,}000 = \frac{1}{\underset{1}{\cancel{16}}} \cdot \frac{\overset{3625}{\cancel{58{,}000}}}{1} = 3625$$

The Gomes family saved $3625 for the year.

25. The error was dividing 21 by 3 and writing 3 instead of 7. The correct solution is

$$\frac{9}{10} \times \frac{20}{21} = \frac{\overset{3}{\cancel{9}}}{\underset{1}{\cancel{10}}} \times \frac{\overset{2}{\cancel{20}}}{\underset{7}{\cancel{21}}} = \frac{6}{7}.$$

27. The amount Pamela earns in 3 hours is $\frac{3}{8}$ of the amount she earns in 8 hours. Multiply 128 times $\frac{3}{8}$.

$$128 \cdot \frac{3}{8} = \frac{\overset{16}{\cancel{128}}}{1} \cdot \frac{3}{\underset{1}{\cancel{8}}} = 48$$

Pamela earned $48 in 3 hours.

29. First multiply $\frac{2}{3}$ and 27,000 to find the number of her votes from senior citizens.

$$\frac{2}{3} \cdot 27{,}000 = \frac{2}{\underset{1}{\cancel{3}}} \cdot \frac{\overset{9000}{\cancel{27{,}000}}}{1} = 18{,}000$$

To find the votes needed from voters other than the senior citizens, subtract:

$$27{,}000 - 18{,}000 = 9000 \text{ votes.}$$

She needed 9000 votes from voters other than the senior citizens.

31. Multiply the remaining $\frac{1}{8}$ of the estate by the fraction going to the American Cancer Society.

$$\frac{1}{4} \cdot \frac{1}{8} = \frac{1}{32}$$

$\frac{1}{32}$ of the estate goes to the American Cancer Society.

2.7 Dividing Fractions

2.7 Margin Exercises

1. **(a)** The reciprocal of $\frac{3}{4}$ is $\frac{4}{3}$ because $\frac{3}{4} \cdot \frac{4}{3} = \frac{12}{12} = 1.$

(b) The reciprocal of $\frac{3}{8}$ is $\frac{8}{3}$ because $\frac{3}{8} \cdot \frac{8}{3} = \frac{24}{24} = 1.$

(c) The reciprocal of $\frac{9}{4}$ is $\frac{4}{9}$ because $\frac{9}{4} \cdot \frac{4}{9} = \frac{36}{36} = 1.$

(d) The reciprocal of 16 is $\frac{1}{16}$ because $\frac{16}{1} \cdot \frac{1}{16} = \frac{16}{16} = 1.$

2. **(a)** $\frac{1}{4} \div \frac{2}{3} = \frac{1}{4} \cdot \frac{3}{2} = \frac{1 \cdot 3}{4 \cdot 2} = \frac{3}{8}$

(b) $\frac{3}{8} \div \frac{5}{8} = \frac{3}{8} \cdot \frac{8}{5} = \frac{3 \cdot \overset{1}{\cancel{8}}}{\underset{1}{\cancel{8}} \cdot 5} = \frac{3}{5}$

(c) $\dfrac{\frac{2}{3}}{\frac{4}{5}} = \frac{2}{3} \div \frac{4}{5} = \frac{\overset{1}{\cancel{2}}}{3} \cdot \frac{5}{\underset{2}{\cancel{4}}} = \frac{1 \cdot 5}{3 \cdot 2} = \frac{5}{6}$

(d) $\dfrac{\frac{5}{6}}{\frac{7}{12}} = \dfrac{5}{6} \div \dfrac{7}{12} = \dfrac{5}{\underset{1}{\cancel{6}}} \cdot \dfrac{\overset{2}{\cancel{12}}}{7} = \dfrac{5 \cdot 2}{1 \cdot 7} = \dfrac{10}{7} = 1\dfrac{3}{7}$

3. **(a)** $10 \div \dfrac{1}{2} = 10 \cdot \dfrac{2}{1} = \dfrac{10}{1}\cdot\dfrac{2}{1} = 20$

(b) $6 \div \dfrac{6}{7} = \dfrac{6}{1} \div \dfrac{6}{7} = \dfrac{\overset{1}{\cancel{6}}}{1} \cdot \dfrac{7}{\underset{1}{\cancel{6}}} = 7$

(c) $\dfrac{4}{5} \div 6 = \dfrac{4}{5} \div \dfrac{6}{1} = \dfrac{\overset{2}{\cancel{4}}}{5} \cdot \dfrac{1}{\underset{3}{\cancel{6}}} = \dfrac{2 \cdot 1}{5 \cdot 3} = \dfrac{2}{15}$

(d) $\dfrac{3}{8} \div 4 = \dfrac{3}{8} \div \dfrac{4}{1} = \dfrac{3}{8} \cdot \dfrac{1}{4} = \dfrac{3 \cdot 1}{8 \cdot 4} = \dfrac{3}{32}$

4. **(a)** Divide the total amount of fuel by the size of the tanks.

$$15 \div \frac{3}{4} = \frac{15}{1} \div \frac{3}{4} = \frac{\overset{5}{\cancel{15}}}{1} \cdot \frac{4}{\underset{1}{\cancel{3}}} = \frac{5 \cdot 4}{1 \cdot 1} = \frac{20}{1} = 20$$

20 fuel tanks can be filled.

(b) Divide the total number of gallons by the size of the sprayer.

$$36 \div \frac{2}{3} = \frac{36}{1} \div \frac{2}{3} = \frac{\overset{18}{\cancel{36}}}{1} \cdot \frac{3}{\underset{1}{\cancel{2}}} = \frac{18 \cdot 3}{1 \cdot 1} = \frac{54}{1} = 54$$

54 sprayers can be filled.

5. **(a)** Divide the fraction of the bonus money by the number of employees.

$$\frac{3}{4} \div 12 = \frac{3}{4} \div \frac{12}{1} = \frac{\overset{1}{\cancel{3}}}{4} \cdot \frac{1}{\underset{4}{\cancel{12}}} = \frac{1 \cdot 1}{4 \cdot 4} = \frac{1}{16}$$

Each employee will receive $\frac{1}{16}$ of the bonus money.

(b) Divide the fraction of the scholarship money by the number of students.

$$\frac{4}{5} \div 8 = \frac{4}{5} \div \frac{8}{1} = \frac{\overset{1}{\cancel{4}}}{5} \cdot \frac{1}{\underset{2}{\cancel{8}}} = \frac{1}{10}$$

Each employee will receive $\frac{1}{10}$ of the prize money.

2.7 Section Exercises

1. The reciprocal of $\frac{3}{8}$ is $\frac{8}{3}$ because $\dfrac{3}{8} \cdot \dfrac{8}{3} = \dfrac{24}{24} = 1.$

3. The reciprocal of $\frac{5}{6}$ is $\frac{6}{5}$ because $\dfrac{5}{6} \cdot \dfrac{6}{5} = \dfrac{30}{30} = 1.$

5. The reciprocal of $\frac{8}{5}$ is $\frac{5}{8}$ because $\dfrac{8}{5} \cdot \dfrac{5}{8} = \dfrac{40}{40} = 1.$

7. The reciprocal of 4 is $\frac{1}{4}$ because $\dfrac{4}{1} \cdot \dfrac{1}{4} = \dfrac{4}{4} = 1.$

9. $\dfrac{1}{2} \div \dfrac{3}{4} = \dfrac{1}{\underset{1}{\cancel{2}}} \cdot \dfrac{\overset{2}{\cancel{4}}}{3} = \dfrac{2}{3}$

11. $\dfrac{7}{8} \div \dfrac{1}{3} = \dfrac{7}{8} \cdot \dfrac{3}{1} = \dfrac{21}{8} = 2\dfrac{5}{8}$

13. $\dfrac{3}{4} \div \dfrac{5}{3} = \dfrac{3}{4} \cdot \dfrac{3}{5} = \dfrac{9}{20}$

15. $\dfrac{7}{9} \div \dfrac{7}{36} = \dfrac{\overset{1}{\cancel{7}}}{\underset{1}{\cancel{9}}} \cdot \dfrac{\overset{4}{\cancel{36}}}{\underset{1}{\cancel{7}}} = \dfrac{4}{1} = 4$

17. $\dfrac{15}{32} \div \dfrac{5}{64} = \dfrac{\overset{3}{\cancel{15}}}{\underset{1}{\cancel{32}}} \cdot \dfrac{\overset{2}{\cancel{64}}}{\underset{1}{\cancel{5}}} = \dfrac{6}{1} = 6$

19. $\dfrac{\frac{13}{20}}{\frac{4}{5}} = \dfrac{13}{20} \div \dfrac{4}{5} = \dfrac{13}{\underset{4}{\cancel{20}}} \cdot \dfrac{\overset{1}{\cancel{5}}}{4} = \dfrac{13}{16}$

21. $\dfrac{\frac{5}{6}}{\frac{25}{24}} = \dfrac{5}{6} \div \dfrac{25}{24} = \dfrac{\overset{1}{\cancel{5}}}{\underset{1}{\cancel{6}}} \cdot \dfrac{\overset{4}{\cancel{24}}}{\underset{5}{\cancel{25}}} = \dfrac{4}{5}$

23. $12 \div \dfrac{2}{3} = \dfrac{12}{1} \div \dfrac{2}{3} = \dfrac{\overset{6}{\cancel{12}}}{1} \cdot \dfrac{3}{\underset{1}{\cancel{2}}} = \dfrac{18}{1} = 18$

25. $\dfrac{18}{\frac{3}{4}} = 18 \div \dfrac{3}{4} = \dfrac{18}{1} \div \dfrac{3}{4} = \dfrac{\overset{6}{\cancel{18}}}{1} \cdot \dfrac{4}{\underset{1}{\cancel{3}}} = 24$

27. $\dfrac{\frac{4}{7}}{8} = \dfrac{4}{7} \div 8 = \dfrac{4}{7} \div \dfrac{8}{1} = \dfrac{\overset{1}{\cancel{4}}}{7} \cdot \dfrac{1}{\underset{2}{\cancel{8}}} = \dfrac{1}{14}$

29. $\frac{8}{9}$ of a liter divided into 4 parts:

$$\frac{8}{9} \div 4 = \frac{8}{9} \div \frac{4}{1} = \frac{\overset{2}{\cancel{8}}}{9} \cdot \frac{1}{\underset{1}{\cancel{4}}} = \frac{2}{9}$$

Each cat will get $\frac{2}{9}$ of a liter.

31. Divide the total number of cups by the size of the measuring cup.

$$5 \div \frac{1}{3} = \frac{5}{1} \div \frac{1}{3} = \frac{5}{1} \cdot \frac{3}{1} = 15$$

They need to fill the measuring cup 15 times.

33. Divide the amount of eye drops by the amount needed for each dispenser.

$$11 \div \frac{1}{8} = \frac{11}{1} \div \frac{1}{8} = \frac{11}{1} \cdot \frac{8}{1} = 88$$

88 dispensers can be filled.

35. Divide the total cords of wood by the amount per trip.

$$40 \div \frac{2}{3} = \frac{\overset{20}{\cancel{40}}}{1} \cdot \frac{3}{\underset{1}{\cancel{2}}} = \frac{60}{1} = 60$$

Pam has to make 60 trips.

37. Answers will vary. A sample answer follows:

You can divide two fractions by using the reciprocal of the second fraction (divisor) and then multiplying.

39. Each loafcake requires $\frac{3}{4}$ pound of jellybeans, so to make 16 loafcakes, use multiplication.

$$16 \cdot \frac{3}{4} = \frac{\overset{4}{\cancel{16}}}{1} \cdot \frac{3}{\underset{1}{\cancel{4}}} = 12$$

12 pounds will be needed.

41. Divide the 156 cans of compound by the fraction of a can needed for each new home.

$$156 \div \frac{3}{4} = \frac{156}{1} \cdot \frac{4}{3} = \frac{\overset{52}{\cancel{156}}}{1} \cdot \frac{4}{\underset{1}{\cancel{3}}} = 208$$

208 homes can be plumbed.

43. $\frac{4}{5}$ of the 380 miles have been completed—use multiplication.

$$\frac{4}{5} \cdot 380 = \frac{4}{\underset{1}{\cancel{5}}} \cdot \frac{\overset{76}{\cancel{380}}}{1} = 304$$

They have gone 304 miles.

45. Divide the 912 yards of fabric by the fraction of a yard needed for each dish towel.

$$912 \div \frac{3}{8} = \frac{912}{1} \cdot \frac{8}{3} = \frac{\overset{304}{\cancel{912}}}{1} \cdot \frac{8}{\underset{1}{\cancel{3}}} = 2432$$

2432 towels can be made.

47. The indicator words for multiplication are underlined below.

more than	per
$\underline{\text{double}}$	$\underline{\text{twice}}$
$\underline{\text{times}}$	$\underline{\text{product}}$
less than	difference
equals	$\underline{\text{twice as much}}$

48. The indicator words for division are underlined below.

fewer	sum of
$\underline{\text{goes into}}$	$\underline{\text{divide}}$
$\underline{\text{per}}$	$\underline{\text{quotient}}$
equals	double
loss of	$\underline{\text{divided by}}$

49. To divide two fractions, multiply the first fraction by the *reciprocal* of the second fraction.

50. The reciprocal of $\frac{3}{4}$ is $\frac{4}{3}$ because $\frac{3}{4} \cdot \frac{4}{3} = \frac{12}{12} = 1$.

The reciprocal of $\frac{7}{8}$ is $\frac{8}{7}$ because $\frac{7}{8} \cdot \frac{8}{7} = \frac{56}{56} = 1$.

The reciprocal of 5 is $\frac{1}{5}$ because $\frac{5}{1} \cdot \frac{1}{5} = \frac{5}{5} = 1$.

The reciprocal of $\frac{12}{19}$ is $\frac{19}{12}$ because $\frac{12}{19} \cdot \frac{19}{12} = \frac{228}{228} = 1$.

51. The stamp is square, so all 4 sides have equal measure.

Perimeter = 4 • side length

$$= 4 \cdot \frac{15}{16} = \frac{\overset{1}{\cancel{4}}}{1} \cdot \frac{15}{\underset{4}{\cancel{16}}} = \frac{15}{4} = 3\frac{3}{4}$$

The perimeter is $3\frac{3}{4}$ in. To find the perimeter of any regular 3-, 4-, 5-, or 6-sided figure, multiply the length of one side by 3, 4, 5, or 6.

52. Area = length • width

$$= \frac{15}{16} \cdot \frac{15}{16} = \frac{225}{256}$$

The area is $\frac{225}{256}$ in^2. Multiply the length by the width to find the area of any square.

2.8 Multiplying and Dividing Mixed Numbers

2.8 Margin Exercises

1. **(a)** $4\frac{2}{3}$

$4\frac{2}{3}$ ← 2 is more than $1\frac{1}{2}$. ← Half of 3 is $1\frac{1}{2}$.

$4\frac{2}{3}$ rounds up to 5.

(b) $3\frac{2}{5}$

$3\frac{2}{5}$ ← 2 is less than $2\frac{1}{2}$. ← Half of 5 is $2\frac{1}{2}$.

$3\frac{2}{5}$ rounds to 3.

(c) $5\frac{3}{4}$

$5\frac{3}{4}$ ← 3 is more than 2. ← Half of 4 is 2.

$5\frac{3}{4}$ rounds up to 6.

(d) $4\frac{7}{12}$

$4\frac{7}{12}$ ← 7 is more than 6. ← Half of 12 is 6.

$4\frac{7}{12}$ rounds up to 5.

(e) $1\frac{1}{2}$

$1\frac{1}{2}$ ← 1 is the same as 1. ← Half of 2 is 1.

$1\frac{1}{2}$ rounds up to 2.

(f) $8\frac{4}{9}$

$8\frac{4}{9}$ ← 4 is less than $4\frac{1}{2}$. ← Half of 9 is $4\frac{1}{2}$.

$8\frac{4}{9}$ rounds to 8.

2. **(a)** $3\frac{1}{2} \cdot 6\frac{1}{3}$

Estimate:

$3\frac{1}{2}$ rounds to 4. $6\frac{1}{3}$ rounds to 6.

$4 \cdot 6 = 24$

Exact:

$$3\frac{1}{2} \cdot 6\frac{1}{3} = \frac{7}{2} \cdot \frac{19}{3} = \frac{133}{6} = 22\frac{1}{6}$$

(b) $4\frac{2}{3} \cdot 2\frac{3}{4}$

Estimate:

$4\frac{2}{3}$ rounds to 5. $2\frac{3}{4}$ rounds to 3.

$5 \cdot 3 = 15$

Exact:

$$4\frac{2}{3} \cdot 2\frac{3}{4} = \frac{14}{3} \cdot \frac{11}{4} = \frac{\overset{7}{\cancel{14}}}{3} \cdot \frac{11}{\underset{2}{\cancel{4}}} = \frac{77}{6} = 12\frac{5}{6}$$

(c) $3\frac{3}{5} \cdot 4\frac{4}{9}$

Estimate:

$3\frac{3}{5}$ rounds to 4. $4\frac{4}{9}$ rounds to 4.

$4 \cdot 4 = 16$

Exact:

$$3\frac{3}{5} \cdot 4\frac{4}{9} = \frac{18}{5} \cdot \frac{40}{9} = \frac{\overset{2}{\cancel{18}}}{\underset{1}{\cancel{5}}} \cdot \frac{\overset{8}{\cancel{40}}}{\underset{1}{\cancel{9}}} = 16$$

(d) $5\frac{1}{4} \cdot 3\frac{2}{5}$

Estimate:

$5\frac{1}{4}$ rounds to 5. $3\frac{2}{5}$ rounds to 3.

$5 \cdot 3 = 15$

Exact:

$$5\frac{1}{4} \cdot 3\frac{2}{5} = \frac{21}{4} \cdot \frac{17}{5} = \frac{357}{20} = 17\frac{17}{20}$$

3. **(a)** $3\frac{1}{8} \div 6\frac{1}{4}$

Estimate:

$3\frac{1}{8}$ rounds to 3. $6\frac{1}{4}$ rounds to 6.

$3 \div 6 = \frac{1}{2}$

Exact:

$$3\frac{1}{8} \div 6\frac{1}{4} = \frac{25}{8} \div \frac{25}{4} = \frac{25}{8} \cdot \frac{4}{25} = \frac{\overset{1}{\cancel{25}}}{\underset{2}{\cancel{8}}} \cdot \frac{\overset{1}{\cancel{4}}}{\underset{1}{\cancel{25}}} = \frac{1}{2}$$

(b) $5\frac{1}{3} \div 1\frac{1}{4}$

Estimate:

$5\frac{1}{3}$ rounds to 5. $1\frac{1}{4}$ rounds to 1.

$5 \div 1 = 5$

Exact:

$$5\frac{1}{3} \div 1\frac{1}{4} = \frac{16}{3} \div \frac{5}{4} = \frac{16}{3} \cdot \frac{4}{5} = \frac{64}{15} = 4\frac{4}{15}$$

(c) $8 \div 5\frac{1}{3}$

Estimate:

8 rounds to 8. $5\frac{1}{3}$ rounds to 5.

$8 \div 5 = 1\frac{3}{5}$

Exact:

$$8 \div 5\frac{1}{3} = \frac{8}{1} \div \frac{16}{3} = \frac{\overset{1}{\cancel{8}}}{1} \cdot \frac{3}{\underset{2}{\cancel{16}}} = \frac{3}{2} = 1\frac{1}{2}$$

(d) $13\frac{1}{2} \div 18$

Estimate:

$13\frac{1}{2}$ rounds to 14. 18 rounds to 18.

$14 \div 18 = \frac{14}{18} = \frac{7}{9}$

Exact:

$$13\frac{1}{2} \div 18 = \frac{27}{2} \div \frac{18}{1} = \frac{\overset{3}{\cancel{27}}}{2} \cdot \frac{1}{\underset{2}{\cancel{18}}} = \frac{3}{4}$$

4. **(a)** Multiply the amount of paint needed for each car by the number of cars.

Estimate:

$2\frac{5}{8}$ rounds to 3. 15 rounds to 15.

$3 \cdot 15 = 45$

Exact:

$$2\frac{5}{8} \cdot 15 = \frac{21}{8} \cdot \frac{15}{1} = \frac{315}{8} = 39\frac{3}{8}$$

$39\frac{3}{8}$ quarts are needed for 15 cars.

(b) Multiply the amount that she earns per hour by the number of hours that she worked.

Estimate:

$9\frac{1}{4}$ rounds to 9. $6\frac{1}{2}$ rounds to 7.

$9 \cdot 7 = 63$

Exact:

$$9\frac{1}{4} \cdot 6\frac{1}{2} = \frac{37}{4} \cdot \frac{13}{2} = \frac{481}{8} = 60\frac{1}{8}$$

Clare would earn $\$60\frac{1}{8}$.

5. **(a)** Divide the total pounds of brass by the number of pounds needed for one engine.

Estimate:

57 rounds to 57. $4\frac{3}{4}$ rounds to 5.

$57 \div 5 \approx 11$

Exact:

$$57 \div 4\frac{3}{4} = \frac{57}{1} \div \frac{19}{4} = \frac{\overset{3}{\cancel{57}}}{1} \cdot \frac{4}{\underset{1}{\cancel{19}}} = \frac{12}{1} = 12$$

12 propellers can be manufactured from 57 pounds of brass.

(b) Divide the total amount of money by the student rate per hour.

Estimate:

315 rounds to 300. $8\frac{3}{4}$ rounds to 9.

$300 \div 9 \approx 33$

Exact:

$$315 \div 8\frac{3}{4} = 315 \div \frac{35}{4} = \overset{9}{\cancel{315}} \cdot \frac{4}{\underset{1}{\cancel{35}}} = 36$$

\$315 can pay for 36 hours of student help.

2.8 Section Exercises

1. $4\frac{1}{2} \cdot 1\frac{3}{4}$

Estimate: $5 \cdot 2 = 10$

Exact: $4\frac{1}{2} \cdot 1\frac{3}{4} = \frac{9}{2} \cdot \frac{7}{4} = \frac{63}{8} = 7\frac{7}{8}$

3. $1\frac{2}{3} \cdot 2\frac{7}{10}$

Estimate: $2 \cdot 3 = 6$

Exact: $1\frac{2}{3} \cdot 2\frac{7}{10} = \frac{5}{3} \cdot \frac{27}{10} = \frac{\overset{1}{\cancel{5}}}{\underset{1}{\cancel{3}}} \cdot \frac{\overset{9}{\cancel{27}}}{\underset{2}{\cancel{10}}} = \frac{9}{2} = 4\frac{1}{2}$

5. $3\frac{1}{9} \cdot 1\frac{2}{7}$

Estimate: $3 \cdot 1 = 3$

Exact: $3\frac{1}{9} \cdot 1\frac{2}{7} = \frac{28}{9} \cdot \frac{9}{7} = \frac{\overset{4}{\cancel{28}}}{\underset{1}{\cancel{9}}} \cdot \frac{\overset{1}{\cancel{9}}}{\underset{1}{\cancel{7}}} = \frac{4}{1} = 4$

7. $8 \cdot 6\frac{1}{4}$

Estimate: $8 \cdot 6 = 48$

Exact: $8 \cdot 6\frac{1}{4} = \frac{8}{1} \cdot \frac{25}{4} = \frac{\overset{2}{\cancel{8}}}{1} \cdot \frac{25}{\underset{1}{\cancel{4}}} = \frac{50}{1} = 50$

9. $4\frac{1}{2}\cdot 2\frac{1}{5}\cdot 5$

Estimate: $5\cdot 2\cdot 5 = 10\cdot 5 = 50$

Exact: $4\frac{1}{2}\cdot 2\frac{1}{5}\cdot 5 = \frac{9}{2}\cdot\frac{11}{\underset{1}{\cancel{5}}}\cdot\frac{\overset{1}{\cancel{5}}}{1} = \frac{99}{2} = 49\frac{1}{2}$

11. $3\cdot 1\frac{1}{2}\cdot 2\frac{2}{3}$

Estimate: $3\cdot 2\cdot 3 = 6\cdot 3 = 18$

Exact: $3\cdot 1\frac{1}{2}\cdot 2\frac{2}{3} = \frac{\overset{1}{\cancel{3}}}{1}\cdot\frac{3}{\underset{1}{\cancel{2}}}\cdot\frac{\overset{4}{\cancel{8}}}{\underset{1}{\cancel{3}}} = \frac{12}{1} = 12$

13. $1\frac{1}{4}\div 3\frac{3}{4}$

Estimate: $1\div 4 = \frac{1}{4}$

Exact: $1\frac{1}{4}\div 3\frac{3}{4} = \frac{5}{4}\div\frac{15}{4} = \frac{\overset{1}{\cancel{5}}}{\underset{1}{\cancel{4}}}\cdot\frac{\overset{1}{\cancel{4}}}{\underset{3}{\cancel{15}}} = \frac{1}{3}$

15. $2\frac{1}{2}\div 3$

Estimate: $3\div 3 = 1$

Exact: $2\frac{1}{2}\div 3 = \frac{5}{2}\div\frac{3}{1} = \frac{5}{2}\cdot\frac{1}{3} = \frac{5}{6}$

17. $9\div 2\frac{1}{2}$

Estimate: $9\div 3 = 3$

Exact: $9\div 2\frac{1}{2} = \frac{9}{1}\div\frac{5}{2} = \frac{9}{1}\cdot\frac{2}{5} = \frac{18}{5} = 3\frac{3}{5}$

19. $\frac{5}{8}\div 1\frac{1}{2}$

Estimate: $1\div 2 = \frac{1}{2}$

Exact: $\frac{5}{8}\div 1\frac{1}{2} = \frac{5}{8}\div\frac{3}{2} = \frac{5}{\underset{4}{\cancel{8}}}\cdot\frac{\overset{1}{\cancel{2}}}{3} = \frac{5}{12}$

21. $1\frac{7}{8}\div 6\frac{1}{4}$

Estimate: $2\div 6 = \frac{2}{6} = \frac{1}{3}$

Exact: $1\frac{7}{8}\div 6\frac{1}{4} = \frac{15}{8}\div\frac{25}{4} = \frac{\overset{3}{\cancel{15}}}{\underset{2}{\cancel{8}}}\cdot\frac{\overset{1}{\cancel{4}}}{\underset{5}{\cancel{25}}} = \frac{3}{10}$

23. $5\frac{2}{3}\div 6$

Estimate: $6\div 6 = 1$

Exact: $5\frac{2}{3}\div 6 = \frac{17}{3}\div\frac{6}{1} = \frac{17}{3}\cdot\frac{1}{6} = \frac{17}{18}$

25. Multiply each amount by $2\frac{1}{2}$.

(a) Applesauce: $\frac{3}{4}$ cup

Estimate: $1\cdot 3 = 3$ cups

Exact: $\frac{3}{4}\cdot 2\frac{1}{2} = \frac{3}{4}\cdot\frac{5}{2} = \frac{15}{8} = 1\frac{7}{8}$ cups

(b) Salt: $\frac{1}{2}$ tsp.

Estimate: $1\cdot 3 = 3$ tsp.

Exact: $\frac{1}{2}\cdot 2\frac{1}{2} = \frac{1}{2}\cdot\frac{5}{2} = \frac{5}{4} = 1\frac{1}{4}$ tsp.

(c) Flour: $1\frac{3}{4}$ cups

Estimate: $2\cdot 3 = 6$ cups

Exact: $1\frac{3}{4}\cdot 2\frac{1}{2} = \frac{7}{4}\cdot\frac{5}{2} = \frac{35}{8} = 4\frac{3}{8}$ cups

27. Divide each amount by 2.

(a) Vanilla extract: $\frac{1}{2}$ tsp.

Estimate: $1\div 2 = \frac{1}{2}$ tsp.

Exact: $\frac{1}{2}\div 2 = \frac{1}{2}\div\frac{2}{1} = \frac{1}{2}\cdot\frac{1}{2} = \frac{1}{4}$ tsp.

(b) Applesauce: $\frac{3}{4}$ cup

Estimate: $1\div 2 = \frac{1}{2}$ cup

Exact: $\frac{3}{4}\div 2 = \frac{3}{4}\div\frac{2}{1} = \frac{3}{4}\cdot\frac{1}{2} = \frac{3}{8}$ cup

(c) Flour: $1\frac{3}{4}$ cups

Estimate: $2\div 2 = 1$ cup

Exact: $1\frac{3}{4}\div 2 = \frac{7}{4}\div\frac{2}{1} = \frac{7}{4}\cdot\frac{1}{2} = \frac{7}{8}$ cup

29. Divide the amount of baseboard available by the amount needed for each home.

Estimate: $1314\div 110\approx 12$ homes

Exact:

$1314\div 109\frac{1}{2} = \frac{1314}{1}\div\frac{219}{2} = \frac{\overset{6}{\cancel{1314}}}{1}\cdot\frac{2}{\underset{1}{\cancel{219}}} = 12$

12 homes can be fitted with the baseboard.

31. Each handle requires $19\frac{1}{2}$ inches of steel tubing. Use multiplication.

Estimate: $135 \cdot 20 = 2700$ in.

Exact:

$135 \cdot 19\frac{1}{2} = \frac{135}{1} \cdot \frac{39}{2} = \frac{5265}{2} = 2632\frac{1}{2}$ in.

$2632\frac{1}{2}$ inches of steel tubing is needed to make 135 jacks.

33. The answer should include:

Step 1
Change mixed numbers to improper fractions.

Step 2
Multiply the fractions.

Step 3
Write the answer in lowest terms, changing to mixed or whole numbers where possible.

35. Divide the total pounds of rubber by the pounds of rubber required for a tire.

Estimate: $51{,}460 \div 21 \approx 2450$ tires

Exact: $51{,}460 \div 20\frac{3}{4} = 51{,}460 \div \frac{83}{4}$

$= \frac{\overset{620}{\cancel{51{,}460}}}{1} \cdot \frac{4}{\underset{1}{\cancel{83}}} = 2480$

2480 tires can be manufactured.

37. Divide the length of the tube by the length of each spacer.

Estimate: $10 \div 1 = 10$ spacers

Exact: $9\frac{3}{4} \div \frac{3}{4} = \frac{39}{4} \div \frac{3}{4} = \frac{\overset{13}{\cancel{39}}}{\underset{1}{\cancel{4}}} \cdot \frac{\overset{1}{\cancel{4}}}{\underset{1}{\cancel{3}}} = 13$

13 spacers can be cut from the tube.

39. Multiply the gallons per acre times the number of acres to get the total number of gallons needed.

Estimate: $7 \cdot 26 = 182$ gallons

Exact: $6\frac{3}{4} \cdot 25\frac{1}{2} = \frac{27}{4} \cdot \frac{51}{2} = \frac{1377}{8} = 172\frac{1}{8}$

$172\frac{1}{8}$ gallons of the chemical are needed.

Chapter 2 Review Exercises

1. $\frac{1}{3}$ There are 3 parts, and 1 is shaded.

2. $\frac{5}{8}$ There are 8 parts, and 5 are shaded.

3. $\frac{2}{4}$ There are 4 parts, and 2 are shaded.

4. Proper fractions have numerator (top) smaller than denominator (bottom).

They are: $\frac{1}{8}, \frac{3}{4}, \frac{2}{3}$.

Improper fractions have numerator (top) larger than or equal to the denominator (bottom).

They are: $\frac{4}{3}, \frac{5}{5}$.

5. Proper fractions have numerator (top) smaller than denominator (bottom).

They are: $\frac{15}{16}, \frac{1}{8}$

Improper fractions have numerator (top) larger than or equal to the denominator (bottom).

They are: $\frac{6}{5}, \frac{16}{13}, \frac{5}{3}$

6. $4\frac{3}{4}$ $4 \cdot 4 = 16$

$16 + 3 = 19$

$4\frac{3}{4} = \frac{19}{4}$

7. $9\frac{5}{6}$ $9 \cdot 6 = 54$

$54 + 5 = 59$

$9\frac{5}{6} = \frac{59}{6}$

8. $\frac{27}{8}$

$$\begin{array}{r} 3 \\ 8\overline{)27} \\ \underline{24} \\ 3 \end{array}$$

3 ← Whole number part; 3 ← Remainder

$\frac{27}{8} = 3\frac{3}{8}$

9. $\frac{63}{5}$

$$\begin{array}{r} 12 \\ 5\overline{)63} \\ \underline{5} \\ 13 \\ \underline{10} \\ 3 \end{array}$$

12 ← Whole number part; 3 ← Remainder

$\frac{63}{5} = 12\frac{3}{5}$

10. Factorizations of 6:

$1 \cdot 6 = 6 \quad 2 \cdot 3 = 6$

The factors of 6 are 1, 2, 3, and 6.

11. Factorizations of 24:

$1 \cdot 24 = 24 \quad 2 \cdot 12 = 24 \quad 3 \cdot 8 = 24 \quad 4 \cdot 6 = 24$

The factors of 24 are 1, 2, 3, 4, 6, 8, 12, and 24.

12. Factorizations of 55:

$1 \cdot 55 = 55 \quad 5 \cdot 11 = 55$

The factors of 55 are 1, 5, 11, and 55.

13. Factorizations of 90:

$1 \cdot 90 = 90 \quad 2 \cdot 45 = 90 \quad 3 \cdot 30 = 90 \quad 5 \cdot 18 = 90$
$6 \cdot 15 = 90 \quad 9 \cdot 10 = 90$

The factors of 90 are 1, 2, 3, 5, 6, 9, 10, 15, 18, 30, 45, and 90.

14.

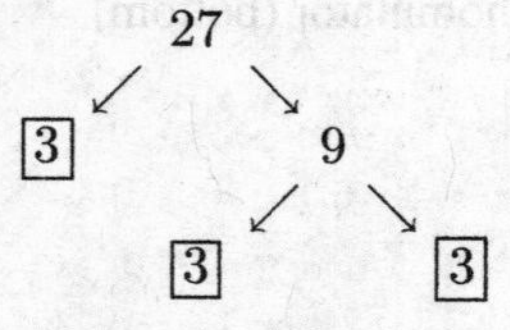

$27 = 3 \cdot 3 \cdot 3 = 3^3$

15.

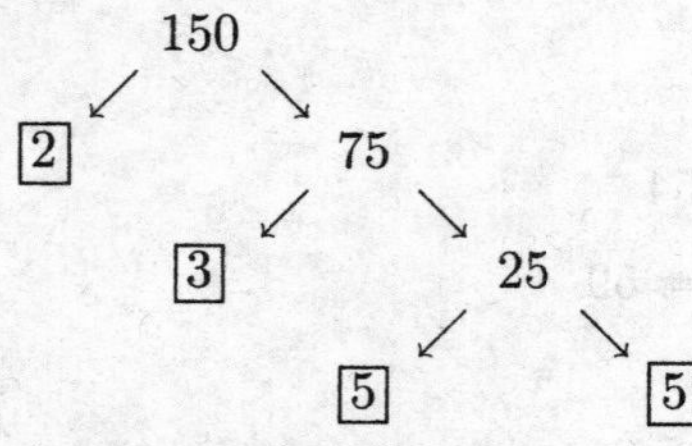

$150 = 2 \cdot 3 \cdot 5 \cdot 5 = 2 \cdot 3 \cdot 5^2$

16.

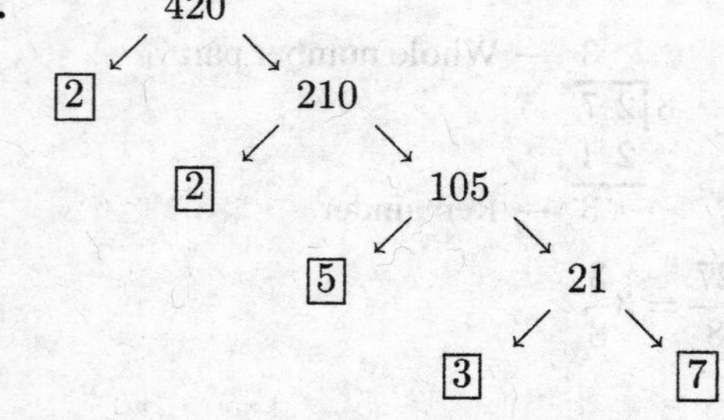

$420 = 2 \cdot 2 \cdot 3 \cdot 5 \cdot 7 = 2^2 \cdot 3 \cdot 5 \cdot 7$

17. $5^2 = 5 \cdot 5 = 25$

18. $6^2 \cdot 2^3 = 6 \cdot 6 \cdot 2 \cdot 2 \cdot 2 = 36 \cdot 8 = 288$

19. $8^2 \cdot 3^3 = 8 \cdot 8 \cdot 3 \cdot 3 \cdot 3 = 64 \cdot 27 = 1728$

20. $4^3 \cdot 2^5 = 4 \cdot 4 \cdot 4 \cdot 2 \cdot 2 \cdot 2 \cdot 2 \cdot 2 = 64 \cdot 32 = 2048$

21. $\frac{16}{24} = \frac{16 \div 8}{24 \div 8} = \frac{2}{3}$

22. $\frac{36}{45} = \frac{36 \div 9}{45 \div 9} = \frac{4}{5}$

23. $\frac{75}{80} = \frac{75 \div 5}{80 \div 5} = \frac{15}{16}$

24. $\frac{25}{60} = \frac{\overset{1}{\cancel{5}} \cdot 5}{2 \cdot 2 \cdot 3 \cdot \underset{1}{\cancel{5}}} = \frac{5}{12}$

25. $\frac{384}{96} = \frac{\overset{1}{\cancel{2}} \cdot \overset{1}{\cancel{2}} \cdot \overset{1}{\cancel{2}} \cdot \overset{1}{\cancel{2}} \cdot \overset{1}{\cancel{2}} \cdot 2 \cdot 2 \cdot \overset{1}{\cancel{3}}}{\underset{1}{\cancel{2}} \cdot \underset{1}{\cancel{2}} \cdot \underset{1}{\cancel{2}} \cdot \underset{1}{\cancel{2}} \cdot \underset{1}{\cancel{2}} \cdot \underset{1}{\cancel{3}}} = \frac{4}{1} = 4$

26. $\frac{3}{4}$ and $\frac{48}{64}$

$\frac{48}{64} = \frac{48 \div 16}{64 \div 16} = \frac{3}{4}$

The fractions are equivalent $\left(\frac{3}{4} = \frac{3}{4}\right)$.

27. $\frac{3}{4}$ and $\frac{42}{58}$

$\frac{42}{58} = \frac{42 \div 2}{58 \div 2} = \frac{21}{29}$

The fractions are not equivalent $\left(\frac{3}{4} \neq \frac{21}{29}\right)$.

28. $\frac{4}{5} \cdot \frac{3}{4} = \frac{\overset{1}{\cancel{4}}}{5} \cdot \frac{3}{\underset{1}{\cancel{4}}} = \frac{1 \cdot 3}{5 \cdot 1} = \frac{3}{5}$

29. $\frac{3}{10} \cdot \frac{5}{8} = \frac{3}{\underset{2}{\cancel{10}}} \cdot \frac{\overset{1}{\cancel{5}}}{8} = \frac{3 \cdot 1}{2 \cdot 8} = \frac{3}{16}$

30. $\frac{70}{175} \cdot \frac{5}{14} = \frac{\overset{\overset{1}{\cancel{5}}}{\cancel{70}}}{\underset{\underset{7}{\cancel{35}}}{\cancel{175}}} \cdot \frac{\overset{1}{\cancel{5}}}{\underset{1}{\cancel{14}}} = \frac{1 \cdot 1}{7 \cdot 1} = \frac{1}{7}$

31. $\frac{44}{63} \cdot \frac{3}{11} = \frac{\overset{4}{\cancel{44}}}{\underset{21}{\cancel{63}}} \cdot \frac{\overset{1}{\cancel{3}}}{\underset{1}{\cancel{11}}} = \frac{4 \cdot 1}{21 \cdot 1} = \frac{4}{21}$

32. $\frac{5}{16} \cdot 48 = \frac{5}{\underset{1}{\cancel{16}}} \cdot \frac{\overset{3}{\cancel{48}}}{1} = \frac{5 \cdot 3}{1 \cdot 1} = \frac{15}{1} = 15$

33. $\frac{5}{8} \cdot 1000 = \frac{5}{\underset{1}{\cancel{8}}} \cdot \frac{\overset{125}{\cancel{1000}}}{1} = \frac{5 \cdot 125}{1 \cdot 1} = \frac{625}{1} = 625$

34. $\frac{2}{3} \div \frac{1}{2} = \frac{2}{3} \cdot \frac{2}{1} = \frac{2 \cdot 2}{3 \cdot 1} = \frac{4}{3} = 1\frac{1}{3}$

35. $\frac{5}{6} \div \frac{1}{2} = \frac{5}{\underset{3}{\cancel{6}}} \cdot \frac{\overset{1}{\cancel{2}}}{1} = \frac{5}{3} = 1\frac{2}{3}$

36. $\dfrac{\frac{15}{18}}{\frac{10}{30}} = \dfrac{15}{18} \div \dfrac{10}{30}$

$= \dfrac{\overset{5}{\cancel{15}}}{\underset{6}{\cancel{18}}} \cdot \dfrac{\overset{3}{\cancel{30}}}{\underset{1}{\cancel{10}}} = \dfrac{5 \cdot \overset{1}{\cancel{3}}}{\underset{2}{\cancel{6}} \cdot 1} = \dfrac{5 \cdot 1}{2 \cdot 1} = \dfrac{5}{2} = 2\dfrac{1}{2}$

37. $\dfrac{\frac{3}{4}}{\frac{3}{8}} = \dfrac{3}{4} \div \dfrac{3}{8} = \dfrac{\overset{1}{\cancel{3}}}{\underset{1}{\cancel{4}}} \cdot \dfrac{\overset{2}{\cancel{8}}}{\underset{1}{\cancel{3}}} = \dfrac{1 \cdot 2}{1 \cdot 1} = 2$

38. $7 \div \dfrac{7}{8} = \dfrac{7}{1} \div \dfrac{7}{8} = \dfrac{\overset{1}{\cancel{7}}}{1} \cdot \dfrac{8}{\underset{1}{\cancel{7}}} = \dfrac{1 \cdot 8}{1 \cdot 1} = 8$

39. $18 \div \dfrac{3}{4} = \dfrac{\overset{6}{\cancel{18}}}{1} \cdot \dfrac{4}{\underset{1}{\cancel{3}}} = \dfrac{6 \cdot 4}{1 \cdot 1} = 24$

40. $\dfrac{5}{8} \div 3 = \dfrac{5}{8} \div \dfrac{3}{1} = \dfrac{5}{8} \cdot \dfrac{1}{3} = \dfrac{5 \cdot 1}{8 \cdot 3} = \dfrac{5}{24}$

41. $\dfrac{2}{3} \div 5 = \dfrac{2}{3} \div \dfrac{5}{1} = \dfrac{2}{3} \cdot \dfrac{1}{5} = \dfrac{2 \cdot 1}{3 \cdot 5} = \dfrac{2}{15}$

42. $\dfrac{\frac{12}{13}}{3} = \dfrac{12}{13} \div 3 = \dfrac{12}{13} \div \dfrac{3}{1} = \dfrac{\overset{4}{\cancel{12}}}{13} \cdot \dfrac{1}{\underset{1}{\cancel{3}}} = \dfrac{4 \cdot 1}{13 \cdot 1} = \dfrac{4}{13}$

43. To find the area, multiply the length and the width.

$\dfrac{9}{10} \cdot \dfrac{1}{4} = \dfrac{9}{40}$

The area is $\frac{9}{40}$ ft^2.

44. To find the area, multiply the length and the width.

$\dfrac{7}{\underset{4}{\cancel{8}}} \cdot \dfrac{\overset{1}{\cancel{2}}}{3} = \dfrac{7 \cdot 1}{4 \cdot 3} = \dfrac{7}{12}$

The area is $\frac{7}{12}$ in.^{2}.

45. Multiply the length and width.

$5 \cdot \dfrac{5}{6} = \dfrac{5}{1} \cdot \dfrac{5}{6} = \dfrac{25}{6} = 4\dfrac{1}{6}$

The area is $4\frac{1}{6}$ ft^2.

46. Multiply the length and width.

$48 \cdot 5\dfrac{3}{4} = \dfrac{\overset{12}{\cancel{48}}}{1} \cdot \dfrac{23}{\underset{1}{\cancel{4}}} = 276$

The area is 276 yd^2.

47. $5\dfrac{1}{2} \cdot 1\dfrac{1}{4}$

Estimate: $6 \cdot 1 = 6$

Exact: $5\dfrac{1}{2} \cdot 1\dfrac{1}{4} = \dfrac{11}{2} \cdot \dfrac{5}{4} = \dfrac{55}{8} = 6\dfrac{7}{8}$

48. $2\dfrac{1}{4} \cdot 7\dfrac{1}{8} \cdot 1\dfrac{1}{3}$

Estimate: $2 \cdot 7 \cdot 1 = 14$

Exact: $2\dfrac{1}{4} \cdot 7\dfrac{1}{8} \cdot 1\dfrac{1}{3} = \dfrac{\overset{3}{\cancel{9}}}{4} \cdot \dfrac{57}{\underset{2}{\cancel{8}}} \cdot \dfrac{\overset{1}{\cancel{4}}}{\underset{1}{\cancel{3}}} = \dfrac{171}{8} = 21\dfrac{3}{8}$

49. $15\dfrac{1}{2} \div 3$

Estimate: $16 \div 3 = \dfrac{16}{3} = 5\dfrac{1}{3}$

Exact: $15\dfrac{1}{2} \div 3 = \dfrac{31}{2} \cdot \dfrac{1}{3} = \dfrac{31}{6} = 5\dfrac{1}{6}$

50. $4\dfrac{3}{4} \div 6\dfrac{1}{3}$

Estimate: $5 \div 6 = \dfrac{5}{6}$

Exact: $4\dfrac{3}{4} \div 6\dfrac{1}{3} = \dfrac{19}{4} \div \dfrac{19}{3} = \dfrac{\overset{1}{\cancel{19}}}{4} \cdot \dfrac{3}{\underset{1}{\cancel{19}}} = \dfrac{3}{4}$

51. Divide the total tons of almonds by the size of the bins.

$320 \div \dfrac{5}{8} = \dfrac{\overset{64}{\cancel{320}}}{1} \cdot \dfrac{8}{\underset{1}{\cancel{5}}} = 512$

512 bins will be needed to store the almonds.

52. $\frac{2}{3}$ of the estate is to be divided into 5 parts.

$\dfrac{2}{3} \div 5 = \dfrac{2}{3} \div \dfrac{5}{1} = \dfrac{2}{3} \cdot \dfrac{1}{5} = \dfrac{2 \cdot 1}{3 \cdot 5} = \dfrac{2}{15}$

Each child will receive $\frac{2}{15}$ of the estate.

53. Divide the total yardage by the amount needed for each pull cord.

Estimate: $158 \div 4 \approx 40$ pull cords

Exact:

$157\dfrac{1}{2} \div 4\dfrac{3}{8} = \dfrac{315}{2} \div \dfrac{35}{8} = \dfrac{\overset{9}{\cancel{315}}}{\underset{1}{\cancel{2}}} \cdot \dfrac{\overset{4}{\cancel{8}}}{\underset{1}{\cancel{35}}} = 36$

36 pull cords can be made.

54. Total wages equal the number of hours worked times the wages per hour.

Estimate: $9 \cdot 38 = \$342$

Exact: $8\frac{1}{2} \cdot 38 = \frac{17}{\underset{1}{\cancel{2}}} \cdot \frac{\overset{19}{\cancel{38}}}{1} = 323$

Neta earned $323.

55. Ebony sold $\frac{1}{4}$ of 100 pounds of rice.

$\frac{1}{4} \cdot 100 = \frac{1}{\underset{1}{\cancel{4}}} \cdot \frac{\overset{25}{\cancel{100}}}{1} = \frac{1 \cdot 25}{1 \cdot 1} = \frac{25}{1} = 25$ pounds

Thus, $100 - 25 = 75$ pounds remain. She gave $\frac{2}{3}$ of 75 to her parents.

$\frac{2}{3} \cdot 75 = \frac{2}{\underset{1}{\cancel{3}}} \cdot \frac{\overset{25}{\cancel{75}}}{1} = \frac{2 \cdot 25}{1 \cdot 1} = \frac{50}{1} = 50$ pounds

Ebony gave 50 pounds to her parents. The amount she has left is $75 - 50 = 25$ pounds.

56. Tiffany paid $\frac{1}{3}$ of $1275 for rent.

$\frac{1}{3} \cdot 1275 = \frac{1}{\underset{1}{\cancel{3}}} \cdot \frac{\overset{425}{\cancel{1275}}}{1} = 425$

She paid $425 for rent.

She paid $\frac{2}{5}$ of the remainder, $\$1275 - \$425 = \$850$, for food, utilities, and transportation.

$\frac{2}{5} \cdot 850 = \frac{2}{\underset{1}{\cancel{5}}} \cdot \frac{\overset{170}{\cancel{850}}}{1} = \frac{2 \cdot 170}{1 \cdot 1} = 340$

She has $\$850 - \$340 = \$510$ left.

57. $\frac{3}{4}$ must be divided by 8.

$\frac{3}{4} \div 8 = \frac{3}{4} \div \frac{8}{1} = \frac{3}{4} \cdot \frac{1}{8} = \frac{3 \cdot 1}{4 \cdot 8} = \frac{3}{32}$

Each park will receive $\frac{3}{32}$ of the budget.

58. $\frac{4}{5}$ of the catch must be divided evenly among 5 fishermen.

$\frac{4}{5} \div 5 = \frac{4}{5} \div \frac{5}{1} = \frac{4}{5} \cdot \frac{1}{5} = \frac{4}{25}$

Each fisherman receives $\frac{4}{25}$ ton.

59. **[2.5]** $\frac{1}{2} \cdot \frac{3}{4} = \frac{1 \cdot 3}{2 \cdot 4} = \frac{3}{8}$

60. **[2.5]** $\frac{2}{3} \cdot \frac{3}{5} = \frac{2}{\underset{1}{\cancel{3}}} \cdot \frac{\overset{1}{\cancel{3}}}{5} = \frac{2 \cdot 1}{1 \cdot 5} = \frac{2}{5}$

61. **[2.8]** $12\frac{1}{2} \cdot 2\frac{1}{2} = \frac{25}{2} \cdot \frac{5}{2} = \frac{125}{4} = 31\frac{1}{4}$

62. **[2.8]** $12\frac{1}{2} \cdot 2\frac{1}{4} = \frac{25}{2} \cdot \frac{9}{4} = \frac{225}{8} = 28\frac{1}{8}$

63. **[2.7]**

$\frac{\frac{4}{5}}{8} = \frac{4}{5} \div 8 = \frac{4}{5} \div \frac{8}{1} = \frac{\overset{1}{\cancel{4}}}{5} \cdot \frac{1}{\underset{2}{\cancel{8}}} = \frac{1 \cdot 1}{5 \cdot 2} = \frac{1}{10}$

64. **[2.7]** $\frac{\frac{5}{8}}{4} = \frac{5}{8} \div \frac{4}{1} = \frac{5}{8} \cdot \frac{1}{4} = \frac{5}{32}$

65. **[2.5]** $\frac{15}{31} \cdot 62 = \frac{15}{\underset{1}{\cancel{31}}} \cdot \frac{\overset{2}{\cancel{62}}}{1} = \frac{15 \cdot 2}{1 \cdot 1} = \frac{30}{1} = 30$

66. **[2.8]**

$3\frac{1}{4} \div 1\frac{1}{2} = \frac{13}{4} \div \frac{3}{2} = \frac{13}{\underset{2}{\cancel{4}}} \cdot \frac{\overset{1}{\cancel{2}}}{3} = \frac{13 \cdot 1}{2 \cdot 3} = \frac{13}{6} = 2\frac{1}{6}$

67. **[2.2]** $\frac{8}{5}$

$$\begin{array}{r l} 1 & \leftarrow \text{Whole number part} \\ 5\overline{)8} & \\ \underline{5} & \\ 3 & \leftarrow \text{Remainder} \end{array}$$

$\frac{8}{5} = 1\frac{3}{5}$

68. **[2.2]** $\frac{153}{4}$

$$\begin{array}{r l} 38 & \leftarrow \text{Whole number part} \\ 4\overline{)153} & \\ \underline{12} & \\ 33 & \\ \underline{32} & \\ 1 & \leftarrow \text{Remainder} \end{array}$$

$\frac{153}{4} = 38\frac{1}{4}$

69. **[2.2]** $5\frac{2}{3}$ $\quad 5 \cdot 3 = 15$

$15 + 2 = 17$

$5\frac{2}{3} = \frac{17}{3}$

70. **[2.2]** $38\frac{3}{8}$ $\quad 38 \cdot 8 = 304$

$304 + 3 = 307$

$38\frac{3}{8} = \frac{307}{8}$

71. [2.4] $\frac{8}{12}=\frac{\overset{1}{\cancel{2}}\cdot\overset{1}{\cancel{2}}\cdot 2}{\underset{1}{\cancel{2}}\cdot\underset{1}{\cancel{2}}\cdot 3}=\frac{1\cdot 1\cdot 2}{1\cdot 1\cdot 3}=\frac{2}{3}$

72. [2.4] $\frac{108}{210}=\frac{\overset{1}{\cancel{2}}\cdot 2\cdot\overset{1}{\cancel{3}}\cdot 3\cdot 3}{\underset{1}{\cancel{2}}\cdot\underset{1}{\cancel{3}}\cdot 5\cdot 7}=\frac{1\cdot 2\cdot 1\cdot 3\cdot 3}{1\cdot 1\cdot 5\cdot 7}=\frac{18}{35}$

73. [2.4] $\frac{75}{90}=\frac{75\div 15}{90\div 15}=\frac{5}{6}$

74. [2.4] $\frac{48}{72}=\frac{48\div 24}{72\div 24}=\frac{2}{3}$

75. [2.4] $\frac{44}{110}=\frac{44\div 22}{110\div 22}=\frac{2}{5}$

76. [2.4] $\frac{87}{261}=\frac{87\div 87}{261\div 87}=\frac{1}{3}$

77. [2.8] Multiply the area by the amount needed for each square yard.

Estimate: $4\cdot 44=176$ ounces

Exact: $3\frac{1}{2}\cdot 43\frac{5}{9}=\frac{7}{\underset{1}{\cancel{2}}}\cdot\frac{\overset{196}{\cancel{392}}}{9}=\frac{7\cdot 196}{1\cdot 9}=\frac{1372}{9}$

```
   1 5 2  ← Whole number part
9 )1 3 7 2
   9
   4 7
   4 5
     2 2
     1 8
       4  ← Remainder
```

$152\frac{4}{9}$ ounces of glue are needed.

78. [2.8] Multiply the number of in-ground tanks by the number of quarts needed for each in-ground tank.

Estimate: $7\cdot 26=182$ qt

Exact: $7\frac{1}{4}\cdot 25\frac{1}{2}=\frac{29}{4}\cdot\frac{51}{2}=\frac{1479}{8}=184\frac{7}{8}$

$184\frac{7}{8}$ quarts are needed.

79. [2.8] To find the area, multiply the length and the width.

$1\frac{3}{4}\cdot\frac{7}{8}=\frac{7}{4}\cdot\frac{7}{8}=\frac{49}{32}=1\frac{17}{32}$

The area is $1\frac{17}{32}$ in.^{2}.

80. [2.5] To find the area, multiply the length and the width.

$\frac{5}{8}\cdot\frac{1}{2}=\frac{5}{16}$

The area is $\frac{5}{16}$ yd^2.

Chapter 2 Test

1. $\frac{5}{6}$ There are 6 parts, and 5 are shaded.

2. $\frac{3}{8}$ There are 8 parts, and 3 are shaded.

3. Proper fractions have the numerator (top) smaller than the denominator (bottom).

$\frac{2}{3},\frac{6}{7},\frac{1}{4},\frac{5}{8}$

4. $3\frac{3}{8}$ $3\cdot 8=24$

$24+3=27$

$3\frac{3}{8}=\frac{27}{8}$

5. $\frac{123}{4}$

```
   3 0  ← Whole number part
4 )1 2 3
   1 2
     3  ← Remainder
```

$\frac{123}{4}=30\frac{3}{4}$

6. Factorizations of 18:

$1\cdot 18=18\quad 2\cdot 9=18\quad 3\cdot 6=18$

The factors of 18 are 1, 2, 3, 6, 9, and 18.

7.

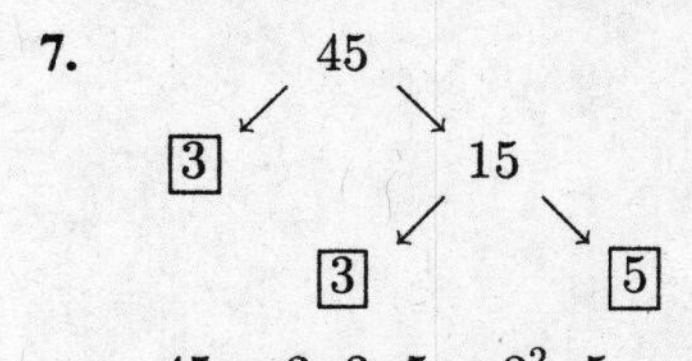

$45=3\cdot 3\cdot 5=3^2\cdot 5$

8.

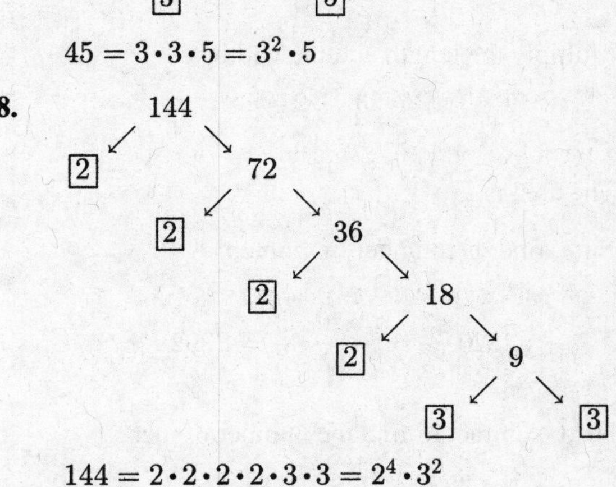

$144=2\cdot 2\cdot 2\cdot 2\cdot 3\cdot 3=2^4\cdot 3^2$

9. Factor tree: 500 → $\boxed{2}$, 250; 250 → $\boxed{2}$, 125; 125 → $\boxed{5}$, 25; 25 → $\boxed{5}$, $\boxed{5}$

$500 = 2 \cdot 2 \cdot 5 \cdot 5 \cdot 5 = 2^2 \cdot 5^3$

10. $\frac{36}{48} = \frac{36 \div 12}{48 \div 12} = \frac{3}{4}$

11. $\frac{60}{72} = \frac{60 \div 12}{72 \div 12} = \frac{5}{6}$

12. Write the prime factorization of both numerator and denominator. Divide the numerator and denominator by any common factors. Multiply the remaining factors in the numerator and denominator.

$$\frac{56}{84} = \frac{\overset{1}{\cancel{2}} \cdot \overset{1}{\cancel{2}} \cdot 2 \cdot \overset{1}{\cancel{7}}}{\underset{1}{\cancel{2}} \cdot \underset{1}{\cancel{2}} \cdot 3 \cdot \underset{1}{\cancel{7}}} = \frac{2}{3}$$

13. Multiply fractions by multiplying the numerators and multiplying the denominators. Divide the two fractions by using the reciprocal of the second fraction (divisor) and multiplying.

14. $\frac{3}{4} \cdot \frac{4}{9} = \frac{\overset{1}{\cancel{3}}}{\underset{1}{\cancel{4}}} \cdot \frac{\overset{1}{\cancel{4}}}{\underset{3}{\cancel{9}}} = \frac{1}{3}$

15. $54 \cdot \frac{2}{3} = \frac{\overset{18}{\cancel{54}}}{1} \cdot \frac{2}{\underset{1}{\cancel{3}}} = 36$

16. Multiply the length and the width.

$$\frac{7}{8} \cdot \frac{3}{4} = \frac{21}{32}$$

The area is $\frac{21}{32}$ yd^2.

17. First, find the number of women.

$$3820 \cdot \frac{3}{5} = \frac{\overset{764}{\cancel{3820}}}{1} \cdot \frac{3}{\underset{1}{\cancel{5}}} = 2292$$

Next, subtract to find the number of men.

$$3820 - 2292 = 1528$$

There are 1528 men.

18. $\frac{3}{4} \div \frac{5}{6} = \frac{3}{\underset{2}{\cancel{4}}} \cdot \frac{\overset{3}{\cancel{6}}}{5} = \frac{3 \cdot 3}{2 \cdot 5} = \frac{9}{10}$

19. $\frac{7}{\frac{4}{9}} = 7 \div \frac{4}{9} = \frac{7}{1} \div \frac{4}{9} = \frac{7}{1} \cdot \frac{9}{4} = \frac{63}{4} = 15\frac{3}{4}$

20. Divide the total quarts by the size of the bottle.

$$120 \div \frac{3}{5} = \frac{\overset{40}{\cancel{120}}}{1} \cdot \frac{5}{\underset{1}{\cancel{3}}} = \frac{40 \cdot 5}{1 \cdot 1} = 200$$

200 sport bottles can be filled.

21. $4\frac{1}{8} \cdot 3\frac{1}{2}$

Estimate: $4 \cdot 4 = 16$

Exact: $4\frac{1}{8} \cdot 3\frac{1}{2} = \frac{33}{8} \cdot \frac{7}{2} = \frac{231}{16} = 14\frac{7}{16}$

22. $1\frac{5}{6} \cdot 4\frac{1}{3}$

Estimate: $2 \cdot 4 = 8$

Exact: $1\frac{5}{6} \cdot 4\frac{1}{3} = \frac{11}{6} \cdot \frac{13}{3} = \frac{143}{18} = 7\frac{17}{18}$

23. $9\frac{3}{5} \div 2\frac{1}{4}$

Estimate: $10 \div 2 = 5$

Exact:

$$9\frac{3}{5} \div 2\frac{1}{4} = \frac{48}{5} \div \frac{9}{4} = \frac{\overset{16}{\cancel{48}}}{5} \cdot \frac{4}{\underset{3}{\cancel{9}}} = \frac{64}{15} = 4\frac{4}{15}$$

24. $\frac{8\frac{1}{2}}{1\frac{2}{3}}$

Estimate: $9 \div 2 = \frac{9}{2} = 4\frac{1}{2}$

Exact: $\frac{8\frac{1}{2}}{1\frac{2}{3}} = 8\frac{1}{2} \div 1\frac{2}{3} = \frac{17}{2} \div \frac{5}{3}$

$$= \frac{17}{2} \cdot \frac{3}{5} = \frac{17 \cdot 3}{2 \cdot 5} = \frac{51}{10} = 5\frac{1}{10}$$

25. If $2\frac{1}{2}$ grams can be synthesized per day, multiply to find the amount synthesized in $12\frac{1}{4}$ days.

Estimate: $3 \cdot 12 = 36$ grams

Exact:

$$2\frac{1}{2} \cdot 12\frac{1}{4} = \frac{5}{2} \cdot \frac{49}{4} = \frac{5 \cdot 49}{2 \cdot 4} = \frac{245}{8} = 30\frac{5}{8}$$

$30\frac{5}{8}$ grams can be synthesized.

Cumulative Review Exercises (Chapters 1–2)

1. 783

hundreds: 7
tens: 8

2. 8,621,785

millions: 8
ten thousands: 2

3.
```
   1
  71
  23
  47
+ 36
 177
```

4.
```
   1 11
  82,121
   5 468
     316
+ 61,294
 149,199
```

5.
```
   413
  6537
- 2085
  4452
```

6.
```
   7118 1510
  4,819,604
- 1,597,783
  3,221,821
```

7.
```
  2
  83
×  9
 747
```

8. $9 \cdot 4 \cdot 2 = (9 \cdot 4) \cdot 2 = 36 \cdot 2 = 72$

9.
```
3784*573
        2168232
```

10.
```
 563      52         563
× 800     563      × 800
        ×   8
         4504     450,400
```
Attach 00.

11. $\frac{63}{7}$ $7\overline{)63}$ = 9 $\frac{63}{7} = 9$

12.
```
136458/18
        7581
```

13. 33,886 ÷ 4
```
   8471 R2
4)33,886
  32
   18
   16
    28
    28
     06
      4
      2
```

Check:
```
  8471
×    4
33,884
+    2
33,886
```

14.
```
        22 R26
492)10,850
     984
     1010
      984
       26
```

Check:
```
   492
×   22
   984
  984
10,824
+   26
10,850
```

15. **To the nearest ten:** $65\underline{8}3$

Next digit is 4 or less.

Tens place does not change.

All digits to the right of the underlined place change to zero. **6580**

To the nearest hundred: $6\underline{5}83$

Next digit is 5 or more.

Hundreds place changes $(5 + 1 = 6)$.

All digits to the right of the underlined place change to zero. **6600**

To the nearest thousand: $\underline{6}583$

Next digit is 5 or more.

Thousands place changes $(6 + 1 = 7)$.

All digits to the right of the underlined place change to zero. **7000**

16. **To the nearest ten:** $76{,}2\underline{7}1$

Next digit is 4 or less.

Tens place does not change.

All digits to the right of the underlined place change to zero. **76,270**

To the nearest hundred: $76{,}\underline{2}71$

Next digit is 5 or more.

Hundreds place changes $(2 + 1 = 3)$.

All digits to the right of the underlined place change to zero. **76,300**

To the nearest thousand: $7\underline{6}{,}271$

Next digit is 4 or less.

Thousands place does not change.

All digits to the right of the underlined place change to zero. **76,000**

17.
$$\begin{aligned} &2^5 - 6(4) && \textit{Exponent} \\ &= 32 - 6(4) && \textit{Multiply} \\ &= 32 - 24 = 8 && \textit{Subtract} \end{aligned}$$

18.
$$\begin{aligned} &\sqrt{36} - 2 \cdot 3 + 5 && \textit{Square root} \\ &= 6 - 2 \cdot 3 + 5 && \textit{Multiply} \\ &= 6 - 6 + 5 && \textit{Subtract} \\ &= 0 + 5 = 5 && \textit{Add} \end{aligned}$$

19. Multiply to find the cost of each order; then add to find the total.

$$\begin{array}{r} \$46 \\ \times\ 8 \\ \hline \$368 \end{array} \qquad \begin{array}{r} \$52 \\ \times\ 12 \\ \hline 104 \\ 52 \\ \hline \$624 \end{array} \qquad \begin{array}{r} \$368 \\ +\ \$624 \\ \hline \$992 \end{array}$$

The total cost is \$992.

20. Subtract to find the difference between the decibel levels.

$$\begin{array}{r} 150 \\ -\ 90 \\ \hline 60 \end{array}$$

The rock concert produces 60 more decibels than a lawnmower.

21. Find the number of hairs lost in 2 years and subtract to find the hairs remaining.

$$\begin{array}{r} 365 \\ \times\ 100 \\ \hline 36{,}500 \end{array} \qquad \begin{array}{r} 36{,}500 \\ \times\ \ \ \ 2 \\ \hline 73{,}000 \end{array} \qquad \begin{array}{r} 120{,}000 \\ -\ \ 73{,}000 \\ \hline 47{,}000 \end{array}$$

47,000 hairs remain.

22. Divide the total number of hours by the number of workers.

$$\begin{array}{r} 188 \\ 22\overline{)4136} \\ 22 \\ \hline 193 \\ 176 \\ \hline 176 \\ 176 \\ \hline 0 \end{array}$$

Each health care worker will work 188 hours.

23. Multiply the length and the width to find the area.

$$\frac{11}{12} \cdot \frac{3}{4} = \frac{11}{\underset{4}{\cancel{12}}} \cdot \frac{\overset{1}{\cancel{3}}}{4} = \frac{11}{16}$$

The area is $\frac{11}{16}$ ft^2.

24. Multiply the cost per minute by the number of minutes the oven is used.

6 hours $= 6 \cdot 60 = 360$ minutes

$$\frac{1}{6} \cdot 360 = \frac{1}{\underset{1}{\cancel{6}}} \cdot \frac{\overset{60}{\cancel{360}}}{1} = 60$$

The cost of operating the oven for 6 hours is 60¢.

25. $\frac{2}{3}$ is *proper* because the numerator (2) is smaller than the denominator (3).

26. $\frac{6}{6}$ is *improper* because the numerator (6) is larger than or the same as the denominator (6).

27. $\frac{9}{18}$ is *proper* because the numerator (9) is smaller than the denominator (18).

28. $3\frac{3}{8}$ $\quad 3 \cdot 8 = 24$

$24 + 3 = 27$

$3\frac{3}{8} = \frac{27}{8}$

29. $6\frac{2}{5}$ $\quad 6 \cdot 5 = 30$

$30 + 2 = 32$

$6\frac{2}{5} = \frac{32}{5}$

30. $\frac{14}{7}$

$$\begin{array}{r} 2 \\ 7\overline{)\,14} \\ \underline{14} \\ 0 \end{array} \quad \begin{array}{l} \leftarrow \text{Whole number part} \\ \\ \\ \leftarrow \text{Remainder} \end{array}$$

$\frac{14}{7} = 2$

31. $\frac{103}{8}$

$$\begin{array}{r} 12 \\ 8\overline{)\,103} \\ \underline{8} \\ 23 \\ \underline{16} \\ 7 \end{array} \quad \begin{array}{l} \leftarrow \text{Whole number part} \\ \\ \\ \\ \\ \leftarrow \text{Remainder} \end{array}$$

$\frac{103}{8} = 12\frac{7}{8}$

32. 72

$$\begin{array}{r} 36 \\ 2\overline{)\,72} \\ 18 \\ 2\overline{)\,36} \\ 9 \\ 2\overline{)\,18} \\ 3 \\ 3\overline{)\,9} \\ 1 \\ 3\overline{)\,3} \end{array}$$

$72 = 2 \cdot 2 \cdot 2 \cdot 3 \cdot 3 = 2^3 \cdot 3^2$

33. 126

126 → [2], 63

63 → [3], 21

21 → [3], [7]

$126 = 2 \cdot 3 \cdot 3 \cdot 7 = 2 \cdot 3^2 \cdot 7$

34. 350

350 → [2], 175

175 → [5], 35

35 → [5], [7]

$350 = 2 \cdot 5 \cdot 5 \cdot 7 = 2 \cdot 5^2 \cdot 7$

35. $4^2 \cdot 2^2 = 4 \cdot 4 \cdot 2 \cdot 2 = 16 \cdot 4 = 64$

36. $2^3 \cdot 6^2 = 2 \cdot 2 \cdot 2 \cdot 6 \cdot 6 = 8 \cdot 36 = 288$

37. $$\begin{aligned} 2^3 \cdot 4^2 \cdot 5 &= 2 \cdot 2 \cdot 2 \cdot 4 \cdot 4 \cdot 5 \\ &= 8 \cdot 16 \cdot 5 \\ &= 128 \cdot 5 \\ &= 640 \end{aligned}$$

38. $\frac{42}{48} = \frac{42 \div 6}{48 \div 6} = \frac{7}{8}$

39. $\frac{24}{36} = \frac{24 \div 12}{36 \div 12} = \frac{2}{3}$

40. $\frac{30}{54} = \frac{30 \div 6}{54 \div 6} = \frac{5}{9}$

41. $\frac{1}{2} \cdot \frac{3}{4} = \frac{1 \cdot 3}{2 \cdot 4} = \frac{3}{8}$

42. $30 \cdot \frac{2}{3} \cdot \frac{3}{5} = \frac{\overset{6}{\cancel{30}} \cdot 2 \cdot \overset{1}{\cancel{3}}}{\underset{1}{\cancel{3}} \cdot \underset{1}{\cancel{5}}} = 12$

43. $7\frac{1}{2} \cdot 3\frac{1}{3} = \frac{\overset{5}{\cancel{15}}}{\underset{1}{\cancel{2}}} \cdot \frac{\overset{5}{\cancel{10}}}{\underset{1}{\cancel{3}}} = \frac{25}{1} = 25$

44. $\frac{3}{5} \div \frac{5}{8} = \frac{3}{5} \cdot \frac{8}{5} = \frac{24}{25}$

45. $\frac{7}{8} \div 1\frac{1}{2} = \frac{7}{8} \div \frac{3}{2} = \frac{7}{\underset{4}{\cancel{8}}} \cdot \frac{\overset{1}{\cancel{2}}}{3} = \frac{7 \cdot 1}{4 \cdot 3} = \frac{7}{12}$

46. $3 \div 1\frac{1}{4} = \frac{3}{1} \div \frac{5}{4} = \frac{3}{1} \cdot \frac{4}{5} = \frac{12}{5} = 2\frac{2}{5}$

CHAPTER 3 ADDING AND SUBTRACTING FRACTIONS

3.1 Adding and Subtracting Like Fractions

3.1 Margin Exercises

1. **(a)** $\frac{2}{5}\ \frac{3}{5}$

The denominators are the same, so $\frac{2}{5}$ and $\frac{3}{5}$ are *like* fractions.

(b) $\frac{2}{3}\ \frac{3}{4}$

The denominators are different, so $\frac{2}{3}$ and $\frac{3}{4}$ are *unlike* fractions.

(c) $\frac{7}{12}\ \frac{11}{12}$

The denominators are the same, so $\frac{7}{12}$ and $\frac{11}{12}$ are *like* fractions.

(d) $\frac{3}{8}\ \frac{3}{16}$

The denominators are different, so $\frac{3}{8}$ and $\frac{3}{16}$ are *unlike* fractions.

2. **(a)** $\frac{3}{8}+\frac{1}{8}=\frac{3+1}{8}=\frac{4}{8}$

$=\frac{4\div 4}{8\div 4}=\frac{1}{2}$ *Lowest terms*

(b) $\frac{2}{9}$

$+\frac{5}{9}$

Add numerators. The denominator stays the same.

$$\frac{2+5}{9}=\frac{7}{9}$$

(c) $\frac{3}{16}+\frac{1}{16}=\frac{3+1}{16}=\frac{4}{16}$

$=\frac{4\div 4}{16\div 4}=\frac{1}{4}$ *Lowest terms*

(d) $\frac{3}{10}+\frac{1}{10}+\frac{4}{10}=\frac{3+1+4}{10}$

$=\frac{8}{10}=\frac{8\div 2}{10\div 2}=\frac{4}{5}$ *Lowest terms*

3. **(a)** $\frac{7}{8}-\frac{3}{8}=\frac{7-3}{8}=\frac{4}{8}=\frac{4\div 4}{8\div 4}=\frac{1}{2}$

(b) $\frac{15}{16}$

$-\frac{3}{16}$

$\frac{15-3}{16}=\frac{12}{16}=\frac{12\div 4}{16\div 4}=\frac{3}{4}$

(c) $\frac{15}{3}-\frac{5}{3}=\frac{15-5}{3}=\frac{10}{3}=3\frac{1}{3}$

(d) $\frac{25}{32}$

$-\frac{6}{32}$

$\frac{25-6}{32}=\frac{19}{32}$

3.1 Section Exercises

1. $\frac{3}{8}+\frac{2}{8}=\frac{3+2}{8}=\frac{5}{8}$

3. $\frac{2}{6}+\frac{3}{6}=\frac{2+3}{6}=\frac{5}{6}$

5. $\frac{1}{4}+\frac{1}{4}=\frac{1+1}{4}=\frac{2}{4}$

$=\frac{2\div 2}{4\div 2}=\frac{1}{2}$ *Lowest terms*

7. $\frac{9}{10}$

$+\frac{3}{10}$

$\frac{9}{10}+\frac{3}{10}=\frac{9+3}{10}=\frac{12}{10}=\frac{12\div 2}{10\div 2}=\frac{6}{5}=1\frac{1}{5}$

9. $\frac{2}{9}$

$+\frac{1}{9}$

$\frac{2}{9}+\frac{1}{9}=\frac{2+1}{9}=\frac{3}{9}=\frac{3\div 3}{9\div 3}=\frac{1}{3}$

11. $\frac{6}{20}+\frac{4}{20}+\frac{3}{20}=\frac{6+4+3}{20}=\frac{13}{20}$

13. $\frac{4}{15}+\frac{2}{15}+\frac{5}{15}=\frac{4+2+5}{15}=\frac{11}{15}$

15. $\frac{3}{8}+\frac{7}{8}+\frac{2}{8}=\frac{3+7+2}{8}=\frac{12}{8}$

$=\frac{12\div 4}{8\div 4}=\frac{3}{2}=1\frac{1}{2}$

17. $\frac{2}{54}+\frac{8}{54}+\frac{12}{54}=\frac{2+8+12}{54}=\frac{22}{54}$

$=\frac{22\div 2}{54\div 2}=\frac{11}{27}$

19. $\frac{7}{8} - \frac{4}{8} = \frac{7-4}{8} = \frac{3}{8}$

21. $\frac{10}{11} - \frac{4}{11} = \frac{10-4}{11} = \frac{6}{11}$

23. $\frac{9}{10} - \frac{3}{10} = \frac{9-3}{10} = \frac{6}{10} = \frac{6 \div 2}{10 \div 2} = \frac{3}{5}$

25.

$$\begin{array}{r} \frac{31}{21} \\ -\frac{7}{21} \\ \hline \end{array}$$

$\frac{31}{21} - \frac{7}{21} = \frac{31-7}{21} = \frac{24}{21} = \frac{24 \div 3}{21 \div 3} = \frac{8}{7} = 1\frac{1}{7}$

27.

$$\begin{array}{r} \frac{27}{40} \\ -\frac{19}{40} \\ \hline \end{array}$$

$\frac{27}{40} - \frac{19}{40} = \frac{27-19}{40} = \frac{8}{40} = \frac{8 \div 8}{40 \div 8} = \frac{1}{5}$

29. $\frac{47}{36} - \frac{5}{36} = \frac{47-5}{36} = \frac{42}{36} = \frac{42 \div 6}{36 \div 6} = \frac{7}{6} = 1\frac{1}{6}$

31. $\frac{73}{60} - \frac{7}{60} = \frac{73-7}{60} = \frac{66}{60} = \frac{66 \div 6}{60 \div 6} = \frac{11}{10} = 1\frac{1}{10}$

33. Three steps to add like fractions are:

1. Add the numerators of the fractions to find the numerator of the sum (the answer).
2. Use the denominator of the fractions as the denominator of the sum.
3. Write the answer in lowest terms.

35. Add the two amounts to find the total fraction saved.

$$\frac{2}{5} + \frac{2}{5} = \frac{2+2}{5} = \frac{4}{5}$$

They saved $\frac{4}{5}$ of the amount.

37. Subtract the two fractions.

$$\frac{5}{9} - \frac{2}{9} = \frac{5-2}{9} = \frac{3}{9} = \frac{3 \div 3}{9 \div 3} = \frac{1}{3}$$

They still owe $\frac{1}{3}$ of the loan.

39. First, add the two fractions of land that were purchased.

$$\frac{9}{10} + \frac{3}{10} = \frac{9+3}{10} = \frac{12}{10}$$

Next, subtract the fraction of land that was planted in carrots.

$$\frac{12}{10} - \frac{7}{10} = \frac{12-7}{10} = \frac{5}{10} = \frac{1}{2}$$

$\frac{1}{2}$ acre is planted in squash.

3.2 Least Common Multiples

3.2 Margin Exercises

1. **(a)** The multiples of 5 are

$$5, 10, 15, 20, 25, 30, 35, 40, \ldots.$$

(b) The multiples of 8 are

$$8, 16, 24, 32, 40, 48, 56, \ldots.$$

(c) Look at the answers for (a) and (b). 40 is the only number found in both lists, so it is the least common multiple.

2. **(a)** 2 and 5

The multiples of 5 are

$$5, 10, 15, 20, \ldots.$$

The first multiple of 5 that is divisible by 2 is 10, so the least common multiple of the numbers 2 and 5 is 10.

(b) 3 and 9

The multiples of 9 are

$$9, 18, 27, 36, 45, \ldots.$$

The first multiple of 9 that is divisible by 3 is 9, so the least common multiple of the numbers 3 and 9 is 9.

(c) 6 and 8

The multiples of 8 are

$$8, 16, 24, 32, 40, 48, \ldots.$$

The first multiple of 8 that is divisible by 6 is 24, so the least common multiple of the numbers 6 and 8 is 24.

(d) 4 and 7

The multiples of 7 are

$$7, 14, 21, 28, 35, 42, \ldots.$$

The first multiple of 7 that is divisible by 4 is 28, so the least common multiple of the numbers 4 and 7 is 28.

3. **(a)** 15 and 18

$15 = 3 \cdot 5$
$18 = 2 \cdot 3 \cdot 3$

To find the LCM of 15 and 18, we'll start with the factors of 18, $2 \cdot 3 \cdot 3$. Now 15 only has 3 and 5 as factors and we already have a 3 in the LCM, so we just need to include a 5.

$\text{LCM} = 2 \cdot 3 \cdot 3 \cdot 5 = 90$

The least common multiple of 15 and 18 is 90.

(b) 12 and 20

$12 = 2 \cdot 2 \cdot 3$
$20 = 2 \cdot 2 \cdot 5$

$\text{LCM} = 2 \cdot 2 \cdot 3 \cdot 5 = 60$

The least common multiple of 12 and 20 is 60.

4. **(a)** $\frac{3}{8}$ and $\frac{6}{5}$

$8 = 2 \cdot 2 \cdot 2$
$5 = 1 \cdot 5$

$\text{LCM} = 2 \cdot 2 \cdot 2 \cdot 5 = 40$

The least common multiple of 8 and 5 is 40.

(b) $\frac{5}{6}$ and $\frac{1}{14}$

$6 = 2 \cdot 3$
$14 = 2 \cdot 7$

$\text{LCM} = 2 \cdot 3 \cdot 7 = 42$

The least common multiple of 6 and 14 is 42.

(c) $\frac{4}{9}, \frac{5}{18},$ and $\frac{7}{24}$

$9 = 3 \cdot 3$
$18 = 2 \cdot 3 \cdot 3$
$24 = 2 \cdot 2 \cdot 2 \cdot 3$

$\text{LCM} = 2 \cdot 2 \cdot 2 \cdot 3 \cdot 3 = 72$

The least common multiple of 9, 18, and 24 is 72.

5. **(a)** 6, 8

$6 = 2 \cdot 3$
$8 = 2 \cdot 2 \cdot 2$

$\text{LCM} = 2 \cdot 2 \cdot 2 \cdot 3 = 24$

The least common multiple of 6 and 8 is 24.

(b) 3, 6, 8

$3 = 1 \cdot 3$
$6 = 2 \cdot 3$
$8 = 2 \cdot 2 \cdot 2$

$\text{LCM} = 2 \cdot 2 \cdot 2 \cdot 3 = 24$

The least common multiple of 3, 6, and 8 is 24. Note that since 3 is a factor of 6, we don't need to consider it; that is, this problem is the same as finding the LCM of 6 and 8, which is part (a).

(c) 9, 36, 48

$9 = 3 \cdot 3$
$36 = 2 \cdot 2 \cdot 3 \cdot 3$
$48 = 2 \cdot 2 \cdot 2 \cdot 2 \cdot 3$

$\text{LCM} = 2 \cdot 2 \cdot 2 \cdot 2 \cdot 3 \cdot 3 = 144$

The least common multiple of 9, 36, and 48 is 144.

(d) 15, 20, 30, 40

$15 = 3 \cdot 5$
$20 = 2 \cdot 2 \cdot 5$
$30 = 2 \cdot 3 \cdot 5$
$40 = 2 \cdot 2 \cdot 2 \cdot 5$

$\text{LCM} = 2 \cdot 2 \cdot 2 \cdot 3 \cdot 5 = 120$

The least common multiple of 15, 20, 30, and 40 is 120.

6. **(a)** The least common multiple of 6 and 15 is the product of the numbers on the left side.

$$\text{LCM} = 2 \cdot 3 \cdot 5 = 30$$

(b) The least common multiple of 20 and 36 is the product of the numbers on the left side.

$$\text{LCM} = 2 \cdot 2 \cdot 3 \cdot 3 \cdot 5 = 180$$

7. **(a)** 3, 6, and 10

$$\begin{array}{r|rrr} 2 & \not{3} & 6 & 10 \\ \hline 3 & 3 & 3 & \not{5} \\ \hline 5 & \not{1} & \not{1} & 5 \\ \hline & 1 & 1 & 1 \end{array}$$

The LCM of 3, 6, and 10 is

$$2 \cdot 3 \cdot 5 = 30.$$

(b) 15 and 40

$$\begin{array}{r|rr} 2 & \not{15} & 40 \\ \hline 2 & \not{15} & 20 \\ \hline 2 & \not{15} & 10 \\ \hline 3 & 15 & \not{5} \\ \hline 5 & 5 & 5 \\ \hline & 1 & 1 \end{array}$$

The LCM of 15 and 40 is

$$2 \cdot 2 \cdot 2 \cdot 3 \cdot 5 = 120.$$

(c) 9 and 24

$$\begin{array}{r|rr} 2 & \not{9} & 24 \\ \hline 2 & \not{9} & 12 \\ \hline 2 & \not{9} & 6 \\ \hline 3 & 9 & 3 \\ \hline 3 & 3 & \not{1} \\ \hline & 1 & 1 \end{array}$$

The LCM of 9 and 24 is

$$2 \cdot 2 \cdot 2 \cdot 3 \cdot 3 = 72.$$

(d) 8, 21, and 24

$$\begin{array}{r|rrr} 2 & 8 & \cancel{21} & 24 \\ 2 & 4 & \cancel{21} & 12 \\ 2 & 2 & \cancel{21} & 6 \\ 3 & \cancel{1} & 21 & 3 \\ 7 & \cancel{1} & 7 & \cancel{1} \\ \hline & 1 & 1 & 1 \end{array}$$

The LCM of 8, 21, and 24 is

$$2 \cdot 2 \cdot 2 \cdot 3 \cdot 7 = 168.$$

8. (a) $\frac{1}{4} = \frac{?}{16} \quad 16 \div 4 = 4$

$\frac{1}{4} = \frac{1 \cdot 4}{4 \cdot 4} = \frac{4}{16}$

(b) $\frac{2}{3} = \frac{?}{15} \quad 15 \div 3 = 5$

$\frac{2}{3} = \frac{2 \cdot 5}{3 \cdot 5} = \frac{10}{15}$

(c) $\frac{7}{16} = \frac{?}{32} \quad 32 \div 16 = 2$

$\frac{7}{16} = \frac{7 \cdot 2}{16 \cdot 2} = \frac{14}{32}$

(d) $\frac{6}{11} = \frac{?}{33} \quad 33 \div 11 = 3$

$\frac{6}{11} = \frac{6 \cdot 3}{11 \cdot 3} = \frac{18}{33}$

3.2 Section Exercises

1. 3 and 6

Multiples of 6:

$$\underline{6}, 12, 18, 24, 30, \ldots$$

6 is the first number divisible by 3. $(6 \div 3 = 2)$

The least common multiple of 3 and 6 is 6.

3. 3 and 5

Multiples of 5:

$$5, 10, \underline{15}, 20, 25, \ldots$$

15 is the first number divisible by 3. $(15 \div 3 = 5)$

The least common multiple of 3 and 5 is 15.

5. 4 and 9

Multiples of 9:

$$9, 18, 27, \underline{36}, 45, \ldots$$

36 is the first number divisible by 4. $(36 \div 4 = 9)$

The least common multiple of 4 and 9 is 36.

7. 2 and 7

Multiples of 7:

$$7, \underline{14}, 21, 28, 35, \ldots$$

14 is the first number divisible by 2. $(14 \div 2 = 7)$

The least common multiple of 2 and 7 is 14.

9. 6 and 10

Multiples of 10:

$$10, 20, \underline{30}, 40, 50, \ldots$$

30 is the first number divisible by 6. $(30 \div 6 = 5)$

The least common multiple of 6 and 10 is 30.

11. 20 and 50

Multiples of 50:

$$50, \underline{100}, 150, 200, 250, \ldots$$

100 is the first number divisible by 20. $(100 \div 20 = 5)$

The least common multiple of 20 and 50 is 100.

13. 4, 10

$$\begin{array}{r|rr} 2 & 4 & 10 \\ 2 & 2 & \cancel{5} \\ 5 & \cancel{1} & 5 \\ \hline & 1 & 1 \end{array}$$

The LCM of 4 and 10 is

$$2 \cdot 2 \cdot 5 = 20.$$

15. 12, 20

$$\begin{array}{r|rr} 2 & 12 & 20 \\ 2 & 6 & 10 \\ 3 & 3 & \cancel{5} \\ 5 & \cancel{1} & 5 \\ \hline & 1 & 1 \end{array}$$

The LCM of 12 and 20 is

$$2 \cdot 2 \cdot 3 \cdot 5 = 60.$$

17. 6, 9, 12

$$\begin{array}{r|rrr} 2 & 6 & \cancel{9} & 12 \\ 2 & \cancel{3} & \cancel{9} & 6 \\ 3 & 3 & 9 & 3 \\ 3 & \cancel{1} & 3 & \cancel{1} \\ \hline & 1 & 1 & 1 \end{array}$$

The LCM of 6, 9, and 12 is

$$2 \cdot 2 \cdot 3 \cdot 3 = 36.$$

19. 4, 6, 8, 10

$4 = 2 \cdot 2$
$6 = 2 \cdot 3$
$8 = 2 \cdot 2 \cdot 2$
$10 = 2 \cdot 5$

$\text{LCM} = 2 \cdot 2 \cdot 2 \cdot 3 \cdot 5 = 120$

The LCM of 4, 6, 8, and 10 is 120.

21. 12, 15, 18, 20

$12 = 2 \cdot 2 \cdot 3$
$15 = 3 \cdot 5$
$18 = 2 \cdot 3 \cdot 3$
$20 = 2 \cdot 2 \cdot 5$

$\text{LCM} = 2 \cdot 2 \cdot 3 \cdot 3 \cdot 5 = 180$

The LCM of 12, 15, 18, and 20 is 180.

23. 8, 12, 16, 36

$8 = 2 \cdot 2 \cdot 2$
$12 = 2 \cdot 2 \cdot 3$
$16 = 2 \cdot 2 \cdot 2 \cdot 2$
$36 = 2 \cdot 2 \cdot 3 \cdot 3$

$\text{LCM} = 2 \cdot 2 \cdot 2 \cdot 2 \cdot 3 \cdot 3 = 144$

The LCM of 8, 12, 16, and 36 is 144.

25. $\frac{2}{3} = \frac{?}{24}$ $\quad 24 \div 3 = 8$
$\frac{2}{3} = \frac{2 \cdot 8}{3 \cdot 8} = \frac{16}{24}$

27. $\frac{3}{4} = \frac{?}{24}$ $\quad 24 \div 4 = 6$
$\frac{3}{4} = \frac{3 \cdot 6}{4 \cdot 6} = \frac{18}{24}$

29. $\frac{5}{6} = \frac{?}{24}$ $\quad 24 \div 6 = 4$
$\frac{5}{6} = \frac{5 \cdot 4}{6 \cdot 4} = \frac{20}{24}$

31. $\frac{1}{2} = \frac{?}{6}$ $\quad 6 \div 2 = 3$
$\frac{1}{2} = \frac{1 \cdot 3}{2 \cdot 3} = \frac{3}{6}$

33. $\frac{3}{4} = \frac{?}{16}$ $\quad 16 \div 4 = 4$
$\frac{3}{4} = \frac{3 \cdot 4}{4 \cdot 4} = \frac{12}{16}$

35. $\frac{7}{8} = \frac{?}{32}$ $\quad 32 \div 8 = 4$
$\frac{7}{8} = \frac{7 \cdot 4}{8 \cdot 4} = \frac{28}{32}$

37. $\frac{3}{16} = \frac{?}{64}$ $\quad 64 \div 16 = 4$
$\frac{3}{16} = \frac{3 \cdot 4}{16 \cdot 4} = \frac{12}{64}$

39. $\frac{8}{5} = \frac{?}{20}$ $\quad 20 \div 5 = 4$
$\frac{8}{5} = \frac{8 \cdot 4}{5 \cdot 4} = \frac{32}{20}$

41. $\frac{9}{7} = \frac{?}{56}$ $\quad 56 \div 7 = 8$
$\frac{9}{7} = \frac{9 \cdot 8}{7 \cdot 8} = \frac{72}{56}$

43. $\frac{7}{4} = \frac{?}{48}$ $\quad 48 \div 4 = 12$
$\frac{7}{4} = \frac{7 \cdot 12}{4 \cdot 12} = \frac{84}{48}$

45. $\frac{8}{11} = \frac{?}{132}$ $\quad 132 \div 11 = 12$
$\frac{8}{11} = \frac{8 \cdot 12}{11 \cdot 12} = \frac{96}{132}$

47. $\frac{3}{16} = \frac{?}{144}$ $\quad 144 \div 16 = 9$
$\frac{3}{16} = \frac{3 \cdot 9}{16 \cdot 9} = \frac{27}{144}$

49. Answers will vary. A sample answer follows: It probably depends on how large the numbers are. If the numbers are small, the method using multiples of the largest number seems best. If the numbers are larger, or there are more than two numbers, then the factorization method will be better.

51. $\frac{25}{400}, \frac{38}{1800}$

Multiples of 1800:

1800, <u>3600</u>, 5400, ...

3600 is the first number divisible by 400. $(3600 \div 400 = 9)$

The LCM of the denominators is 3600.

53. $\frac{109}{1512}, \frac{23}{392}$

2	1512	392
2	756	196
2	378	98
3	189	~~49~~
3	63	~~49~~
3	21	~~49~~
7	7	49
7	~~1~~	7
	1	1

The LCM of the denominators is

$2 \cdot 2 \cdot 2 \cdot 3 \cdot 3 \cdot 3 \cdot 7 \cdot 7 = 10{,}584.$

55. Fractions with the same denominators are *like* fractions and fractions with different denominators are *unlike* fractions.

56. To subtract like fractions, first find the numerator of the difference by subtracting the *numerators*. Write the denominator of the like fractions as the *denominator* of the difference. The answer is then written in *lowest* terms.

57. The *least* common multiple (LCM) of two numbers is the *smallest* whole number divisible by both of those numbers.

58. The smallest number in both lists is 40, so 40 is the least common multiple of 8 and 10.

59. 5, 7, 14, 10

$5 = 1 \cdot 5$
$7 = 1 \cdot 7$
$14 = 2 \cdot 7$
$10 = 2 \cdot 5$

$\text{LCM} = 2 \cdot 5 \cdot 7 = 70$

The LCM of 5, 7, 14, and 10 is 70.

60. 25, 18, 30, 5

$25 = 5 \cdot 5$
$18 = 2 \cdot 3 \cdot 3$
$30 = 2 \cdot 3 \cdot 5$
$5 = 1 \cdot 5$

$\text{LCM} = 2 \cdot 3 \cdot 3 \cdot 5 \cdot 5 = 450$

The LCM of 25, 18, 30, and 15 is 450.

61. $8 = 2 \cdot 2 \cdot 2$
$3 = 1 \cdot 3$
$5 = 1 \cdot 5$
$4 = 2 \cdot 2$
$10 = 2 \cdot 5$

$\text{LCM} = 2 \cdot 2 \cdot 2 \cdot 3 \cdot 5 = 120$

The LCM of 8, 3, 5, 4, and 10 is 120. 240 is a common multiple, but it is not the *least* common multiple.

62. The least common multiple can be no smaller than the largest number in a group and the number 1760 is a multiple of 55 ($1760 \div 55 = 32$).

3.3 Adding and Subtracting Unlike Fractions

3.3 Margin Exercises

1. **(a)** *Step 1* $\frac{1}{2} + \frac{3}{8}$ (LCD = 8)

$\frac{1}{2} = \frac{1 \cdot 4}{2 \cdot 4} = \frac{4}{8}$

Step 2 $\frac{1}{2} + \frac{3}{8} = \frac{4}{8} + \frac{3}{8} = \frac{4+3}{8} = \frac{7}{8}$

(b) *Step 1* $\frac{3}{4} + \frac{1}{8}$ (LCD = 8)

$\frac{3}{4} = \frac{3 \cdot 2}{4 \cdot 2} = \frac{6}{8}$

Step 2 $\frac{3}{4} + \frac{1}{8} = \frac{6}{8} + \frac{1}{8} = \frac{6+1}{8} = \frac{7}{8}$

(c) *Step 1* $\frac{3}{5} + \frac{3}{10}$ (LCD = 10)

$\frac{3}{5} = \frac{3 \cdot 2}{5 \cdot 2} = \frac{6}{10}$

Step 2 $\frac{3}{5} + \frac{3}{10} = \frac{6}{10} + \frac{3}{10} = \frac{6+3}{10} = \frac{9}{10}$

(d) *Step 1* $\frac{1}{12} + \frac{5}{6}$ (LCD = 12)

$\frac{5}{6} = \frac{5 \cdot 2}{6 \cdot 2} = \frac{10}{12}$

Step 2 $\frac{1}{12} + \frac{5}{6} = \frac{1}{12} + \frac{10}{12} = \frac{1+10}{12} = \frac{11}{12}$

2. **(a)** *Step 1* $\frac{3}{10} + \frac{1}{5} = \frac{3}{10} + \frac{2}{10}$

Step 2 $\frac{3}{10} + \frac{2}{10} = \frac{3+2}{10} = \frac{5}{10}$

Step 3 $\frac{5}{10} = \frac{1}{2}$

(b) *Step 1* $\frac{5}{8} + \frac{1}{3} = \frac{15}{24} + \frac{8}{24}$

Step 2 $\frac{15}{24} + \frac{8}{24} = \frac{15+8}{24} = \frac{23}{24}$

(c) *Step 1* $\frac{1}{10} + \frac{1}{3} + \frac{1}{6} = \frac{3}{30} + \frac{10}{30} + \frac{5}{30}$

Step 2 $\frac{3}{30} + \frac{10}{30} + \frac{5}{30} = \frac{3+10+5}{30} = \frac{18}{30}$

Step 3 $\frac{18}{30} = \frac{3}{5}$

3. **(a)**

$$\begin{array}{rcccl} \frac{5}{8} & = & \frac{5 \cdot 3}{8 \cdot 3} & = & \frac{15}{24} \\ +\frac{1}{12} & = & \frac{1 \cdot 2}{12 \cdot 2} & = & \frac{2}{24} \\ \hline & & & & \frac{17}{24} \end{array}$$

(b)
$$\frac{7}{16} = \frac{7}{16}$$
$$+\frac{1}{4} = \frac{1\cdot 4}{4\cdot 4} = \frac{4}{16}$$
$$\frac{11}{16}$$

4. (a) *Step 1* $\quad \frac{5}{8} - \frac{1}{4} = \frac{5}{8} - \frac{2}{8}$

Step 2 $\quad \frac{5}{8} - \frac{2}{8} = \frac{5-2}{8} = \frac{3}{8}$

(b) *Step 1* $\quad \frac{4}{5} - \frac{3}{4} = \frac{16}{20}\ \frac{15}{20}$

Step 2 $\quad \frac{16}{20} - \frac{15}{20} = \frac{16-15}{20} = \frac{1}{20}$

(c)
$$\frac{7}{8} = \frac{7\cdot 3}{8\cdot 3} = \frac{21}{24}$$
$$-\frac{2}{3} = \frac{2\cdot 8}{3\cdot 8} = \frac{16}{24}$$
$$\frac{5}{24}$$

(d)
$$\frac{5}{6} = \frac{5\cdot 2}{6\cdot 2} = \frac{10}{12}$$
$$-\frac{1}{12} = \frac{1}{12}$$
$$\frac{9}{12} = \frac{3}{4}$$

3.3 Section Exercises

1. $\frac{3}{4} + \frac{1}{8} = \frac{6}{8} + \frac{1}{8}$ *LCD is 8*
$$= \frac{6+1}{8}$$
$$= \frac{7}{8}$$

3. $\frac{2}{3} + \frac{2}{9} = \frac{6}{9} + \frac{2}{9}$ *LCD is 9*
$$= \frac{6+2}{9}$$
$$= \frac{8}{9}$$

5. $\frac{9}{20} + \frac{3}{10} = \frac{9}{20} + \frac{6}{20}$ *LCD is 20*
$$= \frac{9+6}{20}$$
$$= \frac{15}{20} = \frac{3}{4} \quad \textit{In lowest terms}$$

7. $\frac{3}{5} + \frac{3}{8} = \frac{24}{40} + \frac{15}{40}$ *LCD is 40*
$$= \frac{24+15}{40}$$
$$= \frac{39}{40}$$

9. $\frac{2}{9} + \frac{5}{12} = \frac{8}{36} + \frac{15}{36}$ *LCD is 36*
$$= \frac{8+15}{36}$$
$$= \frac{23}{36}$$

11. $\frac{1}{3} + \frac{3}{5} = \frac{5}{15} + \frac{9}{15}$ *LCD is 15*
$$= \frac{5+9}{15}$$
$$= \frac{14}{15}$$

13. $\frac{1}{4} + \frac{2}{9} + \frac{1}{3} = \frac{9}{36} + \frac{8}{36} + \frac{12}{36}$ *LCD is 36*
$$= \frac{9+8+12}{36}$$
$$= \frac{29}{36}$$

15. $\frac{3}{10} + \frac{2}{5} + \frac{3}{20} = \frac{6}{20} + \frac{8}{20} + \frac{3}{20}$ *LCD is 20*
$$= \frac{6+8+3}{20}$$
$$= \frac{17}{20}$$

17. $\frac{4}{15} + \frac{1}{6} + \frac{1}{3} = \frac{8}{30} + \frac{5}{30} + \frac{10}{30}$ *LCD is 30*
$$= \frac{8+5+10}{30}$$
$$= \frac{23}{30}$$

19.
$$\frac{1}{4} = \frac{1\cdot 2}{4\cdot 2} = \frac{2}{8} \quad \textit{LCD is 8}$$
$$+\frac{1}{8} = \frac{1}{8}$$
$$\frac{3}{8}$$

21.
$$\frac{5}{12} = \frac{5\cdot 4}{12\cdot 4} = \frac{20}{48} \quad \textit{LCD is 48}$$
$$+\frac{1}{16} = \frac{1\cdot 3}{16\cdot 3} = \frac{3}{48}$$
$$\frac{23}{48}$$

23. $\frac{5}{6} - \frac{1}{3} = \frac{5}{6} - \frac{2}{6}$ *LCD is 6*
$$= \frac{5-2}{6}$$
$$= \frac{3}{6} = \frac{1}{2} \quad \textit{In lowest terms}$$

25. $\frac{2}{3} - \frac{1}{6} = \frac{4}{6} - \frac{1}{6}$ *LCD is 6*
$$= \frac{4-1}{6}$$
$$= \frac{3}{6} = \frac{1}{2} \quad \textit{In lowest terms}$$

27. $\frac{2}{3} - \frac{1}{5} = \frac{10}{15} - \frac{3}{15}$ *LCD is 15*

$= \frac{10-3}{15}$

$= \frac{7}{15}$

29. $\frac{5}{12} - \frac{1}{4} = \frac{5}{12} - \frac{3}{12}$ *LCD is 12*

$= \frac{5-3}{12}$

$= \frac{2}{12} = \frac{1}{6}$ *In lowest terms*

31. $\frac{8}{9} - \frac{7}{15} = \frac{40}{45} - \frac{21}{45}$ *LCD is 45*

$= \frac{40-21}{45}$

$= \frac{19}{45}$

33. $\frac{7}{8} = \frac{7 \cdot 5}{8 \cdot 5} = \frac{35}{40}$ *LCD is 40*

$-\frac{4}{5} = \frac{4 \cdot 8}{5 \cdot 8} = \frac{32}{40}$

$\frac{3}{40}$

35. $\frac{5}{12} = \frac{5 \cdot 4}{12 \cdot 4} = \frac{20}{48}$ *LCD is 48*

$-\frac{1}{16} = \frac{1 \cdot 3}{16 \cdot 3} = \frac{3}{48}$

$\frac{17}{48}$

37. The widest blades have widths 1" and $\frac{3}{4}$". The difference is

$$1 - \frac{3}{4} = \frac{4}{4} - \frac{3}{4} = \frac{4-3}{4} = \frac{1}{4} \text{ inch.}$$

39. The indicator word *remainder* tells you to subtract.

$\frac{3}{4} = \frac{3 \cdot 3}{4 \cdot 3} = \frac{9}{12}$ *LCD is 12*

$-\frac{1}{6} = \frac{1 \cdot 2}{6 \cdot 2} = \frac{2}{12}$

$\frac{7}{12}$

The amount of land that is buildable is $\frac{7}{12}$ acre.

41. Add the fractions to find the total length.

$\frac{1}{8} + \frac{1}{4} + \frac{2}{5} = \frac{5}{40} + \frac{10}{40} + \frac{16}{40}$ *LCD is 40*

$= \frac{5+10+16}{40}$

$= \frac{31}{40}$

The total length of the screw is $\frac{31}{40}$ inch.

43. First add to find the total loss.

$$\frac{1}{6} + \frac{1}{3} = \frac{1}{6} + \frac{2}{6} = \frac{1+2}{6} = \frac{3}{6} = \frac{1}{2}$$

Then subtract to find the amount remaining.

$$\frac{7}{8} - \frac{1}{2} = \frac{7}{8} - \frac{4}{8} = \frac{7-4}{8} = \frac{3}{8}$$

The amount of fluid remaining is $\frac{3}{8}$ gallon.

45. Answers will vary. A sample answer follows: You cannot add or subtract until all the fractional pieces are the same size. For example, halves are larger than fourths, so you cannot add $\frac{1}{2} + \frac{1}{4}$ until you rewrite $\frac{1}{2}$ as $\frac{2}{4}$.

47. Add the time spent in class and study.

$$\frac{1}{6} + \frac{1}{8} = \frac{4}{24} + \frac{3}{24} = \frac{4+3}{24} = \frac{7}{24}$$

$\frac{7}{24}$ of the student's day was spent in class and study.

49. One way to compare fractions accurately is to rewrite each fraction with a common denominator.

$$\frac{1}{3} = \frac{8}{24}, \frac{1}{6} = \frac{4}{24}, \frac{1}{8} = \frac{3}{24}, \frac{1}{12} = \frac{2}{24}, \frac{7}{24}$$

The fraction with the largest numerator is the largest fraction. This is $\frac{8}{24}$ or $\frac{1}{3}$, which is "Work and Travel."

To find the number of hours, multiply.

$$\frac{1}{3} \cdot 24 = \frac{1}{\cancel{3}_1} \cdot \frac{\cancel{24}^8}{1} = \frac{8}{1} = 8$$

8 hours were spent on work and travel. Note that this is just the numerator in the equivalent fraction $\frac{8}{24}$.

51. First add the lengths on either side of the hole.

$$\frac{3}{8} + \frac{3}{8} = \frac{3+3}{8} = \frac{6}{8} = \frac{3}{4}$$

Then subtract this length from the total length.

$$\frac{15}{16} - \frac{3}{4} = \frac{15}{16} - \frac{12}{16} = \frac{15-12}{16} = \frac{3}{16}$$

The diameter is $\frac{3}{16}$ inch.

3.4 Adding and Subtracting Mixed Numbers

3.4 Margin Exercises

1. **(a)** $\frac{9}{2}$

$$\begin{array}{r} 4 \\ 2\overline{)9} \\ \underline{8} \\ 1 \end{array}$$

4 ← Whole number part; 1 ← Remainder

$\frac{9}{2} = 4\frac{1}{2}$

(b) $\frac{8}{3}$

$$\begin{array}{r} 2 \\ 3\overline{)8} \\ \underline{6} \\ 2 \end{array}$$

2 ← Whole number part; 2 ← Remainder

$\frac{8}{3} = 2\frac{2}{3}$

(c) $4\frac{3}{4}$ $\quad 4 \cdot 4 = 16$; $16 + 3 = 19$

$4\frac{3}{4} = \frac{19}{4}$

(d) $3\frac{7}{8}$ $\quad 3 \cdot 8 = 24$; $24 + 7 = 31$

$3\frac{7}{8} = \frac{31}{8}$

2. **(a)** *Estimate:* *Exact:*

$$\begin{array}{r} 7 \\ +2 \\ \hline 9 \end{array} \qquad \begin{array}{rcl} 6\frac{7}{8} & = & 6\frac{7}{8} \\ +2\frac{1}{4} & = & 2\frac{2}{8} \\ \hline & & 8\frac{9}{8} \end{array}$$

$$8 + \frac{9}{8} = 8 + 1\frac{1}{8} = 9\frac{1}{8}$$

(b) $25\frac{3}{5} + 12\frac{3}{10}$

Estimate:
$26 + 12 = 38$

Exact:
$$25\frac{3}{5} + 12\frac{3}{10} = 25\frac{6}{10} + 12\frac{3}{10} = 37\frac{9}{10}$$

(c) *Estimate:* *Exact:*

$$\begin{array}{r} 5 \\ -3 \\ \hline 2 \end{array} \qquad \begin{array}{rcl} 4\frac{7}{9} & = & 4\frac{7}{9} \\ -2\frac{2}{3} & = & 2\frac{6}{9} \\ \hline & & 2\frac{1}{9} \end{array}$$

3. **(a)** *Estimate:* *Exact:*

$$\begin{array}{r} 10 \\ +8 \\ \hline 18 \end{array} \qquad \begin{array}{rcl} 9\frac{3}{4} & = & 9\frac{3}{4} \\ +7\frac{1}{2} & = & 7\frac{2}{4} \\ \hline & & 16\frac{5}{4} \end{array}$$

$$16\frac{5}{4} = 16 + \frac{5}{4} = 16 + 1\frac{1}{4} = 17\frac{1}{4}$$

(b) *Estimate:* *Exact:*

$$\begin{array}{r} 16 \\ +13 \\ \hline 29 \end{array} \qquad \begin{array}{rcl} 15\frac{4}{5} & = & 15\frac{12}{15} \\ +12\frac{2}{3} & = & 12\frac{10}{15} \\ \hline & & 27\frac{22}{15} \end{array}$$

$$27\frac{22}{15} = 27 + \frac{22}{15} = 27 + 1\frac{7}{15} = 28\frac{7}{15}$$

4. **(a)** *Estimate:* *Exact:*

$$\begin{array}{r} 7 \\ -5 \\ \hline 2 \end{array} \qquad \begin{array}{rcl} 7\frac{1}{3} & = & 7\frac{2}{6} \\ -4\frac{5}{6} & = & 4\frac{5}{6} \\ \hline \end{array}$$

Borrow:

$$7\frac{2}{6} = 7 + \frac{2}{6} = 6 + 1 + \frac{2}{6} = 6 + \frac{6}{6} + \frac{2}{6} = 6\frac{8}{6}$$

$$\begin{array}{r} 6\frac{8}{6} \\ -4\frac{5}{6} \\ \hline 2\frac{3}{6} \end{array} = 2\frac{1}{2}$$

(b) *Estimate:* *Exact:*

$$\begin{array}{r} 5 \\ -3 \\ \hline 2 \end{array} \qquad \begin{array}{rcl} 4\frac{5}{8} & = & 4\frac{10}{16} \\ -2\frac{15}{16} & = & 2\frac{15}{16} \\ \hline \end{array}$$

Borrow:

$$4\frac{10}{16} = 4 + \frac{10}{16} = 3 + 1 + \frac{10}{16} = 3 + \frac{16}{16} + \frac{10}{16} = 3\frac{26}{16}$$

$$\begin{array}{r} 3\frac{26}{16} \\ -2\frac{15}{16} \\ \hline 1\frac{11}{16} \end{array}$$

(c) *Estimate:* *Exact:*

$$\begin{array}{r} 15 \\ -\ 6 \\ \hline 9 \end{array} \qquad \begin{array}{r} 15 \\ -\ 6\frac{4}{9} \\ \hline \end{array}$$

Borrow:

$$15 = 14 + 1 = 14 + \frac{9}{9} = 14\frac{9}{9}$$

$$\begin{array}{r} 14\frac{9}{9} \\ -\ 6\frac{4}{9} \\ \hline 8\frac{5}{9} \end{array}$$

5. **(a)**

$$\begin{array}{rcccc} 3\frac{3}{8} & = & \frac{27}{8} & = & \frac{27}{8} \\ +\ 2\frac{1}{2} & = & \frac{5}{2} & = & \frac{20}{8} \\ \hline & & & & \frac{47}{8} = 5\frac{7}{8} \end{array}$$

(b)

$$\begin{array}{rcccc} 6\frac{3}{4} & = & \frac{27}{4} & = & \frac{81}{12} \\ -\ 4\frac{2}{3} & = & \frac{14}{3} & = & \frac{56}{12} \\ \hline & & & & \frac{25}{12} = 2\frac{1}{12} \end{array}$$

3.4 Section Exercises

1. *Estimate:* *Exact:*

$$\begin{array}{r} 6 \\ +\ 3 \\ \hline 9 \end{array} \qquad \begin{array}{rcl} 5\frac{1}{2} & = & 5\frac{3}{6} \\ +\ 3\frac{1}{3} & = & 3\frac{2}{6} \\ \hline & & 8\frac{5}{6} \end{array}$$

3. *Estimate:* *Exact:*

$$\begin{array}{r} 7 \\ +\ 4 \\ \hline 11 \end{array} \qquad \begin{array}{rcl} 7\frac{1}{3} & = & 7\frac{2}{6} \\ +\ 4\frac{1}{6} & = & 4\frac{1}{6} \\ \hline & & 11\frac{3}{6} = 11\frac{1}{2} \end{array}$$

5. *Estimate:* *Exact:*

$$\begin{array}{r} 1 \\ +\ 4 \\ \hline 5 \end{array} \qquad \begin{array}{rcl} \frac{5}{8} & = & \frac{15}{24} \\ +\ 3\frac{7}{12} & = & 3\frac{14}{24} \\ \hline & & 3\frac{29}{24} \end{array}$$

$$3\frac{29}{24} = 3 + \frac{29}{24} = 3 + 1\frac{5}{24} = 4\frac{5}{24}$$

7. *Estimate:* *Exact:*

$$\begin{array}{r} 25 \\ +\ 19 \\ \hline 44 \end{array} \qquad \begin{array}{r} 24\frac{5}{6} \\ +\ 18\frac{5}{6} \\ \hline 42\frac{10}{6} \end{array}$$

$$42\frac{10}{6} = 42 + 1\frac{4}{6} = 43\frac{2}{3}$$

9. *Estimate:* *Exact:*

$$\begin{array}{r} 34 \\ +\ 19 \\ \hline 53 \end{array} \qquad \begin{array}{rcl} 33\frac{3}{5} & = & 33\frac{6}{10} \\ +\ 18\frac{1}{2} & = & 18\frac{5}{10} \\ \hline & & 51\frac{11}{10} \end{array}$$

$$51\frac{11}{10} = 51 + 1\frac{1}{10} = 52\frac{1}{10}$$

11. *Estimate:* *Exact:*

$$\begin{array}{r} 23 \\ +\ 15 \\ \hline 38 \end{array} \qquad \begin{array}{rcl} 22\frac{3}{4} & = & 22\frac{21}{28} \\ +\ 15\frac{3}{7} & = & 15\frac{12}{28} \\ \hline & & 37\frac{33}{28} \end{array}$$

$$37\frac{33}{28} = 37 + 1\frac{5}{28} = 38\frac{5}{28}$$

13. *Estimate:* *Exact:*

$$\begin{array}{r} 13 \\ 19 \\ +\ 15 \\ \hline 47 \end{array} \qquad \begin{array}{rcl} 12\frac{8}{15} & = & 12\frac{16}{30} \\ 18\frac{3}{5} & = & 18\frac{18}{30} \\ +\ 14\frac{7}{10} & = & 14\frac{21}{30} \\ \hline & & 44\frac{55}{30} \end{array}$$

$$44\frac{55}{30} = 44 + 1\frac{25}{30} = 45\frac{5}{6}$$

15. *Estimate:* *Exact:*

$$\begin{array}{r} 15 \\ -\ 12 \\ \hline 3 \end{array} \qquad \begin{array}{rcl} 14\frac{7}{8} & = & 14\frac{7}{8} \\ -\ 12\frac{1}{4} & = & 12\frac{2}{8} \\ \hline & & 2\frac{5}{8} \end{array}$$

17. *Estimate:* *Exact:*

$$\begin{array}{r} 13 \\ -\ 1 \\ \hline 12 \end{array} \qquad \begin{array}{rcl} 12\frac{2}{3} & = & 12\frac{10}{15} \\ -\ 1\frac{1}{5} & = & 1\frac{3}{15} \\ \hline & & 11\frac{7}{15} \end{array}$$

19. *Estimate:* *Exact:*

$$\begin{array}{r} 28 \\ -\ 6 \\ \hline 22 \end{array} \qquad \begin{array}{rcl} 28\frac{3}{10} & = & 28\frac{9}{30} \\ -\ 6\frac{1}{15} & = & 6\frac{2}{30} \\ \hline & & 22\frac{7}{30} \end{array}$$

21. *Estimate:* *Exact:*

$$\begin{array}{r} 17 \\ -\ 7 \\ \hline 10 \end{array} \qquad \begin{array}{r} 17 \\ -\ 6\frac{5}{8} \\ \hline \end{array}$$

Borrow:

$$17 = 16 + 1 = 16 + \frac{8}{8}$$

$$\begin{array}{r} 16\frac{8}{8} \\ -\ 6\frac{5}{8} \\ \hline 10\frac{3}{8} \end{array}$$

23. *Estimate:* *Exact:*

$$\begin{array}{r} 19 \\ -\ 6 \\ \hline 13 \end{array} \qquad \begin{array}{rcl} 18\frac{3}{4} & = & 18\frac{15}{20} \\ -\ 5\frac{4}{5} & = & 5\frac{16}{20} \\ \hline \end{array}$$

Borrow:

$$18\frac{15}{20} = 17 + 1 + \frac{15}{20} = 17 + \frac{20}{20} + \frac{15}{20} = 17\frac{35}{20}$$

$$\begin{array}{r} 17\frac{35}{20} \\ -\ 5\frac{16}{20} \\ \hline 12\frac{19}{20} \end{array}$$

25. *Estimate:* *Exact:*

$$\begin{array}{r} 20 \\ -\ 12 \\ \hline 8 \end{array} \qquad \begin{array}{rcl} 19\frac{2}{3} & = & 19\frac{8}{12} \\ -\ 11\frac{3}{4} & = & 11\frac{9}{12} \\ \hline \end{array}$$

Borrow:

$$19\frac{8}{12} = 18 + 1 + \frac{8}{12} = 18 + \frac{12}{12} + \frac{8}{12} = 18\frac{20}{12}$$

$$\begin{array}{r} 18\frac{20}{12} \\ -\ 11\frac{9}{12} \\ \hline 7\frac{11}{12} \end{array}$$

27.

$$\begin{array}{rcccl} 7\frac{5}{8} & = & \frac{61}{8} & = & \frac{61}{8} \\ +\ 1\frac{1}{2} & = & \frac{3}{2} & = & \frac{12}{8} \\ \hline & & & & \frac{73}{8} = 9\frac{1}{8} \end{array}$$

29.

$$\begin{array}{rcccl} 4\frac{2}{3} & = & \frac{14}{3} & = & \frac{28}{6} \\ +\ 6\frac{5}{6} & = & \frac{41}{6} & = & \frac{41}{6} \\ \hline & & & & \frac{69}{6} = 11\frac{3}{6} = 11\frac{1}{2} \end{array}$$

31.

$$\begin{array}{rcccl} 2\frac{2}{3} & = & \frac{8}{3} & = & \frac{16}{6} \\ +\ 1\frac{1}{6} & = & \frac{7}{6} & = & \frac{7}{6} \\ \hline & & & & \frac{23}{6} = 3\frac{5}{6} \end{array}$$

33.

$$\begin{array}{rcccl} 3\frac{1}{4} & = & \frac{13}{4} & = & \frac{39}{12} \\ +\ 3\frac{2}{3} & = & \frac{11}{3} & = & \frac{44}{12} \\ \hline & & & & \frac{83}{12} = 6\frac{11}{12} \end{array}$$

35.

$$\begin{array}{rcccl} 1\frac{3}{8} & = & \frac{11}{8} & = & \frac{11}{8} \\ +\ 6\frac{3}{4} & = & \frac{27}{4} & = & \frac{54}{8} \\ \hline & & & & \frac{65}{8} = 8\frac{1}{8} \end{array}$$

37.

$$\begin{array}{rcccl} 3\frac{1}{2} & = & \frac{7}{2} & = & \frac{21}{6} \\ -\ 2\frac{2}{3} & = & \frac{8}{3} & = & \frac{16}{6} \\ \hline & & & & \frac{5}{6} \end{array}$$

39.

$$\begin{array}{rcccl} 8\frac{3}{4} & = & \frac{35}{4} & = & \frac{70}{8} \\ -\ 5\frac{7}{8} & = & \frac{47}{8} & = & \frac{47}{8} \\ \hline & & & & \frac{23}{8} = 2\frac{7}{8} \end{array}$$

41.
$$\begin{array}{rcccl} 7\frac{1}{4} & = & \frac{29}{4} & = & \frac{87}{12} \\ -\,4\frac{2}{3} & = & \frac{14}{3} & = & \frac{56}{12} \\ \hline & & & & \frac{31}{12} = 2\frac{7}{12} \end{array}$$

43.
$$\begin{array}{rcccl} 9\frac{1}{5} & = & \frac{46}{5} & = & \frac{184}{20} \\ -\,3\frac{3}{4} & = & \frac{15}{4} & = & \frac{75}{20} \\ \hline & & & & \frac{109}{20} = 5\frac{9}{20} \end{array}$$

45.
$$\begin{array}{rcccl} 6\frac{3}{7} & = & \frac{45}{7} & = & \frac{135}{21} \\ -\,2\frac{2}{3} & = & \frac{8}{3} & = & \frac{56}{21} \\ \hline & & & & \frac{79}{21} = 3\frac{16}{21} \end{array}$$

47. Find the least common denominator. Change the fraction parts so that they have the same denominator. Add the fraction parts. Add the whole number parts. Write the answer as a mixed number.

49. Subtract $14\frac{1}{2}$ from $25\frac{1}{4}$ to determine the difference.

Estimate: $25 - 15 = 10$ feet

Exact:
$$\begin{array}{rcccl} 25\frac{1}{4} & = & \frac{101}{4} & = & \frac{101}{4} \\ -\,14\frac{1}{2} & = & \frac{29}{2} & = & \frac{58}{4} \\ \hline & & & & \frac{43}{4} = 10\frac{3}{4} \end{array}$$

The current world record is $10\frac{3}{4}$ feet taller than the first world record.

51. The longest wrench is $22\frac{3}{4}$" and the second to shortest wrench is $18\frac{5}{8}$". Subtract to find the difference.

Estimate: $23 - 19 = 4$ inches

Exact:
$$\begin{array}{rcl} 22\frac{3}{4} & = & 22\frac{6}{8} \\ -\,18\frac{5}{8} & = & 18\frac{5}{8} \\ \hline & & 4\frac{1}{8} \end{array}$$

The longest wrench is $4\frac{1}{8}$ inches longer than the second to shortest wrench.

53. The three longest wrenches measure $22\frac{3}{4}$", $22\frac{1}{4}$", and 21". Add to find the total length.

Estimate: $23 + 22 + 21 = 66$ inches

Exact:
$$\begin{array}{r} 22\frac{3}{4} \\ 22\frac{1}{4} \\ +\,21 \\ \hline 65\frac{4}{4} = 65 + 1 = 66 \end{array}$$

The total length of the three longest wrenches is 66 inches.

55. Add the hours worked on Monday through Friday.

Estimate: $3 + 6 + 5 + 3 + 6 = 23$ hr

Exact:
$$\begin{array}{rcl} 3\frac{3}{8} & = & 3\frac{3}{8} \\ 5\frac{1}{2} & = & 5\frac{4}{8} \\ 4\frac{3}{4} & = & 4\frac{6}{8} \\ 3\frac{1}{4} & = & 3\frac{2}{8} \\ +\,6 & = & 6 \\ \hline & & 21\frac{15}{8} = 21 + 1\frac{7}{8} = 22\frac{7}{8} \end{array}$$

She worked $22\frac{7}{8}$ hours altogether.

57. Add the lengths of the four sides.

Estimate: $24 + 35 + 24 + 35 = 118$ in.

Exact:
$$\begin{array}{rcl} 23\frac{3}{4} & = & 23\frac{3}{4} \\ 34\frac{1}{2} & = & 34\frac{2}{4} \\ 23\frac{3}{4} & = & 23\frac{3}{4} \\ +\,34\frac{1}{2} & = & 34\frac{2}{4} \\ \hline & & 114\frac{10}{4} = 114 + 2\frac{2}{4} = 116\frac{1}{2} \end{array}$$

The craftsperson needs $116\frac{1}{2}$ inches of lead stripping.

59. Add the amounts he unloads from all three stops. Then subtract this amount from the total amount of concrete in the truck.

Estimate: $9 - 3 - 3 - 2 = 1$ yd^3

Exact:
$$\begin{array}{rcl} 2\frac{1}{2} & = & 2\frac{2}{4} \\ 3 & = & 3 \\ +\,1\frac{3}{4} & = & 1\frac{3}{4} \\ \hline & & 6\frac{5}{4} = 6 + 1\frac{1}{4} = 7\frac{1}{4} \end{array}$$

$$\begin{array}{rcl} 8\frac{7}{8} & = & 8\frac{7}{8} \\ -\,7\frac{1}{4} & = & 7\frac{2}{8} \\ \hline & & 1\frac{5}{8} \end{array}$$

$1\frac{5}{8}$ yd^3 of concrete remains in the truck.

61. Add the lengths of the three sides. Then subtract this amount from the total length to find the length of the fourth side.

Estimate: $527 - 108 - 151 - 139 = 129$ ft

Exact:

$$\begin{array}{rcl} 107\frac{2}{3} & = & 107\frac{16}{24} \\ 150\frac{3}{4} & = & 150\frac{18}{24} \\ +\,138\frac{5}{8} & = & 138\frac{15}{24} \\ \hline & & 395\frac{49}{24} = 395 + 2\frac{1}{24} = 397\frac{1}{24} \end{array}$$

$$\begin{array}{r} 527\frac{1}{24} \\ -\,397\frac{1}{24} \\ \hline 130 \end{array}$$

The length of the fourth side is 130 feet.

63. Add the weights.

Estimate: $3 + 7 + 2 + 3 + 7 = 22$ tons

Exact:

$$\begin{array}{rcl} 2\frac{5}{8} & = & 2\frac{15}{24} \\ 6\frac{1}{2} & = & 6\frac{12}{24} \\ 1\frac{5}{6} & = & 1\frac{20}{24} \\ 3\frac{1}{4} & = & 3\frac{6}{24} \\ +\,7\frac{3}{8} & = & 7\frac{9}{24} \\ \hline & & 19\frac{62}{24} = 19 + 2\frac{14}{24} = 21\frac{7}{12} \end{array}$$

The total weight is $21\frac{7}{12}$ tons.

65. First add the two given portions of the line.

$$\begin{array}{r} 2\frac{3}{8} \\ +\,2\frac{3}{8} \\ \hline 4\frac{6}{8} \end{array}$$

Then subtract to find the unknown length.

$$\begin{array}{rcccl} 9\frac{7}{16} & = & 9\frac{7}{16} & = & 8\frac{23}{16} \\ -\,4\frac{6}{8} & = & 4\frac{12}{16} & = & 4\frac{12}{16} \\ \hline & & & & 4\frac{11}{16} \end{array}$$

The unknown length is $4\frac{11}{16}$ inches.

67. Add the length of the sections at each end.

$$\begin{array}{rcl} 6\frac{1}{4} & = & 6\frac{2}{8} \\ +\,1\frac{7}{8} & = & 1\frac{7}{8} \\ \hline & & 7\frac{9}{8} = 7 + 1\frac{1}{8} = 8\frac{1}{8} \end{array}$$

Then subtract this total from the total length of the arrow.

$$\begin{array}{rcl} 29\frac{1}{2} & = & 29\frac{4}{8} \\ -\,8\frac{1}{8} & = & 8\frac{1}{8} \\ \hline & & 21\frac{3}{8} \end{array}$$

The unknown length is $21\frac{3}{8}$ inches.

69. **(a)** $\frac{5}{9} = \frac{?}{54}$ $\quad 54 \div 9 = 6$

$\frac{5}{9} = \frac{5 \cdot 6}{9 \cdot 6} = \frac{30}{54}$

(b) $\frac{7}{12} = \frac{?}{48}$ $\quad 48 \div 12 = 4$

$\frac{7}{12} = \frac{7 \cdot 4}{12 \cdot 4} = \frac{28}{48}$

(c) $\frac{5}{8} = \frac{?}{40}$ $\quad 40 \div 8 = 5$

$\frac{5}{8} = \frac{5 \cdot 5}{8 \cdot 5} = \frac{25}{40}$

(d) $\frac{11}{5} = \frac{?}{120}$ $\quad 120 \div 5 = 24$

$\frac{11}{5} = \frac{11 \cdot 24}{5 \cdot 24} = \frac{264}{120}$

70. When rewriting unlike fractions as like fractions with the least common multiple as a denominator, the new denominator is called the *least common denominator*, or LCD.

71. **(a)** $\frac{5}{8} + \frac{1}{3} = \frac{15}{24} + \frac{8}{24} = \frac{15 + 8}{24} = \frac{23}{24}$

(b) $\frac{19}{20} - \frac{5}{12} = \frac{57}{60} - \frac{25}{60} = \frac{57 - 25}{60} = \frac{32}{60} = \frac{8}{15}$

(c)
$$\begin{array}{rcl} \dfrac{7}{12} & = & \dfrac{28}{48} \\ \dfrac{3}{16} & = & \dfrac{9}{48} \\ +\dfrac{3}{24} & = & \dfrac{6}{48} \\ \hline & & \dfrac{43}{48} \end{array}$$

(d)
$$\begin{array}{rcl} \dfrac{6}{7} & = & \dfrac{18}{21} \\ -\dfrac{2}{3} & = & \dfrac{14}{21} \\ \hline & & \dfrac{4}{21} \end{array}$$

72. A common method for adding or subtracting mixed numbers is to add or subtract the *fraction parts* and then the whole number parts.

73. Another method for adding or subtracting mixed numbers is to first change the mixed numbers to *improper* fractions. After adding or subtracting, write the answer as a mixed number in lowest terms.

74. (a) *First Method:*
$$\begin{array}{rcl} 4\dfrac{5}{8} & = & 4\dfrac{5}{8} \\ +3\dfrac{1}{4} & = & 3\dfrac{2}{8} \\ \hline & & 7\dfrac{7}{8} \end{array}$$

Second Method:
$$\begin{array}{rcccl} 4\dfrac{5}{8} & = & \dfrac{37}{8} & = & \dfrac{37}{8} \\ +3\dfrac{1}{4} & = & \dfrac{13}{4} & = & \dfrac{26}{8} \\ \hline & & & & \dfrac{63}{8} = 7\dfrac{7}{8} \end{array}$$

(b) *First Method:*
$$\begin{array}{rcccl} 12\dfrac{2}{5} & = & 12\dfrac{16}{40} & = & 11\dfrac{56}{40} \\ -8\dfrac{7}{8} & = & 8\dfrac{35}{40} & = & 8\dfrac{35}{40} \\ \hline & & & & 3\dfrac{21}{40} \end{array}$$

Second Method:
$$\begin{array}{rcccl} 12\dfrac{2}{5} & = & \dfrac{62}{5} & = & \dfrac{496}{40} \\ -8\dfrac{7}{8} & = & \dfrac{71}{8} & = & \dfrac{355}{40} \\ \hline & & & & \dfrac{141}{40} = 3\dfrac{21}{40} \end{array}$$

Both methods give the same answer. Preferences will vary. A sample answer follows: When a problem requires borrowing, it is easier to change all the numbers to improper fractions. Otherwise, adding the whole numbers and then adding the fractions seems easier.

3.5 Order Relations and the Order of Operations

3.5 Margin Exercises

1. (a)–(c)

(a) (b) (c)

0 ⅔ 1 1½ 2 2¾ 3

2. (a) 1 is to the left of $\frac{5}{4} = 1\frac{1}{4}$ on the number line, so 1 is less than $\frac{5}{4}$.

$$1 \boxed{<} \frac{5}{4}$$

(b) $\frac{8}{3} = 2\frac{2}{3}$ is to the right of $\frac{3}{2} = 1\frac{1}{2}$ on the number line, so $\frac{8}{3}$ is greater than $\frac{3}{2}$.

$$\frac{8}{3} \boxed{>} \frac{3}{2}$$

(c) 0 is to the left of 1 on the number line, so 0 is less than 1.

$$0 \boxed{<} 1$$

(d) $\frac{17}{8} = 2\frac{1}{8}$ is to the right of $\frac{8}{4} = 2$ on the number line, so $\frac{17}{8}$ is greater than $\frac{8}{4}$.

$$\frac{17}{8} \boxed{>} \frac{8}{4}$$

3. (a) $\frac{7}{8} \text{ — } \frac{3}{4}$ *LCD is 8*

$\frac{3}{4} = \frac{6}{8}$

$\frac{7}{8} > \frac{6}{8}$, so $\frac{7}{8} \boxed{>} \frac{3}{4}$.

(b) $\frac{13}{8} \text{ — } \frac{15}{9}$ *LCD is 72*

$\frac{13}{8} = \frac{117}{72}$ and $\frac{15}{9} = \frac{120}{72}$

$\frac{117}{72} < \frac{120}{72}$, so $\frac{13}{8} \boxed{<} \frac{15}{9}$.

(c) $\frac{9}{4} \text{ — } \frac{7}{3}$ *LCD is 12*

$\frac{9}{4} = \frac{27}{12}$ and $\frac{7}{3} = \frac{28}{12}$

$\frac{27}{12} < \frac{28}{12}$, so $\frac{9}{4} \boxed{<} \frac{7}{3}$.

(d) $\frac{9}{10} \text{ — } \frac{14}{15}$ *LCD is 30*

$\frac{9}{10} = \frac{27}{30}$ and $\frac{14}{15} = \frac{28}{30}$

$\frac{27}{30} < \frac{28}{30}$, so $\frac{9}{10} \boxed{<} \frac{14}{15}$.

4. (a) $\left(\frac{1}{2}\right)^4 = \frac{1}{2} \cdot \frac{1}{2} \cdot \frac{1}{2} \cdot \frac{1}{2} = \frac{1}{16}$

(b) $\left(\frac{3}{4}\right)^2 = \frac{3}{4}\cdot\frac{3}{4} = \frac{9}{16}$

(c) $\left(\frac{1}{2}\right)^3\cdot\left(\frac{2}{3}\right)^2 = \left(\frac{1}{2}\cdot\frac{1}{2}\cdot\frac{1}{2}\right)\cdot\left(\frac{2}{3}\cdot\frac{2}{3}\right)$

$= \frac{1\cdot 1\cdot 1\cdot \overset{1}{\cancel{2}}\cdot \overset{1}{\cancel{2}}}{2\cdot \underset{1}{\cancel{2}}\cdot \underset{1}{\cancel{2}}\cdot 3\cdot 3}$

$= \frac{1}{18}$

(d) $\left(\frac{1}{5}\right)^2\cdot\left(\frac{5}{3}\right)^2 = \left(\frac{1}{5}\cdot\frac{1}{5}\right)\cdot\left(\frac{5}{3}\cdot\frac{5}{3}\right)$

$= \frac{1\cdot 1\cdot \overset{1}{\cancel{5}}\cdot \overset{1}{\cancel{5}}}{\underset{1}{\cancel{5}}\cdot \underset{1}{\cancel{5}}\cdot 3\cdot 3}$

$= \frac{1}{9}$

5. (a) $\frac{5}{9} - \frac{3}{4}\left(\frac{2}{3}\right) = \frac{5}{9} - \frac{\overset{1}{\cancel{3}}}{\underset{2}{\cancel{4}}}\left(\frac{\overset{1}{\cancel{2}}}{\underset{1}{\cancel{3}}}\right)$

$= \frac{5}{9} - \frac{1}{2}$ *LCD is 18*

$= \frac{10}{18} - \frac{9}{18}$

$= \frac{1}{18}$

(b) $\frac{3}{4}\left(\frac{2}{3}\cdot\frac{3}{5}\right) = \frac{3}{4}\left(\frac{2}{\underset{1}{\cancel{3}}}\cdot\frac{\overset{1}{\cancel{3}}}{5}\right)$

$= \frac{3}{\underset{2}{\cancel{4}}}\cdot\frac{\overset{1}{\cancel{2}}}{5}$

$= \frac{3}{10}$

(c) $\frac{7}{8}\left(\frac{2}{3}\right) - \left(\frac{1}{2}\right)^2 = \frac{7}{\underset{4}{\cancel{8}}}\left(\frac{\overset{1}{\cancel{2}}}{3}\right) - \frac{1}{2}\cdot\frac{1}{2}$

$= \frac{7}{12} - \frac{1}{4}$ *LCD is 12*

$= \frac{7}{12} - \frac{3}{12}$

$= \frac{4}{12} = \frac{1}{3}$

(d) $\frac{\left(\frac{5}{6}\right)^2}{\frac{4}{3}} = \left(\frac{5}{6}\cdot\frac{5}{6}\right) \div \frac{4}{3}$

$= \frac{5\cdot 5}{6\cdot 6}\cdot\frac{3}{4}$

$= \frac{5\cdot 5\cdot \overset{1}{\cancel{3}}}{6\cdot \underset{2}{\cancel{6}}\cdot 4}$

$= \frac{25}{48}$

3.5 Section Exercises

1.–12.

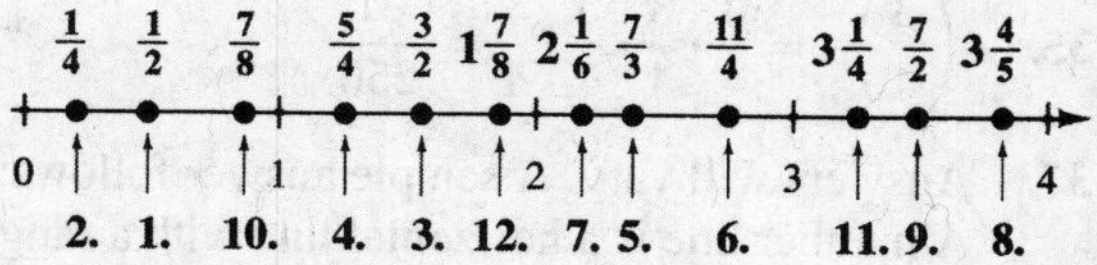

13. $\frac{1}{2}__\frac{3}{8}$ *LCD is 8*

$\frac{1}{2} = \frac{4}{8}$

$\frac{4}{8} > \frac{3}{8}$, so $\frac{1}{2}\ \boxed{>}\ \frac{3}{8}$.

15. $\frac{5}{6}__\frac{11}{12}$ *LCD is 12*

$\frac{5}{6} = \frac{10}{12}$

$\frac{10}{12} < \frac{11}{12}$, so $\frac{5}{6}\ \boxed{<}\ \frac{11}{12}$.

17. $\frac{5}{12}__\frac{3}{8}$ *LCD is 24*

$\frac{5}{12} = \frac{10}{24}$ and $\frac{3}{8} = \frac{9}{24}$

$\frac{10}{24} > \frac{9}{24}$, so $\frac{5}{12}\ \boxed{>}\ \frac{3}{8}$.

19. $\frac{7}{12}__\frac{11}{18}$ *LCD is 36*

$\frac{7}{12} = \frac{21}{36}$ and $\frac{11}{18} = \frac{22}{36}$

$\frac{21}{36} < \frac{22}{36}$, so $\frac{7}{12}\ \boxed{<}\ \frac{11}{18}$.

21. $\frac{11}{18}__\frac{5}{9}$ *LCD is 18*

$\frac{5}{9} = \frac{10}{18}$

$\frac{11}{18} > \frac{10}{18}$, so $\frac{11}{18}\ \boxed{>}\ \frac{5}{9}$.

23. $\frac{37}{50}__\frac{13}{20}$ *LCD is 100*

$\frac{37}{50} = \frac{74}{100}$ and $\frac{13}{20} = \frac{65}{100}$

$\frac{74}{100} > \frac{65}{100}$, so $\frac{37}{50}\ \boxed{>}\ \frac{13}{20}$.

25. $\left(\frac{1}{3}\right)^2 = \frac{1}{3}\cdot\frac{1}{3} = \frac{1\cdot 1}{3\cdot 3} = \frac{1}{9}$

27. $\left(\frac{5}{8}\right)^2 = \frac{5}{8}\cdot\frac{5}{8} = \frac{5\cdot 5}{8\cdot 8} = \frac{25}{64}$

29. $\left(\frac{3}{4}\right)^2 = \frac{3}{4}\cdot\frac{3}{4} = \frac{9}{16}$

31. $\left(\frac{4}{5}\right)^3 = \frac{4}{5}\cdot\frac{4}{5}\cdot\frac{4}{5} = \frac{64}{125}$

33. $\left(\frac{3}{2}\right)^4 = \frac{3}{2}\cdot\frac{3}{2}\cdot\frac{3}{2}\cdot\frac{3}{2} = \frac{81}{16} = 5\frac{1}{16}$

35. $\left(\frac{3}{4}\right)^4 = \frac{3}{4}\cdot\frac{3}{4}\cdot\frac{3}{4}\cdot\frac{3}{4} = \frac{81}{256}$

37. Answers will vary. A sample answer follows: A number line is a horizontal line with a range of numbers placed on it. The lowest number is on the left and the highest number is on the right. It can be used to compare the size or value of numbers.

$0 \quad \frac{1}{2} \quad 1 \quad 1\frac{1}{2} \quad 2 \quad 2\frac{1}{2} \quad 3$

39.
$$\begin{aligned} 2^4 - 4(3) &= 16 - 4(3) \\ &= 16 - 12 \\ &= 4 \end{aligned}$$

41.
$$\begin{aligned} 3\cdot 2^2 - \frac{6}{3} &= 3\cdot 4 - \frac{6}{3} \\ &= 12 - \frac{6}{3} \\ &= 12 - 2 \\ &= 10 \end{aligned}$$

43.
$$\begin{aligned} \left(\frac{1}{2}\right)^2\cdot 4 &= \frac{1}{2}\cdot\frac{1}{2}\cdot 4 \\ &= \frac{1}{4}\cdot 4 \\ &= \frac{1}{4}\cdot\frac{4}{1} = 1 \end{aligned}$$

45.
$$\begin{aligned} \left(\frac{3}{4}\right)^2\cdot\left(\frac{1}{3}\right) &= \frac{3}{4}\cdot\frac{3}{4}\cdot\frac{1}{3} \\ &= \frac{3\cdot \overset{1}{\cancel{3}}\cdot 1}{4\cdot 4\cdot \underset{1}{\cancel{3}}} \\ &= \frac{3}{16} \end{aligned}$$

47.
$$\begin{aligned} \left(\frac{4}{5}\right)^2\cdot\left(\frac{5}{6}\right)^2 &= \left(\frac{4}{5}\cdot\frac{4}{5}\right)\cdot\left(\frac{5}{6}\cdot\frac{5}{6}\right) \\ &= \frac{\overset{2}{\cancel{4}}\cdot\overset{2}{\cancel{4}}\cdot\overset{1}{\cancel{5}}\cdot\overset{1}{\cancel{5}}}{\underset{1}{\cancel{5}}\cdot\underset{1}{\cancel{5}}\cdot\underset{3}{\cancel{6}}\cdot\underset{3}{\cancel{6}}} \\ &= \frac{4}{9} \end{aligned}$$

49.
$$\begin{aligned} 6\left(\frac{2}{3}\right)^2\left(\frac{1}{2}\right)^3 &= 6\left(\frac{2}{3}\cdot\frac{2}{3}\right)\left(\frac{1}{2}\cdot\frac{1}{2}\cdot\frac{1}{2}\right) \\ &= \frac{\overset{\overset{1}{\cancel{2}}}{\cancel{6}}\cdot\overset{1}{\cancel{2}}\cdot\overset{1}{\cancel{2}}\cdot 1\cdot 1\cdot 1}{\underset{1}{\cancel{3}}\cdot 3\cdot\underset{1}{\cancel{2}}\cdot\underset{1}{\cancel{2}}\cdot\underset{1}{\cancel{2}}} \\ &= \frac{1}{3} \end{aligned}$$

51.
$$\begin{aligned} \frac{3}{5}\left(\frac{1}{3}\right) + \frac{2}{5}\left(\frac{3}{4}\right) &= \frac{\overset{1}{\cancel{3}}}{5}\left(\frac{1}{\underset{1}{\cancel{3}}}\right) + \frac{\overset{1}{\cancel{2}}}{5}\left(\frac{3}{\underset{2}{\cancel{4}}}\right) \\ &= \frac{1}{5} + \frac{3}{10} \qquad \textit{LCD is 10} \\ &= \frac{2}{10} + \frac{3}{10} \\ &= \frac{5}{10} = \frac{1}{2} \end{aligned}$$

53.
$$\begin{aligned} \frac{1}{2} + \left(\frac{1}{2}\right)^2 - \frac{3}{8} &= \frac{1}{2} + \frac{1}{2}\cdot\frac{1}{2} - \frac{3}{8} \\ &= \frac{2}{4} + \frac{1}{4} - \frac{3}{8} \\ &= \frac{3}{4} - \frac{3}{8} \qquad \textit{LCD is 8} \\ &= \frac{6}{8} - \frac{3}{8} = \frac{3}{8} \end{aligned}$$

55.
$$\begin{aligned} \left(\frac{1}{3} + \frac{1}{6}\right)\cdot\frac{1}{2} &= \left(\frac{2}{6} + \frac{1}{6}\right)\cdot\frac{1}{2} \\ &= \frac{3}{6}\cdot\frac{1}{2} \\ &= \frac{\overset{1}{\cancel{3}}\cdot 1}{\underset{2}{\cancel{6}}\cdot 2} \\ &= \frac{1}{4} \end{aligned}$$

57.
$$\begin{aligned} \frac{9}{8} \div \left(\frac{2}{3} + \frac{1}{12}\right) &= \frac{9}{8} \div \left(\frac{8}{12} + \frac{1}{12}\right) = \frac{9}{8} \div \frac{9}{12} \\ &= \frac{\overset{1}{\cancel{9}}}{\underset{2}{\cancel{8}}}\cdot\frac{\overset{3}{\cancel{12}}}{\underset{1}{\cancel{9}}} \\ &= \frac{3}{2} = 1\frac{1}{2} \end{aligned}$$

59.
$$\begin{aligned} \left(\frac{7}{8} - \frac{3}{4}\right) \div \frac{3}{2} &= \left(\frac{7}{8} - \frac{6}{8}\right) \div \frac{3}{2} \\ &= \frac{1}{8} \div \frac{3}{2} \\ &= \frac{1}{\underset{4}{\cancel{8}}}\cdot\frac{\overset{1}{\cancel{2}}}{3} \\ &= \frac{1}{12} \end{aligned}$$

61.
$$\begin{aligned} \frac{3}{8}\left(\frac{1}{4} + \frac{1}{2}\right)\cdot\frac{32}{3} &= \frac{3}{8}\left(\frac{1}{4} + \frac{2}{4}\right)\cdot\frac{32}{3} \\ &= \frac{3}{\underset{1}{\cancel{8}}}\cdot\frac{\overset{1}{\cancel{3}}}{\underset{1}{\cancel{4}}}\cdot\frac{\overset{\overset{1}{\cancel{8}}}{\cancel{32}}}{\underset{1}{\cancel{3}}} \\ &= 3 \end{aligned}$$

63.
$$\begin{aligned}\left(\frac{3}{4}\right)^2-\left(\frac{1}{2}-\frac{1}{6}\right)\div\frac{4}{3}&=\left(\frac{3}{4}\right)^2-\left(\frac{3}{6}-\frac{1}{6}\right)\div\frac{4}{3}\\&=\frac{3}{4}\cdot\frac{3}{4}-\left(\frac{2}{6}\right)\cdot\frac{3}{4}\\&=\frac{9}{16}-\frac{\cancel{2}^{1}}{\cancel{6}_{2}}\cdot\frac{\cancel{3}^{1}}{\cancel{4}_{2}}\\&=\frac{9}{16}-\frac{1}{4}\qquad \textit{LCD is 16}\\&=\frac{9}{16}-\frac{4}{16}\\&=\frac{5}{16}\end{aligned}$$

65.
$$\begin{aligned}\left(\frac{7}{8}-\frac{1}{4}\right)-\frac{2}{3}\left(\frac{3}{4}\right)^2&=\left(\frac{7}{8}-\frac{2}{8}\right)-\frac{2}{3}\left(\frac{3}{4}\right)^2\\&=\frac{5}{8}-\frac{\cancel{2}^{1}}{\cancel{3}_{1}}\cdot\frac{\cancel{9}^{3}}{\cancel{16}_{8}}\\&=\frac{5}{8}-\frac{3}{8}\\&=\frac{2}{8}=\frac{1}{4}\end{aligned}$$

67.
$$\begin{aligned}&\left(\frac{3}{4}\right)^2\left(\frac{2}{3}-\frac{5}{9}\right)-\frac{1}{4}\left(\frac{1}{8}\right)\\&=\left(\frac{3}{4}\right)^2\left(\frac{6}{9}-\frac{5}{9}\right)-\frac{1}{4}\left(\frac{1}{8}\right)\\&=\left(\frac{3}{4}\right)^2\cdot\frac{1}{9}-\frac{1}{4}\cdot\frac{1}{8}\\&=\frac{\cancel{3}^{1}}{4}\cdot\frac{\cancel{3}^{1}}{4}\cdot\frac{1}{\cancel{9}_{\cancel{3}_{1}}}-\frac{1}{4}\cdot\frac{1}{8}\\&=\frac{1}{16}-\frac{1}{32}\\&=\frac{2}{32}-\frac{1}{32}=\frac{1}{32}\end{aligned}$$

69. $\frac{9}{25}$ — $\frac{7}{20}$ *LCD is 100*

$\frac{9}{25}=\frac{36}{100}$ and $\frac{7}{20}=\frac{35}{100}$

$\frac{36}{100}>\frac{35}{100}$, so $\frac{9}{25}>\frac{7}{20}$.

$\frac{9}{25}$ of the employees is greater.

The group that thought that more lighting on the grounds and parking lot was most important was greater.

71. When comparing the size of two numbers, we use the symbol $<$ for **is less than** and the symbol $>$ for **is greater than**.

72. **(a)** To identify the greater of two or more fractions, we must first write the fractions as *like* fractions and then compare the *numerators.* The fraction with the greater *numerator* is the greater fraction.

(b) Answers will vary. A sample answer follows: $\frac{1}{4}<\frac{1}{2}, \frac{1}{10}>\frac{1}{16}, \frac{3}{8}<\frac{3}{2}, \frac{5}{6}>\frac{5}{7}$

73.
1. Do all operations inside *parentheses* or other grouping symbols.
2. Simplify any expressions with *exponents* and find any *square* roots.
3. *Multiply* or *divide* proceeding from left to right.
4. *Add* or *subtract* proceeding from left to right.

74.
$$\begin{aligned}&\left(\frac{2}{3}\right)^2-\left(\frac{4}{5}-\frac{3}{10}\right)\div\frac{5}{4}\\&=\left(\frac{2}{3}\right)^2-\left(\frac{8}{10}-\frac{3}{10}\right)\div\frac{5}{4}\\&=\left(\frac{2}{3}\right)^2-\frac{5}{10}\div\frac{5}{4}\\&=\frac{4}{9}-\frac{5}{10}\cdot\frac{4}{5}\\&=\frac{4}{9}-\frac{\cancel{5}^{1}}{\cancel{10}_{5}}\cdot\frac{\cancel{4}^{2}}{\cancel{5}_{1}}\\&=\frac{4}{9}-\frac{2}{5}\qquad \textit{LCD is 45}\\&=\frac{20}{45}-\frac{18}{45}\\&=\frac{2}{45}\end{aligned}$$

For Exercises 75–80, see the number line following Exercise 80.

75. $\left(\frac{2}{3}\right)^2=\frac{2}{3}\cdot\frac{2}{3}=\frac{4}{9}$

76. $\left(\frac{3}{2}\right)^2=\frac{3}{2}\cdot\frac{3}{2}=\frac{9}{4}=2\frac{1}{4}$

77. $\left(\frac{3}{5}\right)^3=\frac{3}{5}\cdot\frac{3}{5}\cdot\frac{3}{5}=\frac{27}{125}$

78. $\left(\frac{5}{4}\right)^2=\frac{5}{4}\cdot\frac{5}{4}=\frac{25}{16}=1\frac{9}{16}$

79.
$$\begin{aligned}4+2-2^2&=4+2-4\\&=6-4\\&=2\end{aligned}$$

80. $\left(\frac{5}{8}\right)^2 + \left(\frac{7}{8} - \frac{1}{4}\right) \div \frac{1}{4} = \left(\frac{5}{8}\right)^2 + \left(\frac{7}{8} - \frac{2}{8}\right) \div \frac{1}{4}$

$$= \left(\frac{5}{8}\right)^2 + \frac{5}{8} \div \frac{1}{4}$$
$$= \frac{25}{64} + \frac{5}{8} \cdot \frac{4}{1}$$
$$= \frac{25}{64} + \frac{5}{\cancel{8}_{2}} \cdot \frac{\cancel{4}^{1}}{1}$$
$$= \frac{25}{64} + \frac{5}{2}$$
$$= \frac{25}{64} + 2 + \frac{1}{2}$$
$$= 2 + \frac{25}{64} + \frac{32}{64}$$
$$= 2\frac{57}{64}$$

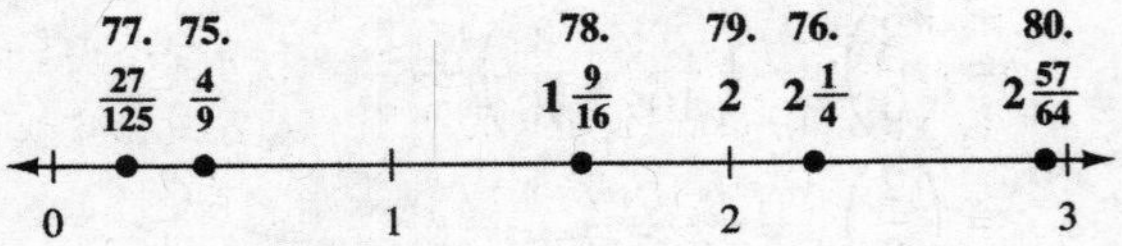

Summary Exercises on Fractions

1. $\frac{3}{4}$ is a proper fraction since the numerator is less than the denominator.

3. $\frac{10}{10}$ is an improper fraction since the numerator is equal to the denominator.

5. $\frac{30}{36} = \frac{30 \div 6}{36 \div 6} = \frac{5}{6}$

7. $\frac{15}{35} = \frac{15 \div 5}{35 \div 5} = \frac{3}{7}$

9. $\frac{3}{4} \cdot \frac{2}{3} = \frac{\cancel{3}^{1} \cdot \cancel{2}^{1}}{\cancel{4}_{2} \cdot \cancel{3}_{1}} = \frac{1 \cdot 1}{2 \cdot 1} = \frac{1}{2}$

11. $56 \cdot \frac{5}{8} = \frac{\cancel{56}^{7}}{1} \cdot \frac{5}{\cancel{8}_{1}} = \frac{7 \cdot 5}{1 \cdot 1} = \frac{35}{1} = 35$

13. $\frac{35}{45} \div \frac{10}{15} = \frac{35}{45} \cdot \frac{15}{10}$

$$= \frac{\cancel{35}^{7}}{\cancel{45}_{3}} \cdot \frac{\cancel{15}^{1}}{\cancel{10}_{2}}$$
$$= \frac{7 \cdot 1}{3 \cdot 2} = \frac{7}{6} = 1\frac{1}{6}$$

15. $\frac{7}{8} + \frac{2}{3} = \frac{7 \cdot 3}{8 \cdot 3} + \frac{2 \cdot 8}{3 \cdot 8} = \frac{21}{24} + \frac{16}{24}$

$$= \frac{21 + 16}{24} = \frac{37}{24} = 1\frac{13}{24}$$

17. $\frac{7}{12} + \frac{5}{6} + \frac{2}{3} = \frac{7}{12} + \frac{10}{12} + \frac{8}{12} = \frac{7 + 10 + 8}{12}$

$$= \frac{25}{12} = 2\frac{1}{12}$$

19. $\frac{7}{8} - \frac{5}{12} = \frac{21}{24} - \frac{10}{24} = \frac{21 - 10}{24} = \frac{11}{24}$

21. $3\frac{1}{2} \cdot 2\frac{1}{4}$

Estimate: $4 \cdot 2 = 8$

Exact: $3\frac{1}{2} \cdot 2\frac{1}{4} = \frac{7}{2} \cdot \frac{9}{4} = \frac{63}{8} = 7\frac{7}{8}$

23. $8 \cdot 5\frac{2}{3} \cdot 2\frac{3}{8}$

Estimate: $8 \cdot 6 \cdot 2 = 48 \cdot 2 = 96$

Exact: $8 \cdot 5\frac{2}{3} \cdot 2\frac{3}{8} = \frac{\cancel{8}^{1}}{1} \cdot \frac{17}{3} \cdot \frac{19}{\cancel{8}_{1}} = \frac{323}{3} = 107\frac{2}{3}$

25. $6\frac{7}{8} \div 2$

Estimate: $7 \div 2 = 3\frac{1}{2}$

Exact: $6\frac{7}{8} \div 2 = \frac{55}{8} \div \frac{2}{1}$

$$= \frac{55}{8} \cdot \frac{1}{2}$$
$$= \frac{55}{16} = 3\frac{7}{16}$$

27. *Estimate:* *Exact:*

$$\begin{array}{r} 6 \\ +\,4 \\ \hline 10 \end{array} \qquad \begin{array}{rcl} 5\frac{2}{3} & = & 5\frac{8}{12} \\ +\,4\frac{1}{4} & = & 4\frac{3}{12} \\ \hline & & 9\frac{11}{12} \end{array}$$

29. *Estimate:* *Exact:*

$$\begin{array}{r} 15 \\ +\,11 \\ \hline 26 \end{array} \qquad \begin{array}{rcl} 14\frac{3}{5} & = & 14\frac{9}{15} \\ +\,10\frac{2}{3} & = & 10\frac{10}{15} \\ \hline & & 24\frac{19}{15} \end{array}$$

$$24\frac{19}{15} = 24 + 1\frac{4}{15} = 25\frac{4}{15}$$

31. *Estimate:* *Exact:*

$$\begin{array}{r} 14 \\ -\,7 \\ \hline 7 \end{array} \qquad \begin{array}{rcl} 14 & = & 13\frac{8}{8} \\ -\,7\frac{3}{8} & = & 7\frac{3}{8} \\ \hline & & 6\frac{5}{8} \end{array}$$

33. $\frac{1}{5}\left(\frac{2}{3}-\frac{1}{4}\right)=\frac{1}{5}\left(\frac{8}{12}-\frac{3}{12}\right)$

$=\frac{1}{5}\cdot\frac{5}{12}=\frac{1\cdot\overset{1}{\cancel{5}}}{\underset{1}{\cancel{5}}\cdot 12}=\frac{1}{12}$

35. $\frac{2}{3}+\left(\frac{2}{3}\right)^2-\frac{5}{6}=\frac{2}{3}+\frac{4}{9}-\frac{5}{6}$

$=\frac{12}{18}+\frac{8}{18}-\frac{15}{18}$

$=\frac{20}{18}-\frac{15}{18}$

$=\frac{5}{18}$

37. 9, 18, 24

$9=3\cdot 3$
$18=2\cdot 3\cdot 3$
$24=2\cdot 2\cdot 2\cdot 3$

$\text{LCM}=2\cdot 2\cdot 2\cdot 3\cdot 3=72$

The LCM of 9, 18, and 24 is 72.

39. $\frac{5}{6}=\frac{?}{42}\quad 42\div 6=7$

$\frac{5}{6}=\frac{5\cdot 7}{6\cdot 7}=\frac{35}{42}$

41. $\frac{11}{12}=\frac{?}{60}\quad 60\div 12=5$

$\frac{11}{12}=\frac{11\cdot 5}{12\cdot 5}=\frac{55}{60}$

43. $\frac{16}{20}$ — $\frac{23}{30}$ *LCD is 60*

$\frac{16}{20}=\frac{48}{60}$ and $\frac{23}{30}=\frac{46}{60}$

$\frac{48}{60}>\frac{46}{60}$, so $\frac{16}{20}\ \boxed{>}\ \frac{23}{30}$.

Chapter 3 Review Exercises

1. $\frac{5}{7}+\frac{1}{7}=\frac{5+1}{7}=\frac{6}{7}$

2. $\frac{4}{9}+\frac{3}{9}=\frac{4+3}{9}=\frac{7}{9}$

3. $\frac{1}{8}+\frac{3}{8}+\frac{2}{8}=\frac{1+3+2}{8}=\frac{6}{8}=\frac{3}{4}$

4. $\frac{5}{16}-\frac{3}{16}=\frac{5-3}{16}=\frac{2}{16}=\frac{1}{8}$

5. $\frac{5}{10}+\frac{3}{10}=\frac{5+3}{10}=\frac{8}{10}=\frac{4}{5}$

6. $\frac{5}{12}-\frac{3}{12}=\frac{5-3}{12}=\frac{2}{12}=\frac{1}{6}$

7. $\frac{36}{62}-\frac{10}{62}=\frac{36-10}{62}=\frac{26}{62}=\frac{13}{31}$

8. $\frac{68}{75}-\frac{43}{75}=\frac{68-43}{75}=\frac{25}{75}=\frac{1}{3}$

9. Add to find what fraction of her patients she screened in two hours.

$\frac{7}{16}+\frac{5}{16}=\frac{7+5}{16}=\frac{12}{16}=\frac{3}{4}$

She screened $\frac{3}{4}$ of her patients in two hours.

10. Subtract to find the answer.

$\frac{5}{8}-\frac{3}{8}=\frac{5-3}{8}=\frac{2}{8}=\frac{1}{4}$

They completed $\frac{1}{4}$ Web page less in the afternoon.

11. 5, 2

Multiples of 5:

$5, \underline{10}, 15, 20, 25, 30, \ldots$

10 is the first number divisible by 2. $(10\div 2=5)$ The least common multiple of 5 and 2 is 10.

12. 3, 4

Multiples of 4:

$4, 8, \underline{12}, 16, 20, 24, \ldots$

12 is the first number divisible by 3. $(12\div 3=4)$ The least common multiple of 3 and 4 is 12.

13.

$$\begin{array}{r|ccc} 2 & 10 & 12 & 20 \\ 2 & \cancel{5} & 6 & 10 \\ 3 & \cancel{5} & 3 & \cancel{5} \\ 5 & 5 & \cancel{1} & 5 \\ \hline & 1 & 1 & 1 \end{array}$$

$\text{LCM}=2\cdot 2\cdot 3\cdot 5=60$

14.

$$\begin{array}{r|ccc} 2 & \cancel{3} & 8 & 4 \\ 2 & \cancel{3} & 4 & 2 \\ 2 & \cancel{3} & 2 & \cancel{1} \\ 3 & 3 & \cancel{1} & \cancel{1} \\ \hline & 1 & 1 & 1 \end{array}$$

$\text{LCM}=2\cdot 2\cdot 2\cdot 3=24$

15. 6, 8, 5, 15

$6=2\cdot 3$
$8=2\cdot 2\cdot 2$
$5=1\cdot 5$
$15=3\cdot 5$

$\text{LCM}=2\cdot 2\cdot 2\cdot 3\cdot 5=120$

The LCM of 6, 8, 5, and 15 is 120.

16.

$$\begin{array}{r|rrr} 2 & \cancel{15} & \cancel{9} & 20 \\ 2 & \cancel{15} & \cancel{9} & 10 \\ 3 & 15 & 9 & \cancel{5} \\ 3 & \cancel{5} & 3 & \cancel{5} \\ 5 & 5 & \cancel{1} & 5 \\ \hline & 1 & 1 & 1 \end{array}$$

$\text{LCM} = 2 \cdot 2 \cdot 3 \cdot 3 \cdot 5 = 180$

17. $\frac{2}{3} = \frac{?}{12} \quad 12 \div 3 = 4$

$\frac{2}{3} = \frac{2 \cdot 4}{3 \cdot 4} = \frac{8}{12}$

18. $\frac{3}{8} = \frac{?}{56} \quad 56 \div 8 = 7$

$\frac{3}{8} = \frac{3 \cdot 7}{8 \cdot 7} = \frac{21}{56}$

19. $\frac{2}{5} = \frac{?}{25} \quad 25 \div 5 = 5$

$\frac{2}{5} = \frac{2 \cdot 5}{5 \cdot 5} = \frac{10}{25}$

20. $\frac{5}{9} = \frac{?}{81} \quad 81 \div 9 = 9$

$\frac{5}{9} = \frac{5 \cdot 9}{9 \cdot 9} = \frac{45}{81}$

21. $\frac{4}{5} = \frac{?}{40} \quad 40 \div 5 = 8$

$\frac{4}{5} = \frac{4 \cdot 8}{5 \cdot 8} = \frac{32}{40}$

22. $\frac{5}{16} = \frac{?}{64} \quad 64 \div 16 = 4$

$\frac{5}{16} = \frac{5 \cdot 4}{16 \cdot 4} = \frac{20}{64}$

23.

$$\begin{aligned} \frac{1}{2} + \frac{1}{3} &= \frac{3}{6} + \frac{2}{6} \\ &= \frac{3+2}{6} \\ &= \frac{5}{6} \end{aligned}$$

24.

$$\begin{aligned} \frac{1}{5} + \frac{3}{10} + \frac{3}{8} &= \frac{8}{40} + \frac{12}{40} + \frac{15}{40} \\ &= \frac{8+12+15}{40} \\ &= \frac{35}{40} = \frac{7}{8} \end{aligned}$$

25.

$$\begin{array}{rcl} \frac{5}{12} &=& \frac{10}{24} \\ +\frac{5}{24} &=& \frac{5}{24} \\ \hline && \frac{15}{24} = \frac{5}{8} \end{array}$$

26. $\frac{2}{3} - \frac{1}{4} = \frac{8}{12} - \frac{3}{12} = \frac{5}{12}$

27.

$$\begin{array}{rcl} \frac{7}{8} &=& \frac{21}{24} \\ -\frac{1}{3} &=& \frac{8}{24} \\ \hline && \frac{13}{24} \end{array}$$

28.

$$\begin{array}{rcl} \frac{11}{12} &=& \frac{33}{36} \\ -\frac{4}{9} &=& \frac{16}{36} \\ \hline && \frac{17}{36} \end{array}$$

29. Add the fractional amounts of fertilizer.

$$\begin{array}{rcl} \frac{3}{8} &=& \frac{9}{24} \\ \frac{1}{4} &=& \frac{6}{24} \\ +\frac{1}{3} &=& \frac{8}{24} \\ \hline && \frac{23}{24} \end{array}$$

He used $\frac{23}{24}$ of the sack of fertilizer.

30. Add the fractional amounts of her budget.

$$\begin{array}{rcl} \frac{1}{4} &=& \frac{15}{60} \\ \frac{1}{3} &=& \frac{20}{60} \\ +\frac{2}{5} &=& \frac{24}{60} \\ \hline && \frac{59}{60} \end{array}$$

The will be spending $\frac{59}{60}$ of the budget on these three categories.

31. *Estimate:* *Exact:*

$$\begin{array}{r} 19 \\ +14 \\ \hline 33 \end{array} \qquad \begin{array}{rcl} 18\frac{5}{8} &=& 18\frac{5}{8} \\ +13\frac{3}{4} &=& 13\frac{6}{8} \\ \hline && 31\frac{11}{8} = 31 + 1\frac{3}{8} = 32\frac{3}{8} \end{array}$$

32. *Estimate:* *Exact:*

$$\begin{array}{r} 23 \\ +15 \\ \hline 38 \end{array} \qquad \begin{array}{rcl} 22\frac{2}{3} &=& 22\frac{6}{9} \\ +15\frac{4}{9} &=& 15\frac{4}{9} \\ \hline && 37\frac{10}{9} = 37 + 1\frac{1}{9} = 38\frac{1}{9} \end{array}$$

33. *Estimate:*

$$\begin{array}{r} 13 \\ 9 \\ +10 \\ \hline 32 \end{array}$$

Exact:

$$\begin{array}{rcl} 12\frac{3}{5} & = & 12\frac{48}{80} \\ 8\frac{5}{8} & = & 8\frac{50}{80} \\ +10\frac{5}{16} & = & 10\frac{25}{80} \\ \hline & & 30\frac{123}{80} \end{array}$$

$$30\frac{123}{80} = 30 + 1\frac{43}{80} = 31\frac{43}{80}$$

34. *Estimate:*

$$\begin{array}{r} 32 \\ -15 \\ \hline 17 \end{array}$$

Exact:

$$\begin{array}{rcl} 31\frac{3}{4} & = & 31\frac{9}{12} \\ -14\frac{2}{3} & = & 14\frac{8}{12} \\ \hline & & 17\frac{1}{12} \end{array}$$

35. *Estimate:*

$$\begin{array}{r} 34 \\ -16 \\ \hline 18 \end{array}$$

Exact:

$$\begin{array}{rcl} 34 & = & 33\frac{3}{3} \\ -15\frac{2}{3} & = & 15\frac{2}{3} \\ \hline & & 18\frac{1}{3} \end{array}$$

36. *Estimate:*

$$\begin{array}{r} 215 \\ -136 \\ \hline 79 \end{array}$$

Exact:

$$\begin{array}{r} 215\frac{7}{16} \\ -136 \\ \hline 79\frac{7}{16} \end{array}$$

37.

$$\begin{array}{rcccl} 5\frac{2}{5} & = & \frac{27}{5} & = & \frac{54}{10} \\ +3\frac{7}{10} & = & \frac{37}{10} & = & \frac{37}{10} \\ \hline & & & & \frac{91}{10} = 9\frac{1}{10} \end{array}$$

38.

$$\begin{array}{rcccl} 4\frac{3}{4} & = & \frac{19}{4} & = & \frac{57}{12} \\ +5\frac{2}{3} & = & \frac{17}{3} & = & \frac{68}{12} \\ \hline & & & & \frac{125}{12} = 10\frac{5}{12} \end{array}$$

39.

$$\begin{array}{rcccl} 5 & = & \frac{5}{1} & = & \frac{20}{4} \\ -1\frac{3}{4} & = & \frac{7}{4} & = & \frac{7}{4} \\ \hline & & & & \frac{13}{4} = 3\frac{1}{4} \end{array}$$

40.

$$\begin{array}{rcccl} 6\frac{1}{2} & = & \frac{13}{2} & = & \frac{39}{6} \\ -4\frac{5}{6} & = & \frac{29}{6} & = & \frac{29}{6} \\ \hline & & & & \frac{10}{6} = 1\frac{4}{6} = 1\frac{2}{3} \end{array}$$

41.

$$\begin{array}{rcccl} 8\frac{1}{3} & = & \frac{25}{3} & = & \frac{50}{6} \\ -2\frac{5}{6} & = & \frac{17}{6} & = & \frac{17}{6} \\ \hline & & & & \frac{33}{6} = 5\frac{3}{6} = 5\frac{1}{2} \end{array}$$

42.

$$\begin{array}{rcccl} 5\frac{5}{12} & = & \frac{65}{12} & = & \frac{130}{24} \\ -2\frac{5}{8} & = & \frac{21}{8} & = & \frac{63}{24} \\ \hline & & & & \frac{67}{24} = 2\frac{19}{24} \end{array}$$

43. Add to find the total gallons used. Then subtract the gallons used from the total gallons to find the amount remaining.

Estimate: $15 - 6 - 7 = 2$ gallons

Exact:

$$\begin{array}{rcl} 5\frac{1}{2} & = & 5\frac{2}{4} \\ +6\frac{3}{4} & = & 6\frac{3}{4} \\ \hline & & 11\frac{5}{4} = 11 + 1\frac{1}{4} = 12\frac{1}{4} \end{array}$$

$$\begin{array}{rcl} 14\frac{2}{3} & = & 14\frac{8}{12} \\ -12\frac{1}{4} & = & 12\frac{3}{12} \\ \hline & & 2\frac{5}{12} \end{array}$$

The lab has $2\frac{5}{12}$ gallons remaining.

44. Add to find the total weight.

Estimate: $29 + 25 = 54$ tons

Exact:

$$\begin{array}{rcl} 28\frac{2}{3} & = & 28\frac{8}{12} \\ +24\frac{3}{4} & = & 24\frac{9}{12} \\ \hline & & 52\frac{17}{12} = 52 + 1\frac{5}{12} = 53\frac{5}{12} \end{array}$$

The total weight was $53\frac{5}{12}$ tons.

45. Add to find the total weight.

Estimate: $9+9+7=25$ pounds

Exact:
$$\begin{aligned} 8\tfrac{7}{8} &= 8\tfrac{21}{24} \\ 9\tfrac{1}{3} &= 9\tfrac{8}{24} \\ +6\tfrac{3}{4} &= 6\tfrac{18}{24} \\ \hline &\ 23\tfrac{47}{24} = 23+1\tfrac{23}{24} = 24\tfrac{23}{24} \end{aligned}$$

The total weight was $24\frac{23}{24}$ pounds.

46. To find the total amount of land that she already has, add the two parcels.

To find how much more land that she needs, subtract the total amount she has from the total amount she needs.

Estimate: $9-(2+3)=4$ acres

Exact:
$$\begin{aligned} 1\tfrac{11}{16} &= 1\tfrac{11}{16} \\ +2\tfrac{3}{4} &= 2\tfrac{12}{16} \\ \hline &\ 3\tfrac{23}{16} = 3+1\tfrac{7}{16} = 4\tfrac{7}{16} \end{aligned}$$

$$\begin{aligned} 8\tfrac{1}{2} &= 8\tfrac{8}{16} \\ -4\tfrac{7}{16} &= 4\tfrac{7}{16} \\ \hline &\ 4\tfrac{1}{16} \end{aligned}$$

She needs to buy an additional $4\frac{1}{16}$ acres.

47.–50.

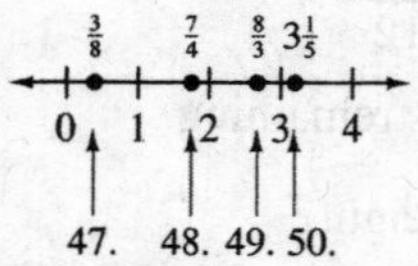

51. $\frac{2}{3}$ — $\frac{3}{4}$ *LCD is 12*

$\frac{2}{3}=\frac{8}{12}$ and $\frac{3}{4}=\frac{9}{12}$

$\frac{8}{12}<\frac{9}{12}$, so $\frac{2}{3} \boxed{<} \frac{3}{4}$.

52. $\frac{3}{4}$ — $\frac{7}{8}$ *LCD is 8*

$\frac{3}{4}=\frac{6}{8}$

$\frac{6}{8}<\frac{7}{8}$, so $\frac{3}{4} \boxed{<} \frac{7}{8}$.

53. $\frac{1}{2}$ — $\frac{7}{15}$ *LCD is 30*

$\frac{1}{2}=\frac{15}{30}$ and $\frac{7}{15}=\frac{14}{30}$

$\frac{15}{30}>\frac{14}{30}$, so $\frac{1}{2} \boxed{>} \frac{7}{15}$.

54. $\frac{7}{10}$ — $\frac{8}{15}$ *LCD is 30*

$\frac{7}{10}=\frac{21}{30}$ and $\frac{8}{15}=\frac{16}{30}$

$\frac{21}{30}>\frac{16}{30}$, so $\frac{7}{10} \boxed{>} \frac{8}{15}$.

55. $\frac{9}{16}$ — $\frac{5}{8}$ *LCD is 16*

$\frac{5}{8}=\frac{10}{16}$

$\frac{9}{16}<\frac{10}{16}$, so $\frac{9}{16} \boxed{<} \frac{5}{8}$.

56. $\frac{7}{20}$ — $\frac{8}{25}$ *LCD is 100*

$\frac{7}{20}=\frac{35}{100}$ and $\frac{8}{25}=\frac{32}{100}$

$\frac{35}{100}>\frac{32}{100}$, so $\frac{7}{20} \boxed{>} \frac{8}{25}$.

57. $\frac{19}{36}$ — $\frac{29}{54}$ *LCD is 108*

$\frac{19}{36}=\frac{57}{108}$ and $\frac{29}{54}=\frac{58}{108}$

$\frac{57}{108}<\frac{58}{108}$, so $\frac{19}{36} \boxed{<} \frac{29}{54}$.

58. $\frac{19}{132}$ — $\frac{7}{55}$ *LCD is 660*

$\frac{19}{132}=\frac{95}{660}$ and $\frac{7}{55}=\frac{84}{660}$

$\frac{95}{660}>\frac{84}{660}$, so $\frac{19}{132} \boxed{>} \frac{7}{55}$.

59. $\left(\frac{1}{2}\right)^2=\frac{1}{2}\cdot\frac{1}{2}=\frac{1}{4}$

60. $\left(\frac{2}{3}\right)^2=\frac{2}{3}\cdot\frac{2}{3}=\frac{4}{9}$

61. $\left(\frac{3}{10}\right)^3=\frac{3}{10}\cdot\frac{3}{10}\cdot\frac{3}{10}=\frac{27}{1000}$

62. $\left(\frac{3}{8}\right)^4=\frac{3}{8}\cdot\frac{3}{8}\cdot\frac{3}{8}\cdot\frac{3}{8}=\frac{81}{4096}$

63.
$$\begin{aligned} 8\left(\frac{1}{4}\right)^2 &= \frac{8}{1}\cdot\frac{1}{4}\cdot\frac{1}{4} \\ &= \frac{\overset{1}{\cancel{\overset{2}{\cancel{8}}}}\cdot 1\cdot 1}{1\cdot \underset{1}{\cancel{4}}\cdot \underset{2}{\cancel{4}}} \\ &= \frac{1}{2} \end{aligned}$$

64. $12\left(\frac{3}{4}\right)^2 = \frac{12}{1}\cdot\frac{3}{4}\cdot\frac{3}{4}$

$= \frac{\overset{3}{\cancel{12}}\cdot 3\cdot 3}{1\cdot\underset{1}{\cancel{4}}\cdot 4}$

$= \frac{27}{4} = 6\frac{3}{4}$

65. $\left(\frac{2}{3}\right)^2\cdot\left(\frac{3}{8}\right)^2 = \frac{2}{3}\cdot\frac{2}{3}\cdot\frac{3}{8}\cdot\frac{3}{8}$

$= \frac{\overset{1}{\cancel{2}}\cdot\overset{1}{\cancel{2}}\cdot\overset{1}{\cancel{3}}\cdot\overset{1}{\cancel{3}}}{\underset{1}{\cancel{3}}\cdot\underset{1}{\cancel{3}}\cdot\underset{4}{\cancel{8}}\cdot\underset{4}{\cancel{8}}}$

$= \frac{1}{16}$

66. $\frac{7}{8}\div\left(\frac{1}{8}+\frac{3}{4}\right) = \frac{7}{8}\div\left(\frac{1}{8}+\frac{6}{8}\right)$

$= \frac{7}{8}\div\frac{7}{8}$

$= \frac{\overset{1}{\cancel{7}}}{\underset{1}{\cancel{8}}}\cdot\frac{\overset{1}{\cancel{8}}}{\underset{1}{\cancel{7}}}$

$= 1$

67. $\left(\frac{1}{2}\right)^2\cdot\left(\frac{1}{4}+\frac{1}{2}\right) = \frac{1}{2}\cdot\frac{1}{2}\cdot\left(\frac{1}{4}+\frac{2}{4}\right)$

$= \frac{1}{2}\cdot\frac{1}{2}\cdot\frac{3}{4}$

$= \frac{3}{16}$

68. $\left(\frac{1}{4}\right)^3 + \left(\frac{5}{8}+\frac{3}{4}\right) = \frac{1}{4}\cdot\frac{1}{4}\cdot\frac{1}{4} + \left(\frac{5}{8}+\frac{6}{8}\right)$

$= \frac{1}{64} + \frac{11}{8}$ *LCD is 64*

$= \frac{1}{64} + \frac{88}{64}$

$= \frac{89}{64} = 1\frac{25}{64}$

69. **[3.1]** $\frac{7}{8} - \frac{1}{8} = \frac{7-1}{8} = \frac{6}{8} = \frac{3}{4}$

70. **[3.1]** $\frac{7}{10} - \frac{3}{10} = \frac{7-3}{10} = \frac{4}{10} = \frac{2}{5}$

71. **[3.3]** $\frac{29}{32} - \frac{5}{16} = \frac{29}{32} - \frac{10}{32} = \frac{29-10}{32} = \frac{19}{32}$

72. **[3.3]**

$\frac{1}{4} + \frac{1}{8} + \frac{5}{16} = \frac{4}{16} + \frac{2}{16} + \frac{5}{16} = \frac{4+2+5}{16} = \frac{11}{16}$

73. **[3.4]**

$$\begin{array}{rcl} 6\frac{2}{3} & = & 6\frac{4}{6} \\ -\,4\frac{1}{2} & = & 4\frac{3}{6} \\ \hline & & 2\frac{1}{6} \end{array}$$

74. **[3.4]**

$$\begin{array}{rcl} 9\frac{1}{2} & = & 9\frac{2}{4} \\ +\,16\frac{3}{4} & = & 16\frac{3}{4} \\ \hline & & 25\frac{5}{4} = 25 + 1\frac{1}{4} = 26\frac{1}{4} \end{array}$$

75. **[3.4]**

$$\begin{array}{rcl} 7 & = & 6\frac{8}{8} \\ -\,1\frac{5}{8} & = & 1\frac{5}{8} \\ \hline & & 5\frac{3}{8} \end{array}$$

76. **[3.4]**

$$\begin{array}{rcl} 2\,\frac{3}{5} & = & 2\,\frac{48}{80} \\ 8\,\frac{5}{8} & = & 8\,\frac{50}{80} \\ +\,\frac{5}{16} & = & \frac{25}{80} \\ \hline & & 10\frac{123}{80} = 10 + 1\frac{43}{80} = 11\frac{43}{80} \end{array}$$

77. **[3.4]**

$$\begin{array}{r} 32\frac{5}{12} \\ -\,17 \\ \hline 15\frac{5}{12} \end{array}$$

78. **[3.3]** $\frac{7}{22} + \frac{3}{22} + \frac{3}{11} = \frac{7}{22} + \frac{3}{22} + \frac{6}{22}$

$= \frac{7+3+6}{22}$

$= \frac{16}{22} = \frac{8}{11}$

79. **[3.5]** $\left(\frac{1}{4}\right)^2\cdot\left(\frac{2}{5}\right)^3 = \frac{1}{\underset{\underset{1}{\cancel{2}}}{\cancel{4}}}\cdot\frac{1}{\underset{2}{\cancel{4}}}\cdot\frac{\overset{1}{\cancel{2}}}{5}\cdot\frac{\overset{1}{\cancel{2}}}{5}\cdot\frac{\overset{1}{\cancel{2}}}{5}$

$= \frac{1}{250}$

80. **[3.5]** $\frac{3}{8}\div\left(\frac{1}{2}+\frac{1}{4}\right) = \frac{3}{8}\div\left(\frac{2}{4}+\frac{1}{4}\right)$

$= \frac{3}{8}\div\frac{3}{4}$

$= \frac{\overset{1}{\cancel{3}}}{\underset{2}{\cancel{8}}}\cdot\frac{\overset{1}{\cancel{4}}}{\underset{1}{\cancel{3}}}$

$= \frac{1}{2}$

81. **[3.5]**

$$\left(\frac{2}{3}\right)^2 \cdot \left(\frac{1}{3}+\frac{1}{6}\right) = \left(\frac{2}{3}\right)^2 \cdot \left(\frac{2}{6}+\frac{1}{6}\right)$$
$$= \frac{\overset{1}{\cancel{2}}}{3}\cdot\frac{2}{3}\cdot\frac{1}{\underset{1}{\cancel{2}}} \qquad \frac{3}{6}=\frac{1}{2}$$
$$= \frac{2}{9}$$

82. **[3.5]**
$$\left(\frac{2}{3}\right)^3 + \left(\frac{2}{3}-\frac{5}{9}\right) = \left(\frac{2}{3}\right)^3 + \left(\frac{6}{9}-\frac{5}{9}\right)$$
$$= \frac{2}{3}\cdot\frac{2}{3}\cdot\frac{2}{3}+\frac{1}{9}$$
$$= \frac{8}{27}+\frac{1}{9}$$
$$= \frac{8}{27}+\frac{3}{27}=\frac{11}{27}$$

83. **[3.5]** $\frac{2}{3} \text{—} \frac{7}{12}$ *LCD is 12*

$\frac{2}{3}=\frac{8}{12}$

$\frac{8}{12} > \frac{7}{12}$, so $\frac{2}{3}\ \boxed{>}\ \frac{7}{12}$.

84. **[3.5]** $\frac{8}{9} \text{—} \frac{15}{8}$ *LCD is 72*

$\frac{8}{9}=\frac{64}{72}$ and $\frac{15}{8}=\frac{135}{72}$

$\frac{64}{72} < \frac{135}{72}$, so $\frac{8}{9}\ \boxed{<}\ \frac{15}{8}$.

85. **[3.5]** $\frac{17}{30} \text{—} \frac{36}{60}$ *LCD is 60*

$\frac{17}{30}=\frac{34}{60}$

$\frac{34}{60} < \frac{36}{60}$, so $\frac{17}{30}\ \boxed{<}\ \frac{36}{60}$.

86. **[3.5]** $\frac{5}{8} \text{—} \frac{17}{30}$ *LCD is 120*

$\frac{5}{8}=\frac{75}{120}$ and $\frac{17}{30}=\frac{68}{120}$

$\frac{75}{120} > \frac{68}{120}$, so $\frac{5}{8}\ \boxed{>}\ \frac{17}{30}$.

87. **[3.2]** 12 and 18

$$\begin{array}{r|ll} 2 & 12 & 18 \\ 2 & 6 & \cancel{9} \\ 3 & 3 & 9 \\ 3 & \cancel{1} & 3 \\ \hline & 1 & 1 \end{array}$$

LCM $= 2\cdot2\cdot3\cdot3 = 36$

88. **[3.2]** 6, 8, 10, and 12

$6 = 2\cdot3$
$8 = 2\cdot2\cdot2$
$10 = 2\cdot5$
$12 = 2\cdot2\cdot3$

LCM $= 2\cdot2\cdot2\cdot3\cdot5 = 120$

89. **[3.2]** 9, 14, and 21

$$\begin{array}{r|lll} 2 & \cancel{9} & 14 & \cancel{21} \\ 3 & 9 & \cancel{7} & 21 \\ 3 & 3 & \cancel{7} & \cancel{7} \\ 7 & \cancel{1} & 7 & 7 \\ \hline & 1 & 1 & 1 \end{array}$$

LCM $= 2\cdot3\cdot3\cdot7 = 126$

90. **[3.2]** $\frac{2}{3}=\frac{?}{27} \quad 27 \div 3 = 9$

$\frac{2}{3}=\frac{2\cdot9}{3\cdot9}=\frac{18}{27}$

91. **[3.2]** $\frac{9}{12}=\frac{?}{144} \quad 144 \div 12 = 12$

$\frac{9}{12}=\frac{9\cdot12}{12\cdot12}=\frac{108}{144}$

92. **[3.2]** $\frac{4}{5}=\frac{?}{75} \quad 75 \div 5 = 15$

$\frac{4}{5}=\frac{4\cdot15}{5\cdot15}=\frac{60}{75}$

93. **[3.4]** Add to find the number of feet needed to complete the two jobs. Then subtract to find the number of feet remaining on the roll.

Estimate: $93 - (14 + 22) = 57$ feet

Exact:
$$\begin{array}{rcr} 13\frac{1}{2} & = & 13\frac{4}{8} \\ +\,22\frac{3}{8} & = & 22\frac{3}{8} \\ \hline & & 35\frac{7}{8} \end{array}$$

$$\begin{array}{rcrcr} 92\frac{3}{4} & = & 92\frac{6}{8} & = & 91\frac{14}{8} \\ -\,35\frac{7}{8} & = & 35\frac{7}{8} & = & 35\frac{7}{8} \\ \hline & & & & 56\frac{7}{8} \end{array}$$

After completing the two jobs, $56\frac{7}{8}$ feet of wire mesh remain.

94. **[3.4]** Add to find the number of cubic yards of mulch used. Then subtract to find the number of cubic yards of mulch that remain.

Estimate: $45 - (11 + 26) = 8$ cubic yards

$$Exact: \quad \begin{array}{r} 10\frac{7}{8} = 10\frac{7}{8} \\ +25\frac{1}{2} = 25\frac{4}{8} \\ \hline 35\frac{11}{8} = 35 + 1\frac{3}{8} = 36\frac{3}{8} \end{array}$$

$$\begin{array}{r} 45 \quad = 44\frac{8}{8} \\ -36\frac{3}{8} = 36\frac{3}{8} \\ \hline 8\frac{5}{8} \end{array}$$

$8\frac{5}{8}$ cubic yards of mulch remain.

Chapter 3 Test

1. $\frac{5}{8} + \frac{1}{8} = \frac{5+1}{8} = \frac{6}{8} = \frac{3}{4}$

2. $\frac{1}{16} + \frac{7}{16} = \frac{1+7}{16} = \frac{8}{16} = \frac{1}{2}$

3. $\frac{7}{10} - \frac{3}{10} = \frac{7-3}{10} = \frac{4}{10} = \frac{2}{5}$

4. $\frac{7}{12} - \frac{5}{12} = \frac{7-5}{12} = \frac{2}{12} = \frac{1}{6}$

5. 2, 3, 4

$$\begin{array}{r|rrr} 2 & 2 & \not{3} & 4 \\ 2 & \not{1} & \not{3} & 2 \\ 3 & \not{1} & 3 & \not{1} \\ \hline & 1 & 1 & 1 \end{array}$$

LCM $= 2 \cdot 2 \cdot 3 = 12$

The LCM of 2, 3, and 4 is 12.

6. 6, 3, 5, 15

$6 = 2 \cdot 3$
$3 = 1 \cdot 3$
$5 = 1 \cdot 5$
$15 = 3 \cdot 5$

LCM $= 2 \cdot 3 \cdot 5 = 30$

The LCM of 6, 3, 5, and 15 is 30.

7. 6, 9, 27, 36

$6 = 2 \cdot 3$
$9 = 3 \cdot 3$
$27 = 3 \cdot 3 \cdot 3$
$36 = 2 \cdot 2 \cdot 3 \cdot 3$

LCM $= 2 \cdot 2 \cdot 3 \cdot 3 \cdot 3 = 108$

The LCM of 6, 9, 27, and 36 is 108.

8. $\frac{3}{8} + \frac{1}{4} = \frac{3}{8} + \frac{2}{8}$ *LCD is 8*
$= \frac{3+2}{8} = \frac{5}{8}$

9. $\frac{2}{9} + \frac{5}{12} = \frac{8}{36} + \frac{15}{36}$ *LCD is 36*
$= \frac{8+15}{36} = \frac{23}{36}$

10. $\frac{7}{8} - \frac{2}{3} = \frac{21}{24} - \frac{16}{24}$ *LCD is 24*
$= \frac{21-16}{24} = \frac{5}{24}$

11. $\frac{2}{5} - \frac{3}{8} = \frac{16}{40} - \frac{15}{40}$ *LCD is 40*
$= \frac{16-15}{40} = \frac{1}{40}$

12. $7\frac{2}{3} + 4\frac{5}{6}$

Estimate: $8 + 5 = 13$

Exact: $7\frac{2}{3} + 4\frac{5}{6} = 7\frac{4}{6} + 4\frac{5}{6}$
$= 11\frac{9}{6} = 11 + 1\frac{3}{6} = 12\frac{1}{2}$

13. $16\frac{2}{5} - 11\frac{2}{3}$

Estimate: $16 - 12 = 4$

$$Exact: \quad \begin{array}{r} 16\frac{2}{5} = 16\frac{6}{15} = 15\frac{21}{15} \\ -11\frac{2}{3} = 11\frac{10}{15} = 11\frac{10}{15} \\ \hline 4\frac{11}{15} \end{array}$$

14. $18\frac{3}{4} + 9\frac{2}{5} + 12\frac{1}{3}$

Estimate: $19 + 9 + 12 = 40$

$$Exact: \quad \begin{array}{r} 18\frac{3}{4} = 18\frac{45}{60} \\ 9\frac{2}{5} = 9\frac{24}{60} \\ +12\frac{1}{3} = 12\frac{20}{60} \\ \hline 39\frac{89}{60} = 39 + 1\frac{29}{60} = 40\frac{29}{60} \end{array}$$

15. $24 - 18\frac{3}{8}$

Estimate: $24 - 18 = 6$

Exact:
$$\begin{array}{rcl} 24 & = & 23\frac{8}{8} \\ -\ 18\frac{3}{8} & = & 18\frac{3}{8} \\ \hline & & 5\frac{5}{8} \end{array}$$

16. Probably addition and subtraction of fractions is more difficult because you have to find the least common denominator and then change the fractions to the same denominator.

17. Round mixed numbers to the nearest whole number. Then add, subtract, multiply, or divide to estimate the answer. The estimate may vary from the exact answer but it lets you know if your answer is reasonable.

18. Add the hours of training Monday through Friday.

Estimate: $6 + 5 + 4 + 7 + 5 = 27$ hours

Exact:
$$\begin{array}{rcl} 5\frac{2}{3} & = & 5\frac{8}{12} \\ 4\frac{3}{4} & = & 4\frac{9}{12} \\ 3\frac{5}{6} & = & 3\frac{10}{12} \\ 7\frac{1}{3} & = & 7\frac{4}{12} \\ +\ 5\frac{1}{6} & = & 5\frac{2}{12} \\ \hline & & 24\frac{33}{12} \end{array}$$

$$24\frac{33}{12} = 24 + 2\frac{9}{12} = 26\frac{3}{4}$$

He trained a total of $26\frac{3}{4}$ hours.

19. Add the number of gallons of paint that were used. Then subtract the number of gallons that were used from the amount the contractor had when he arrived to find the number of gallons remaining.

Estimate: $148 - (69 + 37 + 6) = 36$ gallons

Exact:
$$\begin{array}{rcl} 68\frac{1}{2} & = & 68\frac{4}{8} \\ 37\frac{3}{8} & = & 37\frac{3}{8} \\ +\ 5\frac{3}{4} & = & 5\frac{6}{8} \\ \hline & & 110\frac{13}{8} \end{array}$$

$$110\frac{13}{8} = 110 + 1\frac{5}{8} = 111\frac{5}{8}$$

$$\begin{array}{rcccl} 147\frac{1}{2} & = & 147\frac{4}{8} & = & 146\frac{12}{8} \\ -\ 111\frac{5}{8} & = & 111\frac{5}{8} & = & 111\frac{5}{8} \\ \hline & & & & 35\frac{7}{8} \end{array}$$

The number of gallons remaining is $35\frac{7}{8}$.

20. $\frac{3}{4} \text{ — } \frac{17}{24}$ *LCD is 24*

$$\frac{3}{4} = \frac{18}{24}$$

$$\frac{18}{24} > \frac{17}{24}, \text{ so } \frac{3}{4} \boxed{>} \frac{17}{24}.$$

21. $\frac{19}{24} \text{ — } \frac{17}{36}$ *LCD is 72*

$$\frac{19}{24} = \frac{57}{72} \text{ and } \frac{17}{36} = \frac{34}{72}$$

$$\frac{57}{72} > \frac{34}{72}, \text{ so } \frac{19}{24} \boxed{>} \frac{17}{36}.$$

22.
$$\left(\frac{1}{3}\right)^3 \cdot 54 = \left(\frac{1}{3} \cdot \frac{1}{3} \cdot \frac{1}{3}\right) \cdot 54$$
$$= \frac{1}{\cancel{27}_{1}} \cdot \frac{\cancel{54}^{2}}{1}$$
$$= \frac{2}{1} = 2$$

23.
$$\left(\frac{3}{4}\right)^2 - \left(\frac{7}{8} \cdot \frac{1}{3}\right) = \frac{3}{4} \cdot \frac{3}{4} - \frac{7}{24}$$
$$= \frac{9}{16} - \frac{7}{24} \quad \textit{LCD is 48}$$
$$= \frac{27}{48} - \frac{14}{48} = \frac{13}{48}$$

24.
$$4\left(\frac{7}{8} - \frac{7}{16}\right) = 4\left(\frac{14}{16} - \frac{7}{16}\right)$$
$$= 4 \cdot \frac{7}{16}$$
$$= \frac{\cancel{4}^{1}}{1} \cdot \frac{7}{\cancel{16}_{4}}$$
$$= \frac{7}{4} = 1\frac{3}{4}$$

25.
$$\frac{5}{6} + \frac{4}{3}\left(\frac{3}{8}\right) = \frac{5}{6} + \frac{\cancel{4}^{1}}{\cancel{3}_{1}}\left(\frac{\cancel{3}^{1}}{\cancel{8}_{2}}\right)$$
$$= \frac{5}{6} + \frac{1}{2} \quad \textit{LCD is 6}$$
$$= \frac{5}{6} + \frac{3}{6}$$
$$= \frac{5+3}{6}$$
$$= \frac{8}{6} = 1\frac{2}{6} = 1\frac{1}{3}$$

Cumulative Review Exercises (Chapters 1–3)

1. $\underline{8}7\underline{1}$

8 is in the hundreds place.
1 is in the ones place.

2. $\underline{5},62\underline{9},428$
5 is in the millions place.
9 is in the thousands place.

3. **To the nearest ten:** $14\underline{3}8$

Next digit is 5 or more. Tens place $(3 + 1 = 4)$ changes. All digits to the right of the underlined place are changed to zero. **1440**

To the nearest hundred: $1\underline{4}38$

Next digit is 4 or less. Hundreds place does not change. All digits to the right of the underlined place are changed to zero. **1400**

To the nearest thousand: $\underline{1}438$

Next digit is 4 or less. Thousands place does not change. All digits to the right of the underlined place are changed to zero. **1000**

4. **To the nearest ten:** $59,8\underline{0}3$

Next digit is 4 or less. Tens place does not change. All digits to the right of the underlined place are changed to zero. **59,800**

To the nearest hundred: $59,\underline{8}03$

Next digit is 4 or less. Hundreds place does not change. All digits to the right of the underlined place are changed to zero. **59,800**

To the nearest thousand: $5\underline{9},803$
Next digit is 5 or more. All digits to the right of the underlined place are changed to zero. Add 1 to 9. Write 0 and carry 1. **60,000**

5. *Estimate:* *Exact:*

$$\begin{array}{r} 2000 \\ 400 \\ 50,000 \\ +\,30,000 \\ \hline 82,400 \end{array} \qquad \begin{array}{r} 2361 \\ 386 \\ 47,304 \\ +\,29,728 \\ \hline 79,779 \end{array}$$

6. *Estimate:* *Exact:*

$$\begin{array}{r} 20,000 \\ -\,10,000 \\ \hline 10,000 \end{array} \qquad \begin{array}{r} 24,276 \\ -\,9887 \\ \hline 14,389 \end{array}$$

7. *Estimate:* *Exact:*

$$\begin{array}{r} 4000 \\ \times\,300 \\ \hline 1,200,000 \end{array} \qquad \begin{array}{r} 4468 \\ \times\,280 \\ \hline 1,251,040 \end{array}$$

8. *Estimate:* *Exact:*

$$\begin{array}{r} 2500 \\ 40\overline{)100,000} \end{array} \qquad \begin{array}{r} 3211 \\ 35\overline{)112,385} \\ \underline{105} \\ 73 \\ \underline{70} \\ 38 \\ \underline{35} \\ 35 \\ \underline{35} \\ 0 \end{array}$$

9.

$$\begin{array}{r} 4 \\ 8 \\ 5 \\ +\,9 \\ \hline 26 \end{array}$$

10.

$$\begin{array}{r} {\scriptstyle 1\,1\,1\,2\;\,2\,1} \\ 375,899 \\ 521,742 \\ +\,357,968 \\ \hline 1,255,609 \end{array}$$

11.

$$\begin{array}{r} {\scriptstyle 5\,18} \\ 1\cancel{6}\cancel{8}7 \\ -\,1096 \\ \hline 591 \end{array}$$

12.

$$\begin{array}{r} {\scriptstyle 14} \\ {\scriptstyle 5\;\;\cancel{4}\;9\,12} \\ 3,89\cancel{6},\cancel{5}\cancel{0}\cancel{2} \\ -\,1,094,807 \\ \hline 2,801,695 \end{array}$$

13. $3 \times 8 \times 5 = 24 \times 5 = 120$

14. $6 \cdot 3 \cdot 7 = 18 \cdot 7 = 126$

15. $5(8)(4) = 40(4) = 160$

16.

$$\begin{array}{r} {\scriptstyle 5} \\ 57 \\ \times\,8 \\ \hline 456 \end{array}$$

17.

$$\begin{array}{rl} 962 & \\ \times\,384 & \\ \hline 3848 & \leftarrow 4 \times 962 \\ 7696 & \leftarrow 8 \times 962 \\ 2886 & \leftarrow 3 \times 962 \\ \hline 369,408 & \end{array}$$

18.
$$\begin{array}{r} \overset{2}{3}40 \\ \times \quad 50 \\ \hline 17{,}000 \end{array}$$

19.
$$\begin{array}{r} 1\,3\,5 \\ 8\overline{)1\,0\,8\,0} \\ \underline{8} \\ 2\,8 \\ \underline{2\,4} \\ 4\,0 \\ \underline{4\,0} \\ 0 \end{array}$$

20. $13{,}467 \div 5$

$$\begin{array}{r} 2\ 6\ 9\ 3\ \mathbf{R2} \\ 5\overline{)1\,3,\,{}^{3}4\,{}^{4}6\,{}^{1}7} \end{array}$$

21.
$$\begin{array}{r} 3\,2\ \mathbf{R166} \\ 506\overline{)1\,6,\,3\,5\,8} \\ \underline{1\,5\,1\,8} \\ 1\,1\,7\,8 \\ \underline{1\,0\,1\,2} \\ 1\,6\,6 \end{array}$$

22. To find the perimeter, add the six measurements of the four sides.

Estimate: $20 + 9 + 5 + 20 + 9 + 5 = 68$ ft

Exact: $18 + 9 + 5 + 18 + 9 + 5 = 64$ ft

The perimeter of this parking space is 64 feet.

23. To find the area of the parking space, multiply the length (18) and the width (14).

Estimate: $20 \cdot 10 = 200\text{ ft}^2$

Exact:
$$\begin{array}{r} 18 \\ \times\ 14 \\ \hline 72 \\ 18 \\ \hline 252 \end{array}$$

The area of the parking space is 252 ft^2.

24. To find the number of cartons, divide 40,320 by 32.

Estimate: $40{,}000 \div 30 \approx 1333$ cartons

Exact:
$$\begin{array}{r} 1\,2\,6\,0 \\ 32\overline{)4\,0,\,3\,2\,0} \\ \underline{3\,2} \\ 8\,3 \\ \underline{6\,4} \\ 1\,9\,2 \\ \underline{1\,9\,2} \\ 0 \end{array}$$

1260 32-ounce cartons can be filled with 40,320 ounces of orange juice.

25. Multiply the revolutions made in one minute (3600) by the number of minutes (60).

Estimate: $4000 \times 60 = 240{,}000$ revolutions

Exact:
$$\begin{array}{r} 3600 \\ \times\quad 60 \\ \hline 216{,}000 \end{array}$$

The drill will make 216,000 revolutions in 60 minutes.

26. Multiply the length and the width to find the area.

Estimate: $2 \cdot 3 = 6\text{ yd}^2$

Exact: $1\frac{3}{4} \cdot 2\frac{2}{3} = \frac{7}{\cancel{4}_1} \cdot \frac{\cancel{8}^2}{3} = \frac{14}{3} = 4\frac{2}{3}$

The area of the pool table is $4\frac{2}{3}\text{ yd}^2$.

27. Multiply the length and the width to find the area.

Estimate: $3 \cdot 5 = 15\text{ mi}^2$

Exact: $2\frac{5}{8} \cdot 4\frac{2}{3} = \frac{\cancel{21}^7}{\cancel{8}_4} \cdot \frac{\cancel{14}^7}{\cancel{3}_1} = \frac{49}{4} = 12\frac{1}{4}$

The area is $12\frac{1}{4}\text{ mi}^2$.

28. Multiply the number of cords per load by the number of loads to find the number of cords he could deliver.

Estimate: $5 \cdot 4 = 20$ cords

Exact: $5\frac{1}{4} \cdot 3\frac{1}{2} = \frac{21}{4} \cdot \frac{7}{2} = \frac{147}{8} = 18\frac{3}{8}$

He could deliver $18\frac{3}{8}$ cords.

29. Subtract the height of the flagpole from the height of the building to find the height of the building itself.

Estimate: $1537 - 83 = 1454$ ft

Exact:
$$\begin{array}{rcr} 1536\frac{7}{8} & = & 1536\frac{7}{8} \\ -\ 82\frac{1}{2} & = & 82\frac{4}{8} \\ \hline & & 1454\frac{3}{8} \end{array}$$

The height of the Sears Tower, without the flagpole, is $1454\frac{3}{8}$ feet.

30.

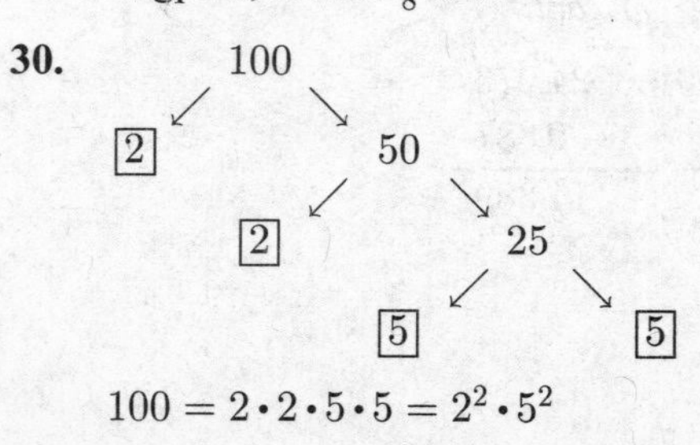

$100 = 2 \cdot 2 \cdot 5 \cdot 5 = 2^2 \cdot 5^2$

31.

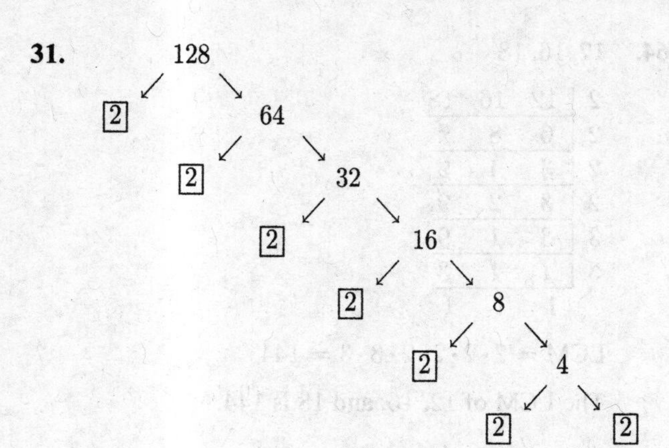

$128 = 2 \cdot 2 \cdot 2 \cdot 2 \cdot 2 \cdot 2 \cdot 2 = 2^7$

32. 1225 → $\boxed{5}$, 245; 245 → $\boxed{5}$, 49; 49 → $\boxed{7}$, $\boxed{7}$

$1225 = 5 \cdot 5 \cdot 7 \cdot 7 = 5^2 \cdot 7^2$

33. $2^4 \cdot 3^2 = (2 \cdot 2 \cdot 2 \cdot 2) \cdot (3 \cdot 3) = 16 \cdot 9 = 144$

34. $4^2 \cdot 2^4 = (4 \cdot 4) \cdot (2 \cdot 2 \cdot 2 \cdot 2) = 16 \cdot 16 = 256$

35. $6^2 \cdot 3^3 = (6 \cdot 6) \cdot (3 \cdot 3 \cdot 3) = 36 \cdot 27 = 972$

36. $\sqrt{36} = 6$ since $6^2 = 36$.

37. $\sqrt{81} = 9$ since $9^2 = 81$.

38. $\sqrt{144} = 12$ since $12^2 = 144$.

39.
$$\begin{aligned} 5^2 - 2(8) &= 25 - 2(8) \\ &= 25 - 16 \\ &= 9 \end{aligned}$$

40.
$$\begin{aligned} \sqrt{25} + 5 \cdot 9 - 6 &= 5 + 5 \cdot 9 - 6 \\ &= 5 + 45 - 6 \\ &= 50 - 6 = 44 \end{aligned}$$

41.
$$\begin{aligned} \frac{2}{3}\left(\frac{4}{5} - \frac{2}{3}\right) &= \frac{2}{3}\left(\frac{12}{15} - \frac{10}{15}\right) \\ &= \frac{2}{3} \cdot \frac{2}{15} \\ &= \frac{4}{45} \end{aligned}$$

42.
$$\begin{aligned} \frac{3}{4} \div \left(\frac{1}{3} + \frac{1}{2}\right) &= \frac{3}{4} \div \left(\frac{2}{6} + \frac{3}{6}\right) \\ &= \frac{3}{4} \div \frac{5}{6} \\ &= \frac{3}{\cancel{4}_2} \cdot \frac{\cancel{6}^3}{5} = \frac{9}{10} \end{aligned}$$

43.
$$\begin{aligned} \frac{7}{8} + \left(\frac{3}{4}\right)^2 - \frac{3}{8} &= \frac{7}{8} + \left(\frac{3}{4} \cdot \frac{3}{4}\right) - \frac{3}{8} \\ &= \frac{7}{8} + \frac{9}{16} - \frac{3}{8} \\ &= \frac{14}{16} + \frac{9}{16} - \frac{6}{16} \\ &= \frac{14 + 9}{16} - \frac{6}{16} \\ &= \frac{23}{16} - \frac{6}{16} \\ &= \frac{23 - 6}{16} \\ &= \frac{17}{16} = 1\frac{1}{16} \end{aligned}$$

44. $\frac{5}{6}$ is proper since the numerator is smaller than the denominator.

45. $\frac{6}{6}$ is improper since the numerator is equal to the denominator.

46. $\frac{11}{10}$ is improper since the numerator is larger than the denominator.

47. $\frac{25}{60} = \frac{25 \div 5}{60 \div 5} = \frac{5}{12}$

48. $\frac{84}{96} = \frac{84 \div 12}{96 \div 12} = \frac{7}{8}$

49. $\frac{63}{70} = \frac{63 \div 7}{70 \div 7} = \frac{9}{10}$

50. $\frac{3}{4} \cdot \frac{2}{3} = \frac{\cancel{3}^1}{\cancel{4}_2} \cdot \frac{\cancel{2}^1}{\cancel{3}_1} = \frac{1 \cdot 1}{2 \cdot 1} = \frac{1}{2}$

51. $\frac{3}{8} \cdot \frac{5}{6} = \frac{\cancel{3}^1}{8} \cdot \frac{5}{\cancel{6}_2} = \frac{1 \cdot 5}{8 \cdot 2} = \frac{5}{16}$

52. $42 \cdot \frac{7}{8} = \frac{\cancel{42}^{21}}{1} \cdot \frac{7}{\cancel{8}_4} = \frac{21 \cdot 7}{1 \cdot 4} = \frac{147}{4} = 36\frac{3}{4}$

53. $\frac{5}{8} \div \frac{3}{8} = \frac{5}{\cancel{8}_1} \cdot \frac{\cancel{8}^1}{3} = \frac{5 \cdot 1}{1 \cdot 3} = \frac{5}{3} = 1\frac{2}{3}$

54. $\frac{25}{40} \div \frac{10}{35} = \frac{\cancel{25}^{\cancel{5}^1}}{\cancel{40}_8} \cdot \frac{35}{\cancel{10}_2} = \frac{35}{16} = 2\frac{3}{16}$

55. $9 \div \frac{2}{3} = \frac{9}{1} \cdot \frac{3}{2} = \frac{27}{2} = 13\frac{1}{2}$

56.
$$\begin{aligned} \frac{3}{4} + \frac{1}{7} &= \frac{21}{28} + \frac{4}{28} \qquad \textit{LCD is 28} \\ &= \frac{21 + 4}{28} = \frac{25}{28} \end{aligned}$$

57. $\frac{3}{8}+\frac{3}{16}+\frac{1}{4}=\frac{6}{16}+\frac{3}{16}+\frac{4}{16}$ *LCD is 16*

$=\frac{6+3+4}{16}=\frac{13}{16}$

58. $\frac{11}{18}-\frac{5}{12}=\frac{22}{36}-\frac{15}{36}$ *LCD is 36*

$=\frac{22-15}{36}=\frac{7}{36}$

59. *Estimate:* *Exact:*

$$\begin{array}{r} 3 \\ +5 \\ \hline 8 \end{array} \qquad \begin{array}{rcr} 3\frac{3}{8} & = & 3\frac{3}{8} \\ +4\frac{1}{2} & = & 4\frac{4}{8} \\ \hline & & 7\frac{7}{8} \end{array}$$

60. *Estimate:* *Exact:*

$$\begin{array}{r} 22 \\ +4 \\ \hline 26 \end{array} \qquad \begin{array}{rcr} 21\frac{7}{8} & = & 21\frac{21}{24} \\ +4\frac{5}{12} & = & 4\frac{10}{24} \\ \hline & & 25\frac{31}{24} \end{array}$$

$25\frac{31}{24}=25+1\frac{7}{24}=26\frac{7}{24}$

61. *Estimate:* *Exact:*

$$\begin{array}{r} 5 \\ -2 \\ \hline 3 \end{array} \qquad \begin{array}{rcr} 5 & = & 4\frac{8}{8} \\ -2\frac{3}{8} & = & 2\frac{3}{8} \\ \hline & & 2\frac{5}{8} \end{array}$$

62. 8, 12

$8=2\cdot2\cdot2$

$12=2\cdot2\cdot3$

LCM $=2\cdot2\cdot2\cdot3=24$

The LCM of 8 and 12 is 24.

63. 3, 8, 15

$$\begin{array}{r|rrr} 2 & \cancel{3} & 8 & \cancel{15} \\ 2 & \cancel{3} & 4 & \cancel{15} \\ 2 & \cancel{3} & 2 & \cancel{15} \\ 3 & 3 & \cancel{1} & 15 \\ 5 & \cancel{1} & \cancel{1} & 5 \\ \hline & 1 & 1 & 1 \end{array}$$

LCM $=2\cdot2\cdot2\cdot3\cdot5=120$

The LCM of 3, 8, and 15 is 120.

64. 12, 16, 18

$$\begin{array}{r|rrr} 2 & 12 & 16 & 18 \\ 2 & 6 & 8 & \cancel{9} \\ 2 & \cancel{3} & 4 & \cancel{9} \\ 2 & \cancel{3} & 2 & \cancel{9} \\ 3 & 3 & \cancel{1} & 9 \\ 3 & \cancel{1} & \cancel{1} & 3 \\ \hline & 1 & 1 & 1 \end{array}$$

LCM $=2\cdot2\cdot2\cdot2\cdot3\cdot3=144$

The LCM of 12, 16, and 18 is 144.

65. $\frac{4}{5}=\frac{?}{45}$ $45\div5=9$

$\frac{4}{5}=\frac{4\cdot9}{5\cdot9}=\frac{36}{45}$

66. $\frac{7}{9}=\frac{?}{72}$ $72\div9=8$

$\frac{7}{9}=\frac{7\cdot8}{9\cdot8}=\frac{56}{72}$

67. $\frac{9}{15}=\frac{?}{135}$ $135\div15=9$

$\frac{9}{15}=\frac{9\cdot9}{15\cdot9}=\frac{81}{135}$

68. $\frac{5}{7}=\frac{?}{84}$ $84\div7=12$

$\frac{5}{7}=\frac{5\cdot12}{7\cdot12}=\frac{60}{84}$

69.–72.

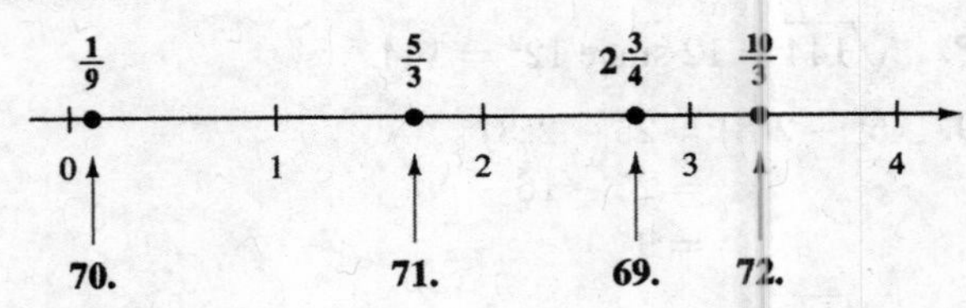

73. $\frac{3}{5}\,\_\_\,\frac{5}{8}$ *LCD is 40*

$\frac{3}{5}=\frac{24}{40}$ and $\frac{5}{8}=\frac{25}{40}$

$\frac{24}{40}<\frac{25}{40}$, so $\frac{3}{5}\ \boxed{<}\ \frac{5}{8}$.

74. $\frac{17}{20}\,\_\_\,\frac{3}{4}$ *LCD is 20*

$\frac{3}{4}=\frac{15}{20}$

$\frac{17}{20}>\frac{15}{20}$, so $\frac{17}{20}\ \boxed{>}\ \frac{3}{4}$.

75. $\frac{7}{12}\,\_\_\,\frac{11}{18}$ *LCD is 36*

$\frac{7}{12}=\frac{21}{36}$ and $\frac{11}{18}=\frac{22}{36}$

$\frac{21}{36}<\frac{22}{36}$, so $\frac{7}{12}\ \boxed{<}\ \frac{11}{18}$.

CHAPTER 4 DECIMALS

4.1 Reading and Writing Decimals

4.1 Margin Exercises

1. **(a)** The figure has 10 equal parts; 1 part is shaded.

$\frac{1}{10}$; 0.1; one tenth

(b) The figure has 10 equal parts; 3 parts are shaded.

$\frac{3}{10}$; 0.3; three tenths

(c) The figure has 10 equal parts; 9 parts are shaded.

$\frac{9}{10}$; 0.9; nine tenths

2. **(a)** The figure has 10 equal parts; 3 parts are shaded.

$\frac{3}{10}$; 0.3; three tenths

(b) The figure has 100 equal parts; 41 parts are shaded.

$\frac{41}{100}$; 0.41; forty-one hundredths

3. **(a)** $0.7 = \frac{7}{10}$

(b) $0.2 = \frac{2}{10}$

(c) $0.03 = \frac{3}{100}$

(d) $0.69 = \frac{69}{100}$

(e) $0.047 = \frac{47}{1000}$

(f) $0.351 = \frac{351}{1000}$

4. **(a)**

hundreds	tens	ones		tenths	hundredths
9	7	1	.	5	4

(b)

ones		tenths
0	.	4

(c)

ones		tenths	hundredths
5	.	6	0

(d)

ones		tenths	hundredths	thousandths	ten-thousandths
0	.	0	8	3	5

5. **(a)** 0.6 is six tenths.

(b) 0.46 is forty-six hundredths.

(c) 0.05 is five hundredths.

(d) 0.409 is four hundred nine thousandths.

(e) 0.0003 is three ten-thousandths.

(f) 0.0703 is seven hundred three ten-thousandths.

(g) 0.088 is eighty-eight thousandths.

6. **(a)** 3.8 is three and eight tenths.

(b) 15.001 is fifteen and one thousandth.

(c) 0.0073 is seventy-three ten-thousandths.

(d) 64.309 is sixty-four and three hundred nine thousandths.

7. **(a)** $0.7 = \frac{7}{10}$

(b) $12.21 = 12\frac{21}{100}$

(c) $0.101 = \frac{101}{1000}$

(d) $0.007 = \frac{7}{1000}$

(e) $1.3717 = 1\frac{3717}{10{,}000}$

8. **(a)** $0.5 = \frac{5}{10} = \frac{5 \div 5}{10 \div 5} = \frac{1}{2}$

(b) $12.6 = 12\frac{6}{10} = 12\frac{6 \div 2}{10 \div 2} = 12\frac{3}{5}$

(c) $0.85 = \frac{85}{100} = \frac{85 \div 5}{100 \div 5} = \frac{17}{20}$

(d) $3.05 = 3\frac{5}{100} = 3\frac{5 \div 5}{100 \div 5} = 3\frac{1}{20}$

(e) $0.225 = \frac{225}{1000} = \frac{225 \div 25}{1000 \div 25} = \frac{9}{40}$

(f) $420.0802 = 420\frac{802}{10{,}000} = 420\frac{802 \div 2}{10{,}000 \div 2}$

$= 420\frac{401}{5000}$

4.1 Section Exercises

1.

tens	ones		tenths		
7	0	.	4	8	9

3.

			hundredths	thousandths	ten-thousandths
0	.	2	5	1	8

5.

			tenths		thousandths	ten-thousandths	
9	3	.	0	1	4	7	2

7.

hundreds	tens			tenths		
3	1	4	.	6	5	8

9.

hundreds		ones			hundredths		
1	4	9	.	0	8	3	2

11.

thousands						hundredths	thousandths	
6	2	8	5	.	7	1	2	5

13. 0 ones, 5 hundredths, 1 ten, 4 hundreds, 2 tenths:

410.25

15. 3 thousandths, 4 hundredths, 6 ones, 2 ten-thousandths, 5 tenths:
6.5432

17. 4 hundredths, 4 hundreds, 0 tens, 0 tenths, 5 thousandths, 5 thousands, 6 ones:
5406.045

19. $0.7 = \frac{7}{10}$

21. $13.4 = 13\frac{4}{10} = 13\frac{4 \div 2}{10 \div 2} = 13\frac{2}{5}$

23. $0.25 = \frac{25}{100} = \frac{25 \div 25}{100 \div 25} = \frac{1}{4}$

25. $0.66 = \frac{66}{100} = \frac{66 \div 2}{100 \div 2} = \frac{33}{50}$

27. $10.17 = 10\frac{17}{100}$

29. $0.06 = \frac{6}{100} = \frac{6 \div 2}{100 \div 2} = \frac{3}{50}$

31. $0.205 = \frac{205}{1000} = \frac{205 \div 5}{1000 \div 5} = \frac{41}{200}$

33. $5.002 = 5\frac{2}{1000} = 5\frac{2 \div 2}{1000 \div 2} = 5\frac{1}{500}$

35. $0.686 = \frac{686}{1000} = \frac{686 \div 2}{1000 \div 2} = \frac{343}{500}$

37. 0.5 is five tenths.

39. 0.78 is seventy-eight hundredths.

41. 0.105 is one hundred five thousandths.

43. 12.04 is twelve and four hundredths.

45. 1.075 is one and seventy-five thousandths.

47. six and seven tenths:

$6\frac{7}{10} = 6.7$

49. thirty-two hundredths:

$\frac{32}{100} = 0.32$

51. four hundred twenty and eight thousandths:

$420\frac{8}{1000} = 420.008$

53. seven hundred three ten-thousandths:

$\frac{703}{10,000} = 0.0703$

55. seventy-five and thirty thousandths:

$75\frac{30}{1000} = 75.030$

57. Anne should not say "and" because that denotes a decimal point.

59. 8-pound test line has a diameter of 0.010 inch. 0.010 inch is read ten thousandths of an inch.

$0.010 = \frac{10}{1000} = \frac{10 \div 10}{1000 \div 10} = \frac{1}{100}$ inch

61. $\frac{13}{1000}$ inch $= 0.013$ inch

A diameter of 0.013 inch corresponds to a test strength of 12 pounds.

63. Six tenths is 0.6, so the correct part number is 3-C.

65. One and six thousandths is 1.006, which is part number 4-A.

67. The size of part number 4-E is 1.602 centimeters, which in words is one and six hundred two thousandths centimeters.

69. Use the whole number place value chart on text page 2 to find the names after hundred thousands. Then change "*s*" to "*ths*" in those names. So millions becomes million*ths* when used on the right side of the decimal point, and so on for ten-millionths, hundred-millionths, and billionths.

70. The first place to the left of the decimal point is ones, so the first place to the right could be one*ths*, like tens and ten*ths*. But anything that is 1 or more is to the left of the decimal point.

71. The eighth place value to the right of the decimal point is hundred-millionths, so 0.72436955 is seventy-two million four hundred thirty-six thousand nine hundred fifty-five hundred-millionths.

72. The ninth place value to the right of the decimal point is billionths, so 0.000678554 is six hundred seventy-eight thousand five hundred fifty-four billionths.

73. 8006.500001 is eight thousand six and five hundred thousand one millionths.

74. 20,060.000505 is twenty thousand, sixty and five hundred five millionths.

75. Three hundred two thousand forty ten-millionths:

0.0302040

76. Nine billion, eight hundred seventy-six million, five hundred forty-three thousand, two hundred ten and one hundred million two hundred thousand three hundred billionths:

9,876,543,210.100200300

4.2 Rounding Decimals

4.2 Margin Exercises

1. Round to the nearest thousandth.

(a) Draw a cut-off line: 0.334|92
The first digit cut is 9, which is 5 or more, so round up.

$$\begin{array}{r} 0.334 \\ +\,0.001 \\ \hline 0.335 \end{array}$$

Answer: ≈ 0.335

(b) Draw a cut-off line: 8.008|51
The first digit cut is 5, which is 5 or more, so round up.

$$\begin{array}{r} 8.008 \\ +\,0.001 \\ \hline 8.009 \end{array}$$

Answer: ≈ 8.009

(c) Draw a cut-off line: 265.420|38
The first digit cut is 3, which is 4 or less. The part you keep stays the same.

Answer: ≈ 265.420

(d) Draw a cut-off line: 10.701|80
The first digit cut is 8, which is 5 or more, so round up.

$$\begin{array}{r} 10.701 \\ +\,0.001 \\ \hline 10.702 \end{array}$$

Answer: ≈ 10.702

2. **(a)** 0.8988 to the nearest hundredth

Draw a cut-off line: 0.89|88
The first digit cut is 8, which is 5 or more, so round up.

$$\begin{array}{r} 0.89 \\ +\,0.01 \\ \hline 0.90 \end{array}$$

Answer: ≈ 0.90

(b) 5.8903 to the nearest hundredth

Draw a cut-off line: 5.89|03
The first digit cut is 0, which is 4 or less. The part you keep stays the same.

Answer: ≈ 5.89

(c) 11.0299 to the nearest thousandth

Draw a cut-off line: 11.029|9
The first digit cut is 9, which is 5 or more, so round up.

$$\begin{array}{r} 11.029 \\ +\,0.001 \\ \hline 11.030 \end{array}$$

Answer: ≈ 11.030

(d) 0.545 to the nearest tenth

Draw a cut-off line: 0.5|45
The first digit cut is 4, which is 4 or less. The part you keep stays the same.

Answer: ≈ 0.5

3. Round to the nearest cent (hundredths).

(a) Draw a cut-off line: \$14.59|5
The first digit cut is 5, which is 5 or more, so round up.

$$\begin{array}{r} \$14.59 \\ +\,0.01 \\ \hline \$14.60 \end{array}$$

Answer: $\approx$ \$14.60

(b) Draw a cut-off line: $578.06|63
The first digit cut is 6, which is 5 or more, so round up.

$$\begin{array}{r} \$578.06 \\ +\ 0.01 \\ \hline \$578.07 \end{array}$$

Answer: ≈ $578.07

(c) Draw a cut-off line: $0.84|9
The first digit cut is 9, which is 5 or more, so round up.

$$\begin{array}{r} \$0.84 \\ +\ 0.01 \\ \hline \$0.85 \end{array}$$

Answer: ≈ $0.85

(d) Draw a cut-off line: $0.05|48
The first digit cut is 4, which is 4 or less. The part you keep stays the same.

Answer: ≈ $0.05

(e) Draw a cut-off line: $49.|60
The first digit cut is 6, which is 5 or more, so round up by adding $1.

$$\begin{array}{r} \$49 \\ +\ 1 \\ \hline \$50 \end{array}$$

Answer: ≈ $50

(f) Draw a cut-off line: $0.|55
The first digit cut is 5, which is 5 or more, so round up by adding $1.

$$\begin{array}{r} \$0 \\ +\ 1 \\ \hline \$1 \end{array}$$

Answer: ≈ $1

(g) Draw a cut-off line: $1.|08
The first digit cut is 0, which is 4 or less. The part you keep stays the same.

Answer: ≈ $1

4. Round to the nearest dollar.

(a) Draw a cut-off line: $29.|10
The first digit cut is 1, which is 4 or less. The part you keep stays the same.

Answer: ≈ $29

(b) Draw a cut-off line: $136.|49
The first digit cut is 4, which is 4 or less. The part you keep stays the same.

Answer: ≈ $136

(c) Draw a cut-off line: $990.|91
The first digit cut is 9, which is 5 or more, so round up by adding $1.

$$\begin{array}{r} \$990 \\ +\ 1 \\ \hline \$991 \end{array}$$

Answer: ≈ $991

(d) Draw a cut-off line: $5949.|88
The first digit cut is 8, which is 5 or more, so round up by adding $1.

$$\begin{array}{r} \$5949 \\ +\ 1 \\ \hline \$5950 \end{array}$$

Answer: ≈ $5950

4.2 Section Exercises

1. 16.8974 to the nearest tenth
Draw a cut-off line after the tenths place:
16.8|974
The first digit cut is 9, which is 5 or more, so round up the tenths place.

$$\begin{array}{r} 16.8 \\ +\ 0.1 \\ \hline 16.9 \end{array}$$

Answer: ≈ 16.9

3. 0.95647 to the nearest thousandth
Draw a cut-off line after the thousandths place:
0.956|47
The first digit cut is 4, which is 4 or less. The part you keep stays the same.

Answer: ≈ 0.956

5. 0.799 to the nearest hundredth
Draw a cut-off line after the hundredths place:
0.79|9
The first digit cut is 9, which is 5 or more, so round up the hundredths place.

$$\begin{array}{r} 0.79 \\ +\ 0.01 \\ \hline 0.80 \end{array}$$

Answer: ≈ 0.80

7. 3.66062 to the nearest thousandth
Draw a cut-off line after the thousandths place:
3.660|62
The first digit cut is 6, which is 5 or more, so round up the thousandths place.

$$\begin{array}{r} 3.660 \\ +0.001 \\ \hline 3.661 \end{array}$$

Answer: ≈ 3.661

9. 793.988 to the nearest tenth
Draw a cut-off line after the tenths place:
793.9|88
The first digit cut is 8, which is 5 or more, so round up the tenths place.

$$\begin{array}{r} 793.9 \\ +\ 0.1 \\ \hline 794.0 \end{array}$$

Answer: ≈ 794.0

11. 0.09804 to the nearest ten-thousandth
Draw a cut-off line after the ten-thousandths place:
0.0980|4
The first digit cut is 4, which is 4 or less. The part you keep stays the same.

Answer: ≈ 0.0980

13. 48.512 to the nearest one
Draw a cut-off line after the ones place: 48.|512
The first digit cut is 5, which is 5 or more, so round up the ones place.

$$\begin{array}{r} 48 \\ +1 \\ \hline 49 \end{array}$$

Answer: ≈ 49

15. 9.0906 to the nearest hundredth
Draw a cut-off line after the hundredths place:
9.09|06
The first digit cut is 0, which is 4 or less. The part you keep stays the same.

Answer: ≈ 9.09

17. 82.000151 to the nearest ten-thousandth
Draw a cut-off line after the ten-thousandths place:
82.0001|51
The first digit cut is 5, which is 5 or more, so round up the ten-thousandths place.

$$\begin{array}{r} 82.0001 \\ +0.0001 \\ \hline 82.0002 \end{array}$$

Answer: ≈ 82.0002

19. Round \$0.81666 to the nearest cent.
Draw a cut-off line: \$0.81|666
The first digit cut is 6, which is 5 or more, so round up.

$$\begin{array}{r} \$0.81 \\ +0.01 \\ \hline \$0.82 \end{array}$$

Nardos pays \$0.82.

21. Round \$1.2225 to the nearest cent.
Draw a cut-off line: \$1.22|25
The first digit cut is 2, which is 4 or less. The part you keep stays the same.

Nardos pays \$1.22.

23. Round \$0.4983 to the nearest cent.
Draw a cut-off line: \$0.49|83
The first digit cut is 8, which is 5 or more, so round up.

$$\begin{array}{r} \$0.49 \\ +0.01 \\ \hline \$0.50 \end{array}$$

Nardos pays \$0.50.

25. Round \$48,649.60 to the nearest dollar.
Draw a cut-off line: \$48,649.|60
The first digit cut is 6, which is 5 or more, so round up.

$$\begin{array}{r} \$48{,}649 \\ +\quad 1 \\ \hline \$48{,}650 \end{array}$$

Income from job: $\approx$ \$48,650

27. Round \$310.08 to the nearest dollar.
Draw a cut-off line: \$310.|08
The first digit cut is 0, which is 4 or less. The part you keep stays the same.

Union dues: $\approx$ \$310

29. Round \$848.91 to the nearest dollar.
Draw a cut-off line: \$848.|91
The first digit cut is 9, which is 5 or more, so round up.

$$\begin{array}{r} \$848 \\ +\ 1 \\ \hline \$849 \end{array}$$

Donations to charity: $\approx$ \$849

31. \$499.98 to the nearest dollar
Draw a cut-off line: \$499.|98
The first digit cut is 9, which is 5 or more, so round up.

$$\begin{array}{r} \$499 \\ +\ 1 \\ \hline \$500 \end{array}$$

Answer: ≈ \$500

33. \$0.996 to the nearest cent
Draw a cut-off line: \$0.99|6
The first digit cut is 6, which is 5 or more, so round up.

$$\begin{array}{r} \$0.99 \\ +\ 0.01 \\ \hline \$1.00 \end{array}$$

Answer: ≈ \$1.00

35. \$999.73 to the nearest dollar
Draw a cut-off line: \$999.|73
The first digit cut is 7, which is 5 or more, so round up.

$$\begin{array}{r} \$999 \\ +\ 1 \\ \hline \$1000 \end{array}$$

Answer: ≈ \$1000

37. **(a)** Round 322.15 to the nearest whole number.
Draw a cut-off line: 322.|15
The first digit cut is 1, which is 4 or less. The part you keep stays the same.

The record speed for a motorcycle is about 322 miles per hour.

(b) Round 106.9 to the nearest whole number.
Draw a cut-off line: 106.|9
The first digit cut is 9, which is 5 or more, so round up.

$$\begin{array}{r} 106 \\ +1 \\ \hline 107 \end{array}$$

The fastest roller-coaster speed record is about 107 miles per hour.

39. **(a)** Round 185.981 to the nearest tenth.
Draw a cut-off line: 185.9|81
The first digit cut is 8, which is 5 or more, so round up.

$$\begin{array}{r} 185.9 \\ +\ 0.1 \\ \hline 186.0 \end{array}$$

The Indianapolis 500 fastest average winning speed is about 186.0 miles per hour.

(b) Round 763.04 to the nearest tenth.
Draw a cut-off line: 763.0|4
The first digit cut is 4, which is 4 or less. The part you keep stays the same.

The land speed record is about 763.0 miles per hour.

41. If you round \$0.499 to the nearest dollar, it will round to \$0 (zero dollars) because \$0.499 is closer to \$0 than to \$1.

42. Round amounts less than \$1.00 to the nearest cent instead of the nearest dollar.

43. If you round \$0.0015 to the nearest cent, it will round to \$0.00 (zero cents) because \$0.0015 is closer to \$0.00 than to \$0.01.

44. Both \$0.5968 and \$0.6014 round to \$0.60. Rounding to nearest thousandth (tenth of a cent) would allow you to identify \$0.597 as less than \$0.601.

4.3 Adding and Subtracting Decimals

4.3 Margin Exercises

1. **(a)** 2.86 + 7.09

$$\begin{array}{r} \overset{1}{2.86} \\ +\ 7.09 \\ \hline 9.95 \end{array}$$ *Line up decimal points.*

(b) 13.761 + 8.325

$$\begin{array}{r} \overset{11}{13.761} \\ +\ 8.325 \\ \hline 22.086 \end{array}$$ *Line up decimal points.*

(c) 0.319 + 56.007 + 8.252

$$\begin{array}{r} 1\ \ 1\ \ \ \ \\ 0.319 \\ 56.007 \\ +\ 8.252 \\ \hline 64.578 \end{array}$$ *Line up decimal points.*

(d) 39.4 + 0.4 + 177.2

$$\begin{array}{r} 1\,11\ \ \\ 39.4 \\ 0.4 \\ +\ 177.2 \\ \hline 217.0 \end{array}$$ *Line up decimal points.*

2. **(a)** $6.54 + 9.8$

$$\begin{array}{r} \scriptstyle 1 \\ 6.54 \\ +\ 9.80 \\ \hline 16.34 \end{array}$$

Line up decimal points.
Write in zeros.

(b) $0.831 + 222.2 + 10$

$$\begin{array}{r} \scriptstyle 1 \\ 0.831 \\ 222.200 \\ +\ 10.000 \\ \hline 233.031 \end{array}$$

Line up decimal points.
Write in zeros.

(c) $8.64 + 39.115 + 3.0076$

$$\begin{array}{r} \scriptstyle 2\ \ 1 \\ 8.6400 \\ 39.1150 \\ +\ 3.0076 \\ \hline 50.7626 \end{array}$$

Line up decimal points.
Write in zeros.

(d) $5 + 429.823 + 0.76$

$$\begin{array}{r} \scriptstyle 11 \\ 5.000 \\ 429.823 \\ +\ 0.760 \\ \hline 435.583 \end{array}$$

Line up decimal points.
Write in zeros.

3. **(a)** Subtract 22.7 from 72.9.

$$\begin{array}{r} 72.9 \\ -\ 22.7 \\ \hline 50.2 \end{array} \quad \text{Check:} \quad \begin{array}{r} 22.7 \\ +\ 50.2 \\ \hline 72.9 \end{array}$$

(b) Subtract 6.425 from 11.813.

$$\begin{array}{r} \scriptstyle 7\,10\,13 \\ 11.\not{8}\not{1}\not{3} \\ -\ 6.4\,2\,5 \\ \hline 5.3\,8\,8 \end{array} \quad \text{Check:} \quad \begin{array}{r} 6.425 \\ +\ 5.388 \\ \hline 11.813 \end{array}$$

(c) $20.15 - 19.67$

$$\begin{array}{r} \scriptstyle 1\,9\,10\,15 \\ \not{2}\not{0}.\not{1}\not{5} \\ -\ 1\,9.6\,7 \\ \hline 0.4\,8 \end{array} \quad \text{Check:} \quad \begin{array}{r} 19.67 \\ +\ 0.48 \\ \hline 20.15 \end{array}$$

4. **(a)** Subtract 18.651 from 25.3.

$$\begin{array}{r} \scriptstyle 1\,14\,12\,9\,10 \\ \not{2}\not{5}.\not{3}\not{0}\not{0} \\ -\ 1\,8.6\,5\,1 \\ \hline 6.6\,4\,9 \end{array}$$

Write two zeros.

$$\text{Check:} \quad \begin{array}{r} 18.651 \\ +\ 6.649 \\ \hline 25.300 \end{array}$$

(b) $5.816 - 4.98$

$$\begin{array}{r} \scriptstyle 4\ 17\,11 \\ \not{5}.\not{8}\not{1}6 \\ -\ 4.9\,8\,0 \\ \hline 0.8\,3\,6 \end{array} \quad \text{Check:} \quad \begin{array}{r} 4.980 \\ +\ 0.836 \\ \hline 5.816 \end{array}$$

Write one zero.

(c) 40 less 3.66

$$\begin{array}{r} \scriptstyle 3\,9\,9\,10 \\ \not{4}\not{0}.\not{0}\not{0} \\ -\ 3.6\,6 \\ \hline 3\,6.3\,4 \end{array} \quad \text{Check:} \quad \begin{array}{r} 36.34 \\ +\ 3.66 \\ \hline 40.00 \end{array}$$

(d) $1 - 0.325$

$$\begin{array}{r} \scriptstyle 0\,9\,9\,10 \\ \not{1}.\not{0}\not{0}\not{0} \\ -\ 0.3\,2\,5 \\ \hline 0.6\,7\,5 \end{array} \quad \text{Check:} \quad \begin{array}{r} 0.325 \\ +\ 0.675 \\ \hline 1.000 \end{array}$$

5. **(a)** $2.83 + 5.009 + 76.1$

Estimate: $\begin{array}{r} 3 \\ 5 \\ +\ 80 \\ \hline 88 \end{array}$ *Exact:* $\begin{array}{r} 2.830 \\ 5.009 \\ +\ 76.100 \\ \hline 83.939 \end{array}$

(b) 11.365 from 58

Estimate: $\begin{array}{r} 60 \\ -\ 10 \\ \hline 50 \end{array}$ *Exact:* $\begin{array}{r} 58.000 \\ -\ 11.365 \\ \hline 46.635 \end{array}$

(c) $398.81 + 47.658 + 4158.7$

Estimate: $\begin{array}{r} 400 \\ 50 \\ +\ 4000 \\ \hline 4450 \end{array}$ *Exact:* $\begin{array}{r} 398.810 \\ 47.658 \\ +\ 4158.700 \\ \hline 4605.168 \end{array}$

(d) difference between 12.837 meters and 46.091 meters

Estimate: $\begin{array}{r} 50 \\ -\ 10 \\ \hline 40 \end{array}$ *Exact:* $\begin{array}{r} 46.091 \\ -\ 12.837 \\ \hline 33.254 \end{array}$ meters

(e) \$19.28 plus \$1.53

Estimate: $\begin{array}{r} \$20 \\ +\ 2 \\ \hline \$22 \end{array}$ *Exact:* $\begin{array}{r} \$19.28 \\ +\ 1.53 \\ \hline \$20.81 \end{array}$

4.3 Section Exercises

1. 5.69 + 11.79

$$\begin{array}{r} 5.69 \\ +\,11.79 \\ \hline 17.48 \end{array}$$ *Line up decimal points.*

3. 24.008 − 0.995

$$\begin{array}{r} 24.008 \\ -\,0.995 \\ \hline 23.013 \end{array}$$ *Line up decimal points.*

5. 8.263 − 0.5

$$\begin{array}{r} 8.263 \\ -\,0.500 \\ \hline 7.763 \end{array}$$ *Line up decimal points. Write in zeros.*

7. 76.5 + 0.506

$$\begin{array}{r} 76.500 \\ +\,0.506 \\ \hline 77.006 \end{array}$$ *Line up decimal points. Write in zeros.*

9. 21 − 0.896

$$\begin{array}{r} 21.000 \\ -\,0.896 \\ \hline 20.104 \end{array}$$ *Line up decimal points. Write in zeros.*

11. 0.4 − 0.291

$$\begin{array}{r} 0.400 \\ -\,0.291 \\ \hline 0.109 \end{array}$$ *Line up decimal points. Write in zeros.*

13. 39.76005 + 182 + 4.799 + 98.31 + 5.9999

$$\begin{array}{r} 39.76005 \\ 182.00000 \\ 4.79900 \\ 98.31000 \\ +\,5.99990 \\ \hline 330.86895 \end{array}$$ *Line up decimal points. Write in zeros.*

15. (a) Humerus 14.35 in., radius 10.4 in.

$$\begin{array}{r} 14.35 \\ +\,10.40 \\ \hline 24.75 \end{array}$$ *Line up decimal points. Write in zeros.*

The combined length is 24.75 inches.

(b)

$$\begin{array}{r} 14.35 \\ -\,10.40 \\ \hline 3.95 \end{array}$$

The difference is 3.95 inches.

17. (a) Humerus 14.35 in., ulna 11.1 in., femur 19.88 in., tibia 16.94 in.

$$\begin{array}{r} 14.35 \\ 11.10 \\ 19.88 \\ +\,16.94 \\ \hline 62.27 \end{array}$$ *Line up decimal points. Write in zeros.*

The sum of the lengths is 62.27 inches.

(b) 8th rib 9.06 in., 7th rib 9.45 in.

$$\begin{array}{r} 9.45 \\ -\,9.06 \\ \hline 0.39 \end{array}$$

The 8th rib is 0.39 in. shorter than the 7th rib.

19. 6 should be written 6.00. The sum should be

$$\begin{array}{r} 0.72 \\ 6.00 \\ +\,39.50 \\ \hline 46.22 \end{array}$$

21. *Estimate:*

$$\begin{array}{r} \$20 \\ -\,7 \\ \hline \$13 \end{array}$$

Exact:

$$\begin{array}{r} \$19.74 \\ -\,6.58 \\ \hline \$13.16 \end{array}$$

23. *Estimate:*

$$\begin{array}{r} 400 \\ 1 \\ +\,20 \\ \hline 421 \end{array}$$

Exact:

$$\begin{array}{r} 392.700 \\ 0.865 \\ +\,21.080 \\ \hline 414.645 \end{array}$$

25. 8.6 less 3.751

Estimate:

$$\begin{array}{r} 9 \\ -\,4 \\ \hline 5 \end{array}$$

Exact:

$$\begin{array}{r} 8.600 \\ -\,3.751 \\ \hline 4.849 \end{array}$$

27. *Estimate:*

$$\begin{array}{r} 60 \\ 500 \\ +\,6 \\ \hline 566 \end{array}$$

Exact:

$$\begin{array}{r} 62.8173 \\ 539.9900 \\ +\,5.6290 \\ \hline 608.4363 \end{array}$$

29. 12 − 11.725

An estimate is 12 − 12 = 0, so 0.275 is the most reasonable answer.

31. 6.5 + 0.007

An estimate is 7 + 0 = 7, so 6.507 is the most reasonable answer.

33. 456.71 − 454.9

An estimate is 457 − 455 = 2, so 1.81 is the most reasonable answer.

35. 6004.003 + 52.7172

An estimate is 6000 + 50 = 6050, so 6056.7202 is the most reasonable answer.

37. Canada: 15.4 million; South Korea: 19 million

Estimate:	*Exact:*
19	19.0
− 15	− 15.4
4	3.6

There are 3.6 million fewer Internet users in Canada than there are in South Korea.

39. Add the number of users from all the countries.

Estimate:	*Exact:*
100	134.6
30	33.9
20	22.5
20	19.0
20	15.4
+ 9	+ 9.0
199	234.4

There are 234.4 million Internet users in all the countries listed in the table.

41. Add the 3 heights.

Estimate:	*Exact:*
2	1.83
2	2.16
+ 2	+ 2.11
6 meters	6.10 meters

The NBA stars' combined height is 0.3 meter less than the rhino's height of 6.4 meters.

43. Subtract the weight of the lighter box from the weight of the heavier box.

Estimate:	*Exact:*
11	10.50
− 10	− 9.85
1 ounce	0.65 ounce

The difference in weight is 0.65 ounce.

45. Add the lengths of the four sides.

Estimate:	*Exact:*
20	19.75
20	19.75
6	6.30
+ 6	+ 6.30
52 inches	52.10 inches

The perimeter is 52.1 inches.

47. Subtract the price of the regular fishing line, $4.84, from the price of the fluorescent fishing line, $5.14.

Estimate:	*Exact:*
$5	$5.14
− 5	− 4.84
$0	$0.30

The fluorescent fishing line costs $0.30 more than the regular fishing line.

49. Add the price of the four items and the sales tax.

Estimate:	*Exact:*
$19	$18.84 *spinning reel*
2	2.07 *tin split shot*
2	2.07 *tin split shot*
10	9.96 *tackle box*
+ 2	+ 2.31 *sales tax*
$35	$35.25

The total cost of the four items and the sales tax is $35.25.

51. Add the monthly expenses.

$994.00
190.78
205.00
39.95
19.95
40.80
57.32
186.81
97.75
+ 107.00
$1939.36

Olivia's total expenses are $1939.36 per month.

53. Subtract to find the difference.

$$\begin{array}{r} \$190.78 \\ -\ 186.81 \\ \hline \$3.97 \end{array}$$

The difference in the amounts spent for groceries and for the car payment is \$3.97.

55. Add the car expenses; then subtract this amount from the rent.

$$\begin{array}{r} 190.78 \\ +\ 205.00 \\ \hline \$395.78 \end{array} \qquad \begin{array}{r} 994.00 \\ -\ 395.78 \\ \hline \$598.22 \end{array}$$

She spends \$598.22 more on rent than on all her car expenses.

57. Subtract the two inside measurements from the total length.

$$\begin{array}{rl} 3.00 & \textit{Total length} \\ -\ 0.91 & \textit{Leftmost measurement} \\ \hline 2.09 & \\ -\ 0.70 & \textit{Middle measurement} \\ \hline 1.39 & \end{array}$$

$b = 1.39$ cm.

59. Add the given lengths and subtract the sum from the total length.

$$\begin{array}{r} 2.981 \\ +\ 2.981 \\ \hline 5.962 \text{ feet} \end{array} \qquad \begin{array}{r} 29.805 \\ -\ 5.962 \\ \hline 23.843 \text{ feet} \end{array}$$

$q = 23.843$ feet

4.4 Multiplying Decimals

4.4 Margin Exercises

1. **(a)**

$$\begin{array}{rcl} 2.6 & \leftarrow & \textit{1 decimal place} \\ \times\ 0.4 & \leftarrow & \textit{1 decimal place} \\ \hline 1.04 & \leftarrow & \textit{2 decimal places} \end{array}$$

(b)

$$\begin{array}{rcl} 45.2 & \leftarrow & \textit{1 decimal place} \\ \times\ 0.25 & \leftarrow & \textit{2 decimal places} \\ \hline 2\,260 & & \\ 9\,04 & & \\ \hline 11.3\,00 & \leftarrow & \textit{3 decimal places} \end{array}$$

(c)

$$\begin{array}{rcl} 0.104 & \leftarrow & \textit{3 decimal places} \\ \times\ \ 7 & \leftarrow & \textit{0 decimal places} \\ \hline 0.728 & \leftarrow & \textit{3 decimal places} \end{array}$$

(d)

$$\begin{array}{rcl} 3.18 & \leftarrow & \textit{2 decimal places} \\ \times\ 2.23 & \leftarrow & \textit{2 decimal places} \\ \hline 9\,54 & & \\ 63\,6 & & \\ 6\,36 & & \\ \hline 7.09\,14 & \leftarrow & \textit{4 decimal places} \end{array}$$

(e)

$$\begin{array}{rcl} 611 & \leftarrow & \textit{0 decimal places} \\ \times\ 3.7 & \leftarrow & \textit{1 decimal place} \\ \hline 427\,7 & & \\ 1833 & & \\ \hline 2260.7 & \leftarrow & \textit{1 decimal place} \end{array}$$

2. **(a)** 0.04×0.09

$$\begin{array}{rcl} 0.04 & \leftarrow & \textit{2 decimal places} \\ \times\ 0.09 & \leftarrow & \textit{2 decimal places} \\ \hline 36 & \leftarrow & \textit{4 decimal places} \end{array}$$

Count 4 places from the right. Write in the decimal point and zeros. 0.0036

(b) $(0.2)(0.008)$

$$\begin{array}{rcl} 0.008 & \leftarrow & \textit{3 decimal places} \\ \times\ 0.2 & \leftarrow & \textit{1 decimal place} \\ \hline 16 & \leftarrow & \textit{4 decimal places} \end{array}$$

Count 4 places from the right. Write in the decimal point and zeros. 0.0016

(c) $(0.003)^2$

$$\begin{array}{rcl} 0.003 & \leftarrow & \textit{3 decimal places} \\ \times\ 0.003 & \leftarrow & \textit{3 decimal places} \\ \hline 9 & \leftarrow & \textit{6 decimal places} \end{array}$$

Count 6 places from the right. Write in the decimal point and zeros. 0.000009

(d) $(0.0081)(0.003)$

$$\begin{array}{rcl} 0.0081 & \leftarrow & \textit{4 decimal places} \\ \times\ 0.003 & \leftarrow & \textit{3 decimal places} \\ \hline 243 & \leftarrow & \textit{7 decimal places} \end{array}$$

Count 7 places from the right. Write in the decimal point and zeros. 0.0000243

(e) $(0.11)(0.0005)$

$$\begin{array}{rcl} 0.11 & \leftarrow & \textit{2 decimal places} \\ \times\ 0.0005 & \leftarrow & \textit{4 decimal places} \\ \hline 55 & \leftarrow & \textit{6 decimal places} \end{array}$$

Count 6 places from the right. Write in the decimal point and zeros. 0.000055

3. (a) (11.62)(4.01)

Estimate:	*Exact:*	
10	11.62	← 2 *decimal places*
× 4	× 4.01	← 2 *decimal places*
40	1162	
	46 480	
	46.5962	← 4 *decimal places*

(b) (5.986)(33)

Estimate:	*Exact:*	
6	5.986	← 3 *decimal places*
× 30	× 33	← 0 *decimal places*
180	17 958	
	179 58	
	197.538	← 3 *decimal places*

(c) (8.31)(4.2)

Estimate:	*Exact:*	
8	8.31	← 2 *decimal places*
× 4	× 4.2	← 1 *decimal place*
32	1 662	
	33 24	
	34.902	← 3 *decimal places*

(d) 58.6 × 17.4

Estimate:	*Exact:*	
60	58.6	← 1 *decimal place*
× 20	× 17.4	← 1 *decimal place*
1200	23 44	
	4 10 2	
	5 86	
	10 19.64	← 2 *decimal places*

4.4 Section Exercises

1.

0.042	← 3 *decimal places*
× 3.2	← 1 *decimal place*
84	
126	
0.1344	← 4 *decimal places*

3.

21.5	← 1 *decimal place*
× 7.4	← 1 *decimal place*
8 60	
150 5	
159.10	← 2 *decimal places*

5.

23.4	← 1 *decimal place*
× 0.666	← 3 *decimal places*
1404	
1 404	
14 04	
15.5844	← 4 *decimal places*

7.

$51.88	← 2 *decimal places*
× 665	
259 40	
3 112 8	
31 12 8	
$34,500.20	← 2 *decimal places*

9. 72 × 0.6

72	72	
× 6	× 0.6	← 1 *decimal place*
432	43.2	← 1 *decimal place*

11. (7.2)(0.06)

72	7.2	← 1 *decimal place*
× 6	× 0.06	← 2 *decimal places*
432	0.432	← 3 *decimal places*

13. 0.72(0.06)

72	0.72	← 2 *decimal places*
× 6	× 0.06	← 2 *decimal places*
432	0.0432	← 4 *decimal places*

15. 0.0072(0.6)

72	0.0072	← 4 *decimal places*
× 6	× 0.6	← 1 *decimal place*
432	0.00432	← 5 *decimal places*

17. (0.006)(0.0052)

0.0052	← 4 *decimal places*
× 0.006	← 3 *decimal places*
0.0000312	← 7 *decimal places*

19. $(0.005)^2 = 0.005 \cdot 0.005$

0.005	← 3 *decimal places*
× 0.005	← 3 *decimal places*
0.000025	← 6 *decimal places*

21. (5.96)(10) = 59.6 (3.2)(10) = 32
(0.476)(10) = 4.76 (80.35)(10) = 803.5
(722.6)(10) = 7226 (0.9)(10) = 9

When you multiply a number by 10, the decimal point moves one place to the right. Multiplying by 100 moves the decimal point two places to the right. Multiplying by 1000 moves the decimal point three places to the right.

22. $(59.6)(0.1) = \underline{5.96}$ $(3.2)(0.1) = \underline{0.32}$
$(0.476)(0.1) = \underline{0.0476}$ $(80.35)(0.1) = \underline{8.035}$
$(65)(0.1) = \underline{6.5}$ $(523)(0.1) = \underline{52.3}$

Multiplying by 0.1 moves the decimal point one place to the left. Multiplying by 0.01 moves the decimal point two places to the left. Multiplying by 0.001 moves the decimal point three places to the left.

23. *Estimate:* *Exact:*

$$\begin{array}{r} 40 \\ \times 5 \\ \hline 200 \end{array} \qquad \begin{array}{rl} 39.6 & \leftarrow \textit{1 decimal place} \\ \times\ 4.8 & \leftarrow \textit{1 decimal place} \\ \hline 31\,68 & \\ 158\,4 & \\ \hline 190.08 & \leftarrow \textit{2 decimal places} \end{array}$$

25. *Estimate:* *Exact:*

$$\begin{array}{r} 40 \\ \times 40 \\ \hline 1600 \end{array} \qquad \begin{array}{rl} 37.1 & \leftarrow \textit{1 decimal place} \\ \times\ 42 & \leftarrow \textit{0 decimal places} \\ \hline 74\,2 & \\ 1484 & \\ \hline 1558.2 & \leftarrow \textit{1 decimal place} \end{array}$$

27. *Estimate:* *Exact:*

$$\begin{array}{r} 7 \\ \times 5 \\ \hline 35 \end{array} \qquad \begin{array}{rl} 6.53 & \leftarrow \textit{2 decimal places} \\ \times\ 4.6 & \leftarrow \textit{1 decimal place} \\ \hline 3\,918 & \\ 26\,12 & \\ \hline 30.038 & \leftarrow \textit{3 decimal places} \end{array}$$

29. *Estimate:* *Exact:*

$$\begin{array}{r} 3 \\ \times 7 \\ \hline 21 \end{array} \qquad \begin{array}{rl} 2.809 & \leftarrow \textit{3 decimal places} \\ \times\ 6.85 & \leftarrow \textit{2 decimal places} \\ \hline 14045 & \\ 2\,2472 & \\ 16\,854 & \\ \hline 19.24165 & \leftarrow \textit{5 decimal places} \end{array}$$

31. An \$18.90 car payment is *unreasonable.* A reasonable answer would be \$189.00.

33. A height of 60.5 inches (about 5 feet) is *reasonable.*

35. A gallon of milk for \$319 is *unreasonable.* A reasonable answer would be \$3.19.

37. 0.095 pounds for a baby's weight is *unreasonable.* A reasonable answer would be 9.5 pounds.

39. Multiply her pay per hour times the hours she worked.

$$\begin{array}{rl} \$18.73 & \leftarrow \textit{2 decimal places} \\ \times\ 50.5 & \leftarrow \textit{1 decimal place} \\ \hline 9\,365 & \\ 936\,50 & \\ \hline \$945.865 & \leftarrow \textit{3 decimal places} \end{array}$$

Round \$945.865 to the nearest cent. LaTasha made \$945.87.

41. Multiply the cost of one meter of canvas by the number of meters needed.

$$\begin{array}{r} \$4.09 \\ \times\ 0.6 \\ \hline \$2.454 \end{array}$$

\$2.454 rounds to \$2.45. Sid will spend \$2.45 on the canvas.

43. Multiply the number of gallons that she pumped into her pickup truck by the price per gallon.

$$\begin{array}{r} 18.605 \\ \times\ \ 1.749 \\ \hline 167445 \\ 74420 \\ 13\,0235 \\ 18\,605 \\ \hline 32.540145 \end{array}$$

Round 32.540145 to the nearest cent. Michelle paid \$32.54 for the gas.

45. Multiply the cost of the home by 0.07.

$$\begin{array}{r} \$229{,}500 \\ \times\ \ 0.07 \\ \hline \$16{,}065.00 \end{array}$$

Ms. Rolack's fee was \$16,065.

47. **(a) Area before 1929:**

$$\begin{array}{r} 7.4218 \\ \times\ \ 3.125 \\ \hline 371090 \\ 148436 \\ 74218 \\ 22\,2654 \\ \hline 23.1931250 \end{array}$$

Rounding to the nearest tenth gives us 23.2 in.^{2}.

Area after 1929:

```
    6.14
 ×  2.61
     614
   3 684
  12 28
 16.0254
```

Rounding to the nearest tenth gives us 16.0 in.^{2}.

(b) Subtract to find the difference.

```
  23.2
– 16.0
   7.2
```

The difference in rounded areas is 7.2 in.^{2}.

49. **(a)** Multiply the thickness of one bill times the number of bills.

```
 0.0043
 ×  100
 0.4300
```

A pile of 100 bills would be 0.43 inch high.

(b)

```
 0.0043
 × 1000
 4.3000
```

A pile of 1000 bills would be 4.3 inches high.

51. Multiply the monthly cost for cable times 24 months (two years).

```
  $38.96  basic per month
 ×    24
  155 84
  779 2
 $935.04  monthly total
 + 49.00  one-time installation
 $984.04  two-year total
```

The total cost for basic cable is $984.04.

```
   $89.95  deluxe per month
 ×     24
   359 80
  1799 0
 $2158.80  monthly total
 +  49.00  one-time installation
 $2207.80  two-year total
```

The total cost for deluxe cable is $2207.80.

53. Multiply the number of sheets by the cost per sheet.

```
     5100
 × $0.015
   25 500
   51 00
  $76.500
```

The library will pay $76.50 for the paper.

55. Multiply to find the cost of the rope, then multiply to find the cost of the wire. Add the results to find Barry's total purchases. Subtract the purchases from $15 (three $5 bills is $15).

Cost of rope	**Cost of wire**
16.5	$1.05
× $0.47	× 3
$7.755	$3.15

The cost of the rope rounds to $7.76.

Purchases	**Change**
$7.76 *rope*	$15.00
+ 3.15 *wire*	– 10.91
$10.91	$4.09

Barry received $4.09 in change.

57. Find the cost of the 4 long-sleeve, solid color shirts.

```
 $18.95
 ×    4
 $75.80
```

Then find the cost of the 2 short-sleeve, striped shirts.

```
 $16.75
 ×    2
 $33.50
```

Add these two amounts and the $2 per shirt charge for the XXL size.

```
  $2      $75.80
 × 6       33.50
 $12     + 12.00
         $121.30
```

The total cost is $121.30 plus shipping, or $121.30 + $7.95 = $129.25.

59. **(a)** Find the cost for 3 short-sleeved, solid-color shirts.

```
  $14.75
×      3
  $44.25
```

Based on this subtotal, shipping is $5.95. Find the cost of the 3 monograms.

```
  $4.95
×     3
 $14.85
```

Add these amounts, plus $5.00 for a gift box.

```
  $44.25
    5.95
   14.85
+   5.00
  $70.05
```

The total cost is $70.05.

(b) Subtract the cost of the shirts to find the difference.

```
  $70.05
– 44.25
  $25.80
```

The monograms, gift box, and shipping added $25.80 to the cost of the gift.

4.5 Dividing Decimals

4.5 Margin Exercises

1. **(a)**

```
   2 3. 4
 4)9 3. 6
   8
   1 3
   1 2
     1 6
     1 6
       0
```

Check:

```
  23.4
×    4
  93.6
```

(b)

```
   1. 1 3 4
 6)6. 8 0 4
   6
   0 8
     6
     2 0
     1 8
       2 4
       2 4
         0
```

Check:

```
  1.134
×     6
  6.804
```

(c)

```
     2 5. 3
 11)2 7 8. 3
    2 2
      5 8
      5 5
        3 3
        3 3
          0
```

Check:

```
  25.3
×   11
  25 3
  253
 278.3
```

(d) $0.51835 \div 5$

```
   0. 1 0 3 6 7
 5)0. 5 1 8 3 5
      5
        1
        0
        1 8
        1 5
          3 3
          3 0
            3 5
            3 5
              0
```

Check:

```
  0.10367
×       5
  0.51835
```

(e) $213.45 \div 15$

```
     1 4. 2 3
 15)2 1 3. 4 5
    1 5
      6 3
      6 0
        3 4
        3 0
          4 5
          4 5
            0
```

Check:

```
  14.23
×    15
  71 15
 142 3
 213.45
```

2. **(a)**

```
   1. 2 8
 5)6. 4 0  ← Write one zero.
   5
   1 4
   1 0
     4 0
     4 0
       0
```

Check:

```
  1.28
×    5
  6.40 or 6.4
```

(b) $30.87 \div 14$

```
     2. 2 0 5
14 | 3 0. 8 7 0  ← Write one zero.
     2 8
       2 8
       2 8
         0 7
           0
           7 0
           7 0
             0
```

Check:
```
   2.205
 ×    14
   8 820
  22 05
  30.870 or 30.87
```

(c) $\dfrac{259.5}{30}$

```
       8. 6 5
30 | 2 5 9. 5 0  ← Write one zero.
     2 4 0
       1 9 5
       1 8 0
         1 5 0
         1 5 0
             0
```

Check:
```
   8.65
 ×   30
 259.50 or 259.5
```

(d) $0.3 \div 8$

```
   0. 0 3 7 5
8 | 0. 3 0 0 0  ← Write three zeros.
       0
       3 0
       2 4
         6 0
         5 6
           4 0
           4 0
             0
```

Check:
```
   0.0375
 ×      8
   0.3000 or 0.3
```

3. **(a)** $13\overline{)267.01}$

$267.01 \div 13 \approx 20.539231$

There are no repeating digits visible on the calculator.

20.539|231 rounds to 20.539.

Check: $(20.539)(13) = 267.007 \approx 267.01$

(b) $6\overline{)20.5}$

$20.5 \div 6 \approx 3.416666$

There is a repeating decimal: $3.41\overline{6}$.

3.416|666 rounds to 3.417.

Check: $(3.417)(6) = 20.502 \approx 20.5$

(c) $\dfrac{10.22}{9} = 10.22 \div 9 \approx 1.135555$

There is a repeating decimal: $1.13\overline{5}$.

1.135|555 rounds to 1.136.

Check: $(1.136)(9) = 10.224 \approx 10.22$

(d) $16.15 \div 3 \approx 5.383333$

There is a repeating decimal: $5.38\overline{3}$.

5.383|333 rounds to 5.383.

Check: $(5.383)(3) = 16.149 \approx 16.15$

(e) $116.3 \div 11 \approx 10.5727272$

The answer has a repeating decimal that starts repeating in the fourth place: $10.5\overline{72}$.

10.572|72 rounds to 10.573.

Check: $(10.573)(11) = 116.303 \approx 116.3$

4. **(a)**

```
          5. 2
0.2∧ | 1. 0∧ 4
       1 0
         0 4
           4
           0
```

(b)

```
            3 0. 1 2
0.06∧ | 1. 8 0∧ 7 2
        1 8
          0 0
            0
            0 7
              6
              1 2
              1 2
                0
```

(c)

```
              6 4 0 0
0.005∧ | 3 2. 0 0 0∧
         3 0
           2 0
           2 0
             0 0
               0
               0 0
                 0
                 0
```

(d) $8.1 \div 0.025$

$$\begin{array}{r} 3\,2\,4 \\ 0.025_\wedge\overline{)8.\,1\,0\,0_\wedge} \\ \underline{7\,5} \\ 6\,0 \\ \underline{5\,0} \\ 1\,0\,0 \\ \underline{1\,0\,0} \\ 0 \end{array}$$

(e) $\dfrac{7}{1.3}$

$$\begin{array}{r} 5.\,3\,8\,4 \approx 5.38 \\ 1.3_\wedge\overline{)7.\,0_\wedge\,0\,0\,0} \\ \underline{6\,5} \\ 5\,0 \\ \underline{3\,9} \\ 1\,1\,0 \\ \underline{1\,0\,4} \\ 6\,0 \\ \underline{5\,2} \\ 8 \end{array}$$

(f) $5.3091 \div 6.2$

$$\begin{array}{r} 0.\,8\,5\,6 \approx 0.86 \\ 6.2_\wedge\overline{)5.\,3_\wedge\,0\,9\,1} \\ \underline{4\,9\,6} \\ 3\,4\,9 \\ \underline{3\,1\,0} \\ 3\,9\,1 \\ \underline{3\,7\,2} \\ 1\,9 \end{array}$$

5. **(a)** $42.75 \div 3.8 = 1.125$

Estimate: $40 \div 4 = 10$

The answer is not reasonable.

$$\begin{array}{r} 1\,1.\,2\,5 \\ 3.8_\wedge\overline{)4\,2.\,7_\wedge\,5\,0} \\ \underline{3\,8} \\ 4\,7 \\ \underline{3\,8} \\ 9\,5 \\ \underline{7\,6} \\ 1\,9\,0 \\ \underline{1\,9\,0} \\ 0 \end{array}$$

The answer should be 11.25.

(b) $807.1 \div 1.76 = 458.580$

Estimate: $800 \div 2 = 400$

The answer is reasonable.

(c) $48.63 \div 52 = 93.519$

Estimate: $50 \div 50 = 1$

The answer is not reasonable.

$$\begin{array}{r} 0.\,9\,3\,5\,1 \approx 0.935 \\ 52\overline{)4\,8.\,6\,3\,0\,0} \\ \underline{4\,6\,8} \\ 1\,8\,3 \\ \underline{1\,5\,6} \\ 2\,7\,0 \\ \underline{2\,6\,0} \\ 1\,0\,0 \\ \underline{5\,2} \\ 4\,8 \end{array}$$

The answer should be 0.935.

(d) $9.0584 \div 2.68 = 0.338$

Estimate: $9 \div 3 = 3$

The answer is not reasonable.

$$\begin{array}{r} 3.\,3\,8 \\ 2.68_\wedge\overline{)9.\,0\,5_\wedge\,8\,4} \\ \underline{8\,0\,4} \\ 1\,0\,1\,8 \\ \underline{8\,0\,4} \\ 2\,1\,4\,4 \\ \underline{2\,1\,4\,4} \\ 0 \end{array}$$

The answer should be 3.38.

6. **(a)** $4.6 - 0.79 + 1.5^2$ *Exponent*
$4.6 - 0.79 + 2.25$ *Subtract*
$3.81 + 2.25 = 6.06$ *Add*

(b) $3.64 \div 1.3 \times 3.6$ *Divide*
$2.8 \times 3.6 = 10.08$ *Multiply*

(c) $0.08 + 0.6(3 - 2.99)$ *Parentheses*
$0.08 + 0.6(0.01)$ *Multiply*
$0.08 + 0.006 = 0.086$ *Add*

(d) $10.85 - 2.3(5.2) \div 3.2$ *Multiply*
$10.85 - 11.96 \div 3.2$ *Divide*
$10.85 - 3.7375 = 7.1125$ *Subtract*

4.5 Section Exercises

1.
$$\begin{array}{r} 3.\,9 \\ 7\overline{)2\,7.\,3} \\ \underline{2\,1} \\ 6\,3 \\ \underline{6\,3} \\ 0 \end{array}$$
Line up decimal points.

3. $\frac{4.23}{9}$

```
    0. 4 7    Line up decimal points.
 9 ) 4. 2 3
     3 6
     ---
       6 3
       6 3
       ---
         0
```

5.

```
           4 0 0. 2
 0.05^ ) 2 0. 0 1^ 0    Move decimal point
         2 0            in divisor and dividend
         ---            2 places; write one zero.
           0 0 1 0
               1 0
               ---
                 0
```

7.

```
          3 6
 1.5^ ) 5 4. 0^    Move decimal point
        4 5        in divisor and dividend
        ---        1 place; write 0.
          9 0
          9 0
          ---
            0
```

9. Given: $108 \div 18 = 6$

Find: $0.108 \div 1.8$

The new dividend, 0.108, has the effect of moving the decimal place in the answer *left* three places. The new divisor, 1.8, has the effect of moving the decimal place in the answer *right* one place. So the new answer has the decimal point moved to the left two places.

$0.108 \div 1.8 = 0.06$

11. Given: $108 \div 18 = 6$

Find: $0.018\overline{)108}$ or $108 \div 0.018$

The new divisor, 0.018, has the effect of moving the decimal place in the answer *right* three places.

$108 \div 0.018 = 6000$

13.

```
            2 5. 3
 4.6^ ) 1 1 6. 3^ 8    Move decimal point
          9 2          in divisor and dividend
          ---          1 place.
          2 4 3
          2 3 0
          -----
            1 3 8
            1 3 8
            -----
                0
```

15. $\frac{3.1}{0.006}$

```
              5 1 6. 6 6 6    Line up decimal points.
 0.006^ ) 3. 1 0 0^ 0 0 0
          3 0                 Move decimal point in
          ---                 divisor and dividend 3
            1 0               places. Write 00
              6               in dividend.
            ---
              4 0
              3 6
              ---
                4 0           Write 0 in dividend.
                3 6
                ---
                  4 0         Write 0 in dividend.
                  3 6
                  ---
                    4 0       Write 0 in dividend.
                    3 6
                    ---
                      4       Stop and round answer
                              to the nearest hundredth.
```

516.666 rounds to 516.67.

17. Enter on calculator:

240 [÷] 9.88 [=]

Round 24.29149798 to 24.291.

19. $0.034\overline{)342.81}$

Enter on calculator:

342.81 [÷] 0.034 [=]

Round 10,082.64706 to 10,082.647.

21.

$3.77 \div 10 = \underline{0.377}$	$9.1 \div 10 = \underline{0.91}$
$0.886 \div 10 = \underline{0.0886}$	$30.19 \div 10 = \underline{3.019}$
$406.5 \div 10 = \underline{40.65}$	$6625.7 \div 10 = \underline{662.57}$

(a) Dividing by 10, decimal point moves one place to the left; by 100, two places to the left; by 1000, three places to the left.

(b) The decimal point moved to the *right* when *multiplying* by 10 or 100 or 1000. Here the decimal point moves to the *left* when *dividing* by those numbers.

22.

$40.2 \div 0.1 = \underline{402}$	$7.1 \div 0.1 = \underline{71}$
$0.339 \div 0.1 = \underline{3.39}$	$15.77 \div 0.1 = \underline{157.7}$
$46 \div 0.1 = \underline{460}$	$873 \div 0.1 = \underline{8730}$

(a) Dividing by 0.1, decimal point moves one place to the right; by 0.01, two places to the right; by 0.001, three places to the right.

(b) The decimal point moved to the *left* when multiplying by 0.1 or 0.01 or 0.001. Here the decimal point moves to the *right* when *dividing* by those numbers.

23. $37.8 \div 8 = 47.25$

Estimate: $40 \div 8 = 5$

The answer 47.25 is *unreasonable.*

```
    4. 7 2 5
 8|3 7. 8 0 0
   3 2
   ---
     5 8
     5 6
     ---
       2 0
       1 6
       ---
         4 0
         4 0
         ---
           0
```

The correct answer is 4.725.

25. $54.6 \div 48.1 = 1.135$

Estimate: $50 \div 50 = 1$

The answer 1.135 is *reasonable.*

27. $307.02 \div 5.1 = 6.2$

Estimate: $300 \div 5 = 60$

The answer 6.2 is *unreasonable.*

```
          6 0. 2
 5.1^|3 0 7. 0^ 2
      3 0 6
      -----
          1 0
            0
          -----
          1 0 2
          1 0 2
          -----
              0
```

The correct answer is 60.2.

29. $9.3 \div 1.25 = 0.744$

Estimate: $9 \div 1 = 9$

The answer 0.744 is *unreasonable.*

```
             7. 4 4
 1.25^|9. 3 0^ 0 0
       8 7 5
       -----
         5 5 0
         5 0 0
         -----
           5 0 0
           5 0 0
           -----
               0
```

The correct answer is 7.44.

31. Divide the cost by the number of pairs of tights.

```
     3. 9 9 6
 6|2 3. 9 8 0
   1 8
   ---
     5 9
     5 4
     ---
       5 8
       5 4
       ---
         4 0
         3 6
         ---
           4
```

$3.996 rounds to $4.00.

One pair costs $4.00.

33. Divide the balance by the number of months.

```
         6 7. 0 7 9
 21|1 4 0 8. 6 6 0
    1 2 6
    -----
      1 4 8
      1 4 7
      -----
          1 6 6
          1 4 7
          -----
            1 9 0   Write one zero.
            1 8 9
            -----
                1
```

Aimee is paying $67.08 (rounded) per month.

35. Divide the total cost by the number of bricks to find the cost per brick.

```
           0. 3 0
 619|1 8 5. 7 0
     1 8 5 7
     -------
           0 0
             0
           ---
             0
```

One brick costs $0.30.

37. Divide the total earnings by the number of hours.

```
      1 1. 9 2
 40|4 7 6. 8 0
    4 0
    ---
      7 6
      4 0
      ---
      3 6 8
      3 6 0
      -----
          8 0
          8 0
          ---
            0
```

Darren earns $11.92 per hour.

39. Divide the miles driven by the gallons of gas.

$346.2 \div 16.35 \approx 21.17$

She got 21.2 miles per gallon (rounded).

41. First find the sum of lengths for Jackie Joyner-Kersee.

$$\begin{array}{r} 7.49 \\ 7.45 \\ 7.40 \\ 7.32 \\ +\,7.20 \\ \hline 36.86 \end{array}$$

Then divide by the number of jumps, namely 5.

$$\begin{array}{r} 7.372 \\ 5\overline{)36.860} \\ 35 \\ \hline 1\,8 \\ 1\,5 \\ \hline 36 \\ 35 \\ \hline 10 \\ 10 \\ \hline 0 \end{array}$$

7.372 rounds to 7.37.

The average length of the long jumps made by Jackie Joyner-Kersee is 7.37 meters (rounded).

43. Subtract to find the difference.

7.40	*fifth longest jump*
− 7.32	*sixth longest jump*
0.08	*difference*

The fifth longest jump was 0.08 meter longer than the sixth longest jump.

45. Add the lengths of the top three athletes.

$$\begin{array}{r} 7.52 \\ 7.49 \\ +\,7.48 \\ \hline 22.49 \end{array}$$

The total length jumped by the top three athletes was 22.49 meters.

47.

$7.2 - 5.2 + 3.5^2$	*Exponent*
$7.2 - 5.2 + 12.25$	*Subtract*
$2 + 12.25 = 14.25$	*Add*

49.

$38.6 + 11.6(13.4 - 10.4)$	*Parentheses*
$38.6 + 11.6(3)$	*Multiply*
$38.6 + 34.8 = 73.4$	*Add*

51.

$8.68 - 4.6(10.4) \div 6.4$	*Multiply*
$8.68 - 47.84 \div 6.4$	*Divide*
$8.68 - 7.475 = 1.205$	*Subtract*

53.

$33 - 3.2(0.68 + 9) - 1.3^2$	*Parentheses; Exponent*
$33 - 3.2(9.68) - 1.69$	*Multiply*
$33 - 30.976 - 1.69$	*Subtract*
$2.024 - 1.69 = 0.334$	*Subtract*

55. Multiply the price per can by the number of cans.

$$\begin{array}{r} \$0.57 \\ \times\ \ 6 \\ \hline \$3.42 \end{array}$$

Subtract to find total savings.

$$\begin{array}{r} \$3.42 \\ -\,3.25 \\ \hline \$0.17 \end{array}$$

There are six cans, so divide by 6 to find the savings per can.

$$\begin{array}{r} 0.028 \\ 6\overline{)0.170} \\ 12 \\ \hline 50 \\ 48 \\ \hline 2 \end{array}$$

\$0.028 rounds to \$0.03.

You will save \$0.03 per can (rounded).

57. **(a)** $\frac{37{,}000{,}000}{24} \approx 1{,}541{,}667$ pieces each hour

(b) There are $24 \times 60 = 1440$ minutes in a day.

$\frac{37{,}000{,}000}{1440} \approx 25{,}694$ pieces each minute

(c) There are $24 \times 60 \times 60 = 86{,}400$ seconds in a day.

$\frac{37{,}000{,}000}{86{,}400} \approx 428$ pieces each second

59. Divide \$10,000 by 10¢.

$$\begin{array}{r} 1\,0\,0{,}0\,0\,0. \\ 0.10_\wedge\overline{)1\,0{,}0\,0\,0.\,0\,0_\wedge} \\ 1\,0\phantom{{,}000.00} \\ \hline 0\,0\,0\,0\,0\,0 \end{array}$$

The school would need to collect 100,000 box tops.

61. Divide 100,000 (the answer from Exercise 59) by 38.

```
        2 6 3 1. 5
 38 | 1 0 0, 0 0 0. 0
      7 6
      2 4 0
      2 2 8
        1 2 0
        1 1 4
            6 0
            3 8
            2 2 0
            1 9 0
              3 0
```

2631.5 rounds to 2632.

The school needs to collect 2632 box tops (rounded) during each of the 38 weeks.

Summary Exercises on Decimals

1. $0.8 = \frac{8}{10} = \frac{8 \div 2}{10 \div 2} = \frac{4}{5}$

3. $0.35 = \frac{35}{100} = \frac{35 \div 5}{100 \div 5} = \frac{7}{20}$

5. 2.0003 is two and three ten-thousandths.

7. five hundredths:

$\frac{5}{100} = 0.05$

9. ten and seven tenths:

$10\frac{7}{10} = 10.7$

11. 0.95 to the nearest tenth

Draw a cut-off line: 0.9|5

The first digit cut is 5, which is 5 or more, so round up.

$$\begin{array}{r} 0.9 \\ +0.1 \\ \hline 1.0 \end{array}$$

Answer: ≈ 1.0

13. \$0.893 to the nearest cent

Draw a cut-off line: \$0.89|3

The first digit cut is 3, which is 4 or less. The part you keep stays the same.

Answer: $\approx$ \$0.89

15. \$99.64 to the nearest dollar

Draw a cut-off line: \$99.|64

The first digit cut is 6, which is 5 or more, so round up.

$$\begin{array}{r} \$99 \\ +1 \\ \hline \$100 \end{array}$$

Answer: $\approx$ \$100

17. $50 - 0.3801$

$$\begin{array}{r} \overset{4}{\not5}\overset{9}{\not0}.\overset{9}{\not0}\overset{9}{\not0}\overset{9}{\not0}\overset{10}{\not0} \\ -\ 0.3801 \\ \hline 49.6199 \end{array}$$ *Write four zeros.*

Check: $$\begin{array}{r} 0.3801 \\ +49.6199 \\ \hline 50.0000 \end{array}$$

19. $\frac{90.18}{6}$

```
     1 5. 0 3
 6 | 9 0. 1 8
     6
     3 0
     3 0
       0 1
         0
         1 8
         1 8
           0
```

21. $3.7 - 1.55$

$$\begin{array}{r} 3.\overset{6}{\not7}\overset{10}{\not0} \\ -1.55 \\ \hline 2.15 \end{array}$$ Check: $$\begin{array}{r} 1.55 \\ +2.15 \\ \hline 3.70 \end{array}$$

23. $3.6 + 0.718 + 9 + 5.0829$

$$\begin{array}{r} \overset{1\ 11}{3.6000} \\ 0.7180 \\ 9.0000 \\ +5.0829 \\ \hline 18.4009 \end{array}$$ *Line up decimal points. Write in zeros.*

25. $8.9 - 0.4^2 \div 0.02$ *Exponent*
$8.9 - 0.16 \div 0.02$ *Divide*
$8.9 - 8 = 0.9$ *Subtract*

27. $0.64 \div 16.3 \approx 0.0392638037 \approx 0.04$

29. Perimeter $= 0.875 + 0.75 + 0.875 + 0.75$ *Add*
$= 1.625 + 0.875 + 0.75$ *Add*
$= 2.5 + 0.75 = 3.25$ inches *Add*

Area $= (0.875)(0.75)$ *Multiply*
$= 0.65625 \approx 0.66$ inch^2

31.

$$\begin{array}{rl} \overset{21\,2}{68.563} & \textit{radish} \\ 7.750 & \textit{tomato} \\ +\,15.490 & \textit{onion} \\ \hline 91.803 & \textit{combined weight} \end{array}$$

The combined weight is 91.803 pounds.

33. Divide 262 by 3.5.

$$\begin{array}{r} 74.8 \\ 3.5_\wedge \overline{)262.0_\wedge 0} \\ 245 \\ \hline 170 \\ 140 \\ \hline 300 \\ 280 \\ \hline 20 \end{array}$$

It would take 75 apples (rounded) to match the weight of the watermelon.

35. Find her total revenue.

$$\begin{array}{rl} \overset{27\,4}{\$12.95} & \textit{price per blanket} \\ \times \quad 8 & \textit{number of blankets} \\ \hline \$103.60 & \textit{total revenue} \end{array}$$

Subtract her cost from the revenue.

$$\begin{array}{rl} \overset{510}{\$103.\not{6}\not{0}} & \textit{total revenue} \\ -\,40.32 & \textit{cost} \\ \hline \$63.28 & \textit{profit} \end{array}$$

Madison made a profit of $63.28.

4.6 Writing Fractions as Decimals

4.6 Margin Exercises

1. **(a)** $\frac{1}{9}$ is written $9\overline{)1}$.

(b) $\frac{2}{3}$ is written $3\overline{)2}$.

(c) $\frac{5}{4}$ is written $4\overline{)5}$.

(d) $\frac{3}{10}$ is written $10\overline{)3}$.

(e) $\frac{21}{16}$ is written $16\overline{)21}$.

(f) $\frac{1}{50}$ is written $50\overline{)1}$.

2. **(a)** $\frac{1}{4}$

$$\begin{array}{r} 0.25 \\ 4\overline{)1.00} \\ 8 \\ \hline 20 \\ 20 \\ \hline 0 \end{array}$$

$\frac{1}{4} = 0.25$

(b) $2\frac{1}{2} = \frac{5}{2}$

$$\begin{array}{r} 2.5 \\ 2\overline{)5.0} \\ 4 \\ \hline 10 \\ 10 \\ \hline 0 \end{array}$$

$2\frac{1}{2} = 2.5$

(c) $\frac{5}{8}$

$$\begin{array}{r} 0.625 \\ 8\overline{)5.000} \\ 48 \\ \hline 20 \\ 16 \\ \hline 40 \\ 40 \\ \hline 0 \end{array}$$

$\frac{5}{8} = 0.625$

(d) $4\frac{3}{5}$

$$\begin{array}{r} 0.6 \\ 5\overline{)3.0} \\ 30 \\ \hline 0 \end{array}$$

$4 + 0.6 = 4.6$
$4\frac{3}{5} = 4.6$

(e) $\frac{7}{8}$

$$\begin{array}{r} 0.875 \\ 8\overline{)7.000} \\ 64 \\ \hline 60 \\ 56 \\ \hline 40 \\ 40 \\ \hline 0 \end{array}$$

$\frac{7}{8} = 0.875$

3. **(a)** $\frac{1}{3}$

$$\begin{array}{r} 0.3333 \\ 3\overline{)1.0000} \\ 9 \\ \hline 10 \\ 9 \\ \hline 10 \\ 9 \\ \hline 10 \\ 9 \\ \hline 1 \end{array}$$

Rounded to the nearest thousandth, $\frac{1}{3} \approx 0.333$.

(b) $2\frac{7}{9} = \frac{25}{9}$

$$\begin{array}{r} 2.7777 \\ 9\overline{)25.0000} \\ \underline{18} \\ 7\,0 \\ \underline{6\,3} \\ 70 \\ \underline{63} \\ 70 \\ \underline{63} \\ 70 \\ \underline{63} \\ 7 \end{array}$$

Rounded to the nearest thousandth, $2\frac{7}{9} \approx 2.778$.

(c) $\frac{10}{11}$

$$\begin{array}{r} 0.9090 \\ 11\overline{)10.0000} \\ \underline{9\,9} \\ 10 \\ \underline{0} \\ 100 \\ \underline{99} \\ 10 \\ \underline{0} \\ 10 \end{array}$$

Rounded to the nearest thousandth, $\frac{10}{11} \approx 0.909$.

(d) $\frac{3}{7}$

$$\begin{array}{r} 0.4285 \\ 7\overline{)3.0000} \\ \underline{2\,8} \\ 20 \\ \underline{14} \\ 60 \\ \underline{56} \\ 40 \\ \underline{35} \\ 5 \end{array}$$

Rounded to the nearest thousandth, $\frac{3}{7} \approx 0.429$.

(e) $3\frac{5}{6} = \frac{23}{6}$

$$\begin{array}{r} 3.8333 \\ 6\overline{)23.0000} \\ \underline{18} \\ 5\,0 \\ \underline{4\,8} \\ 20 \\ \underline{18} \\ 20 \\ \underline{18} \\ 20 \\ \underline{18} \\ 2 \end{array}$$

Rounded to the nearest thousandth, $3\frac{5}{6} \approx 3.833$.

4. **(a)** On the first number line, 0.4375 is to the *left* of 0.5, so use the $\boxed{<}$ symbol: $0.4375 < 0.5$

(b) On the first number line, 0.75 is to the *right* of 0.6875, so use the $\boxed{>}$ symbol: $0.75 > 0.6875$

(c) $0.625 > 0.0625$

(d) $\frac{2}{8} = \frac{1}{4} = 0.25$

$0.25 < 0.375$, so $\frac{2}{8} < 0.375$.

(e) On the second number line, $0.8\overline{3}$ and $\frac{5}{6}$ are at the *same point*, so use the $\boxed{=}$ symbol:

$0.8\overline{3} = \frac{5}{6}$

(f) $\frac{1}{2} < 0.\overline{5}$ $(\frac{1}{2} = 0.5)$

(g) $0.\overline{1} < 0.1\overline{6}$

(h) $\frac{8}{9} = 0.\overline{8}$

(i) $\frac{4}{6} = \frac{2}{3} = 0.\overline{6}$

$0.\overline{7} > 0.\overline{6}$, so $0.\overline{7} > \frac{4}{6}$.

(j) $\frac{1}{4} = 0.25$

5. **(a)**

0.7	0.703	0.7029
↓	↓	↓
0.7000	0.7030	0.7029

From smallest to largest:
0.7000, 0.7029, 0.7030.
or
0.7, 0.7029, 0.703

(b)

6.39	6.309	6.4	6.401
↓	↓	↓	↓
6.390	6.309	6.400	6.401

From smallest to largest:
6.309, 6.390, 6.400, 6.401
or
6.309, 6.39, 6.4, 6.401

(c)

1.085	$1\frac{3}{4}$	0.9
↓	↓	↓
1.085	1.750	0.900

From smallest to largest:
0.900, 1.085, 1.750
or
0.9, 1.085, $1\frac{3}{4}$

(d) $\frac{1}{4}, \frac{2}{5}, \frac{3}{7}, 0.428$

To compare, change fractions to decimals.

$\frac{1}{4} = 0.250 \quad \frac{2}{5} = 0.400 \quad \frac{3}{7} \approx 0.429$

From smallest to largest:
0.250, 0.400, 0.428, 0.429
or
$\frac{1}{4}, \frac{2}{5}, 0.428, \frac{3}{7}$

4.6 Section Exercises

1. $\frac{1}{2} = 0.5$

$$\begin{array}{r}0.5\\2\overline{)1.0}\\\underline{1\,0}\\0\end{array}$$

3. $\frac{3}{4} = 0.75$

$$\begin{array}{r}0.75\\4\overline{)3.00}\\\underline{2\,8}\\2\,0\\\underline{2\,0}\\0\end{array}$$

5. $\frac{3}{10} = 0.3$

$$\begin{array}{r}0.3\\10\overline{)3.0}\\\underline{3\,0}\\0\end{array}$$

7. $\frac{9}{10} = 0.9$

$$\begin{array}{r}0.9\\10\overline{)9.0}\\\underline{9\,0}\\0\end{array}$$

9. $\frac{3}{5} = 0.6$

$$\begin{array}{r}0.6\\5\overline{)3.0}\\\underline{3\,0}\\0\end{array}$$

11. $\frac{7}{8} = 0.875$

$$\begin{array}{r}0.875\\8\overline{)7.000}\\\underline{6\,4}\\6\,0\\\underline{5\,6}\\4\,0\\\underline{4\,0}\\0\end{array}$$

13. $2\frac{1}{4} = \frac{9}{4} = 2.25$

$$\begin{array}{r}2.25\\4\overline{)9.00}\\\underline{8}\\1\,0\\\underline{8}\\2\,0\\\underline{2\,0}\\0\end{array}$$

15. $14\frac{7}{10} = 14 + \frac{7}{10} = 14 + 0.7 = 14.7$

17. $3\frac{5}{8} = \frac{29}{8} = 3.625$

$$\begin{array}{r}3.625\\8\overline{)29.000}\\\underline{2\,4}\\5\,0\\\underline{4\,8}\\2\,0\\\underline{1\,6}\\4\,0\\\underline{4\,0}\\0\end{array}$$

19. $\frac{1}{3} \approx 0.3333333 \approx 0.333$

21. $\frac{5}{6} \approx 0.8333333 \approx 0.833$

23. $1\frac{8}{9} = \frac{17}{9} \approx 1.8888889 \approx 1.889$

25. **(a)** In $\frac{5}{9}$, the numerator is less than the denominator, so the decimal must be less than 1 (not 1.8).

(b) $\frac{5}{9}$ means $5 \div 9$ or $9\overline{)5}$, so the correct answer is 0.556 (rounded). This makes sense because both the fraction and the decimal are less than 1.

$$\begin{array}{r}0.5555\\9\overline{)5.0000}\\\underline{4\,5}\\5\,0\\\underline{4\,5}\\5\,0\\\underline{4\,5}\\5\,0\\\underline{4\,5}\\5\end{array}$$

26. **(a)** She changed 0.35 to 0.035 when adding to 2.

(b) Adding the whole number part gives $2 + 0.35$, which is 2.35, not 2.035. To check, $2.35 = 2\frac{35}{100} = 2\frac{7}{20}$, but $2.035 = 2\frac{35}{1000} = 2\frac{7}{200}$.

27. Just add the whole number part to 0.375. So $1\frac{3}{8} = 1.375$; $3\frac{3}{8} = 3.375$; $295\frac{3}{8} = 295.375$.

28. It works only when the fraction part has a one-digit numerator and a denominator of 10, a two-digit numerator and a denominator of 100, and so on.

29. $0.4 = \frac{4}{10} = \frac{4 \div 2}{10 \div 2} = \frac{2}{5}$

31. $0.625 = \frac{625}{1000} = \frac{625 \div 125}{1000 \div 125} = \frac{5}{8}$

33. $0.35 = \frac{35}{100} = \frac{35 \div 5}{100 \div 5} = \frac{7}{20}$

35. $\frac{7}{20} = 0.35$

$$\begin{array}{r} 0.35 \\ 20\overline{)7.00} \\ \underline{6\,0} \\ 1\,00 \\ \underline{1\,00} \\ 0 \end{array}$$

37. $0.04 = \frac{4}{100} = \frac{4 \div 4}{100 \div 4} = \frac{1}{25}$

39. $0.15 = \frac{15}{100} = \frac{15 \div 5}{100 \div 5} = \frac{3}{20}$

41. $\frac{1}{5} = 0.2$

$$\begin{array}{r} 0.2 \\ 5\overline{)1.0} \\ \underline{1\,0} \\ 0 \end{array}$$

43. $0.09 = \frac{9}{100}$

45. Compare the two lengths.
average length → 20.80 *longer*
Charlene's baby → 20.08 *shorter*

$$\begin{array}{r} 20.80 \\ -\,20.08 \\ \hline 0.72 \end{array}$$

Her baby is 0.72 inch *shorter* than the average length.

47. Write two zeros to the right of 0.5 so it has the same number of decimal places as 0.505. Then you can compare the numbers: $0.505 > 0.500$. There was too much calcium in each capsule. Subtract to find the difference.

$$\begin{array}{r} 0.505 \\ -\,0.500 \\ \hline 0.005 \end{array}$$

There was 0.005 gram too much.

49. Write two zeros so that all the numbers have four decimal places.

$1.0100 > 1.0020$ *unacceptable*
$0.9991 > 0.9980$ and $0.9991 < 1.0020$ *acceptable*
$1.0007 > 0.9980$ and $1.0007 < 1.0020$ *acceptable*
$0.9900 < 0.9980$ *unacceptable*

The lengths of 0.9991 cm and 1.0007 cm are acceptable.

51. Compare the two amounts.

$3\frac{3}{4} \to 3.75$ *less*
$3.8 \to 3.80$ *more*

$$\begin{array}{r} 3.80 \\ -\,3.75 \\ \hline 0.05 \end{array}$$

3.8 inches is 0.05 inch *more* than Ginny hoped for.

53. 0.54, 0.5455, 0.5399

$0.54 = 0.5400$
$0.5455 = 0.5455$ *largest*
$0.5399 = 0.5399$ *smallest*

From smallest to largest: 0.5399, 0.54, 0.5455

55. 5.8, 5.79, 5.0079, 5.804

$5.8 = 5.8000$
$5.79 = 5.7900$
$5.0079 = 5.0079$ *smallest*
$5.804 = 5.8040$ *largest*

From smallest to largest: 5.0079, 5.79, 5.8, 5.804

57. 0.628, 0.62812, 0.609, 0.6009

$0.628 = 0.62800$
$0.62812 = 0.62812$ *largest*
$0.609 = 0.60900$
$0.6009 = 0.60090$ *smallest*

From smallest to largest:
0.6009, 0.609, 0.628, 0.62812

59. 5.8751, 4.876, 2.8902, 3.88

The numbers after the decimal places are irrelevant since the whole number parts are all different.

From smallest to largest:
2.8902, 3.88, 4.876, 5.8751

61. 0.043, 0.051, 0.006, $\frac{1}{20}$

$0.043 = 0.043$
$0.051 = 0.051$ *largest*
$0.006 = 0.006$ *smallest*

$\frac{1}{20} = 0.050$

From smallest to largest: 0.006, 0.043, $\frac{1}{20}$, 0.051

63. $\frac{3}{8}$, $\frac{2}{5}$, 0.37, 0.4001

$\frac{3}{8} = 0.3750$
$\frac{2}{5} = 0.4000$
$0.37 = 0.3700$ *smallest*
$0.4001 = 0.4001$ *largest*

From smallest to largest: 0.37, $\frac{3}{8}$, $\frac{2}{5}$, 0.4001

65. Find the largest of:
0.018, 0.01, 0.008, 0.010

$0.018 = 0.018$ *largest*
$0.01 = 0.010$
$0.008 = 0.008$ *smallest*
$0.010 = 0.010$

List from smallest to largest:

0.008, 0.01 = 0.010, 0.018

The red box, labeled 0.018 in. diameter, has the strongest line.

67. Subtract 0.008 from 0.018.

$$\begin{array}{r} 0.018 \\ -\,0.008 \\ \hline 0.010 \end{array}$$

The difference in diameter is 0.01 inch.

69. $1\frac{7}{16} = \frac{23}{16}$

$$\begin{array}{r} 1.\,4\,3 \\ 16\overline{)\,2\,3.\,0\,0} \\ \underline{1\,6} \\ 7\,0 \\ \underline{6\,4} \\ 6\,0 \\ \underline{4\,8} \\ 1\,2 \end{array}$$

1.43 rounded to the nearest tenth is 1.4.
Length (a) is 1.4 inches (rounded).

71. $\frac{1}{4} = 0.25$

0.25 rounded to the nearest tenth is 0.3.
Length (c) is 0.3 inch (rounded).

73. $\frac{3}{8}$

$$\begin{array}{r} 0.\,3\,7 \\ 8\overline{)\,3.\,0\,0} \\ \underline{2\,4} \\ 6\,0 \\ \underline{5\,6} \\ 4 \end{array}$$

0.37 rounded to the nearest tenth is 0.4.
Length (e) is 0.4 inch (rounded).

Chapter 4 Review Exercises

1. 2 4 3 . 0 5 9 (tenths: 0; hundredths: 5)

2. 0 . 6 8 1 7 (ones: 0; tenths: 6)

3. \$5 8 2 4 . 3 9 (hundreds: 8; hundredths: 9)

4. 8 9 6 . 5 0 3 (tens: 8; tenths: 5)

5. 2 0 . 7 3 8 6 1 (tenths: 7; ten-thousandths: 6)

6. $0.5 = \frac{5}{10} = \frac{1}{2}$

7. $0.75 = \frac{75}{100} = \frac{75 \div 25}{100 \div 25} = \frac{3}{4}$

8. $4.05 = 4\frac{5}{100} = 4\frac{5 \div 5}{100 \div 5} = 4\frac{1}{20}$

9. $0.875 = \frac{875}{1000} = \frac{875 \div 125}{1000 \div 125} = \frac{7}{8}$

10. $0.027 = \frac{27}{1000}$

11. $27.8 = 27\frac{8}{10} = 27\frac{8 \div 2}{10 \div 2} = 27\frac{4}{5}$

12. 0.8 is eight tenths.

13. 400.29 is four hundred and twenty-nine hundredths.

14. 12.007 is twelve and seven thousandths.

15. 0.0306 is three hundred six ten-thousandths.

16. eight and three tenths:

$8\frac{3}{10} = 8.3$

17. two hundred five thousandths:

$\frac{205}{1000} = 0.205$

18. seventy and sixty-six ten-thousandths:

$70\frac{66}{10{,}000} = 70.0066$

19. thirty hundredths:

$\frac{30}{100} = 0.30$

20. 275.635 to the nearest tenth
Draw a cut-off line: 275.6|35
The first digit cut is 3, which is 4 or less. The part you keep stays the same.
Answer: ≈ 275.6

21. 72.789 to the nearest hundredth
Draw a cut-off line: 72.78|9
The first digit cut is 9, which is 5 or more, so round up.
Answer: ≈ 72.79

22. 0.1604 to the nearest thousandth
Draw a cut-off line: 0.160|4
The first digit cut is 4, which is 4 or less. The part you keep stays the same.
Answer: ≈ 0.160

23. 0.0905 to the nearest thousandth
Draw a cut-off line: 0.090|5
The first digit cut is 5, which is 5 or more, so round up.
Answer: ≈ 0.091

24. 0.98 to the nearest tenth
Draw a cut-off line: 0.9|8
The first digit cut is 8, which is 5 or more, so round up.
Answer: ≈ 1.0

25. \$15.8333 to the nearest cent
Draw a cut-off line: \$15.83|33
The first digit cut is 3, which is 4 or less. The part you keep stays the same.
Answer: $\approx$ \$15.83

26. \$0.698 to the nearest cent
Draw a cut-off line: \$0.69|8
The first digit cut is 8, which is 5 or more, so round up.
Answer: $\approx$ \$0.70

27. \$17,625.7906 to the nearest cent
Draw a cut-off line: \$17,625.79|06
The first digit cut is 0, which is 4 or less. The part you keep stays the same.
Answer: $\approx$ \$17,625.79

28. \$350.48 to the nearest dollar
Draw a cut-off line: \$350.|48
The first digit cut is 4, which is 4 or less. The part you keep stays the same.
Answer: $\approx$ \$350

29. \$129.50 to the nearest dollar
Draw a cut-off line: \$129.|50
The first digit cut is 5, which is 5 or more, so round up.
Answer: $\approx$ \$130

30. \$99.61 to the nearest dollar
Draw a cut-off line: \$99.|61
The first digit cut is 6, which is 5 or more, so round up.
Answer: $\approx$ \$100

31. \$29.37 to the nearest dollar
Draw a cut-off line: \$29.|37
The first digit cut is 3, which is 4 or less. The part you keep stays the same.
Answer: $\approx$ \$29

32. *Estimate:*

$$\begin{array}{r} 6 \\ 400 \\ +\ 20 \\ \hline 426 \end{array}$$

Exact:

$$\begin{array}{r} 5.81 \\ 423.96 \\ +\ 15.09 \\ \hline 444.86 \end{array}$$

33. *Estimate:*

$$\begin{array}{r} 80 \\ 1 \\ 100 \\ 1 \\ +\ 30 \\ \hline 212 \end{array}$$

Exact:

$$\begin{array}{r} 75.600 \\ 1.290 \\ 122.045 \\ 0.880 \\ +\ 33.700 \\ \hline 233.515 \end{array}$$

34. *Estimate:*

$$\begin{array}{r} 300 \\ -\ 20 \\ \hline 280 \end{array}$$

Exact:

$$\begin{array}{r} 308.5 \\ -\ 17.8 \\ \hline 290.7 \end{array}$$

35. *Estimate:*

$$\begin{array}{r} 9 \\ -\ 8 \\ \hline 1 \end{array}$$

Exact:

$$\begin{array}{r} 9.2000 \\ -\ 7.9316 \\ \hline 1.2684 \end{array}$$

36. *Estimate:*

$$\begin{array}{r} 80 \\ -\ 50 \\ \hline 30 \end{array} \text{ million}$$

Exact:

$$\begin{array}{r} 81.3 \\ -\ 49.9 \\ \hline 31.4 \end{array}$$

31.4 million more people go walking than camping.

37. Add the amounts of the two checks.

Estimate:

$$\begin{array}{r} \$200 \\ +\ 40 \\ \hline \$240 \end{array}$$

Exact:

$$\begin{array}{r} \$215.53 \\ +\ 44.67 \\ \hline \$260.20 \end{array}$$

The total amount of the two checks was \$260.20. Now subtract from her balance of \$306.

Estimate:	*Exact:*
$300	$306.00
− 240	− 260.20
$60	$45.80

The new balance is $45.80.

38. First total the money that Joey spent.

Estimate:	*Exact:*
$2	$1.59
5	5.33
+ 20	+ 18.94
$27	$25.86

Then subtract to find the change.

Estimate:	*Exact:*
$30	$30.00
− 27	− 25.86
$3	$4.14

Joey's change was $4.14.

39. Add the kilometers that she raced each day.

Estimate:	*Exact:*
2	2.30
4	4.00
+ 5	+ 5.25
11 km	11.55 km

Roseanne raced 11.55 kilometers.

40.

Estimate:	*Exact:*
6	6.138
× 4	× 3.7
24	4 2966
	18 414
	22.7106

41.

Estimate:	*Exact:*
40	42.9
× 3	× 3.3
120	12 87
	128 7
	141.57

42. (5.6)(0.002)

5.6	←	*1 decimal place*
× 0.002	←	*3 decimal places*
0.0112	←	*4 decimal places*

43. 0.071(0.005)

0.071	←	*3 decimal places*
× 0.005	←	*3 decimal places*
0.000355	←	*6 decimal places*

44. $706.2 \div 12 = 58.85$

Estimate: $700 \div 10 = 70$

58.85 is *reasonable.*

45. $26.6 \div 2.8 = 0.95$

Estimate: $30 \div 3 = 10$

0.95 is *not reasonable.*

```
            9. 5
2.8^ | 2 6. 6^ 0   Move decimal point 1
       2 5 2        place in divisor and
         1 4  0     dividend; write one zero.
         1 4  0
              0
```

The correct answer is 9.5.

46.

```
   1 4. 4 6 6 6
3 | 4 3. 4 0 0 0
    3
    1 3
    1 2
      1 4
      1 2
        2 0      Write one zero.
        1 8
          2 0    Write one zero.
          1 8
            2 0  Write one zero.
            1 8
              2
```

14.4666 rounds to 14.467.

47. $\dfrac{72}{0.06} = \dfrac{7200}{6} = 1200$

48. $0.00048 \div 0.0012 = 4.8 \div 12$

```
     0. 4
12 | 4. 8
     4 8
       0
```

49. Multiply the hourly wage times the hours worked.

Pay for first 40 hours	*Pay rate after 40 hours*	*Pay for over 40 hours*
14.24	14.24	21.36
× 40	× 1.5	× 6.5
569.60	7 120	10 680
	14 24	128 16
	21.360	138.840

Add the two amounts.

continued

$$\begin{array}{r} \$569.60 \\ 138.84 \\ \hline \$708.44 \end{array}$$

Adrienne's total earnings were $708 (rounded).

50. Divide the cost of the book by the number of tickets in the book.

$$\begin{array}{r} 2.990 \\ 12\overline{)\$35.890} \\ 24 \\ \hline 118 \\ 108 \\ \hline 109 \\ 108 \\ \hline 10 \end{array}$$

Round to the nearest cent.
Draw a cut-off line: $2.99|0
The first digit cut is 0, which is 4 or less. The part you keep stays the same.

Each ticket costs $2.99 (rounded).

51. Divide the amount of the investment by the price per share.

$$\begin{array}{r} 133.3 \\ 3.75_\wedge\overline{)500.00_\wedge 0} \\ 375 \\ \hline 1250 \\ 1125 \\ \hline 1250 \\ 1125 \\ \hline 1250 \\ 1125 \\ \hline 125 \end{array}$$

Round 133.3 down. Kenneth could buy 133 shares.

52. Multiply the price per pound by the amount to be purchased.

$$\begin{array}{r} \$0.99 \\ \times\ 3.5 \\ \hline 495 \\ 2\,97 \\ \hline \$3.465 \end{array}$$

$3.465 rounds to $3.47.
Ms. Lee will pay $3.47.

53. $3.5^2 + 8.7(1.95)$ *Exponent*
$12.25 + 8.7(1.95)$ *Multiply*
$12.25 + 16.965 = 29.215$ *Add*

54. $11 - 3.06 \div (3.95 - 0.35)$ *Parentheses*
$11 - 3.06 \div 3.6$ *Divide*
$11 - 0.85 = 10.15$ *Subtract*

55. $3\frac{4}{5} = \frac{19}{5} = 3.8$

$$\begin{array}{r} 3.8 \\ 5\overline{)19.0} \\ 15 \\ \hline 40 \\ 40 \\ \hline 0 \end{array}$$

56. $\frac{16}{25} = 0.64$

$$\begin{array}{r} 0.64 \\ 25\overline{)16.00} \\ 150 \\ \hline 100 \\ 100 \\ \hline 0 \end{array}$$

57. $1\frac{7}{8} = \frac{15}{8} = 1.875$

$$\begin{array}{r} 1.875 \\ 8\overline{)15.000} \\ 8 \\ \hline 70 \\ 64 \\ \hline 60 \\ 56 \\ \hline 40 \\ 40 \\ \hline 0 \end{array}$$

58. $\frac{1}{9}$

$$\begin{array}{r} 0.1111 \\ 9\overline{)1.0000} \\ 9 \\ \hline 10 \\ 9 \\ \hline 10 \\ 9 \\ \hline 10 \\ 9 \\ \hline 1 \end{array}$$

0.1111 rounds to 0.111.
$\frac{1}{9} \approx 0.111$

59. 3.68, 3.806, 3.6008

$3.68 = 3.6800$
$3.806 = 3.8060$ *largest*
$3.6008 = 3.6008$ *smallest*

From smallest to largest:
3.6008, 3.68, 3.806

60. 0.215, 0.22, 0.209, 0.2102

$0.215 = 0.2150$
$0.22 = 0.2200$ *largest*
$0.209 = 0.2090$ *smallest*
$0.2102 = 0.2102$

From smallest to largest:
0.209, 0.2102, 0.215, 0.22

61. $0.17, \frac{3}{20}, \frac{1}{8}, 0.159$

$0.17 = 0.170$ *largest*

$\frac{3}{20} = 0.150$

$\frac{1}{8} = 0.125$ *smallest*

$0.159 = 0.159$

From smallest to largest:

$\frac{1}{8}, \frac{3}{20}, 0.159, 0.17$

62. **[4.3]** $89.19 + 0.075 + 310.6 + 5$

$$\begin{array}{r} {}^{11\ 1}\\ 89.190 \\ 0.075 \\ 310.600 \\ +\ 5.000 \\ \hline 404.865 \end{array}$$

Line up decimal points. *Write in zeros.*

63. **[4.4]** 72.8×3.5

$$\begin{array}{rll} 72.8 & \leftarrow & \textit{1 decimal place} \\ \times\ 3.5 & \leftarrow & \textit{1 decimal place} \\ \hline 36\,40 & & \\ 218\,4\ \ & & \\ \hline 254.80 & \leftarrow & \textit{2 decimal places} \end{array}$$

64. **[4.5]** $1648.3 \div 0.46 \approx 3583.2609$

3583.2609 rounds to 3583.261.

65. **[4.3]**

$$\begin{array}{r} {}^{2\ 9\ \ 9\ 9\ 9\ 10} \\ \not{3}\,\not{0}.\not{0}\,\not{0}\,\not{0}\,\not{0} \\ -\ 0.9\ 1\ 0\ 2 \\ \hline 2\ 9.0\ 8\ 9\ 8 \end{array}$$

66. **[4.4]** $4.38(0.007)$

$$\begin{array}{rll} 4.38 & \leftarrow & \textit{2 decimal places} \\ \times\ 0.007 & \leftarrow & \textit{3 decimal places} \\ \hline 0.03066 & \leftarrow & \textit{5 decimal places} \end{array}$$

67. **[4.5]**

$$\begin{array}{r} 9.\ 4 \\ 0.005_{\wedge}\overline{)0.\ 0\ 4\ 7_{\wedge}\ 0} \\ 4\ 5\ \ \ \\ \hline 2\ \ 0 \\ 2\ \ 0 \\ \hline 0 \end{array}$$

Move decimal point 3 places in divisor and dividend; write 0.

68. **[4.3]** $72.105 + 8.2 + 95.37$

$$\begin{array}{r} 72.105 \\ 8.200 \\ +\ 95.370 \\ \hline 175.675 \end{array}$$

Line up decimal points. *Write in zeros.*

69. **[4.5]** $81.36 \div 9$

$$\begin{array}{r} 9.\ 0\ 4 \\ 9\overline{)8\ 1.\ 3\ 6} \\ 8\ 1\ \ \ \ \ \ \\ \hline 0\ \ 3\ \ \\ 0\ \ \ \\ \hline 3\ 6 \\ 3\ 6 \\ \hline 0 \end{array}$$

70. **[4.5]** $(5.6 - 1.22) + 4.8(3.15)$ *Parentheses*

$4.38 + 4.8(3.15)$ *Multiply*

$4.38 + 15.12 = 19.50$ *Add*

71. **[4.4]** 0.455×18

$$\begin{array}{rll} 0.455 & \leftarrow & \textit{3 decimal places} \\ \times\ \ 18 & & \\ \hline 3\ 640 & & \\ 4\ 55\ \ & & \\ \hline 8.190 & \leftarrow & \textit{3 decimal places} \end{array}$$

72. **[4.4]** $(1.6)(0.58)$

$$\begin{array}{rll} 0.58 & \leftarrow & \textit{2 decimal places} \\ \times\ 1.6 & \leftarrow & \textit{1 decimal place} \\ \hline 348 & & \\ 58\ \ & & \\ \hline 0.928 & \leftarrow & \textit{3 decimal places} \end{array}$$

73. **[4.5]** $0.218\overline{)7.\ 6\ 3}$

$7.63 \div 0.218 = 35$

74. **[4.3]** $21.059 - 20.8$

$$\begin{array}{r} 21.059 \\ -\ 20.800 \\ \hline 0.259 \end{array}$$

Line up decimal points. *Write in zeros.*

75. **[4.5]** $18.3 - 3^2 \div 0.5$ *Exponent*

$18.3 - 9 \div 0.5$ *Divide*

$18.3 - 18 = 0.3$ *Subtract*

76. **[4.5]** Men's socks are 3 pairs for \$8.99. Divide the price by 3 to find the cost of one pair.

$$\begin{array}{r} 2.\ 9\ 9\ 6 \\ 3\overline{)8.\ 9\ 9\ 0} \\ 6\ \ \ \ \ \ \ \ \ \\ \hline 2\ 9\ \ \ \ \ \ \\ 2\ 7\ \ \ \ \ \ \\ \hline 2\ 9\ \ \ \\ 2\ 7\ \ \ \\ \hline 2\ 0 \\ 1\ 8 \\ \hline 2 \end{array}$$

\$2.996 rounds to \$3.00.

One pair of men's socks costs \$3.00 (rounded).

77. **[4.5]** Children's socks cost 6 pairs for \$5.00. Find the cost for one pair.

$$\begin{array}{r} 0.833 \\ 6\overline{)5.000} \\ \underline{4\,8} \\ 20 \\ \underline{18} \\ 20 \\ \underline{18} \\ 2 \end{array}$$

\$0.833 rounds to \$0.83.

Subtract to find how much more men's socks cost.

$$\begin{array}{r} \$3.00 \\ -\ 0.83 \\ \hline \$2.17 \end{array}$$

Men's socks cost \$2.17 (rounded) more.

78. **[4.4]** A dozen pair of socks is $4 \cdot 3 = 12$ pairs of socks. So multiply 4 times \$8.99.

$$\begin{array}{r} \$8.99 \\ \times\ 4 \\ \hline \$35.96 \end{array}$$

The cost for a dozen pair of men's socks is \$35.96.

79. **[4.4]** Five pairs of teen jeans cost \$19.95 times 5.

$$\begin{array}{r} \$19.95 \\ \times\ 5 \\ \hline \$99.75 \end{array}$$

Four pairs of women's jeans cost \$24.99 times 4.

$$\begin{array}{r} \$24.99 \\ \times\ 4 \\ \hline \$99.96 \end{array}$$

Add the two amounts.

$$\begin{array}{r} \$99.75 \\ +\ 99.96 \\ \hline \$199.71 \end{array}$$

Akiko would pay \$199.71.

80. **[4.3]** The highest regular price for athletic shoes is \$149.50. The cheapest sale price is \$71.

Subtract to find the difference.

$$\begin{array}{r} \$149.50 \\ -\ 71.00 \\ \hline \$78.50 \end{array}$$

The difference in price is \$78.50.

81. **[4.6]** $0.11 = 0.11$
$0.66 = 0.66$
$0.7 = 0.70$
$0.45 = 0.45$
$0.5 = 0.50$
$0.3 = 0.30$
$0.57 = 0.57$

List from smallest to largest.
0.11, 0.3, 0.45, 0.5, 0.57, 0.66, 0.7

(a) 0.7 mg is the largest.

A baked potato with skin has the highest amount of Vitamin B-6.

(b) 0.11 is the smallest.

A cup of orange juice has the lowest amount of Vitamin B-6.

(c) Subtract to find difference.

$$\begin{array}{r} 0.70 \\ -\ 0.11 \\ \hline 0.59 \end{array}$$

The difference is 0.59 milligram.

82. **[4.6]** **(a)** Banana: 0.66 mg
3 ounces of skinless chicken: 0.5 mg
1 cup of strawberries: 0.45

$$\begin{array}{r} \times\ 2 \\ \hline 0.90 \text{ mg} \end{array}$$

Add the amounts.

$$\begin{array}{rl} 0.66 & \textit{banana} \\ 0.50 & \textit{chicken} \\ +\ 0.90 & \textit{strawberries} \\ \hline 2.06 & \end{array}$$

You would get 2.06 milligrams of Vitamin B-6.

(b) Subtract the recommended amount, 2 mg, from the amount you got.

$$\begin{array}{r} 2.06 \\ -\ 2.00 \\ \hline 0.06 \end{array}$$

You got 0.06 milligram more than the recommended amount.

Chapter 4 Test

1. $18.4 = 18\frac{4}{10} = 18\frac{4 \div 2}{10 \div 2} = 18\frac{2}{5}$

2. $0.075 = \frac{75}{1000} = \frac{75 \div 25}{1000 \div 25} = \frac{3}{40}$

3. 60.007 is sixty and seven thousandths.

4. 0.0208 is two hundred eight ten-thousandths.

5. 725.6089 to the nearest tenth
Draw a cut-off line: 725.6|089
The first digit cut is 0, which is 4 or less. The part you keep stays the same.

Answer: ≈ 725.6

6. 0.62951 to the nearest thousandth
Draw a cut-off line: 0.629|51
The first digit cut is 5, which is 5 or more, so round up.

Answer: ≈ 0.630

7. \$1.4945 to the nearest cent
Draw a cut-off line: \$1.49|45
The first digit cut is 4, which is 4 or less. The part you keep stays the same.

Answer: $\approx$ \$1.49

8. \$7859.51 to the nearest dollar
Draw a cut-off line: \$7859.|51
The first digit cut is 5, which is 5 or more, so round up.

Answer: $\approx$ \$7860

9. 7.6 + 82.0128 + 39.59

Estimate: 8 + 80 + 40 = 128

Exact:
7.6000 *Line up decimals.*
82.0128 *Write in zeros.*
+ 39.5900
129.2028

10. 79.1 − 3.602

Estimate: 80 − 4 = 76

Exact:
79.100 *Line up decimals.*
− 3.602 *Write in zeros.*
75.498

11. 5.79(1.2)

Estimate: 6 × 1 = 6

Exact:
5.79
× 1.2
1 158
5 79
6.948

12. 20.04 ÷ 4.8

Estimate: $20 \div 5 = 4$

Exact:
$4.8_\wedge\overline{)\,20.0_\wedge400}$ = 4.175
192
84
48
360
336
240
240
0

13. 53.1 + 4.631 + 782 + 0.031

53.100 *Line up decimals.*
4.631 *Write in zeros.*
782.000
+ 0.031
839.762

14. 670 − 0.996

670.000 *Line up decimals.*
− 0.996 *Write in zeros.*
669.004

15. (0.0069)(0.007)

0.0069 ← *4 decimal places*
× 0.007 ← *3 decimal places*
0.0000483 ← *7 decimal places*

16. $0.15_\wedge\overline{)\,72.00_\wedge}$ = 480.
60
120
120
00
0
0

Move decimal point 2 places in dividend and divisor; write two zeros.

17. $2\frac{5}{8} = \frac{21}{8}$

$8\overline{)\,21.000}$ = 2.625
16
50
48
20
16
40
40
0

$2\frac{5}{8} = 2.625$

18. $0.44, 0.451, \frac{9}{20}, 0.4506$

Change to ten-thousandths, if necessary, and compare.

$0.44 = 0.4400$ *smallest*
$0.451 = 0.4510$ *largest*
$\frac{9}{20} = 0.4500$
$0.4506 = 0.4506$

From smallest to largest: $0.44, \frac{9}{20}, 0.4506, 0.451$

19. $6.3^2 - 5.9 + 3.4(0.5)$ *Exponent*
$39.69 - 5.9 + 3.4(0.5)$ *Multiply*
$39.69 - 5.9 + 1.7$ *Subtract*
$33.79 + 1.7 = 35.49$ *Add*

20. To find the balance, add the interest and the deposit to the balance and then subtract the bank charge.

$$\begin{array}{r} \$71.15 \\ 0.95 \\ +\ 390.77 \\ \hline \$462.87 \end{array} \qquad \begin{array}{r} \$462.87 \\ -\ 16.00 \\ \hline \$446.87 \end{array}$$

The new balance is $446.87.

21. 2.500 million *gadwalls*
2.551 million *wigeons*
2.560 million *pintails*
From highest number to lowest:
pintails, wigeons, gadwalls

22. Multiply the price per pound by the amount purchased.

$$\begin{array}{r} \$2.89 \\ \times\ 1.85 \\ \hline 1445 \\ 2\ 312 \\ 2\ 89 \\ \hline \$5.3465 \end{array}$$

Round $5.3465 to the nearest cent.
Draw a cut-off line: $5.34|65
The first digit cut is 6, which is 5 or more, so round up.

Mr. Yamamoto paid $5.35 for the cheese.

23. Subtract the two temperatures.

$$\begin{array}{r} 102.7 \\ -\ 99.9 \\ \hline 2.8 \end{array}$$

The temperature dropped 2.8 degrees.

24. Divide to find the cost per meter.

$$\begin{array}{r} 4.\ 5\ 5 \\ 3.4_\wedge \overline{)\ \$1\ 5.\ 4_\wedge\ 7\ 0} \\ 1\ 3\ 6 \\ \hline 1\ 8\ 7 \\ 1\ 7\ 0 \\ \hline 1\ 7\ 0 \\ 1\ 7\ 0 \\ \hline 0 \end{array}$$

Move decimal point 1 place in dividend and divisor; write one zero.

The fabric costs $4.55 per meter.

25. Answers will vary.

Cumulative Review Exercises (Chapters 1–4)

1. 1 9,0 7 6,5 4 2 (millions: 9; hundreds: 5; ones: 2)

2. 8 3. 0 7 5 4 (tens: 8; tenths: 0; thousandths: 5)

3. 499,501 to the nearest thousand: 49<u>9</u>,501

Next digit is 5 or more. 499 becomes 500. All digits to the right of the underlined place are changed to zero. **500,000**

4. 602.4937 to the nearest hundredth
Draw a cut-off line: 602.49|37
The first digit cut is 3, which is 4 or less. The part you keep stays the same.

Answer: ≈ 602.49

5. $709.60 to the nearest dollar
Draw a cut-off line: $709.|60
The first digit cut is 6, which is 5 or more, so round up.

Answer: $\approx \$710$

6. $0.0528 to the nearest cent
Draw a cut-off line: $0.05|28
The first digit cut is 2, which is 4 or less. The part you keep stays the same.

Answer: $\approx \$0.05$

7. *Estimate:* *Exact:*

$$\begin{array}{r} 4{,}000 \\ 600 \\ +\ 9{,}000 \\ \hline 13{,}600 \end{array} \qquad \begin{array}{r} 3{,}672 \\ 589 \\ +\ 9{,}078 \\ \hline 13{,}339 \end{array}$$

8. *Estimate:*

$$\begin{array}{r} 4 \\ 20 \\ +1 \\ \hline 25 \end{array}$$

Exact:

$$\begin{array}{r} 4.060 \\ 15.700 \\ +0.923 \\ \hline 20.683 \end{array}$$

9. *Estimate:*

$$\begin{array}{r} 5000 \\ -2000 \\ \hline 3000 \end{array}$$

Exact:

$$\begin{array}{r} \scriptstyle 4\,10\,0\,18 \\ \not{5}\,\not{0}\,1\,\not{8} \\ -1\,8\,0\,9 \\ \hline 3\,2\,0\,9 \end{array}$$

10. *Estimate:*

$$\begin{array}{r} 50 \\ -7 \\ \hline 43 \end{array}$$

Exact:

$$\begin{array}{r} \scriptstyle 9 \\ \scriptstyle 4\,11\;\;5\,\not{10}\,10 \\ \not{5}\,\not{1}.\not{6}\,\not{0}\,\not{0} \\ -\quad 7.0\,9\,4 \\ \hline 4\,4.5\,0\,6 \end{array}$$

11. *Estimate:*

$$\begin{array}{r} 3{,}000 \\ \times\ 200 \\ \hline 600{,}000 \end{array}$$

Exact:

$$\begin{array}{r} 3{,}317 \\ \times\ 166 \\ \hline 19\,902 \\ 199\,02 \\ 331\,7 \\ \hline 550{,}622 \end{array}$$

12. *Estimate:*

$$\begin{array}{r} 7 \\ \times\ 7 \\ \hline 49 \end{array}$$

Exact:

$$\begin{array}{r} 6.82 \\ \times\ 7.3 \\ \hline 2\,046 \\ 47\,74 \\ \hline 49.786 \end{array}$$

13. *Estimate:*

$$50\overline{)100{,}000}\quad = 2{,}000$$

Exact:

$$\begin{array}{r} 2{,}690 \\ 46\overline{)123{,}740} \\ 92 \\ \hline 317 \\ 276 \\ \hline 414 \\ 414 \\ \hline 00 \\ 0 \\ \hline 0 \end{array}$$

14. *Estimate:*

$$8\overline{)40}\quad = 5$$

Exact:

$$\begin{array}{r} 4.5 \\ 8.4_\wedge\overline{)37.8_\wedge 0} \\ 33\,6 \\ \hline 4\,2\,0 \\ 4\,2\,0 \\ \hline 0 \end{array}$$

15. *Estimate:* $2 \cdot 4 = 8$

Exact: $1\frac{9}{10} \cdot 3\frac{3}{4} = \frac{19}{\cancel{10}_2} \cdot \frac{\cancel{15}^3}{4} = \frac{57}{8} = 7\frac{1}{8}$

16. *Estimate:* $2 \div 1 = 2$

Exact: $2\frac{1}{3} \div \frac{5}{6} = \frac{7}{3} \div \frac{5}{6} = \frac{7}{\cancel{3}_1} \cdot \frac{\cancel{6}^2}{5} = \frac{14}{5} = 2\frac{4}{5}$

17. *Estimate:* $2 + 2 = 4$

Exact:

$1\frac{4}{5} + 1\frac{2}{3} = \frac{9}{5} + \frac{5}{3} = \frac{27}{15} + \frac{25}{15} = \frac{52}{15} = 3\frac{7}{15}$

18. *Estimate:* $5 - 2 = 3$

Exact:

$4\frac{1}{2} - 1\frac{7}{8} = \frac{9}{2} - \frac{15}{8} = \frac{36}{8} - \frac{15}{8} = \frac{21}{8} = 2\frac{5}{8}$

19. $10 - 0.329$

$$\begin{array}{r} 10.000 \\ -\ 0.329 \\ \hline 9.671 \end{array}$$

Line up decimals. Write in zeros.

20. $2\frac{3}{5} \cdot \frac{5}{9} = \frac{13}{5} \cdot \frac{5}{9} = \frac{13}{\cancel{5}_1} \cdot \frac{\cancel{5}^1}{9} = \frac{13}{9} = 1\frac{4}{9}$

21. $9 + 72{,}417 + 799$

$$\begin{array}{r} 9 \\ 72{,}417 \\ +\ 799 \\ \hline 73{,}225 \end{array}$$

22. $11\frac{1}{5} \div 8 = \frac{56}{5} \div \frac{8}{1} = \frac{\cancel{56}^7}{5} \cdot \frac{1}{\cancel{8}_1} = \frac{7}{5} = 1\frac{2}{5}$

23. $5006 - 92$

$$\begin{array}{r} 5006 \\ -\ 92 \\ \hline 4914 \end{array}$$

24. $0.7 + 85 + 7.903$

$$\begin{array}{r} 0.700 \\ 85.000 \\ +\ 7.903 \\ \hline 93.603 \end{array}$$

Line up decimals. Write in zeros.

25.
$$\begin{array}{r} 404 \text{ R3} \\ 7\overline{)2831} \\ \underline{28} \\ 03 \\ \underline{0} \\ 31 \\ \underline{28} \\ 3 \end{array}$$

26.
$$\begin{array}{rcl} \frac{5}{6} &=& \frac{20}{24} \\ +\frac{7}{8} &=& \frac{21}{24} \\ \hline && \frac{41}{24} = 1\frac{17}{24} \end{array}$$

27. 332×704

$$\begin{array}{r} 332 \\ \times\ 704 \\ \hline 1\,328 \\ 232\,40 \\ \hline 233{,}728 \end{array}$$

28. $(0.006)(5.44)$

$$\begin{array}{rl} 0.006 & \leftarrow \textit{3 decimal places} \\ \times\ 5.44 & \leftarrow \textit{2 decimal places} \\ \hline 24 & \\ 24 & \\ 30 & \\ \hline 0.03264 & \leftarrow \textit{5 decimal places} \end{array}$$

29. $3.2(2.5)$

$$\begin{array}{rl} 3.2 & \leftarrow \textit{1 decimal place} \\ \times\ 2.5 & \leftarrow \textit{1 decimal place} \\ \hline 1\,60 & \\ 6\,4 & \\ \hline 8.00 & \leftarrow \textit{2 decimal places} \end{array}$$

30. $25.2 \div 0.56$

$$\begin{array}{r} 45 \\ 0.56_\wedge\overline{)25.20_\wedge} \\ \underline{224} \\ 280 \\ \underline{280} \\ 0 \end{array}$$

31. $\frac{2}{3} \div 5\frac{1}{6} = \frac{2}{3} \div \frac{31}{6} = \frac{2}{\cancel{3}_1} \cdot \frac{\cancel{6}^2}{31} = \frac{4}{31}$

32.
$$\begin{array}{rcrcr} 5\frac{1}{4} &=& 5\frac{3}{12} &=& 4\frac{15}{12} \\ -4\frac{7}{12} &=& 4\frac{7}{12} &=& 4\frac{7}{12} \\ \hline &&&& \frac{8}{12} = \frac{2}{3} \end{array}$$

33.
$$\begin{array}{r} 0.505 \\ 9.3_\wedge\overline{)4.7_\wedge000} \\ \underline{465} \\ 500 \\ \underline{465} \\ 35 \end{array}$$

Round 0.505 to the nearest hundredth.
Draw a cut-off line: 0.50|5
The first digit cut is 5, which is 5 or more, so round up.

Answer: ≈ 0.51

34.
$10 - 4 \div 2 \cdot 3$ *Divide*
$10 - 2 \cdot 3$ *Multiply*
$10 - 6 = 4$ *Subtract*

35.
$\sqrt{36} + 3(8) - 4^2$ *Exponent*
$\sqrt{36} + 3(8) - 16$ *Square Root*
$6 + 3(8) - 16$ *Multiply*
$6 + 24 - 16$ *Add*
$30 - 16 = 14$ *Subtract*

36. $\frac{2}{3}\left(\frac{7}{8} - \frac{1}{2}\right) = \frac{2}{3}\left(\frac{7}{8} - \frac{4}{8}\right) = \frac{2}{3} \cdot \left(\frac{3}{8}\right) = \frac{\cancel{2}^1}{\cancel{3}_1} \cdot \frac{\cancel{3}^1}{\cancel{8}_4} = \frac{1}{4}$

37.
$0.9^2 + 10.6 \div 0.53$ *Exponent*
$0.81 + 10.6 \div 0.53$ *Divide*
$0.81 + 20 = 20.81$ *Add*

38. $4^3 \cdot 3^2 = (4 \cdot 4 \cdot 4) \cdot (3 \cdot 3) = 64 \cdot 9 = 576$

39. $14^2 = 196$, so $\sqrt{196} = 14$.

40.

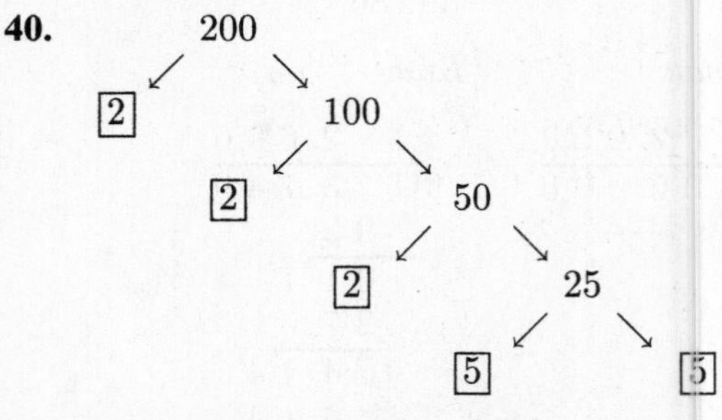

$200 = 2 \cdot 2 \cdot 2 \cdot 5 \cdot 5 = 2^3 \cdot 5^2$

41. 40.035 is forty and thirty-five thousandths.

42. Three hundred six ten-thousandths is 0.0306.

43. $0.125 = \frac{125}{1000} = \frac{125 \div 125}{1000 \div 125} = \frac{1}{8}$

44. $3.08 = 3\frac{8}{100} = 3\frac{8 \div 4}{100 \div 4} = 3\frac{2}{25}$

45. $2\frac{3}{5} = \frac{13}{5} = 2.6$

$$\begin{array}{r} 2.6 \\ 5\overline{)13.0} \\ \underline{10} \\ 3\,0 \\ \underline{3\,0} \\ 0 \end{array}$$

46. $\frac{7}{11}$

$$\begin{array}{r} 0.6363 \\ 11\overline{)7.0000} \\ \underline{6\,6} \\ 4\,0 \\ \underline{3\,3} \\ 7\,0 \\ \underline{6\,6} \\ 4\,0 \\ \underline{3\,3} \\ 7 \end{array}$$

0.6363 rounded to the nearest thousandth is 0.636.

$\frac{7}{11} \approx 0.636$

47. $\frac{5}{8} \_\_ \frac{4}{9}$

$\frac{5}{8} = \frac{45}{72}$ and $\frac{4}{9} = \frac{32}{72}$

$\frac{45}{72} > \frac{32}{72}$, so $\frac{5}{8}\ \boxed{>}\ \frac{4}{9}$.

48. 7.005, 7.5005, 7.5, 7.505

$7.005 = 7.0050$ *smallest*
$7.5005 = 7.5005$
$7.5 = 7.5000$
$7.505 = 7.5050$ *largest*

From smallest to largest:
7.005, 7.5, 7.5005, 7.505

49. $\frac{7}{8}$, 0.8, $\frac{21}{25}$, 0.8015

$\frac{7}{8} = 0.8750$ *largest*
$0.8 = 0.8000$ *smallest*
$\frac{21}{25} = 0.8400$
$0.8015 = 0.8015$

From smallest to largest:
0.8, 0.8015, $\frac{21}{25}$, $\frac{7}{8}$

50. $20\frac{1}{4} < 21\frac{1}{16} < 21\frac{1}{8}$
Order size M (medium).

51. **Size XXS:**

$$\begin{array}{rcr} 18 & = & 17\frac{2}{2} \\ -\,16\frac{1}{2} & = & 16\frac{1}{2} \\ \hline & & 1\frac{1}{2} \end{array}$$

The difference for size XXS is $1\frac{1}{2}$ inches.

Size XS:

$$\begin{array}{rcr} 19 & = & 18\frac{4}{4} \\ -\,18\frac{1}{4} & = & 18\frac{1}{4} \\ \hline & & \frac{3}{4} \end{array}$$

The difference for size XS is $\frac{3}{4}$ inch.

Size S:

$$\begin{array}{rcr} 20 & = & 19\frac{4}{4} \\ -\,19\frac{1}{4} & = & 19\frac{1}{4} \\ \hline & & \frac{3}{4} \end{array}$$

The difference for size S is $\frac{3}{4}$ inch.

Size M:

$$\begin{array}{rcrcr} 21\frac{1}{8} & = & 21\frac{1}{8} & = & 20\frac{9}{8} \\ -\,20\frac{1}{4} & = & 20\frac{2}{8} & = & 20\frac{2}{8} \\ \hline & & & & \frac{7}{8} \end{array}$$

The difference for size M is $\frac{7}{8}$ inch.

Size L:

$$\begin{array}{rcrcr} 22\frac{1}{4} & = & 22\frac{1}{4} & = & 21\frac{5}{4} \\ -\,21\frac{1}{2} & = & 21\frac{2}{4} & = & 21\frac{2}{4} \\ \hline & & & & \frac{3}{4} \end{array}$$

The difference for size L is $\frac{3}{4}$ inch.

52. $\frac{1}{8} = 0.125$

$$\begin{array}{r} 0.125 \\ 8\overline{)1.000} \\ \underline{8} \\ 20 \\ \underline{16} \\ 40 \\ \underline{40} \\ 0 \end{array}$$

$\frac{1}{4} = 0.25$

$$\begin{array}{r} 0.25 \\ 4\overline{)1.00} \\ \underline{8} \\ 20 \\ \underline{20} \\ 0 \end{array}$$

$$\begin{array}{rcl} 21\frac{1}{8} & = & 21.125 \\ \underline{-\ 20\frac{1}{4}} & = & \underline{20.250} \\ & & 0.875 \text{ in.} \end{array}$$

$0.875 = \frac{875}{1000} = \frac{875 \div 125}{1000 \div 125} = \frac{7}{8}$ inch, which is the same as the answer in Exercise 51.

53. Answers will vary. Many people prefer using decimals because you do not need to find a common denominator or rewrite answers in lowest terms.

54. He started with two \$20 bills, or \$40. He spent \$17.96 and \$0.87.

Estimate: \$40 – \$18 – \$1 = \$21

First find how much money Lameck spent.

$$\begin{array}{r} \$17.96 \\ \underline{+\ 0.87} \\ \$18.83 \end{array}$$

Now subtract from \$40.

$$\begin{array}{r} \$40.00 \\ \underline{-\ 18.83} \\ \$21.17 \end{array}$$

He has \$21.17 left.

55. Subtract the two measurements.

Estimate: 50 – 47 = 3 inches

Exact:
$$\begin{array}{rcl} 50 & = & 49\frac{8}{8} \\ \underline{-\ 46\frac{5}{8}} & = & \underline{46\frac{5}{8}} \\ & & 3\frac{3}{8} \end{array}$$

She has grown $3\frac{3}{8}$ inches.

56. Multiply the hourly wage, \$11.63, by the hours worked, 16.5.

Estimate: \$10 × 20 = \$200

Exact:
$$\begin{array}{r} \$11.63 \\ \underline{\times\ 16.5} \\ 5\,815 \\ 69\,78 \\ \underline{116\,3} \\ 191.895 \end{array}$$

\$191.895 rounded to the nearest cent is \$191.90. Sharon earned \$191.90 (rounded).

57. Find the number of students in each size of classroom (8 have 22 students and 12 have 26 students). Then add the two sums.

Estimate: (8 × 20) + (10 × 30) = 160 + 300 = 460 students

Exact:
$$\begin{array}{r} 22 \\ \underline{\times\ 8} \\ 176 \end{array} \qquad \begin{array}{r} 26 \\ \underline{\times\ 12} \\ 52 \\ \underline{26} \\ 312 \end{array} \qquad \begin{array}{r} 176 \\ \underline{+\ 312} \\ 488 \end{array}$$

There are 488 students attending the school.

58. Add the two purchases.

Estimate: 2 + 4 = 6 yards

Exact:
$$\begin{array}{rcl} 2\frac{1}{3} & = & 2\frac{8}{24} \\ \underline{+\ 3\frac{7}{8}} & = & \underline{3\frac{21}{24}} \\ & & 5\frac{29}{24} = 5 + 1 + \frac{5}{24} \\ & & = 6\frac{5}{24} \end{array}$$

Toshihiro bought $6\frac{5}{24}$ yards of fabric.

59. Add the deposit to the amount Kimberly had in her account. Then subtract the amount for the check that she wrote and the overdraft charge.

Estimate: \$30 + \$200 – \$40 – \$20 = \$170

Exact:

\$29.44	*Balance on hand*
+ \$220.06	*Deposit*
\$249.50	
– 40.00	*Check*
\$209.50	
– 18.00	*Overdraft*
\$191.50	*New Balance*

The new balance in her account is \$191.50.

60. Divide the price of the grapes, \$6.18, by the number of pounds of grapes, 2.7, that Paulette bought.

Estimate: \$6 ÷ \$3 = \$2 per pound

Exact:

```
            2. 2 8 8
 2.7^ ) $6. 1^ 8 0 0   Move decimal point
        5 4            one place in divisor
        ---            and dividend.
          7 8
          5 4
          ---
          2 4 0
          2 1 6
          -----
            2 4 0
            2 1 6
            -----
              2 4
```

2.288 rounds to 2.29.

The cost of the grapes is \$2.29 per pound (rounded).

61. Divide the amount of the grant by the number of students.

Estimate: \$80,000 ÷ 100 = \$800

Exact:

```
           7 2 8. 9
 107 ) $7 8, 0 0 0. 0
         7 4 9
         -----
           3 1 0
           2 1 4
           -----
             9 6 0
             8 5 6
             -----
             1 0 4 0
               9 6 3
               -----
                 7 7
```

728.9 rounds to 729.

Each student could be given \$729 (rounded).

62. **(a)** Divide 3.5 by 0.4.

```
          8. 7 5
 0.4^ ) 3. 5^ 0 0
        3 2
        ---
          3 0
          2 8
          ---
            2 0
            2 0
            ---
              0
```

8.75 rounded to the nearest whole number is 9. It will take 9 days (rounded) for a hamster to eat enough food to equal its body weight.

(b) Divide 140 by 8.75.

```
                1 6.
 8.75^ ) 1 4 0. 0 0^
           8 7 5
           -----
           5 2 5 0
           5 2 5 0
           -------
                 0
```

The woman would need to eat 16 pounds of food each day. If we use 9 days instead of 8.75 days, we get 15.6 pounds (rounded) per day.

63. Multiply 0.004 and 80.

$$\begin{array}{r} 0.004 \\ \times\ 80 \\ \hline 0.320 \end{array}$$

The average weight of food eaten each day by a queen bee is 0.32 ounce.

64. Multiply $1\frac{3}{5} = 1.6$ and 0.07.

$$\begin{array}{rcl} 1.6 & \leftarrow & \textit{1 decimal place} \\ \times\ 0.07 & \leftarrow & \textit{2 decimal places} \\ \hline 0.112 & \leftarrow & \textit{3 decimal places} \end{array}$$

The hummingbird's body weight is 0.112 ounce.

$0.07 = \frac{7}{100}$

$$1\frac{3}{5} \times \frac{7}{100} = \frac{\overset{2}{\cancel{8}}}{5} \times \frac{7}{\underset{25}{\cancel{100}}} = \frac{14}{125}$$

```
         0. 1 1 2
 125 ) 1 4. 0 0 0
         1 2 5
         -----
           1 5 0
           1 2 5
           -----
             2 5 0
             2 5 0
             -----
                 0
```

Using decimal numbers and using fractions lead to the same answer.

CHAPTER 5 RATIO AND PROPORTION

5.1 Ratios

5.1 Margin Exercises

1. **(a)** \$7 spent on fresh fruit to \$5 spent on milk

$$\frac{7}{5}$$

(b) \$5 spent on milk to \$14 spent on meat

$$\frac{5}{14}$$

(c) \$14 spent on meat to \$5 spent on milk

$$\frac{14}{5}$$

2. **(a)** 9 hours to 12 hours

$$\frac{9}{12} = \frac{9 \div 3}{12 \div 3} = \frac{3}{4}$$

(b) 100 meters to 50 meters

$$\frac{100}{50} = \frac{100 \div 50}{50 \div 50} = \frac{2}{1}$$

(c) width to length or 24 feet to 48 feet

$$\frac{24}{48} = \frac{24 \div 24}{48 \div 24} = \frac{1}{2}$$

3. **(a)** The increase in price is

\$7.00 − \$5.50 = \$1.50.

The ratio of increase in price to original price is

$$\frac{1.50}{5.50}.$$

Rewrite as two whole numbers.

$$\frac{1.50}{5.50} = \frac{1.50 \times 100}{5.50 \times 100} = \frac{150}{550}$$

Write in lowest terms.

$$\frac{150 \div 50}{550 \div 50} = \frac{3}{11}$$

(b) The decrease in hours is

4.5 hours − 3 hours = 1.5 hours.

The ratio of decrease in hours to the original number of hours is

$$\frac{1.5}{4.5}.$$

Rewrite as two whole numbers.

$$\frac{1.5 \cdot 10}{4.5 \cdot 10} = \frac{15}{45}$$

Write in lowest terms.

$$\frac{15 \div 15}{45 \div 15} = \frac{1}{3}$$

4. **(a)** $3\frac{1}{2}$ to 4

$$\frac{3\frac{1}{2}}{4} = \frac{\frac{7}{2}}{\frac{4}{1}} = \frac{7}{2} \div \frac{4}{1} = \frac{7}{2} \cdot \frac{1}{4} = \frac{7}{8}$$

(b) $5\frac{5}{8}$ pounds to $3\frac{3}{4}$ pounds

$$\frac{5\frac{5}{8}}{3\frac{3}{4}} = \frac{\frac{45}{8}}{\frac{15}{4}} = \frac{45}{8} \div \frac{15}{4} = \frac{\overset{3}{\cancel{45}}}{\underset{2}{\cancel{8}}} \cdot \frac{\overset{1}{\cancel{4}}}{\underset{1}{\cancel{15}}} = \frac{3}{2}$$

(c) $3\frac{1}{2}$ inches to $\frac{7}{8}$ inch

$$\frac{3\frac{1}{2}}{\frac{7}{8}} = \frac{\frac{7}{2}}{\frac{7}{8}} = \frac{7}{2} \div \frac{7}{8} = \frac{\overset{1}{\cancel{7}}}{\underset{1}{\cancel{2}}} \cdot \frac{\overset{4}{\cancel{8}}}{\underset{1}{\cancel{7}}} = \frac{4}{1}$$

5. **(a)** 9 inches to 6 feet

6 feet $= 6 \cdot 12$ inches
$= 72$ inches

$$\frac{9 \text{ inches}}{72 \text{ inches}} = \frac{9}{72} = \frac{9 \div 9}{72 \div 9} = \frac{1}{8}$$

(b) 2 days to 8 hours

2 days $= 2 \cdot 24$ hours $= 48$ hours

$$\frac{48 \text{ hours}}{8 \text{ hours}} = \frac{48}{8} = \frac{48 \div 8}{8 \div 8} = \frac{6}{1}$$

(c) 7 yards to 14 feet

7 yards $= 7 \cdot 3$ feet $= 21$ feet

$$\frac{21 \text{ feet}}{14 \text{ feet}} = \frac{21}{14} = \frac{21 \div 7}{14 \div 7} = \frac{3}{2}$$

(d) 3 quarts to 3 gallons

3 gallons $= 3 \cdot 4$ quarts $= 12$ quarts

$$\frac{3 \text{ quarts}}{12 \text{ quarts}} = \frac{3}{12} = \frac{3 \div 3}{12 \div 3} = \frac{1}{4}$$

(e) 25 minutes to 2 hours

2 hours $= 2 \cdot 60$ minutes
$= 120$ minutes

$$\frac{25 \text{ minutes}}{120 \text{ minutes}} = \frac{25}{120} = \frac{25 \div 5}{120 \div 5} = \frac{5}{24}$$

(f) 4 pounds to 12 ounces

4 pounds $= 4 \cdot 16$ ounces
$= 64$ ounces

$$\frac{64 \text{ ounces}}{12 \text{ ounces}} = \frac{64}{12} = \frac{64 \div 4}{12 \div 4} = \frac{16}{3}$$

5.1 Section Exercises

1. 8 to 9 is written $\frac{8}{9}$

3. \$100 to \$50

$$\frac{\$100}{\$50} = \frac{100}{50} = \frac{100 \div 50}{50 \div 50} = \frac{2}{1}$$

5. 30 minutes to 90 minutes

$$\frac{30 \text{ minutes}}{90 \text{ minutes}} = \frac{30}{90} = \frac{30 \div 30}{90 \div 30} = \frac{1}{3}$$

7. 80 miles to 50 miles

$$\frac{80 \text{ miles}}{50 \text{ miles}} = \frac{80}{50} = \frac{80 \div 10}{50 \div 10} = \frac{8}{5}$$

9. 6 hours to 16 hours

$$\frac{6 \text{ hours}}{16 \text{ hours}} = \frac{6 \div 2}{16 \div 2} = \frac{3}{8}$$

11. \$4.50 to \$3.50

$$\frac{\$4.50}{\$3.50} = \frac{4.50}{3.50} = \frac{4.50 \cdot 10}{3.50 \cdot 10} = \frac{45}{35} = \frac{45 \div 5}{35 \div 5} = \frac{9}{7}$$

13. 15 to $2\frac{1}{2}$

$$\frac{15}{2\frac{1}{2}} = \frac{\frac{15}{1}}{\frac{5}{2}} = \frac{15}{1} \div \frac{5}{2} = \frac{\overset{3}{\cancel{15}}}{1} \cdot \frac{2}{\underset{1}{\cancel{5}}} = \frac{6}{1}$$

15. $1\frac{1}{4}$ to $1\frac{1}{2}$

$$\frac{1\frac{1}{4}}{1\frac{1}{2}} = \frac{\frac{5}{4}}{\frac{3}{2}} = \frac{5}{4} \div \frac{3}{2} = \frac{5}{\underset{2}{\cancel{4}}} \cdot \frac{\overset{1}{\cancel{2}}}{3} = \frac{5}{6}$$

17. 4 feet to 30 inches

4 feet $= 4 \cdot 12$ inches $= 48$ inches

$$\frac{4 \text{ feet}}{30 \text{ inches}} = \frac{48 \text{ inches}}{30 \text{ inches}} = \frac{48}{30} = \frac{48 \div 6}{30 \div 6} = \frac{8}{5}$$

(Do not write the ratio as $1\frac{3}{5}$.)

19. 5 minutes to 1 hour

1 hour $= 60$ minutes

$$\frac{5 \text{ minutes}}{1 \text{ hour}} = \frac{5 \text{ minutes}}{60 \text{ minutes}} = \frac{5}{60} = \frac{5 \div 5}{60 \div 5} = \frac{1}{12}$$

21. 15 hours to 2 days

2 days $= 2 \cdot 24$ hours $= 48$ hours

$$\frac{15 \text{ hours}}{2 \text{ days}} = \frac{15 \text{ hours}}{48 \text{ hours}} = \frac{15}{48} = \frac{15 \div 3}{48 \div 3} = \frac{5}{16}$$

23. 5 gallons to 5 quarts

5 gallons $= 5 \cdot 4$ quarts $= 20$ quarts

$$\frac{5 \text{ gallons}}{5 \text{ quarts}} = \frac{20 \text{ quarts}}{5 \text{ quarts}} = \frac{20}{5} = \frac{20 \div 5}{5 \div 5} = \frac{4}{1}$$

25. Thanksgiving cards to graduation cards

$$\frac{30 \text{ million}}{60 \text{ million}} = \frac{30}{60} = \frac{30 \div 30}{60 \div 30} = \frac{1}{2}$$

27. Valentine's Day cards to Halloween cards

$$\frac{900 \text{ million}}{25 \text{ million}} = \frac{900}{25} = \frac{900 \div 25}{25 \div 25} = \frac{36}{1}$$

29. Answers will vary. One possibility is stocking cards of various types in the same ratios as those in the table.

31. *Candle in the Wind* to *White Christmas*

$$\frac{36 \text{ million}}{30 \text{ million}} = \frac{36}{30} = \frac{36 \div 6}{30 \div 6} = \frac{6}{5}$$

Candle in the Wind to *Rock Around the Clock*

$$\frac{36 \text{ million}}{17 \text{ million}} = \frac{36}{17}$$

33. To get a ratio of $\frac{3}{1}$, we can start with the smallest value in the table and see if there is a value that is three times the smallest. $3 \times 9 = 27$ and 27 million is not in the table. $3 \times 10 = 30$, so there are at least 2 pairs of songs that give a ratio of $\frac{3}{1}$: *White Christmas* to *It's Now or Never* and *White Christmas* to *I Will Always Love You.*
$3 \times 12 = 36$, so there is another pair: *Candle in the Wind* to *I Want to Hold Your Hand.*
$3 \times 17 = 51$, which is greater than the largest value in the table, so we do not need to look for any more pairs.

35. **(a)** rent to utilities

$$\frac{\$750}{\$125} = \frac{750}{125} = \frac{750 \div 125}{125 \div 125} = \frac{6}{1}$$

(b) rent to food

$$\frac{\$750}{\$300} = \frac{750}{300} = \frac{750 \div 150}{300 \div 150} = \frac{5}{2}$$

(c) rent and utilities to total expenses

rent and utilities: $\$750 + \$125 = \$875$

total expenses:
$\$225 + \$125 + \$200 + \$400 + \$300 + \$750 = \$2000$

$$\frac{\$875}{\$2000} = \frac{875}{2000} = \frac{875 \div 125}{2000 \div 125} = \frac{7}{16}$$

37. $$\frac{\text{longest side}}{\text{shortest side}} = \frac{7 \text{ feet}}{5 \text{ feet}} = \frac{7}{5}$$

The ratio of the length of the longest side to the length of the shortest side is $\frac{7}{5}$.

39. $\dfrac{\text{longest side}}{\text{shortest side}} = \dfrac{1.8 \text{ meters}}{0.3 \text{ meters}} = \dfrac{1.8}{0.3}$
$= \dfrac{1.8 \cdot 10}{0.3 \cdot 10} = \dfrac{18}{3} = \dfrac{18 \div 3}{3 \div 3} = \dfrac{6}{1}$

The ratio of the length of the longest side to the length of the shortest side is $\frac{6}{1}$.

41. $\dfrac{9\frac{1}{2} \text{ inches}}{4\frac{1}{4} \text{ inches}} = \dfrac{9\frac{1}{2}}{4\frac{1}{4}} = \dfrac{\frac{19}{2}}{\frac{17}{4}} = \dfrac{19}{2} \div \dfrac{17}{4}$
$= \dfrac{19}{\cancel{2}_1} \cdot \dfrac{\cancel{4}^2}{17} = \dfrac{38}{17}$

The ratio of the length of the longest side to the length of the shortest side is $\frac{38}{17}$.

43. The increase in price is

$\$13.50 - \$10.80 = \$2.70.$

$\dfrac{\$2.70}{\$10.80} = \dfrac{2.70}{10.80} = \dfrac{2.70 \cdot 10}{10.80 \cdot 10} = \dfrac{27}{108}$
$= \dfrac{27 \div 27}{108 \div 27} = \dfrac{1}{4}$

The ratio of the increase in price to the original price is $\frac{1}{4}$.

45. $59\dfrac{1}{2} \text{ days} \div 7 = \dfrac{\cancel{119}^{17}}{2} \cdot \dfrac{1}{\cancel{7}_1} = \dfrac{17}{2} \text{ weeks}$

$\dfrac{\frac{17}{2} \text{ weeks}}{8\frac{3}{4} \text{ weeks}} = \dfrac{\frac{17}{2}}{\frac{35}{4}} = \dfrac{17}{2} \div \dfrac{35}{4} = \dfrac{17}{\cancel{2}_1} \cdot \dfrac{\cancel{4}^2}{35} = \dfrac{34}{35}$

The ratio of $59\frac{1}{2}$ days to $8\frac{3}{4}$ weeks is $\frac{34}{35}$.

47. $\dfrac{\text{longest side}}{\text{shortest side}} = \dfrac{2\frac{7}{12} \text{ feet}}{2\frac{7}{12} \text{ feet}} = \dfrac{1}{1}$

The ratio of the length of the longest side to the length of the shortest side is $\frac{1}{1}$.

As long as the sides all have the same length, any measurement you choose will maintain the ratio.

48. Answers will vary. Some possibilities are:

$\dfrac{4}{5} = \dfrac{8}{10} = \dfrac{12}{15} = \dfrac{16}{20} = \dfrac{20}{25} = \dfrac{24}{30} = \dfrac{28}{35}.$

49. It is not possible. Amelia would have to be older than her mother to have a ratio of 5 to 3.

50. Answers will vary, but a ratio of 3 to 1 means your income is 3 times your friend's income.

5.2 Rates

5.2 Margin Exercises

1. **(a)** $6 for 30 packages

$\dfrac{\$6 \div 6}{30 \text{ packages} \div 6} = \dfrac{\$1}{5 \text{ packages}}$

(b) 500 miles in 10 hours

$\dfrac{500 \text{ miles} \div 10}{10 \text{ hours} \div 10} = \dfrac{50 \text{ miles}}{1 \text{ hour}}$

(c) 4 teachers for 90 students

$\dfrac{4 \text{ teachers} \div 2}{90 \text{ students} \div 2} = \dfrac{2 \text{ teachers}}{45 \text{ students}}$

(d) 1270 bushels on 30 acres

$\dfrac{1270 \text{ bushels} \div 10}{30 \text{ acres} \div 10} = \dfrac{127 \text{ bushels}}{3 \text{ acres}}$

2. **(a)** $4.35 for 3 pounds of cheese

$\dfrac{\$4.35 \div 3}{3 \text{ pounds} \div 3} = \dfrac{\$1.45}{1 \text{ pound}}$

The unit rate is $1.45/pound.

(b) 304 miles on 9.5 gallons of gas

$\dfrac{304 \text{ miles} \div 9.5}{9.5 \text{ gallons} \div 9.5} = \dfrac{32 \text{ miles}}{1 \text{ gallon}}$

The unit rate is 32 miles/gallon.

(c) $850 in 5 days

$\dfrac{\$850 \div 5}{5 \text{ days} \div 5} = \dfrac{\$170}{1 \text{ day}}$

The unit rate is $170/day.

(d) 24-pound turkey for 15 people

$\dfrac{24 \text{ pounds} \div 15}{15 \text{ people} \div 15} = \dfrac{1.6 \text{ pounds}}{1 \text{ person}}$

The unit rate is 1.6 pounds/person.

3. **(a)**

Size	Cost per Unit
2 quarts	$\dfrac{\$3.25}{2 \text{ quarts}} = \$1.625$
3 quarts	$\dfrac{\$4.95}{3 \text{ quarts}} = \$1.65$
4 quarts	$\dfrac{\$6.48}{4 \text{ quarts}} = \$1.62$

The lowest cost per quart is $1.62, so the best buy is 4 quarts for $6.48.

(b) <u>Size</u> <u>Cost per Unit</u>

6 cans $\frac{\$1.99}{6 \text{ cans}} \approx \$0.332$

12 cans $\frac{\$3.49}{12 \text{ cans}} \approx \$0.291$

24 cans $\frac{\$7}{24 \text{ cans}} \approx \$0.292$

The lowest price per can is about $0.291, so the best buy is 12 cans of cola for $3.49.

4. **(a)** One battery that lasts twice as long is like getting two batteries.

$$\frac{\$1.19 \div 2}{2 \text{ batteries} \div 2} = \$0.595/\text{battery}.$$

The cost of the four-pack is

$$\frac{\$2.79 \div 4}{4 \text{ batteries} \div 4} = \$0.6975/\text{battery}.$$

The best buy is the single AA-size battery.

(b) Brand C:

$\$3.89 - \$0.85 = \$3.04$

$$\frac{\$3.04 \div 6}{6 \text{ ounces} \div 6} \approx \$0.507/\text{ounce}$$

Brand D:

$\$1.59 - \$0.20 = \$1.39$

$$\frac{\$1.39 \div 2.5}{2.5 \text{ ounces} \div 2.5} = \$0.556/\text{ounce}$$

Brand C with the 85¢ coupon is the better buy at $0.507 per ounce (rounded).

5.2 Section Exercises

1. 10 cups for 6 people

$$\frac{10 \text{ cups} \div 2}{6 \text{ people} \div 2} = \frac{5 \text{ cups}}{3 \text{ people}}$$

3. 15 feet in 35 seconds

$$\frac{15 \text{ feet} \div 5}{35 \text{ seconds} \div 5} = \frac{3 \text{ feet}}{7 \text{ seconds}}$$

5. 72 miles on 4 gallons

$$\frac{72 \text{ miles} \div 4}{4 \text{ gallons} \div 4} = \frac{18 \text{ miles}}{1 \text{ gallon}}$$

7. $60 in 5 hours

$$\frac{\$60 \div 5}{5 \text{ hours} \div 5} = \frac{\$12}{1 \text{ hour}}$$

The unit rate is $12 per hour or $12/hour.

9. 7.5 pounds for 6 people

$$\frac{7.5 \text{ pounds} \div 6}{6 \text{ people} \div 6} = \frac{1.25 \text{ pounds}}{1 \text{ person}}$$

The unit rate is 1.25 pounds/person.

11. $413.20 for 4 days

$$\frac{\$413.20 \div 4}{4 \text{ days} \div 4} = \frac{\$103.30}{1 \text{ day}}$$

The unit rate is $103.30/day.

13. Miles traveled:

$27{,}758.2 - 27{,}432.3 = 325.9$

Miles per gallon:

$$\frac{325.9 \text{ miles} \div 15.5}{15.5 \text{ gallons} \div 15.5} \approx 21.02 \approx 21.0$$

15. Miles traveled:

$28{,}396.7 - 28{,}058.1 = 338.6$

Miles per gallon:

$$\frac{338.6 \text{ miles} \div 16.2}{16.2 \text{ gallons} \div 16.2} \approx 20.90 \approx 20.9$$

17. <u>Size</u> <u>Cost per Unit</u>

2 ounces $\frac{\$1.79}{2 \text{ ounces}} = \$0.895$

4 ounces $\frac{\$3.09}{4 \text{ ounces}} = \$0.7725$ (*)

The best buy is 4 ounces for $3.09.

19. <u>Size</u> <u>Cost per Unit</u>

13 ounces $\frac{\$2.80}{13 \text{ ounces}} \approx \$0.215$

15 ounces $\frac{\$3.15}{15 \text{ ounces}} = \$0.210$ (*)

18 ounces $\frac{\$3.98}{18 \text{ ounces}} \approx \$0.221$

The best buy is 15 ounces for $3.15.

21. <u>Size</u> <u>Cost per Unit</u>

12 ounces $\frac{\$1.29}{12 \text{ ounces}} \approx \$0.108$

18 ounces $\frac{\$1.79}{18 \text{ ounces}} \approx \$0.099$ (*)

28 ounces $\frac{\$3.39}{28 \text{ ounces}} \approx \$0.121$

40 ounces $\frac{\$4.39}{40 \text{ ounces}} \approx \$0.110$

The best buy is 18 ounces for $1.79.

23. Answers will vary. For example, you might choose Brand B because you like more chicken, so the cost per chicken chunk may actually be the same or less than Brand A.

25. 10.5 pounds in 6 weeks

$$\frac{10.5 \text{ pounds} \div 6}{6 \text{ weeks} \div 6} = \frac{1.75 \text{ pounds}}{1 \text{ week}}$$

Her rate of loss was 1.75 pounds/week.

27. 7 hours to earn \$85.82

$$\frac{\$85.82 \div 7}{7 \text{ hours} \div 7} = \frac{\$12.26}{1 \text{ hour}}$$

His pay rate is \$12.26/hour.

29. **(a)** Add the connection fee and five times the cost per minute.

Radiant Penny:
$\$0.39 + 5(\$0.01) = \$0.39 + \$0.05 = \$0.44$

IDT Special:
$\$0.14 + 5(\$0.022) = \$0.14 + \$0.11 = \$0.25$

Access America:
$\$0.00 + 5(\$0.047) = \$0.235$

(b) Divide the total costs by five minutes.

Radiant Penny: $\frac{\$0.44 \div 5}{5 \text{ minutes} \div 5} = \$0.088\text{/minute}$

IDT Special: $\frac{\$0.25 \div 5}{5 \text{ minutes} \div 5} = \$0.05\text{/minute}$

Access America: $\frac{\$0.235 \div 5}{5 \text{ minutes} \div 5} = \$0.047\text{/minute}$

Access America is the best buy.

31. Add the connection fee and fifteen (and then twenty-five) times the cost per minute.

Radiant Penny:
$\$0.39 + 15(\$0.01) = \$0.39 + \$0.15 = \$0.54$
$\$0.39 + 25(\$0.01) = \$0.39 + \$0.25 = \$0.64$

IDT Special:
$\$0.14 + 15(\$0.022) = \$0.14 + \$0.33 = \$0.47$
$\$0.14 + 25(\$0.022) = \$0.14 + \$0.55 = \$0.69$

Access America:
$\$0.00 + 15(\$0.047) = \$0.705$
$\$0.00 + 25(\$0.047) = \$1.175$

For the 15-minute call, IDT Special is the best buy at \$0.47. Alternatively, we could calculate the cost per minute to determine the best buy.
For the 25-minute call, Radiant Penny is the best buy at \$0.64.

33. One battery for \$1.79 is like getting 3 batteries for $\$1.79 \div 3 \approx \$0.597$ per battery.

An eight-pack of AAA batteries for \$4.99 is $\$4.99 \div 8 \approx \$0.624$ per battery.

The better buy is the one battery package.

35. **Brand G:** $\$2.39 - \$0.60 \text{ coupon} = \$1.79$

$$\frac{\$1.79}{10 \text{ ounces}} = \$0.179 \text{ per ounce}$$

Brand K: $\frac{\$3.99}{20.3 \text{ ounces}} \approx \$0.197 \text{ per ounce}$

Brand P: $\$3.39 - \$0.50 \text{ coupon} = \$2.89$

$$\frac{\$2.89}{16.5 \text{ ounces}} \approx \$0.175 \text{ per ounce}$$

Brand P with the 50¢ coupon has the lowest cost per ounce and is the best buy.

37. The one-time activation fee is \$35 (or \$36) for 12 months.

$$\frac{\$35 \div 12}{12 \text{ months} \div 12} \approx \$2.92\text{/month}$$

$(\$36/12 = \$3\text{/month})$

The activation fee will add \$2.92 (rounded) per month for Verizon, T-Mobile, and Nextel; \$3 per month for Sprint.

38. Assume there are exactly 52 weeks in a year. Then there are $52 \cdot 5 = 260$ weekdays in a year.

$$\frac{260 \text{ weekdays} \div 12}{12 \text{ months} \div 12} \approx 21.7 \text{ weekdays/month}$$

For Verizon:
$$\frac{400 \text{ anytime minutes} \div 21.7}{21.7 \text{ weekdays} \div 21.7} \approx 18 \text{ minutes/weekday}$$

For T-Mobile:
$$\frac{600 \text{ anytime minutes} \div 21.7}{21.7 \text{ weekdays} \div 21.7} \approx 28 \text{ minutes/weekday}$$

For Nextel and Sprint:
$$\frac{500 \text{ anytime minutes} \div 21.7}{21.7 \text{ weekdays} \div 21.7} \approx 23 \text{ minutes/weekday}$$

39. To find the actual average cost per "anytime minute," add the monthly charge for the activation fee (from Exercise 37) to the monthly charge add then divide by the number of anytime minutes.

For Verizon:
$$\frac{\$2.92 + \$59.99}{400 \text{ minutes}} = \frac{\$62.91}{400 \text{ minutes}} \approx \$0.16\text{/minute}$$

For T-Mobile:
$$\frac{\$2.92 + \$39.99}{600 \text{ minutes}} = \frac{\$42.91}{600 \text{ minutes}} \approx \$0.07\text{/minute}$$

continued

For Nextel:

$$\frac{\$2.92 + \$45.99}{500 \text{ minutes}} = \frac{\$48.91}{500 \text{ minutes}} \approx \$0.10/\text{minute}$$

For Sprint:

$$\frac{\$3 + \$55}{500 \text{ minutes}} = \frac{\$58}{500 \text{ minutes}} \approx \$0.12/\text{minute}$$

T-Mobile is the best buy at about $0.07 per "anytime minute."

40. The total cost is the sum of the activation fee, two monthly charges, and the termination fee. The number of "anytime minutes" in each case is $2 \times 100 = 200$.

For Verizon:

$$\frac{\$35 + 2(\$59.99) + \$175}{200 \text{ minutes}} = \frac{\$329.98}{200 \text{ minutes}} \approx \$1.65/\text{minute}$$

For T-Mobile:

$$\frac{\$35 + 2(\$39.99) + \$200}{200 \text{ minutes}} = \frac{\$314.98}{200 \text{ minutes}} \approx \$1.57/\text{minute}$$

For Nextel:

$$\frac{\$35 + 2(\$45.99) + \$200}{200 \text{ minutes}} = \frac{\$326.98}{200 \text{ minutes}} \approx \$1.63/\text{minute}$$

For Sprint:

$$\frac{\$36 + 2(\$55) + \$150}{200 \text{ minutes}} = \frac{\$296}{200 \text{ minutes}} = \$1.48/\text{minute}$$

5.3 Proportions

5.3 Margin Exercises

1. **(a)** $7 is to 3 cans as $28 is to 12 cans

$$\frac{\$7}{3 \text{ cans}} = \frac{\$28}{12 \text{ cans}}$$

(b) 9 meters is to 16 meters as 18 meters is to 32 meters

$$\frac{9}{16} = \frac{18}{32} \quad \textit{Common units cancel}$$

(c) 5 is to 7 as 35 is to 49

$$\frac{5}{7} = \frac{35}{49}$$

(d) 10 is to 30 as 60 is to 180

$$\frac{10}{30} = \frac{60}{180}$$

2. **(a)** $\frac{6}{12} = \frac{15}{30}$

$$\frac{6 \div 6}{12 \div 6} = \frac{1}{2} \text{ and } \frac{15 \div 15}{30 \div 15} = \frac{1}{2}$$

Both ratios are equivalent to $\frac{1}{2}$, so the proportion is *true.*

(b) $\frac{20}{24} = \frac{3}{4}$

$$\frac{20}{24} = \frac{20 \div 4}{24 \div 4} = \frac{5}{6} \text{ and } \frac{3}{4}$$

Because $\frac{5}{6}$ is *not* equivalent to $\frac{3}{4}$, the proportion is *false.*

(c) $\frac{25}{40} = \frac{30}{48}$

$$\frac{25 \div 5}{40 \div 5} = \frac{5}{8} \text{ and } \frac{30 \div 6}{48 \div 6} = \frac{5}{8}$$

Both ratios are equivalent to $\frac{5}{8}$, so the proportion is *true.*

(d) $\frac{35}{45} = \frac{12}{18}$

$$\frac{35 \div 5}{45 \div 5} = \frac{7}{9} \text{ and } \frac{12 \div 6}{18 \div 6} = \frac{2}{3}$$

Because $\frac{7}{9}$ is *not* equivalent to $\frac{2}{3}$, the proportion is *false.*

3. **(a)** $\frac{5}{9} = \frac{10}{18}$

Cross products: $5 \cdot 18 = 90$; $9 \cdot 10 = 90$

The cross products are *equal,* so the proportion is *true.*

(b) $\frac{32}{15} = \frac{16}{8}$

Cross products: $32 \cdot 8 = 256$; $15 \cdot 16 = 240$

The cross products are *unequal,* so the proportion is *false.*

(c) $\frac{10}{17} = \frac{20}{34}$

Cross products: $10 \cdot 34 = 340$; $17 \cdot 20 = 340$

The cross products are *equal,* so the proportion is *true.*

(d) $\frac{2.4}{6} = \frac{5}{12}$

Cross products: $(2.4)(12) = 28.8$; $6 \cdot 5 = 30$

The cross products are *unequal,* so the proportion is *false.*

(e) $\frac{3}{4.25} = \frac{24}{34}$

Cross products: $3 \cdot 34 = 102$; $(4.25)(24) = 102$

The cross products are *equal,* so the proportion is *true.*

(f) $\dfrac{1\frac{1}{6}}{2\frac{1}{3}} = \dfrac{4}{8}$

Cross products:

$$1\frac{1}{6} \cdot 8 = \frac{7}{\cancel{6}_3} \cdot \frac{\cancel{8}^4}{1} = \frac{28}{3}$$

$$2\frac{1}{3} \cdot 4 = \frac{7}{3} \cdot \frac{4}{1} = \frac{28}{3}$$

The cross products are *equal*, so the proportion is *true*.

5.3 Section Exercises

1. \$9 is to 12 cans as \$18 is to 24 cans.

$$\frac{\$9}{12 \text{ cans}} = \frac{\$18}{24 \text{ cans}}$$

3. 200 adults is to 450 children as 4 adults is to 9 children.

$$\frac{200 \text{ adults}}{450 \text{ children}} = \frac{4 \text{ adults}}{9 \text{ children}}$$

5. 120 feet is to 150 feet as 8 feet is to 10 feet.

$\dfrac{120}{150} = \dfrac{8}{10}$ *The common units (feet) cancel.*

7. $\dfrac{6}{10} = \dfrac{3}{5}$

$\dfrac{6 \div 2}{10 \div 2} = \dfrac{3}{5}$ and $\dfrac{3}{5}$

Both ratios are equivalent to $\frac{3}{5}$, so the proportion is *true*.

9. $\dfrac{5}{8} = \dfrac{25}{40}$

$\dfrac{5}{8}$ and $\dfrac{25 \div 5}{40 \div 5} = \dfrac{5}{8}$

Both ratios are equivalent to $\frac{5}{8}$, so the proportion is *true*.

11. $\dfrac{150}{200} = \dfrac{200}{300}$

$\dfrac{150 \div 50}{200 \div 50} = \dfrac{3}{4}$ and $\dfrac{200 \div 100}{300 \div 100} = \dfrac{2}{3}$

Because $\frac{3}{4}$ is *not* equivalent to $\frac{2}{3}$, the proportion is *false*.

13. $\dfrac{42}{15} = \dfrac{28}{10}$

$\dfrac{42 \div 3}{15 \div 3} = \dfrac{14}{5}$ and $\dfrac{28 \div 2}{10 \div 2} = \dfrac{14}{5}$

Both ratios are equivalent to $\frac{14}{5}$, so the proportion is *true*.

15. $\dfrac{32}{18} = \dfrac{48}{27}$

$\dfrac{32 \div 2}{18 \div 2} = \dfrac{16}{9}$ and $\dfrac{48 \div 3}{27 \div 3} = \dfrac{16}{9}$

Both ratios are equivalent to $\frac{16}{9}$, so the proportion is *true*.

17. $\dfrac{7}{6} = \dfrac{54}{48}$

$\dfrac{7}{6}$ and $\dfrac{54 \div 6}{48 \div 6} = \dfrac{9}{8}$

Because $\frac{7}{6}$ is *not* equivalent to $\frac{9}{8}$, the proportion is *false*.

19. $\dfrac{2}{9} = \dfrac{6}{27}$

Cross products: $2 \cdot 27 = 54$; $9 \cdot 6 = 54$

The cross products are *equal*, so the proportion is *true*.

21. $\dfrac{20}{28} = \dfrac{12}{16}$

Cross products: $20 \cdot 16 = 320$; $28 \cdot 12 = 336$

The cross products are *unequal*, so the proportion is *false*.

23. $\dfrac{110}{18} = \dfrac{160}{27}$

Cross products: $110 \cdot 27 = 2970$; $18 \cdot 160 = 2880$

The cross products are *unequal*, so the proportion is *false*.

25. $\dfrac{3.5}{4} = \dfrac{7}{8}$

Cross products: $(3.5)(8) = 28$; $4 \cdot 7 = 28$

The cross products are *equal*, so the proportion is *true*.

27. $\dfrac{18}{16} = \dfrac{2.8}{2.5}$

Cross products: $18(2.5) = 45$; $16(2.8) = 44.8$

The cross products are *unequal*, so the proportion is *false*.

29. $\dfrac{6}{3\frac{2}{3}} = \dfrac{18}{11}$

Cross products: $6 \cdot 11 = 66$;

$$3\frac{2}{3} \cdot 18 = \frac{11}{\cancel{3}_1} \cdot \frac{\cancel{18}^6}{1} = 66$$

The cross products are *equal*, so the proportion is *true*.

31. $\dfrac{2\frac{5}{8}}{3\frac{1}{4}} = \dfrac{21}{26}$

Cross products:

$$2\frac{5}{8} \cdot 26 = \frac{21}{\cancel{8}_4} \cdot \frac{\cancel{26}^{13}}{1} = \frac{273}{4}$$

$$3\frac{1}{4} \cdot 21 = \frac{13}{4} \cdot \frac{21}{1} = \frac{273}{4}$$

The cross products are *equal,* so the proportion is *true.*

33. $\dfrac{\frac{2}{3}}{2} = \dfrac{2.7}{8}$

Cross products:

$$\frac{2}{3} \cdot 8 = \frac{2}{3} \cdot \frac{8}{1} = \frac{16}{3} \approx 5.3$$

$$2(2.7) = 5.4$$

The cross products are *unequal,* so the proportion is *false.*

35. $\dfrac{2\frac{3}{10}}{8.05} = \dfrac{\frac{1}{4}}{0.9}$

Cross products:

$$2\frac{3}{10}(0.9) = \frac{23}{10} \cdot \frac{9}{10} = \frac{207}{100} = 2.07$$

$$(8.05)\frac{1}{4} = (8.05)(0.25) = 2.0125$$

The cross products are *unequal,* so the proportion is *false.*

37. (Answers may vary: proportion may also be set up with "at bats" in both numerators and "hits" in both denominators. Cross product will be the same.)

Ichiro Suzuki		Sammy Sosa
$\dfrac{16 \text{ hits}}{50 \text{ at bats}}$	=	$\dfrac{128 \text{ hits}}{400 \text{ at bats}}$

Cross products: $16 \cdot 400 = 6400$; $50 \cdot 128 = 6400$

The cross products are *equal,* so the proportion is *true.* Paul is correct; the two men hit equally well.

5.4 Solving Proportions

5.4 Margin Exercises

1. (a) $\dfrac{1}{2} = \dfrac{x}{12}$

$2 \cdot x = 1 \cdot 12$ *Cross products are equivalent*

$$\frac{\cancel{2}^1 \cdot x}{\cancel{2}_1} = \frac{1 \cdot \cancel{12}^6}{\cancel{2}_1}$$

$$x = 6$$

We will check our answers by showing that the cross products are equal.

Check: $1 \cdot 12 = 12 = 2 \cdot 6$

(b) $\dfrac{6}{10} = \dfrac{15}{x}$

$6 \cdot x = 10 \cdot 15$ *Cross products are equivalent*

$$\frac{\cancel{6}^1 \cdot x}{\cancel{6}_1} = \frac{\cancel{150}^{25}}{\cancel{6}_1}$$

$$x = 25$$

Check: $6 \cdot 25 = 150 = 10 \cdot 15$

(c) $\dfrac{28}{x} = \dfrac{21}{9}$

$x \cdot 21 = 28 \cdot 9$ *Cross products are equivalent*

$$\frac{x \cdot \cancel{21}^1}{\cancel{21}_1} = \frac{\cancel{28}^4 \cdot 9}{\cancel{21}_3}$$

$$x = 12$$

Check: $28 \cdot 9 = 252 = 12 \cdot 21$

(d) $\dfrac{x}{8} = \dfrac{3}{5}$

$x \cdot 5 = 8 \cdot 3$ *Cross products are equivalent*

$$\frac{x \cdot \cancel{5}^1}{\cancel{5}_1} = \frac{24}{5}$$

$$x = \frac{24}{5} = 4.8$$

Check: $(4.8)(5) = 24 = 8 \cdot 3$

(e) $\dfrac{14}{11} = \dfrac{x}{3}$

$11 \cdot x = 14 \cdot 3$ *Cross products are equivalent*

$$\frac{\cancel{11}^1 \cdot x}{\cancel{11}_1} = \frac{14 \cdot 3}{11}$$

$x = \dfrac{42}{11} \approx 3.82$ *(rounded to the nearest hundredth)*

Check: $14 \cdot 3 = 42 = 11 \cdot \frac{42}{11}$

2. **(a)** $\frac{3\frac{1}{4}}{2} = \frac{x}{8}$

$2 \cdot x = 3\frac{1}{4} \cdot 8$ *Cross products are equivalent*

$2 \cdot x = \frac{13}{4} \cdot \frac{8}{1}$

$\frac{2 \cdot x}{2} = \frac{26}{2}$

$x = 13$

Check: $3\frac{1}{4} \cdot 8 = 26 = 2 \cdot 13$

(b) $\frac{x}{3} = \frac{1\frac{2}{3}}{5}$

$5 \cdot x = 3 \cdot 1\frac{2}{3}$ *Cross products are equivalent*

$5 \cdot x = 3 \cdot \frac{5}{3}$

$\frac{5 \cdot x}{5} = \frac{1 \cdot 5}{5}$

$x = 1$

Check: $1 \cdot 5 = 5 = 3 \cdot 1\frac{2}{3}$

(c) $\frac{0.06}{x} = \frac{0.3}{0.4}$

$0.3 \cdot x = 0.06 \cdot 0.4$ *Cross products are equivalent*

$\frac{0.3 \cdot x}{0.3} = \frac{0.024}{0.3}$

$x = 0.08$

Check: $(0.06)(0.4) = 0.024 = (0.08)(0.3)$

(d) $\frac{2.2}{5} = \frac{13}{x}$

$2.2 \cdot x = 5 \cdot 13$ *Cross products are equivalent*

$\frac{2.2 \cdot x}{2.2} = \frac{65}{2.2}$

$x = \frac{65}{2.2} \approx 29.55$ *(rounded to the nearest hundredth)*

Check: $2.2\left(\frac{65}{2.2}\right) = 65 = 5 \cdot 13$

(e) $\frac{x}{6} = \frac{0.5}{1.2}$

$1.2 \cdot x = 6(0.5)$ *Cross products are equivalent*

$\frac{1.2 \cdot x}{1.2} = \frac{3}{1.2} = \frac{30}{12} = \frac{5}{2}$

$x = 2.5$

Check: $(2.5)(1.2) = 3 = 6(0.5)$

(f) $\frac{0}{2} = \frac{x}{7.092}$

Since $\frac{0}{2} = 0$, x *must* equal 0. The denominator 7.092 could be any number (except 0).

5.4 Section Exercises

1. $\frac{1}{3} = \frac{x}{12}$

$3 \cdot x = 1 \cdot 12$ *Cross products are equivalent*

$\frac{3 \cdot x}{3} = \frac{12}{3}$

$x = 4$

We will check our answers by showing that the cross products are equal.

Check: $1 \cdot 12 = 12 = 3 \cdot 4$

3. $\frac{15}{10} = \frac{3}{x}$ **OR** $\frac{3}{2} = \frac{3}{x}$

Since the numerators are equal, the denominators must be equal, so $x = 2$.

5. $\frac{x}{11} = \frac{32}{4}$ **OR** $\frac{x}{11} = \frac{8}{1}$

$1 \cdot x = 11 \cdot 8$ *Cross products are equivalent*

$x = 88$

Check: $88 \cdot 1 = 88 = 11 \cdot 8$

7. $\frac{42}{x} = \frac{18}{39}$

$x \cdot 18 = 42 \cdot 39$ *Cross products are equivalent*

$\frac{x \cdot 18}{18} = \frac{42 \cdot 39}{18}$

$x = 91$

Check: $42 \cdot 39 = 1638 = 91 \cdot 18$

9. $\frac{x}{25} = \frac{4}{20}$ **OR** $\frac{x}{25} = \frac{1}{5}$

$5 \cdot x = 25 \cdot 1$ *Cross products are equivalent*

$\frac{\overset{1}{\cancel{5}} \cdot x}{\underset{1}{\cancel{5}}} = \frac{\overset{5}{\cancel{25}}}{\underset{1}{\cancel{5}}}$

$x = 5$

Check: $5 \cdot 5 = 25 = 25 \cdot 1$

11. $\frac{8}{x} = \frac{24}{30}$ **OR** $\frac{8}{x} = \frac{4}{5}$

$4 \cdot x = 8 \cdot 5$ *Cross products are equivalent*

$\frac{\overset{1}{\cancel{4}} \cdot x}{\underset{1}{\cancel{4}}} = \frac{\overset{2}{\cancel{8}} \cdot 5}{\underset{1}{\cancel{4}}}$

$x = 10$

Check: $8 \cdot 5 = 40 = 10 \cdot 4$

13. $\frac{99}{55} = \frac{44}{x}$ **OR** $\frac{9}{5} = \frac{44}{x}$

$9 \cdot x = 5 \cdot 44$ *Cross products are equivalent*

$\frac{\overset{1}{\cancel{9}} \cdot x}{\underset{1}{\cancel{9}}} = \frac{220}{9}$

$x = \frac{220}{9}$

≈ 24.44

Check: $9 \cdot \frac{220}{9} = 220 = 5 \cdot 44$

15. $\frac{0.7}{9.8} = \frac{3.6}{x}$

$0.7 \cdot x = 9.8\,(3.6)$ *Cross products are equivalent*

$\frac{\overset{1}{\cancel{0.7}} \cdot x}{\underset{1}{\cancel{0.7}}} = \frac{\overset{50.4}{\cancel{35.28}}}{\underset{1}{\cancel{0.7}}}$

$x = 50.4$

Check: $0.7(50.4) = 35.28 = 9.8(3.6)$

17. $\frac{250}{24.8} = \frac{x}{1.75}$

$24.8 \cdot x = 250(1.75)$ *Cross products are equivalent*

$\frac{\overset{1}{\cancel{24.8}} \cdot x}{\underset{1}{\cancel{24.8}}} = \frac{437.5}{24.8}$

≈ 17.64

Check: $250(1.75) = 437.5 = 24.8 \cdot \frac{437.5}{24.8}$

19. $\frac{15}{1\frac{2}{3}} = \frac{9}{x}$

$15 \cdot x = 1\frac{2}{3} \cdot 9$ *Cross products are equivalent*

$15 \cdot x = \frac{5}{\underset{1}{\cancel{3}}} \cdot \frac{\overset{3}{\cancel{9}}}{1}$

$\frac{\overset{1}{\cancel{15}} \cdot x}{\underset{1}{\cancel{15}}} = \frac{15}{15}$

$x = 1$

21. $\frac{2\frac{1}{3}}{1\frac{1}{2}} = \frac{x}{2\frac{1}{4}}$

$1\frac{1}{2} \cdot x = 2\frac{1}{3} \cdot 2\frac{1}{4}$

$\frac{3}{2} \cdot x = \frac{7}{\underset{1}{\cancel{3}}} \cdot \frac{\overset{3}{\cancel{9}}}{4}$

$\frac{3}{2} \cdot x = \frac{21}{4}$

$\frac{\frac{3}{2} \cdot x}{\frac{3}{2}} = \frac{\frac{21}{4}}{\frac{3}{2}}$

$x = \frac{21}{4} \div \frac{3}{2} = \frac{\overset{7}{\cancel{21}}}{\underset{2}{\cancel{4}}} \cdot \frac{\overset{1}{\cancel{2}}}{\underset{1}{\cancel{3}}} = \frac{7}{2} = 3\frac{1}{2}$

23. Change $\frac{1}{2}$ to a decimal by dividing: $1 \div 2 = 0.5$

$\frac{\frac{1}{2}}{x} = \frac{2}{0.8}$

$\frac{0.5}{x} = \frac{2}{0.8}$

$x \cdot 2 = 0.5(0.8)$

$x \cdot 2 = 0.4$

$\frac{x \cdot \overset{1}{\cancel{2}}}{\underset{1}{\cancel{2}}} = \frac{0.4}{2}$

$x = 0.2$

Now change 0.8 to a fraction and write it in lowest terms:

$0.8 = \frac{8 \div 2}{10 \div 2} = \frac{4}{5}$

$\frac{\frac{1}{2}}{x} = \frac{2}{\frac{4}{5}}$

$x \cdot 2 = \frac{1}{\underset{1}{\cancel{2}}} \cdot \frac{\overset{2}{\cancel{4}}}{5}$

$x \cdot 2 = \frac{2}{5}$

$$\frac{x \cdot \overset{1}{\cancel{2}}}{\underset{1}{\cancel{2}}} = \frac{\frac{2}{5}}{2}$$

$$x = \frac{2}{5} \div 2 = \frac{\overset{1}{\cancel{2}}}{5} \cdot \frac{1}{\underset{1}{\cancel{2}}} = \frac{1}{5}$$

25. $\dfrac{x}{\frac{3}{50}} = \dfrac{0.15}{1\frac{4}{5}}$

Change to decimals.

$$\frac{x}{0.06} = \frac{0.15}{1.8}$$
$$x \cdot 1.8 = 0.15(0.06)$$
$$\frac{x \cdot 1.8}{1.8} = \frac{0.009}{1.8}$$
$$x = 0.005$$

Change to fractions.

$$\frac{x}{\frac{3}{50}} = \frac{\frac{3}{20}}{1\frac{4}{5}}$$
$$1\frac{4}{5} \cdot x = \frac{3}{50} \cdot \frac{3}{20}$$
$$\frac{1\frac{4}{5} \cdot x}{1\frac{4}{5}} = \frac{\frac{9}{1000}}{1\frac{4}{5}}$$
$$x = \frac{9}{1000} \div 1\frac{4}{5} = \frac{\overset{1}{\cancel{9}}}{\underset{200}{\cancel{1000}}} \cdot \frac{\overset{1}{\cancel{5}}}{\underset{1}{\cancel{9}}} = \frac{1}{200}$$

27. $\dfrac{10}{4} = \dfrac{5}{3}$

Find the cross products.

$10 \cdot 3 = 30$
$4 \cdot 5 = 20$

The cross products are *unequal,* so the proportion is *not* true.

Replace 10 with x.

$$\frac{x}{4} = \frac{5}{3}$$
$$3 \cdot x = 4 \cdot 5$$
$$x = \frac{20}{3} = 6\frac{2}{3}, \text{ so}$$

$\dfrac{6\frac{2}{3}}{4} = \dfrac{5}{3}$ is a true proportion.

Replace 4 with x.

$$\frac{10}{x} = \frac{5}{3}$$
$$5 \cdot x = 10 \cdot 3$$
$$x = \frac{30}{5} = 6, \text{ so}$$

$\dfrac{10}{6} = \dfrac{5}{3}$ is a true proportion.

Replace 5 with x.

$$\frac{10}{4} = \frac{x}{3}$$
$$4 \cdot x = 10 \cdot 3$$
$$x = \frac{30}{4} = \frac{15}{2} = 7.5, \text{ so}$$

$\dfrac{10}{4} = \dfrac{7.5}{3}$ is a true proportion.

Replace 3 with x.

$$\frac{10}{4} = \frac{5}{x}$$
$$10 \cdot x = 4 \cdot 5$$
$$x = \frac{20}{10} = 2, \text{ so}$$

$\dfrac{10}{4} = \dfrac{5}{2}$ is a true proportion.

28. $\dfrac{6}{8} = \dfrac{24}{30}$

Find the cross products.

$6 \cdot 30 = 180$
$8 \cdot 24 = 192$

The cross products are *unequal,* so the proportion is *not* true.

Replace 6 with x.

$$\frac{x}{8} = \frac{24}{30}$$
$$30 \cdot x = 8 \cdot 24$$
$$x = \frac{8 \cdot \overset{4}{\cancel{24}}}{\underset{5}{\cancel{30}}} = \frac{32}{5} = 6.4, \text{ so}$$

$\dfrac{6.4}{8} = \dfrac{24}{30}$ is a true proportion.

Replace 8 with x.

$$\frac{6}{x} = \frac{24}{30}$$
$$24 \cdot x = 6 \cdot 30$$
$$x = \frac{6 \cdot \overset{5}{\cancel{30}}}{\underset{4}{\cancel{24}}} = \frac{30}{4} = 7.5, \text{ so}$$

$\dfrac{6}{7.5} = \dfrac{24}{30}$ is a true proportion.

continued

Replace 24 with x.

$$\frac{6}{8}=\frac{x}{30}$$
$$8\cdot x=6\cdot 30$$
$$x=\frac{\overset{3}{\cancel{6}}\cdot\overset{15}{\cancel{30}}}{\underset{\underset{2}{\cancel{4}}}{\cancel{8}}}=\frac{45}{2}=22.5\text{, so}$$

$\frac{6}{8}=\frac{22.5}{30}$ is a true proportion.

Replace 30 with x.

$$\frac{6}{8}=\frac{24}{x}$$
$$6\cdot x=8\cdot 24$$
$$x=\frac{8\cdot\overset{4}{\cancel{24}}}{\underset{1}{\cancel{6}}}=32\text{, so}$$

$\frac{6}{8}=\frac{24}{32}$ is a true proportion.

Summary Exercises on Ratios, Rates, and Proportions

1. 2400 freshmen to 2000 juniors

$$\frac{2400}{2000}=\frac{2400\div 400}{2000\div 400}=\frac{6}{5}$$

3. 1850 seniors and 2150 sophomores to 2000 juniors

$$\frac{1850+2150}{2000}=\frac{4000}{2000}=\frac{4000\div 2000}{2000\div 2000}=\frac{2}{1}$$

5. **(i)** 2 million violin players to 22 million piano players

$$\frac{2\text{ million}}{22\text{ million}}=\frac{2\div 2}{22\div 2}=\frac{1}{11}$$

(ii) 2 million violin players to 20 million guitar players

$$\frac{2\text{ million}}{20\text{ million}}=\frac{2\div 2}{20\div 2}=\frac{1}{10}$$

(iii) 2 million violin players to 6 million organ players

$$\frac{2\text{ million}}{6\text{ million}}=\frac{2\div 2}{6\div 2}=\frac{1}{3}$$

(iv) 2 million violin players to 4 million clarinet players

$$\frac{2\text{ million}}{4\text{ million}}=\frac{2\div 2}{4\div 2}=\frac{1}{2}$$

(v) 2 million violin players to 3 million drum players

$$\frac{2\text{ million}}{3\text{ million}}=\frac{2}{3}$$

7.
$$\frac{100\text{ points}}{48\text{ minutes}}=\frac{100\div 48\text{ points}}{48\div 48\text{ minutes}}=2.08\overline{3}\approx 2.1\text{ points/minute}$$

$$\frac{48\text{ minutes}}{100\text{ points}}=\frac{48\div 100\text{ minutes}}{100\div 100\text{ points}}=0.48\approx 0.5\text{ minute/point}$$

9.
$$\frac{\$652.80}{40\text{ hours}}=\frac{\$652.80\div 40}{40\div 40\text{ hours}}=\$16.32/\text{hour}$$

Her regular hourly pay rate is \$16.32/hour.

$$\frac{\$195.84}{8\text{ hours}}=\frac{\$195.84\div 8}{8\div 8\text{ hours}}=\$24.48/\text{hour}$$

Her overtime hourly pay rate is \$24.48/hour.

11.

<u>Size</u>	<u>Cost per Unit</u>
26 ounces	$\frac{\$4.78}{26\text{ ounces}}\approx\$0.18$
34 ounces	$\frac{\$5.44}{34\text{ ounces}}=\$0.16\ (*)$
39 ounces	$\frac{\$7.49}{39\text{ ounces}}\approx\$0.19$

The best buy is 34 ounces for \$5.44.

13.
$$\frac{28}{21}=\frac{44}{33}$$
$$\frac{28\div 7}{21\div 7}=\frac{4}{3}\text{ and }\frac{44\div 11}{33\div 11}=\frac{4}{3}$$

Both ratios are equivalent to $\frac{4}{3}$, so the proportion is *true*.

15.
$$\frac{2\frac{5}{8}}{3\frac{1}{4}}=\frac{21}{26}$$

Cross products:

$$2\frac{5}{8}\cdot 26=\frac{21}{\underset{4}{\cancel{8}}}\cdot\frac{\overset{13}{\cancel{26}}}{1}=\frac{273}{4}$$

$$3\frac{1}{4}\cdot 21=\frac{13}{4}\cdot\frac{21}{1}=\frac{273}{4}$$

The cross products are *equal*, so the proportion is *true*.

17. $\frac{15}{8} = \frac{6}{x}$

$15 \cdot x = 8 \cdot 6$ *Cross products are equivalent*

$\frac{\overset{1}{\cancel{15}} \cdot x}{\underset{1}{\cancel{15}}} = \frac{8 \cdot \overset{2}{\cancel{6}}}{\underset{5}{\cancel{15}}} = \frac{16}{5} = 3.2$

Check: $15(3.2) = 48 = 8 \cdot 6$

19. $\frac{10}{11} = \frac{x}{4}$

$11 \cdot x = 10 \cdot 4$ *Cross products are equivalent*

$\frac{\overset{1}{\cancel{11}} \cdot x}{\underset{1}{\cancel{11}}} = \frac{40}{11} \approx 3.64$

Check: $10 \cdot 4 = 40 = 11 \cdot \frac{40}{11}$

21. $\frac{2.6}{x} = \frac{13}{7.8}$

$13 \cdot x = 2.6(7.8)$ *Cross products are equivalent*

$\frac{\overset{1}{\cancel{13}} \cdot x}{\underset{1}{\cancel{13}}} = \frac{20.28}{13} = 1.56$

Check: $2.6(7.8) = 20.28 = 1.56(13)$

23. $\frac{\frac{1}{3}}{8} = \frac{x}{24}$

$8 \cdot x = \frac{1}{3} \cdot 24$ *Cross products are equivalent*

$\frac{\overset{1}{\cancel{8}} \cdot x}{\underset{1}{\cancel{8}}} = \frac{8}{8} = 1$

Check: $\frac{1}{3}(24) = 8 = 8 \cdot 1$

5.5 Solving Application Problems with Proportions

5.5 Margin Exercises

1. (a) $\frac{2 \text{ pounds}}{50 \text{ square feet}} = \frac{x \text{ pounds}}{225 \text{ square feet}}$

$50 \cdot x = 2 \cdot 225$ *Cross products are equivalent*

$\frac{50 \cdot x}{50} = \frac{450}{50}$ *Divide both sides by 50*

$x = 9$

9 pounds of fertilizer are needed for 225 square feet.

(b) $\frac{1 \text{ inch}}{75 \text{ miles}} = \frac{4.75 \text{ inches}}{x \text{ miles}}$

$1 \cdot x = 75(4.75)$ *Cross products are equivalent*

$x = 356.25 \approx 356 \text{ miles}$

The lake's actual length is about 356 miles.

(c) $\frac{30 \text{ milliliters}}{100 \text{ pounds}} = \frac{x \text{ milliliters}}{34 \text{ pounds}}$

$100 \cdot x = 30 \cdot 34$ *Cross products are equivalent*

$\frac{100 \cdot x}{100} = \frac{1020}{100}$ *Divide both sides by 100*

$x = 10.2 \approx 10 \text{ milliliters}$

A 34-pound child should be given about 10 milliliters of cough syrup.

2. (a) $\frac{2 \text{ people}}{3 \text{ people}} = \frac{x}{150 \text{ people}}$

$3 \cdot x = 2 \cdot 150$ *Cross products are equivalent*

$\frac{3 \cdot x}{3} = \frac{300}{3}$ *Divide both sides by 3*

$x = 100$

In a group of 150, 100 people want to lose weight. This is a reasonable answer because it's more than half the people, but less than all the people.

Incorrect setup

$\frac{3 \text{ people}}{2 \text{ people}} = \frac{x}{150 \text{ people}}$

$2 \cdot x = 3 \cdot 150$

$\frac{2 \cdot x}{2} = \frac{450}{2}$

$x = 225$

The incorrect setup gives an unreasonable answer of 225 people since there are only 150 people in the group.

(b) $\frac{3 \text{ FA students}}{5 \text{ students}} = \frac{x}{4500 \text{ students}}$

$5 \cdot x = 3 \cdot 4500$ *Cross products are equivalent*

$\frac{5 \cdot x}{5} = \frac{13{,}500}{5}$ *Divide both sides by 5*

$x = 2700$

At Central Community College 2700 students receive financial aid. This is a reasonable answer because it's more than half the people, but less than all the people.

continued

Incorrect setup

$$\frac{5 \text{ students}}{3 \text{ FA students}} = \frac{x}{4500 \text{ students}}$$
$$3 \cdot x = 5 \cdot 4500$$
$$\frac{3 \cdot x}{3} = \frac{22{,}500}{3}$$
$$x = 7500$$

The incorrect setup gives an unreasonable answer of 7500 students since there are only 4500 students at the college.

(c) $$\frac{9 \text{ sugarless}}{10 \text{ total}} = \frac{x}{60 \text{ total}}$$

$$10 \cdot x = 9 \cdot 60$$ *Cross products are equivalent*

$$\frac{10 \cdot x}{10} = \frac{540}{10}$$ *Divide both sides by 10*

$$x = 54$$

If the advertisement is true, 54 of the 60 dentists in the city would recommend sugarless gum. This is a reasonable answer since it is almost all of the dentists.

Incorrect setup

$$\frac{10}{9} = \frac{x}{60}$$
$$9 \cdot x = 10 \cdot 60$$
$$\frac{9 \cdot x}{9} = \frac{600}{9}$$
$$x = 66.\overline{6} \approx 67$$

The incorrect setup gives an unreasonable answer of 67 dentists since there are only 60 dentists in the city.

5.5 Section Exercises

1. $$\frac{5 \text{ hours}}{4 \text{ cartoon strips}} = \frac{x \text{ hours}}{18 \text{ cartoon strips}}$$
$$\frac{5}{4} = \frac{x}{18}$$
$$4 \cdot x = 5 \cdot 18$$
$$4 \cdot x = 90$$
$$\frac{\overset{1}{\cancel{4}} \cdot x}{\underset{1}{\cancel{4}}} = \frac{90}{4}$$
$$x = 22.5$$

It will take 22.5 hours to sketch 18 cartoon strips.

3. $$\frac{60 \text{ newspapers}}{\$27} = \frac{16 \text{ newspapers}}{x}$$
$$60 \cdot x = 27 \cdot 16$$
$$\frac{60 \cdot x}{60} = \frac{432}{60}$$
$$x = 7.2$$

The cost of 16 newspapers is $7.20.

5. $$\frac{3 \text{ pounds}}{350 \text{ square feet}} = \frac{x}{4900 \text{ square feet}}$$
$$350 \cdot x = 3 \cdot 4900$$
$$\frac{350 \cdot x}{350} = \frac{14{,}700}{350}$$
$$x = 42$$

42 pounds are needed to cover 4900 square feet.

7. $$\frac{\$455.75}{5 \text{ days}} = \frac{x}{3 \text{ days}}$$
$$5 \cdot x = (455.75)(3)$$
$$\frac{5 \cdot x}{5} = \frac{1367.25}{5}$$
$$x = 273.45$$

In 3 days Tom makes $273.45.

9. $$\frac{6 \text{ ounces}}{7 \text{ servings}} = \frac{x}{12 \text{ servings}}$$
$$7 \cdot x = 6 \cdot 12$$
$$\frac{7 \cdot x}{7} = \frac{72}{7}$$
$$x = 10\frac{2}{7} \approx 10$$

You need about 10 ounces for 12 servings.

11. $$\frac{3 \text{ quarts}}{270 \text{ square feet}} = \frac{x}{(350 + 100) \text{ square feet}}$$
$$\frac{3}{270} = \frac{x}{450}$$
$$270 \cdot x = 3 \cdot 450$$
$$\frac{270 \cdot x}{270} = \frac{1350}{270}$$
$$x = 5$$

You will need 5 quarts.

13. First find the length.

$$\frac{1 \text{ inch}}{4 \text{ feet}} = \frac{3.5 \text{ inches}}{x}$$
$$1 \cdot x = 4(3.5)$$
$$x = 14$$

The kitchen is 14 feet long.

Then find the width.

$$\frac{1 \text{ inch}}{4 \text{ feet}} = \frac{2.5 \text{ inches}}{x \text{ feet}}$$
$$1 \cdot x = 4(2.5)$$
$$x = 10$$

The kitchen is 10 feet wide.

15. The length of the dining area is the same as the length of the kitchen, which is 14 feet by Exercise 13.

Find the width of the dining area.

$4.5 - 2.5 \text{ inches} = 2 \text{ inches}$

$$\frac{1 \text{ inch}}{4 \text{ feet}} = \frac{2 \text{ inches}}{x}$$
$$1 \cdot x = 4 \cdot 2$$
$$x = 8$$

The dining area is 8 feet wide.

17. Set up and solve a proportion with pieces of chicken and number of guests.

$$\frac{40 \text{ pieces}}{25 \text{ guests}} = \frac{x}{60 \text{ guests}}$$
$$25 \cdot x = 40 \cdot 60$$
$$\frac{25 \cdot x}{25} = \frac{40 \cdot 60}{25}$$
$$x = \frac{40 \cdot 60}{25} = 96 \text{ pieces of chicken}$$

For the other food items, the proportions are similar, so we simply replace "40" in the last step with the appropriate value.

$$\frac{14 \cdot 60}{25} = 33.6 \text{ pounds of lasagna}$$
$$\frac{4.5 \cdot 60}{25} = 10.8 \text{ pounds of deli meats}$$
$$\frac{\frac{7}{3} \cdot 60}{25} = \frac{28}{5} = 5\frac{3}{5} \text{ pounds of cheese}$$
$$\frac{3 \text{ dozen} \cdot 60}{25} = 7.2 \text{ dozen (about 86) buns}$$
$$\frac{6 \cdot 60}{25} = 14.4 \text{ pounds of salad}$$

19. $$\frac{7 \text{ refresher}}{10 \text{ entering}} = \frac{x}{2950 \text{ entering}}$$
$$10 \cdot x = 7 \cdot 2950$$
$$\frac{10 \cdot x}{10} = \frac{20{,}650}{10}$$
$$x = 2065$$

2065 students will probably need a refresher course. This is a reasonable answer because it's more than half the students, but not all the students.

Incorrect setup

$$\frac{10 \text{ entering}}{7 \text{ refresher}} = \frac{x}{2950 \text{ entering}}$$
$$7 \cdot x = 10 \cdot 2950$$
$$\frac{7 \cdot x}{7} = \frac{29{,}500}{7}$$
$$x \approx 4214$$

The incorrect setup gives an unreasonable estimate of 4214 entering students since there are only 2950 entering students.

21. $$\frac{1 \text{ chooses vanilla}}{8 \text{ people}} = \frac{x \text{ choose vanilla}}{238 \text{ people}}$$
$$8 \cdot x = 1 \cdot 238$$
$$\frac{8 \cdot x}{8} = \frac{238}{8}$$
$$x = 29.75 \approx 30$$

You would expect about 30 people to choose vanilla ice cream. This is a reasonable answer.

Incorrect setup

$$\frac{8 \text{ people}}{1 \text{ chooses vanilla}} = \frac{x \text{ choose vanilla}}{238 \text{ people}}$$
$$1 \cdot x = 8 \cdot 238$$
$$x = 1904$$

With an incorrect setup, 1904 people choose vanilla ice cream. This is unreasonable because only 238 people attended the ice cream social.

23. $$\frac{98}{100} = \frac{x}{107{,}500{,}000}$$
$$100 \cdot x = 98 \cdot 107{,}500{,}000$$
$$\frac{100 \cdot x}{100} = \frac{10{,}535{,}000{,}000}{100}$$
$$x = 105{,}350{,}000$$

105,350,000 U.S. households have one or more TVs.

Incorrect setup

$$\frac{100}{98} = \frac{x}{107{,}500{,}000}$$
$$98 \cdot x = 100 \cdot 107{,}500{,}000$$
$$\frac{98 \cdot x}{98} = \frac{10{,}750{,}000{,}000}{98}$$
$$x \approx 109{,}693{,}878$$

The incorrect setup gives about 109,693,878 U.S. households with one or more TVs, but there are only 107,500,000 U.S. households.

25. $$\frac{5 \text{ stocks up}}{6 \text{ stocks down}} = \frac{x \text{ stocks up}}{750 \text{ stocks down}}$$
$$\frac{5}{6} = \frac{x}{750}$$
$$6 \cdot x = 5 \cdot 750$$
$$\frac{6 \cdot x}{6} = \frac{3750}{6}$$
$$x = 625$$

625 stocks went up.

27. $\frac{8 \text{ length}}{1 \text{ width}} = \frac{32.5 \text{ meters length}}{x \text{ meters width}}$

$8 \cdot x = 1 \cdot 32.5$

$\frac{\overset{1}{\cancel{8}} \cdot x}{\underset{1}{\cancel{8}}} = \frac{32.5}{8}$

$x = 4.0625 \approx 4.06$

The wing must be about 4.06 meters wide.

29. $\frac{150 \text{ pounds}}{222 \text{ calories}} = \frac{210 \text{ pounds}}{x}$

$150 \cdot x = 222 \cdot 210$

$\frac{150 \cdot x}{150} = \frac{46{,}620}{150}$

$x = 310.8 \approx 311$

A 210-pound person would burn about 311 calories.

31. $\frac{1.05 \text{ meters}}{1.68 \text{ meters}} = \frac{6.58 \text{ meters}}{x}$

$1.05 \cdot x = 1.68(6.58)$

$\frac{1.05 \cdot x}{1.05} = \frac{11.0544}{1.05}$

$x = 10.528 \approx 10.53$

The height of the tree is about 10.53 meters.

33. You cannot solve this problem using a proportion because the ratio of age to weight is not constant. As Jim's age increases from 25 to 50 years old, his weight may decrease, stay the same, or increase.

35. First find the number of coffee drinkers.

$\frac{4}{5} = \frac{x}{48{,}000}$

$5 \cdot x = 4 \cdot 48{,}000$

$\frac{5 \cdot x}{5} = \frac{192{,}000}{5}$

$x = 38{,}400$

The survey showed that 38,400 students drink coffee. Now find the number of coffee drinkers who use cream.

$\frac{1}{8} = \frac{x}{38{,}400}$

$8 \cdot x = 38{,}400$

$\frac{8 \cdot x}{8} = \frac{38{,}400}{8}$

$x = 4800$

According to the survey, 4800 students use cream.

37. First find the number of calories in a $\frac{1}{2}$-cup serving of bran cereal.

$\frac{\frac{1}{3} \text{ cup}}{80 \text{ calories}} = \frac{\frac{1}{2} \text{ cup}}{x}$

$\frac{1}{3} \cdot x = 80 \cdot \frac{1}{2}$

$\frac{\frac{1}{3} \cdot x}{\frac{1}{3}} = \frac{40}{\frac{1}{3}} = \frac{40}{1} \cdot \frac{3}{1}$

$x = 120$

Then find the number of grams of fiber in a $\frac{1}{2}$-cup serving of bran cereal.

$\frac{\frac{1}{3} \text{ cup}}{8 \text{ grams of fiber}} = \frac{\frac{1}{2} \text{ cup}}{x}$

$\frac{1}{3} \cdot x = 8 \cdot \frac{1}{2}$

$\frac{\frac{1}{3} \cdot x}{\frac{1}{3}} = \frac{4}{\frac{1}{3}} = \frac{4}{1} \cdot \frac{3}{1}$

$x = 12$

A $\frac{1}{2}$-cup serving of bran cereal provides 120 calories and 12 grams of fiber.

39. *Use proportions.*

Water: $\frac{3\frac{1}{2} \text{ cups}}{12 \text{ servings}} = \frac{x}{6 \text{ servings}}$

$12 \cdot x = 3\frac{1}{2} \cdot 6$

$12 \cdot x = \frac{7}{2} \cdot 6$

$\frac{12 \cdot x}{12} = \frac{21}{12}$

$x = \frac{7}{4} = 1\frac{3}{4}$

Margarine: $\frac{6 \text{ Tbsp}}{12 \text{ servings}} = \frac{x}{6 \text{ servings}}$

$12 \cdot x = 6 \cdot 6$

$\frac{12 \cdot x}{12} = \frac{36}{12}$

$x = 3$

Milk: $\frac{1\frac{1}{2} \text{ cups}}{12 \text{ servings}} = \frac{x}{6 \text{ servings}}$

$12 \cdot x = 1\frac{1}{2} \cdot 6$

$12 \cdot x = \frac{3}{2} \cdot 6$

$\frac{12 \cdot x}{12} = \frac{9}{12}$

$x = \frac{3}{4}$

Potato flakes: $\dfrac{4 \text{ cups}}{12 \text{ servings}} = \dfrac{x}{6 \text{ servings}}$

$$12 \cdot x = 4 \cdot 6$$
$$\frac{12 \cdot x}{12} = \frac{24}{12}$$
$$x = 2$$

Multiply the quantities by $\frac{1}{2}$ (or divide by 2), since *6 servings is $\frac{1}{2}$ of 12 servings.*

Water: $\dfrac{3\frac{1}{2}}{2} = \dfrac{\frac{7}{2}}{2} = \dfrac{7}{2} \div 2 = \dfrac{7}{2} \cdot \dfrac{1}{2}$
$= \dfrac{7}{4} = 1\dfrac{3}{4}$

Margarine: $\dfrac{6}{2} = 3$

Milk: $\dfrac{1\frac{1}{2}}{2} = \dfrac{\frac{3}{2}}{2} = \dfrac{3}{2} \div 2$
$= \dfrac{3}{2} \cdot \dfrac{1}{2} = \dfrac{3}{4}$

Potato flakes: $\dfrac{4}{2} = 2$

For 6 servings, use $1\frac{3}{4}$ cups water, 3 Tbsp margarine, $\frac{3}{4}$ cup milk, and 2 cups potato flakes.

40. *Use proportions.*

Water: $\dfrac{3\frac{1}{2} \text{ cups}}{12 \text{ servings}} = \dfrac{x}{18 \text{ servings}}$

$$12 \cdot x = 3\frac{1}{2} \cdot 18$$
$$12 \cdot x = \frac{7}{2} \cdot 18$$
$$\frac{12 \cdot x}{12} = \frac{63}{12}$$
$$x = 5\frac{1}{4}$$

Margarine: $\dfrac{6 \text{ Tbsp}}{12 \text{ servings}} = \dfrac{x}{18 \text{ servings}}$

$$12 \cdot x = 6 \cdot 18$$
$$\frac{12 \cdot x}{12} = \frac{108}{12}$$
$$x = 9$$

Milk: $\dfrac{1\frac{1}{2} \text{ cups}}{12 \text{ servings}} = \dfrac{x}{18 \text{ servings}}$

$$12 \cdot x = 1\frac{1}{2} \cdot 18$$
$$12 \cdot x = \frac{3}{2} \cdot 18$$
$$\frac{12 \cdot x}{12} = \frac{27}{12}$$
$$x = 2\frac{1}{4}$$

Potato flakes: $\dfrac{4 \text{ cups}}{12 \text{ servings}} = \dfrac{x}{18 \text{ servings}}$

$$12 \cdot x = 4 \cdot 18$$
$$\frac{12 \cdot x}{12} = \frac{72}{12}$$
$$x = 6$$

Multiply the quantities in Exercise 39 by 3, since 18 servings is 3 times 6 servings.

Water: $1\dfrac{3}{4} \cdot 3 = \dfrac{7}{4} \cdot \dfrac{3}{1} = \dfrac{21}{4} = 5\dfrac{1}{4}$

Margarine: $3 \cdot 3 = 9$

Milk: $\dfrac{3}{4} \cdot 3 = \dfrac{3}{4} \cdot \dfrac{3}{1} = \dfrac{9}{4} = 2\dfrac{1}{4}$

Potato flakes: $2 \cdot 3 = 6$

For 18 servings, use $5\frac{1}{4}$ cups water, 9 Tbsp margarine, $2\frac{1}{4}$ cups milk, and 6 cups potato flakes.

41. Since 3 servings is $\frac{1}{2}$ of 6 servings, multiply each quantity in Exercise 39 by $\frac{1}{2}$.

Water: $1\dfrac{3}{4} \cdot \dfrac{1}{2} = \dfrac{7}{4} \cdot \dfrac{1}{2} = \dfrac{7}{8}$

Margarine: $3 \cdot \dfrac{1}{2} = \dfrac{3}{1} \cdot \dfrac{1}{2} = \dfrac{3}{2} = 1\dfrac{1}{2}$

Milk: $\dfrac{3}{4} \cdot \dfrac{1}{2} = \dfrac{3}{8}$

Potato flakes: $2 \cdot \dfrac{1}{2} = \dfrac{2}{1} \cdot \dfrac{1}{2} = 1$

For 3 servings, use $\frac{7}{8}$ cup water, $1\frac{1}{2}$ Tbsp margarine, $\frac{3}{8}$ cup milk, and 1 cup potato flakes.

42. Since 9 servings is 3 times 3 servings, multiply each quantity in Exercise 41 by 3.

Water: $\dfrac{7}{8} \cdot 3 = \dfrac{7}{8} \cdot \dfrac{3}{1} = \dfrac{21}{8} = 2\dfrac{5}{8}$

Margarine: $\dfrac{3}{2} \cdot 3 = \dfrac{3}{2} \cdot \dfrac{3}{1} = \dfrac{9}{2} = 4\dfrac{1}{2}$

Milk: $\dfrac{3}{8} \cdot 3 = \dfrac{3}{8} \cdot \dfrac{3}{1} = \dfrac{9}{8} = 1\dfrac{1}{8}$

Potato flakes: $1 \cdot 3 = 3$

For 9 servings, use $2\frac{5}{8}$ cups water, $4\frac{1}{2}$ Tbsp margarine, $1\frac{1}{8}$ cups milk, and 3 cups potato flakes.

Chapter 5 Review Exercises

1. orca whale's length to whale shark's length

$$\frac{30 \text{ ft}}{40 \text{ ft}} = \frac{30}{40} = \frac{30 \div 10}{40 \div 10} = \frac{3}{4}$$

2. blue whale's length to great white shark's length

$$\frac{80 \text{ ft}}{20 \text{ ft}} = \frac{80}{20} = \frac{80 \div 20}{20 \div 20} = \frac{4}{1}$$

3. To get a ratio of $\frac{1}{2}$, we can start with the smallest value in the table and see if there is a value that is two times the smallest. $2 \times 20 = 40$, so the ratio of the *great white shark's* length to the *whale shark's* length is $\frac{20}{40} = \frac{1}{2}$. The length of the orca whale is 30 ft, but $2 \times 30 = 60$ is not in the table. $2 \times 40 = 80$, so the ratio of the *whale shark's* length to the *blue whale's* length is $\frac{40}{80} = \frac{1}{2}$.

4. \$2.50 to \$1.25

$$\frac{\$2.50}{\$1.25} = \frac{2.50}{1.25} = \frac{2.50 \div 1.25}{1.25 \div 1.25} = \frac{2}{1}$$

5. \$0.30 to \$0.45

$$\frac{\$0.30}{\$0.45} = \frac{0.30}{0.45} = \frac{0.30 \div .15}{0.45 \div .15} = \frac{2}{3}$$

6. $1\frac{2}{3}$ cups to $\frac{2}{3}$ cup

$$\frac{1\frac{2}{3} \text{ cups}}{\frac{2}{3} \text{ cup}} = \frac{1\frac{2}{3}}{\frac{2}{3}} = \frac{5}{3} \div \frac{2}{3}$$

$$= \frac{5}{\cancel{3}_1} \cdot \frac{\cancel{3}^1}{2} = \frac{5}{2}$$

7. $2\frac{3}{4}$ miles to $16\frac{1}{2}$ miles

$$\frac{2\frac{3}{4} \text{ miles}}{16\frac{1}{2} \text{ miles}} = \frac{2\frac{3}{4}}{16\frac{1}{2}} = \frac{\frac{11}{4}}{\frac{33}{2}} = \frac{11}{4} \div \frac{33}{2}$$

$$= \frac{\cancel{11}^1}{\cancel{4}_2} \cdot \frac{\cancel{2}^1}{\cancel{33}_3} = \frac{1}{6}$$

8. 5 hours to 100 minutes

$5 \text{ hours} = 5 \cdot 60 \text{ minutes} = 300 \text{ minutes}$

$$\frac{300 \text{ minutes}}{100 \text{ minutes}} = \frac{300}{100} = \frac{300 \div 100}{100 \div 100} = \frac{3}{1}$$

9. 9 inches to 2 feet

$2 \text{ feet} = 24 \text{ inches}$

$$\frac{9 \text{ inches}}{24 \text{ inches}} = \frac{9}{24} = \frac{9 \div 3}{24 \div 3} = \frac{3}{8}$$

10. 1 ton to 1500 pounds

$1 \text{ ton} = 2000 \text{ pounds}$

$$\frac{2000 \text{ pounds}}{1500 \text{ pounds}} = \frac{2000}{1500} = \frac{2000 \div 500}{1500 \div 500} = \frac{4}{3}$$

11. 8 hours to 3 days

$3 \text{ days} = 3 \cdot 24 \text{ hours} = 72 \text{ hours}$

$$\frac{8 \text{ hours}}{72 \text{ hours}} = \frac{8}{72} = \frac{8 \div 8}{72 \div 8} = \frac{1}{9}$$

12. \$500 to \$350

$$\frac{\$500}{\$350} = \frac{500 \div 50}{350 \div 50} = \frac{10}{7}$$

The ratio of her sales to his sales is $\frac{10}{7}$.

13. $$\frac{35 \text{ miles per gallon}}{25 \text{ miles per gallon}} = \frac{35}{25} = \frac{35 \div 5}{25 \div 5} = \frac{7}{5}$$

The ratio of the new car's mileage to the old car's mileage is $\frac{7}{5}$.

14. $$\frac{6000 \text{ students}}{7200 \text{ students}} = \frac{6000}{7200} = \frac{6000 \div 1200}{7200 \div 1200} = \frac{5}{6}$$

The ratio of the math students to the English students is $\frac{5}{6}$.

15. \$88 for 8 dozen

$$\frac{\$88 \div 8}{8 \text{ dozen} \div 8} = \frac{\$11}{1 \text{ dozen}}$$

16. 96 children in 40 families

$$\frac{96 \text{ children} \div 8}{40 \text{ families} \div 8} = \frac{12 \text{ children}}{5 \text{ families}}$$

17. 4 pages in 20 minutes

(i) $$\frac{4 \text{ pages} \div 20}{20 \text{ minutes} \div 20} = \frac{0.2 \text{ page}}{1 \text{ minute}} = 0.2 \text{ page/minute} \text{ or } \frac{1}{5} \text{ page/minute}$$

(ii) $$\frac{20 \text{ minutes} \div 4}{4 \text{ pages} \div 4} = \frac{5 \text{ minutes}}{1 \text{ page}} = 5 \text{ minutes/page}$$

18. \$24 in 3 hours

(i) $$\frac{\$24 \div 3}{3 \text{ hours} \div 3} = \frac{\$8}{1 \text{ hour}} = \$8/\text{hour}$$

(ii) $$\frac{3 \text{ hours} \div 24}{\$24 \div 24} = \frac{0.125 \text{ hour}}{\$1} = 0.125 \text{ hour/dollar} \text{ or } \frac{1}{8} \text{ hour/dollar}$$

19.

<u>Size</u>	<u>Cost per Unit</u>
13 ounces	$\frac{\$2.29}{13 \text{ ounces}} \approx \$0.176$ (*)
8 ounces	$\frac{\$1.45}{8 \text{ ounces}} \approx \$0.181$
3 ounces	$\frac{\$0.95}{3 \text{ ounces}} \approx \$0.317$

The best buy is 13 ounces for \$2.29.

20. 50 pounds for \$19.95 − \$1.00 (coupon) = \$18.95

$\frac{\$18.95}{50 \text{ pounds}} = \$0.379$

25 pounds for \$10.40 − \$1.00 (coupon) = \$9.40

$\frac{\$9.40}{25 \text{ pounds}} = \$0.376\ (*)$

8 pounds for \$3.40

$\frac{\$3.40}{8 \text{ pounds}} = \$0.425$

The best buy is 25 pounds for \$10.40 with the \$1 coupon.

21. $\frac{6}{10} = \frac{9}{15}$

$\frac{6 \div 2}{10 \div 2} = \frac{3}{5}$ and $\frac{9 \div 3}{15 \div 3} = \frac{3}{5}$

Both ratios are equivalent to $\frac{3}{5}$, so the proportion is *true.*

22. $\frac{6}{48} = \frac{9}{36}$

Cross products: $6 \cdot 36 = 216$; $48 \cdot 9 = 432$

The cross products are *unequal,* so the proportion is *false.*

23. $\frac{47}{10} = \frac{98}{20}$

Cross products: $47 \cdot 20 = 940$; $10 \cdot 98 = 980$

The cross products are *unequal,* so the proportion is *false.*

24. $\frac{64}{36} = \frac{96}{54}$

$\frac{64 \div 4}{36 \div 4} = \frac{16}{9}$ and $\frac{96 \div 6}{54 \div 6} = \frac{16}{9}$

Both ratios are equivalent to $\frac{16}{9}$, so the proportion is *true.*

25. $\frac{1.5}{2.4} = \frac{2}{3.2}$

Cross products: $1.5(3.2) = 4.8$; $2.4 \cdot 2 = 4.8$

The cross products are *equal,* so the proportion is *true.*

26. $\frac{3\frac{1}{2}}{2\frac{1}{3}} = \frac{6}{4}$

Cross products:

$3\frac{1}{2} \cdot 4 = \frac{7}{\cancel{2}_1} \cdot \frac{\cancel{4}^2}{1} = 14$

$2\frac{1}{3} \cdot 6 = \frac{7}{\cancel{3}_1} \cdot \frac{\cancel{6}^2}{1} = 14$

The cross products are *equal,* so the proportion is *true.*

27. $\frac{4}{42} = \frac{150}{x}$ **OR** $\frac{2}{21} = \frac{150}{x}$

$2 \cdot x = 21 \cdot 150$ *Cross products are equivalent*

$\frac{2 \cdot x}{2} = \frac{3150}{2}$

$x = 1575$

Check: $2 \cdot 1575 = 3150 = 21 \cdot 150$

28. $\frac{16}{x} = \frac{12}{15}$ **OR** $\frac{16}{x} = \frac{4}{5}$

$4 \cdot x = 5 \cdot 16$ *Cross products are equivalent*

$\frac{4 \cdot x}{4} = \frac{80}{4}$

$x = 20$

Check: $16 \cdot 5 = 80 = 20 \cdot 4$

29. $\frac{100}{14} = \frac{x}{56}$ **OR** $\frac{50}{7} = \frac{x}{56}$

$7 \cdot x = 50 \cdot 56$ *Cross products are equivalent*

$\frac{7 \cdot x}{7} = \frac{2800}{7}$

$x = 400$

Check: $50 \cdot 56 = 2800 = 7 \cdot 400$

30. $\frac{5}{8} = \frac{x}{20}$

$8 \cdot x = 5 \cdot 20$ *Cross products are equivalent*

$\frac{8 \cdot x}{8} = \frac{100}{8}$

$x = 12.5$

Check: $5 \cdot 20 = 100 = 8(12.5)$

31. $\frac{x}{24} = \frac{11}{18}$

$18 \cdot x = 24 \cdot 11$ *Cross products are equivalent*

$\frac{18 \cdot x}{18} = \frac{264}{18}$

$x = \frac{44}{3} \approx 14.67$

Check: $\frac{44}{3} \cdot 18 = 264 = 24 \cdot 11$

32. $\frac{7}{x} = \frac{18}{21}$ **OR** $\frac{7}{x} = \frac{6}{7}$

$6 \cdot x = 7 \cdot 7$ *Cross products are equivalent*

$\frac{6 \cdot x}{6} = \frac{49}{6}$

$x = \frac{49}{6} \approx 8.17$

Check: $7 \cdot 7 = 49 = \frac{49}{6} \cdot 6$

33. $\frac{x}{3.6} = \frac{9.8}{0.7}$

$0.7 \cdot x = 9.8(3.6)$ *Cross products are equivalent*

$\frac{0.7 \cdot x}{0.7} = \frac{35.28}{0.7}$

$x = 50.4$

Check: $50.4(0.7) = 35.28 = 3.6(9.8)$

34. $\frac{13.5}{1.7} = \frac{4.5}{x}$

$13.5 \cdot x = 1.7(4.5)$ *Cross products are equivalent*

$\frac{13.5 \cdot x}{13.5} = \frac{7.65}{13.5}$

$x \approx 0.57$

Check: $13.5\left(\frac{7.65}{13.5}\right) = 7.65 = 1.7(4.5)$

35. $\frac{0.82}{1.89} = \frac{x}{5.7}$

$1.89 \cdot x = 0.82(5.7)$ *Cross products are equivalent*

$\frac{1.89 \cdot x}{1.89} = \frac{4.674}{1.89}$

$x \approx 2.47$

Check: $0.82(5.7) = 4.674 = 1.89\left(\frac{4.674}{1.89}\right)$

36. $\frac{3 \text{ cats}}{5 \text{ dogs}} = \frac{x}{45 \text{ dogs}}$

$5 \cdot x = 3 \cdot 45$

$\frac{5 \cdot x}{5} = \frac{135}{5}$

$x = 27$

There are 27 cats.

37. $\frac{8 \text{ hits}}{28 \text{ at bats}} = \frac{x}{161 \text{ at bats}}$

$28 \cdot x = 8 \cdot 161$

$\frac{28 \cdot x}{28} = \frac{1288}{28}$

$x = 46$

She will get 46 hits.

38. $\frac{3.5 \text{ pounds}}{\$9.77} = \frac{5.6 \text{ pounds}}{x}$

$3.5 \cdot x = 9.77(5.6)$

$\frac{3.5 \cdot x}{3.5} = \frac{54.712}{3.5}$

$x \approx 15.63$

The cost for 5.6 pounds of ground beef is $15.63 (rounded).

39. $\frac{4 \text{ voting students}}{10 \text{ students}} = \frac{x}{8247 \text{ students}}$

$10 \cdot x = 4 \cdot 8247$

$\frac{10 \cdot x}{10} = \frac{32{,}988}{10}$

$x \approx 3299$

They should expect about 3299 students to vote.

40. $\frac{1 \text{ inch}}{16 \text{ feet}} = \frac{4.25 \text{ inches}}{x}$

$1 \cdot x = 16(4.25)$

$x = 68$

The length of the real boxcar is 68 feet.

41. 2 dozen necklaces $= 2 \cdot 12 = 24$ necklaces

$\frac{24 \text{ necklaces}}{16\frac{1}{2} \text{ hours}} = \frac{40 \text{ necklaces}}{x}$

$24 \cdot x = 16\frac{1}{2} \cdot 40 = \frac{33}{2} \cdot \frac{40}{1}$

$\frac{24 \cdot x}{24} = \frac{660}{24}$

$x = 27.5 = 27\frac{1}{2}$

It will take Marvette $27\frac{1}{2}$ hours or 27.5 hours to make 40 necklaces.

42. $\frac{284 \text{ calories}}{25 \text{ minutes}} = \frac{x}{45 \text{ minutes}}$

$25 \cdot x = 284 \cdot 45$

$\frac{25 \cdot x}{25} = \frac{12{,}780}{25}$

$x = 511.2 \approx 511$

A 180-pound person would burn about 511 calories in 45 minutes.

43. $\frac{3.5 \text{ milligrams}}{50 \text{ pounds}} = \frac{x}{210 \text{ pounds}}$

$50 \cdot x = 3.5 \cdot 210$

$\frac{50 \cdot x}{50} = \frac{735}{50}$

$x = 14.7$

A patient who weighs 210 pounds should be given 14.7 milligrams of the medicine.

44. **[5.4]** $\frac{x}{45}=\frac{70}{30}$ **OR** $\frac{x}{45}=\frac{7}{3}$

$3\cdot x=7\cdot 45$ *Cross products are equivalent*

$\frac{3\cdot x}{3}=\frac{315}{3}$

$x=105$

Check: $105\cdot 3=315=45\cdot 7$

45. **[5.4]** $\frac{x}{52}=\frac{0}{20}$

Since $\frac{0}{20}=0$, x *must* equal 0. The denominator 52 could be any number (except 0).

46. **[5.4]** $\frac{64}{10}=\frac{x}{20}$ **OR** $\frac{32}{5}=\frac{x}{20}$

$5\cdot x=32\cdot 20$ *Cross products are equivalent*

$\frac{5\cdot x}{5}=\frac{640}{5}$

$x=128$

Check: $32\cdot 20=640=5\cdot 128$

47. **[5.4]** $\frac{15}{x}=\frac{65}{100}$ **OR** $\frac{15}{x}=\frac{13}{20}$

$13\cdot x=15\cdot 20$ *Cross products are equivalent*

$\frac{13\cdot x}{13}=\frac{300}{13}$

$x=\frac{300}{13}\approx 23.08$

Check: $15\cdot 20=300=\frac{300}{13}\cdot 13$

48. **[5.4]** $\frac{7.8}{3.9}=\frac{13}{x}$ **OR** $\frac{2}{1}=\frac{13}{x}$

$2\cdot x=1\cdot 13$

$\frac{2\cdot x}{2}=\frac{13}{2}$

$x=6.5$

Check: $2(6.5)=13=1\cdot 13$

49. **[5.4]** $\frac{34.1}{x}=\frac{0.77}{2.65}$

$0.77\cdot x=34.1(2.65)$

$\frac{0.77\cdot x}{0.77}=\frac{90.365}{0.77}$

$x\approx 117.36$

Check: $34.1(2.65)=90.365=\left(\frac{90.365}{0.77}\right)(0.77)$

50. **[5.3]** $\frac{55}{18}=\frac{80}{27}$

Cross products: $55\cdot 27=1485$; $18\cdot 80=1440$

The cross products are *unequal,* so the proportion is *false.*

51. **[5.3]** $\frac{5.6}{0.6}=\frac{18}{1.94}$

Cross products:
$5.6(1.94)=10.864$; $0.6\cdot 18=10.8$

The cross products are *unequal,* so the proportion is *false.*

52. **[5.3]** $\frac{\frac{1}{5}}{2}=\frac{1\frac{1}{6}}{11\frac{2}{3}}$

Cross products:

$\frac{1}{5}\cdot 11\frac{2}{3}=\frac{1}{\cancel{5}_1}\cdot\frac{\cancel{35}^7}{3}=\frac{7}{3}=2\frac{1}{3}$

$2\cdot 1\frac{1}{6}=\frac{\cancel{2}^1}{1}\cdot\frac{7}{\cancel{6}_3}=\frac{7}{3}=2\frac{1}{3}$

The cross products are *equal,* so the proportion is *true.*

53. **[5.1]** 4 dollars to 10 quarters

4 dollars $=4\cdot 4=16$ quarters

$\frac{16\text{ quarters}\div 2}{10\text{ quarters}\div 2}=\frac{8}{5}$

54. **[5.1]** $4\frac{1}{8}$ inches to 10 inches

$\frac{4\frac{1}{8}\text{ inches}}{10\text{ inches}}=\frac{\frac{33}{8}}{10}=\frac{33}{8}\div 10$

$=\frac{33}{8}\cdot\frac{1}{10}=\frac{33}{80}$

55. **[5.1]** 10 yards to 8 feet

10 yards $=10\cdot 3=30$ feet

$\frac{30\text{ feet}}{8\text{ feet}}=\frac{30\div 2}{8\div 2}=\frac{15}{4}$

56. **[5.1]** \$3.60 to \$0.90

$\frac{\$3.60}{\$0.90}=\frac{3.60\div 0.90}{0.90\div 0.90}=\frac{4}{1}$

57. **[5.1]** 12 eggs to 15 eggs

$\frac{12\text{ eggs}}{15\text{ eggs}}=\frac{12\div 3}{15\div 3}=\frac{4}{5}$

58. **[5.1]** 37 meters to 7 meters

$\frac{37\text{ meters}}{7\text{ meters}}=\frac{37}{7}$

59. **[5.1]** 3 pints to 4 quarts

4 quarts $=4\cdot 2=8$ pints

$\frac{3\text{ pints}}{8\text{ pints}}=\frac{3}{8}$

60. **[5.1]** 15 minutes to 3 hours

3 hours $= 3 \cdot 60 = 180$ minutes

$$\frac{15 \text{ minutes}}{180 \text{ minutes}} = \frac{15 \div 15}{180 \div 15} = \frac{1}{12}$$

61. **[5.1]** $4\frac{1}{2}$ miles to $1\frac{3}{10}$ miles

$$\frac{4\frac{1}{2} \text{ miles}}{1\frac{3}{10} \text{ miles}} = \frac{4\frac{1}{2}}{1\frac{3}{10}} = 4\frac{1}{2} \div 1\frac{3}{10}$$

$$= \frac{9}{2} \div \frac{13}{10} = \frac{9}{\cancel{2}_1} \cdot \frac{\cancel{10}^5}{13}$$

$$= \frac{45}{13}$$

62. **[5.5]**

$$\frac{7 \text{ buying fans}}{8 \text{ fans}} = \frac{x}{28{,}500 \text{ fans}}$$

$$8 \cdot x = 7 \cdot 28{,}500$$

$$\frac{8 \cdot x}{8} = \frac{199{,}500}{8}$$

$$x \approx 24{,}937.5 \text{ or } 24{,}900$$

(*rounded to the nearest hundred*)

At today's concert, about 24,900 fans can be expected to buy a beverage.

63. **[5.1]**

$$\frac{\$400 \text{ spent on car insurance}}{\$150 \text{ spent on repairs}} = \frac{400 \div 50}{150 \div 50} = \frac{8}{3}$$

The ratio of the amount spent on insurance to the amount spent on repairs is $\frac{8}{3}$.

64. **[5.2]** 25 feet for \$0.78

$$\frac{\$0.78}{25 \text{ feet}} \approx \$0.031 \text{ per foot}$$

75 feet for \$1.99 – \$0.50 (coupon) = \$1.49

$$\frac{\$1.49}{75 \text{ feet}} \approx \$0.020 \text{ per foot } (*)$$

100 feet for \$2.59 – \$0.50 (coupon) = \$2.09

$$\frac{\$2.09}{100 \text{ feet}} \approx \$0.021 \text{ per foot}$$

The best buy is 75 feet for \$1.99 with a 50¢ coupon.

65. **[5.5]** First find the length.

$$\frac{0.5 \text{ inch}}{6 \text{ feet}} = \frac{1.75 \text{ inches}}{x}$$

$$0.5 \cdot x = 6(1.75)$$

$$\frac{0.5 \cdot x}{0.5} = \frac{10.5}{0.5}$$

$$x = 21$$

When it is built, the actual length of the patio will be 21 feet.

Then find the width.

$$\frac{0.5 \text{ inch}}{6 \text{ feet}} = \frac{1.25 \text{ inches}}{x}$$

$$0.5 \cdot x = 6(1.25)$$

$$\frac{0.5 \cdot x}{0.5} = \frac{7.5}{0.5}$$

$$x = 15$$

When it is built, the actual width of the patio will be 15 feet.

66. **[5.5]** $\frac{0.8 \text{ gallons}}{3 \text{ hours}} = \frac{2 \text{ gallons}}{x}$

$$0.8 \cdot x = 3 \cdot 2$$

$$\frac{0.8 \cdot x}{0.8} = \frac{6}{0.8}$$

$$x = 7.5 \text{ or } 7\frac{1}{2}$$

The lawn mower should run 7.5 or $7\frac{1}{2}$ hours on a full tank.

67. **[5.5]** $\frac{1\frac{1}{2} \text{ teaspoons}}{24 \text{ pounds}} = \frac{x}{8 \text{ pounds}}$

$$24 \cdot x = 1\frac{1}{2} \cdot 8 = \frac{3}{2} \cdot \frac{8}{1} = \frac{24}{2} = 12$$

$$\frac{24 \cdot x}{24} = \frac{12}{24}$$

$$x = \frac{1}{2} \text{ or } 0.5$$

The infant should be given $\frac{1}{2}$ or 0.5 teaspoon.

68. **[5.5]** $\frac{251 \text{ points}}{169 \text{ minutes}} = \frac{x}{14 \text{ minutes}}$

$$169 \cdot x = 251 \cdot 14$$

$$\frac{169 \cdot x}{169} = \frac{3514}{169}$$

$$x \approx 20.792899$$

Charles should score 21 points (rounded).

69. **[5.5]** Set up the proportion to compare teaspoons to pounds on both sides.

$$\frac{1.5 \text{ tsp}}{24 \text{ pounds}} = \frac{x \text{ tsp}}{8 \text{ pounds}}$$

Show that the cross products are equal.

$$(24)(x) = (1.5)(8)$$

Divide both sides by 24.

$$\frac{(\cancel{24}^1)(x)}{\cancel{24}_1} = \frac{12}{24}, \text{ so } x = \frac{1}{2} \text{ tsp or } 0.5 \text{ tsp}$$

70. **[5.5] (a)** $\frac{1000 \text{ milligrams}}{5 \text{ pounds}} = \frac{x}{7 \text{ pounds}}$

$$5 \cdot x = 1000 \cdot 7$$
$$\frac{5 \cdot x}{5} = \frac{7000}{5}$$
$$x = 1400$$

A 7-pound cat should be given 1400 milligrams.

(b) 8 ounces $= \frac{8}{16}$ pound $= 0.5$ pound

$$\frac{1000 \text{ milligrams}}{5 \text{ pounds}} = \frac{x}{0.5 \text{ pounds}}$$
$$5 \cdot x = 1000(0.5)$$
$$\frac{5 \cdot x}{5} = \frac{500}{5}$$
$$x = 100$$

An 8-ounce kitten should be given 100 milligrams.

Chapter 5 Test

1. 16 fish to 20 fish

$$\frac{16 \text{ fish}}{20 \text{ fish}} = \frac{16 \div 4}{20 \div 4} = \frac{4}{5}$$

2. 300 miles on 15 gallons

$$\frac{300 \text{ miles} \div 15}{15 \text{ gallons} \div 15} = \frac{20 \text{ miles}}{1 \text{ gallon}}$$

3. \$15 for 75 minutes

$$\frac{\$15 \div 15}{75 \text{ minutes} \div 15} = \frac{\$1}{5 \text{ minutes}}$$

4. 3 hours to 40 minutes

3 hours $= 3 \cdot 60$ minutes $= 180$ minutes

$$\frac{180 \text{ minutes}}{40 \text{ minutes}} = \frac{180 \div 20}{40 \div 20} = \frac{9}{2}$$

5. $\frac{1200 \text{ seats}}{320 \text{ seats}} = \frac{1200 \div 80}{320 \div 80} = \frac{15}{4}$

6.

<u>Size</u>	<u>Cost per Unit</u>
Small	$\frac{\$4.29}{4.5 \text{ inches}} \approx \$0.953/\text{inch}$
Regular	$\frac{\$5.49}{8 \text{ inches}} \approx \$0.686/\text{inch}$ (*)
Large	$\frac{\$8.29}{12 \text{ inches}} \approx \$0.691/\text{inch}$

The best buy is the 8-inch sub, which costs about \$0.686 per inch.

7. 28 ounces of Brand X for
\$3.89 – \$0.75 (coupon) = \$3.14

$$\frac{\$3.14}{28 \text{ ounces}} \approx \$0.112 \text{ per ounce}$$

18 ounces of Brand Y for
\$1.89 – \$0.25 (coupon) = \$1.64

$$\frac{\$1.64}{18 \text{ ounces}} \approx \$0.091 \text{ per ounce } (*)$$

13 ounces of Brand Z for \$1.29.

$$\frac{\$1.29}{13 \text{ ounces}} \approx \$0.099 \text{ per ounce}$$

The best buy is 18 ounces of Brand Y for \$1.89 with a 25¢ coupon.

8. You earned less this year.

An example is:

Last year → \$15,000
This year → \$10,000 $= \frac{3}{2}$

9. $\frac{6}{14} = \frac{18}{45}$

Cross products: $6 \cdot 45 = 270$; $14 \cdot 18 = 252$

The cross products are *unequal,* so the proportion is *false.* We could also write the fractions in lowest terms to see that they are not equal $\left(\frac{3}{7} \neq \frac{2}{5}\right)$.

10. $\frac{8.4}{2.8} = \frac{2.1}{0.7}$

$\frac{3}{1} = \frac{3}{1}$ *Write in lowest terms*

The proportion is *true.* We could also show that the cross products are equal (5.88).

11. $\frac{5}{9} = \frac{x}{45}$

$9 \cdot x = 5 \cdot 45$ *Cross products are equivalent*

$$\frac{9 \cdot x}{9} = \frac{225}{9}$$
$$x = 25$$

Check: $5 \cdot 45 = 225 = 9 \cdot 25$

12. $\frac{3}{1} = \frac{8}{x}$

$3 \cdot x = 8 \cdot 1$ *Cross products are equivalent*

$$\frac{3 \cdot x}{3} = \frac{8}{3}$$
$$x \approx 2.67$$

Check: $3\left(\frac{8}{3}\right) = 8 = 1 \cdot 8$

13. $\frac{x}{20} = \frac{6.5}{0.4}$

$0.4 \cdot x = 20(6.5)$ *Cross products are equivalent*

$$\frac{0.4 \cdot x}{0.4} = \frac{130}{0.4}$$
$$x = 325$$

Check: $325(0.4) = 130 = 20(6.5)$

14. $\dfrac{2\frac{1}{3}}{x} = \dfrac{\frac{8}{9}}{4}$

$\dfrac{8}{9} \cdot x = 2\dfrac{1}{3} \cdot 4 = \dfrac{7}{3} \cdot \dfrac{4}{1} = \dfrac{28}{3}$

$\dfrac{\frac{8}{9} \cdot x}{\frac{8}{9}} = \dfrac{\frac{28}{3}}{\frac{8}{9}}$

$x = \dfrac{28}{3} \div \dfrac{8}{9} = \dfrac{\overset{7}{\cancel{28}}}{\underset{1}{\cancel{3}}} \cdot \dfrac{\overset{3}{\cancel{9}}}{\underset{2}{\cancel{8}}} = \dfrac{21}{2} = 10\dfrac{1}{2}$

Check: $2\frac{1}{3} \cdot 4 = \frac{28}{3} = \frac{21}{2} \cdot \frac{8}{9}$

15. $\dfrac{240 \text{ words}}{5 \text{ minutes}} = \dfrac{x}{12 \text{ minutes}}$

$5 \cdot x = 240 \cdot 12$

$\dfrac{5 \cdot x}{5} = \dfrac{2880}{5}$

$x = 576$

Pedro could type 576 words in 12 minutes.

16. $\dfrac{0.8 \text{ ounce}}{50 \text{ square feet}} = \dfrac{x}{225 \text{ square feet}}$

$50 \cdot x = 0.8(225)$

$\dfrac{50 \cdot x}{50} = \dfrac{180}{50}$

$x = 3.6$

3.6 ounces of wildflower seeds are needed for a garden with 225 square feet.

17. $\dfrac{2 \text{ left-handed people}}{15 \text{ people}} = \dfrac{x}{650 \text{ students}}$

$15 \cdot x = 2 \cdot 650$

$\dfrac{15 \cdot x}{15} = \dfrac{1300}{15}$

$x = \dfrac{260}{3} \approx 87$

You could expect about 87 students to be left-handed.

18. No, 4875 cannot be correct because there are only 650 students in the whole school.

19. $\dfrac{8.2 \text{ grams}}{50 \text{ pounds}} = \dfrac{x}{145 \text{ pounds}}$

$50 \cdot x = 8.2(145)$

$\dfrac{50 \cdot x}{50} = \dfrac{1189}{50}$

$x = 23.78 \approx 23.8$

A 145-pound person should be given 23.8 grams (rounded).

20. $\dfrac{1 \text{ inch}}{8 \text{ feet}} = \dfrac{7.5 \text{ inches}}{x}$

$1 \cdot x = 8(7.5)$

$x = 60$

The actual height of the building is 60 feet.

Cumulative Review Exercises (Chapters 1–5)

1.

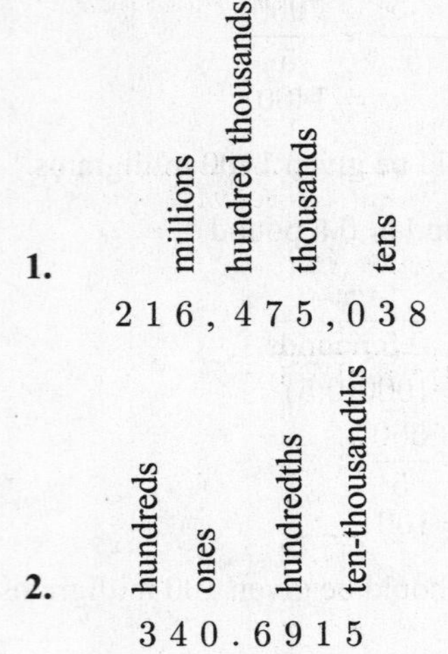

2. hundreds, ones, hundredths, ten-thousandths

3 4 0 . 6 9 1 5

3. 9903 to the nearest hundred: 9<u>9</u>03

Next digit is less than 5. Hundreds place does not change. All digits to the right of the underlined place change to 0. **9900**

4. 617.0519 to the nearest tenth:
Draw a cut-off line: 617.0|519
The first digit cut is 5 or more, so round up.
617.1

5. \$99.81 to the nearest dollar:
Draw a cut-off line: \$99.|81
The first digit cut is 5 or more, so round up. **\$100**

6. \$3.0555 to the nearest cent:
Draw a cut-off line: \$3.05|55
The first digit cut is 5 or more, so round up.
\$3.06

7. *Estimate:*

$$\begin{array}{r} 30 \\ 5000 \\ +\ 400 \\ \hline 5430 \end{array}$$

Exact:

$$\begin{array}{r} 28 \\ 5206 \\ +\ 351 \\ \hline 5585 \end{array}$$

8. *Estimate:*

$$\begin{array}{r} 60 \\ -\ 6 \\ \hline 54 \end{array}$$

Exact:

$$\begin{array}{r} 63.100 \\ -\ 5.692 \\ \hline 57.408 \end{array}$$

9. *Estimate:*

$$\begin{array}{r} 5000 \\ \times\ \ 800 \\ \hline 4{,}000{,}000 \end{array}$$

Exact:

$$\begin{array}{r} 4716 \\ \times\ \ 804 \\ \hline 18\ 864 \\ 3\ 772\ 80 \\ \hline 3{,}791{,}664 \end{array}$$

10. *Estimate:*

$$\begin{array}{r} 1 \\ \times\ 20 \\ \hline 20 \end{array}$$

Exact:

$$\begin{array}{rl} 0.982 & \leftarrow \text{ 3 decimal places} \\ \times\ 17.8 & \leftarrow \text{ 1 decimal place} \\ \hline 7856 & \\ 6\,874\ \ & \\ 9\,82\ \ \ \ & \\ \hline 17.4796 & \leftarrow \text{ 4 decimal places} \end{array}$$

11. *Estimate:*

$$50\overline{)50{,}000} = 1\,000$$

Exact:

$$\begin{array}{r} 907 \\ 53\overline{)48{,}071} \\ 477\ \ \\ \hline 37\ \\ 0\ \\ \hline 371 \\ 371 \\ \hline 0 \end{array}$$

12. *Estimate:*

$$5\overline{)2000} = 400$$

Exact:

$$\begin{array}{r} 364 \\ 4.5_\wedge\overline{)1638.0_\wedge} \\ 135\ \ \ \\ \hline 288\ \ \\ 270\ \ \\ \hline 180 \\ 180 \\ \hline 0 \end{array}$$

13. *Estimate:* $2 \cdot 4 = 8$

Exact: $1\frac{5}{6} \times 3\frac{3}{5} = \frac{11}{\not{6}_1} \cdot \frac{\not{18}^{3}}{5} = \frac{33}{5} = 6\frac{3}{5}$

14. *Estimate:* $5 \div 1 = 5$

Exact: $5\frac{1}{4} \div \frac{7}{8} = 5\frac{1}{4} \cdot \frac{8}{7} = \frac{\not{21}^{3}}{\not{4}_1} \cdot \frac{\not{8}^{2}}{\not{7}_1} = 6$

15. *Estimate:* $3 - 2 = 1$

Exact: $2\frac{4}{5} - 1\frac{5}{6} = \frac{14}{5} - \frac{11}{6} = \frac{84}{30} - \frac{55}{30} = \frac{29}{30}$

16. *Estimate:* $3 + 11 = 14$

Exact: $2\frac{9}{10} + 10\frac{1}{2} = 2\frac{9}{10} + 10\frac{5}{10}$

$= 12\frac{14}{10} = 13\frac{4}{10} = 13\frac{2}{5}$

17. $988 + 373{,}422 + 6$

$$\begin{array}{r} {}^{1\,11}\ \ \\ 988 \\ 373{,}422 \\ +\ \ \ \ \ \ \ \ \ 6 \\ \hline 374{,}416 \end{array}$$

18. $30 - 0.66$

$$\begin{array}{r} {}^{2\,9\,10\,10} \\ \not{3}\not{0}.\not{0}\not{0} \\ -\ \ 0.66 \\ \hline 29.34 \end{array}$$

19.

$$\begin{array}{r} 610\ \textbf{R27} \\ 33\overline{)20{,}157} \\ 198\ \ \ \\ \hline 35\ \ \\ 33\ \ \\ \hline 27\ \\ 0\ \\ \hline 27\ \end{array}$$

20. $(1.9)(0.004)$

$$\begin{array}{rl} 1.9 & \leftarrow \text{ 1 decimal place} \\ \times\ 0.004 & \leftarrow \text{ 3 decimal places} \\ \hline 0.0076 & \leftarrow \text{ 4 decimal places} \end{array}$$

21. $3020 - 708$

$$\begin{array}{r} {}^{2\,10\,1\,10} \\ \not{3}\not{0}\not{2}\not{0} \\ -\ \ 708 \\ \hline 2312 \end{array}$$

22. $0.401 + 62.98 + 5$

$$\begin{array}{rl} 0.401 & \\ 62.980 & \\ +\ 5.000 & \textit{Write in 0's} \\ \hline 68.381 & \end{array}$$

23. $1.39 \div 0.025$

$$\begin{array}{rl} 55.6 & \\ 0.025_\wedge\overline{)1.390_\wedge0} & \textit{Move decimal point in} \\ 125\ \ \ \ & \textit{divisor and dividend} \\ \hline 140\ \ & \\ 125\ \ & \\ \hline 150 & \textit{Write 0 in} \\ 150 & \textit{the dividend} \\ \hline 0 & \end{array}$$

24. $(6392)(5609) = 6392 \cdot 5609 = 35{,}852{,}728$

25. $36 + 18 \div 6 = 36 + (18 \div 6) = 36 + 3 = 39$

26. $8 \div 4 + (10 - 3^2) \cdot 4^2 = 8 \div 4 + (10 - 9) \cdot 16$

$= 8 \div 4 + 1 \cdot 16$

$= 2 + 1 \cdot 16$

$= 2 + 16 = 18$

27. $88 \div \sqrt{121} \cdot 2^3 = 88 \div \sqrt{121} \cdot 8$

$= 88 \div 11 \cdot 8$

$= 8 \cdot 8 = 64$

28. $(16.2 - 5.85) - 2.35(4) = 10.35 - 2.35(4)$
$= 10.35 - 9.4 = 0.95$

29. 0.0105 is one hundred five ten-thousandths.

30. Sixty and seventy-one thousandths is 60.071.

31. $1\frac{9}{16} = \frac{25}{16}$

$$\begin{array}{r} 1.5625 \\ 16\overline{)25.0000} \\ \underline{16} \\ 9\,0 \\ \underline{8\,0} \\ 1\,00 \\ \underline{96} \\ 40 \\ \underline{32} \\ 80 \\ \underline{80} \\ 0 \end{array}$$

$1\frac{9}{16} \approx 1.563$

$1\frac{3}{8} = \frac{11}{8}$

$$\begin{array}{r} 1.375 \\ 8\overline{)11.000} \\ \underline{8} \\ 3\,0 \\ \underline{2\,4} \\ 60 \\ \underline{56} \\ 40 \\ \underline{40} \\ 0 \end{array}$$

$1\frac{3}{8} = 1.375$

$1\frac{1}{8} = \frac{9}{8}$

$$\begin{array}{r} 1.125 \\ 8\overline{)9.000} \\ \underline{8} \\ 1\,0 \\ \underline{8} \\ 20 \\ \underline{16} \\ 40 \\ \underline{40} \\ 0 \end{array}$$

$1\frac{1}{8} = 1.125$

32. Use the answers from Exercise 31.

$1.5 = 1.500$

$1\frac{9}{16} \approx 1.563$ *largest*

$1.25 = 1.250$

$1\frac{3}{8} = 1.375$

$1\frac{1}{8} = 1.125$ *smallest*

The nest box openings (in inches) in order from smallest to largest are:

$1\frac{1}{8},\ 1.25,\ 1\frac{3}{8},\ 1.5,\ 1\frac{9}{16}$

33. $1\frac{9}{16} - 1\frac{1}{8} = 1\frac{9}{16} - 1\frac{2}{16} = \frac{25}{16} - \frac{18}{16}$
$= \frac{25 - 18}{16} = \frac{7}{16}$

The difference is $\frac{7}{16}$ inch.

34. $1\frac{9}{16} - 1.5 = 1\frac{9}{16} - 1\frac{1}{2} = 1\frac{9}{16} - 1\frac{8}{16}$
$= \frac{25}{16} - \frac{24}{16} = \frac{25 - 24}{16} = \frac{1}{16}$

$$\begin{array}{r} 1\frac{9}{16} = 1.5625 \\ \underline{-\ 1.5 = 1.5000} \\ 0.0625 \end{array}$$

The difference is $\frac{1}{16}$ inch or about 0.063 inch.

35. drink coffee daily to drink coffee occasionally

$\frac{550}{250} = \frac{550 \div 50}{250 \div 50} = \frac{11}{5}$

36. don't drink coffee to total

Total = 250 + 200 + 550 = 1000

$\frac{200}{1000} = \frac{200 \div 200}{1000 \div 200} = \frac{1}{5}$

37. drink coffee to don't drink coffee

$\frac{550 + 250}{200} = \frac{800}{200} = \frac{800 \div 200}{200 \div 200} = \frac{4}{1}$

38. 20 minutes to 4 hours

4 hours = $4 \cdot 60$ = 240 minutes

$\frac{20 \text{ minutes}}{240 \text{ minutes}} = \frac{20 \div 20}{240 \div 20} = \frac{1}{12}$

39. 2 ft to 8 in.

2 feet = $2 \cdot 12$ = 24 inches

$\frac{24 \text{ inches}}{8 \text{ inches}} = \frac{24 \div 8}{8 \div 8} = \frac{3}{1}$

40. 20 servings for \$1.59

$\frac{\$1.59}{20 \text{ servings}} \approx \$0.0795$ per serving

36 servings for \$3.24 – \$0.50 (coupon) = \$2.74

$\frac{\$2.74}{36 \text{ servings}} \approx \$0.0761$ per serving (*)

48 servings for \$4.99 – \$0.50 (coupon) = \$4.49

$\frac{\$4.49}{48 \text{ servings}} \approx \$0.0935$ per serving

The best buy is 36 servings for \$3.24 with a 50¢ coupon.

41. $\frac{9}{12} = \frac{x}{28}$ **OR** $\frac{3}{4} = \frac{x}{28}$

$4 \cdot x = 3 \cdot 28$ *Cross products are equivalent*

$\frac{4 \cdot x}{4} = \frac{84}{4}$ *Divide both sides by 4*

$x = 21$

Check: $3 \cdot 28 = 84 = 4 \cdot 21$

42. $\frac{7}{12} = \frac{10}{x}$

$7 \cdot x = 10 \cdot 12$ *Cross products are equivalent*

$\frac{7 \cdot x}{7} = \frac{120}{7}$ *Divide both sides by 7*

$x \approx 17.14$

Check: $7(\frac{120}{7}) = 120 = 12 \cdot 10$

43. $\frac{x}{\frac{3}{4}} = \frac{2\frac{1}{2}}{\frac{1}{6}}$

$\frac{1}{6} \cdot x = \frac{3}{4} \cdot 2\frac{1}{2} = \frac{3}{4} \cdot \frac{5}{2}$

$\frac{\frac{1}{6} \cdot x}{\frac{1}{6}} = \frac{\frac{15}{8}}{\frac{1}{6}}$

$x = \frac{15}{8} \div \frac{1}{6} = \frac{15}{\cancel{8}_4} \cdot \frac{\cancel{6}^3}{1} = \frac{45}{4} = 11\frac{1}{4}$

Check: $\frac{45}{4} \cdot \frac{1}{6} = \frac{45}{24} = \frac{15}{8} = \frac{3}{4} \cdot \frac{5}{2}$

44. $\frac{6.7}{x} = \frac{62.8}{9.15}$

$62.8 \cdot x = 6.7(9.15)$

$\frac{62.8 \cdot x}{62.8} = \frac{61.305}{62.8}$

$x \approx 0.98$

Check: $6.7(9.15) = 61.305 = (\frac{61.305}{62.8})62.8$

45. Find the amount already collected.

$\frac{5}{\cancel{6}_1} \cdot \frac{\cancel{1500}^{250}}{1} = 1250$

$1500 - 1250 = 250$

They need to collect 250 pounds.

46. $\frac{10 \text{ cm wide}}{15 \text{ cm long}} = \frac{x}{40 \text{ cm long}}$

$15 \cdot x = 10 \cdot 40$

$\frac{15 \cdot x}{15} = \frac{400}{15} = \frac{80}{3}$

$x \approx 26.7$

The enlarged width of the photo is 26.7 centimeters (rounded).

47. First add the number of times Norma ran in the morning and in the afternoon.

$4 + 2\frac{1}{2} = 6\frac{1}{2}$

Then multiply the sum by the distance around Dunning Pond.

$1\frac{1}{10} \cdot 6\frac{1}{2} = \frac{11}{10} \cdot \frac{13}{2} = \frac{143}{20} = 7\frac{3}{20}$

Norma ran $7\frac{3}{20}$ miles.

48. Divide the number of miles driven by the number of gallons of gas purchased.

```
          1 8. 0 0
49.8^)8 9 6. 5^ 0 0
      4 9 8
      3 9 8 5
      3 9 8 4
            1 0
              0
            1 0 0
              0
            1 0 0
```

Rodney's SUV got 18.0 miles per gallon (rounded) on the vacation.

49. $\frac{5 \text{ bothered residents}}{8 \text{ residents}} = \frac{x}{224 \text{ residents}}$

$8 \cdot x = 5 \cdot 224$

$\frac{8 \cdot x}{8} = \frac{1120}{8}$

$x = 140$

You would expect 140 residents to be bothered by noise.

50. $\frac{\frac{1}{2}\text{ teaspoon}}{2\text{ quarts water}} = \frac{x}{5\text{ quarts water}}$

$$2 \cdot x = \frac{1}{2} \cdot 5$$

$$\frac{2 \cdot x}{2} = \frac{\frac{5}{2}}{2}$$

$$x = \frac{5}{2} \div 2 = \frac{5}{2} \cdot \frac{1}{2} = \frac{5}{4} = 1\frac{1}{4}$$

Use $1\frac{1}{4}$ teaspoons of plant food.

51. The cost per minute rates (in dollars) from least to greatest are:
0.08, 0.10, 0.104, 0.111, 0.14, 0.172, 0.333

52. $\frac{\$10}{\$0.14/\text{minute}} = \frac{10}{0.14} \approx 71.4$ minutes

You could make a call to Japan for 71 minutes (rounded).

53. $\frac{\$10}{\$0.333/\text{minute}} \approx 30.03$ minutes

You get 30 minutes (rounded) per $10 card, so buy 4 cards for 120 minutes (2 hours).

54. **For Canada:**

$\frac{\$10}{\$0.08/\text{minute}} = 125$ minutes

For Mexico City:

$\frac{\$10}{\$0.10/\text{minute}} = 100$ minutes

minutes to Canada to minutes to Mexico City

$\frac{125 \div 25}{100 \div 25} = \frac{5}{4}$

The ratio of *minutes for a $10 call to Canada* to *minutes for a $10 call to Mexico* is $\frac{5}{4}$.

55. **(a)** $\frac{\$595}{25{,}000\text{ homes}} = \$0.0238/\text{home}$

For a full page ad, it costs $0.024 (rounded) per home.

(b) $\frac{\$375}{25{,}000\text{ homes}} = \$0.015/\text{home}$

For a half page ad, it costs $0.015 per home.

56. **(a)** If you're a new customer, you get $\frac{1}{2}$ off on your second page. The cost of a two-page ad is

$$\$595 + \frac{1}{2}(\$595) = \$595 + \$297.50$$
$$= \$892.50$$

(b) $\frac{\$892.50}{25{,}000\text{ homes}} = \$0.0357/\text{home}$

For a two-page ad, it costs $0.036 (rounded) per home.

CHAPTER 6 PERCENT

6.1 Basics of Percent

6.1 Margin Exercises

1. **(a)** 46 out of 100 is $\frac{46}{100}$ or 46%, so 46% of the people are homeowners.

(b) The tax rate is \$6 per \$100. The ratio is $\frac{6}{100}$ and the percent of tax is 6%.

(c) 68 out of 100 is $\frac{68}{100}$ or 68%, so 68% of the students are working full or part-time.

2. **(a)** $68\% = 68 \div 100 = 0.68$

(b) $34\% = 34 \div 100 = 0.34$

(c) $58.5\% = 58.5 \div 100 = 0.585$

(d) $175\% = 175 \div 100 = 1.75$

(e) $200\% = 200 \div 100 = 2.00$ or 2

3. **(a)** $96\% = 96.\% = 0.96$

Drop the percent sign and move the decimal point two places to the left.

(b) $6\% = 6.\% = 0.06$

Drop the percent sign and move the decimal point two places to the left.

(c) $24.8\% = 0.248$

Drop the percent sign and move the decimal point two places to the left.

(d) $0.9\% = 0.009$

Drop the percent sign and move the decimal point two places to the left.

4. **(a)** $0.74 = 74\%$

Move the decimal point two places to the right. (The decimal point is not written with whole number percents.) Attach a percent sign.

(b) $0.15 = 15\%$

Move the decimal point two places to the right and attach a percent sign.

(c) $0.09 = 9\%$

Move the decimal point two places to the right and attach a percent sign.

(d) $0.617 = 61.7\%$

Move the decimal point two places to the right and attach a percent sign.

(e) $0.834 = 83.4\%$

Move the decimal point two places to the right and attach a percent sign.

(f) $5.34 = 534\%$

Move the decimal point two places to the right and attach a percent sign.

(g) $2.8 = 280\%$

One zero is attached so the decimal point can be moved two places to the right. Attach a percent sign.

(h) $4 = 400\%$

Two zeros are attached so the decimal point can be moved two places to the right. Attach a percent sign.

5. **(a)** 100% is all of the money.
So, 100% of \$7.80 is \$7.80.

(b) 100% is all of the workers.
So, 100% of 1850 workers is 1850 workers.

(c) 200% is twice (2 times) as many photographs.
So, 200% of 24 photographs is $2 \cdot 24 = 48$ photographs.

(d) 300% is 3 times as many miles.
So, 300% of 8 miles is $3 \cdot 8 = 24$ miles.

6. **(a)** 50% is half of the patients.
So, 50% of 200 patients is $\frac{1}{2} \cdot 200 = 100$ patients.

(b) 50% is half of the e-mails.
So, 50% of 64 e-mails is 32 e-mails.

(c) 10% is $\frac{1}{10}$ of the elm trees. Move the decimal point *one* place to the left.
So, 10% of 3850. elm trees is 385 elm trees.

(d) 1% is $\frac{1}{100}$ of the feet. Move the decimal point *two* places to the left.
So, 1% of 240. feet is 2.4 feet.

6.1 Section Exercises

1. $12\% = 12.\% = 0.12$

Drop the percent sign and move the decimal point two places to the left.

3. $70\% = 70.\% = 0.70$ or 0.7

Drop the percent sign and move the decimal point two places to the left.

5. $25\% = 25.\% = 0.25$

Drop the percent sign and move the decimal point two places to the left.

7. $140\% = 140.\% = 1.40$ or 1.4

Drop the percent sign and move the decimal point two places to the left.

9. $5.5\% = 0.055$

Drop the percent sign. Attach one zero so the decimal point can be moved two places to the left.

11. $100\% = 100.\% = 1.00$ or 1

Drop the percent sign and move the decimal point two places to the left.

13. $0.5\% = 0.005$

Drop the percent sign. Attach two zeros so the decimal point can be moved two places to the left.

15. $0.35\% = 0.0035$

Drop the percent sign. Attach two zeros so the decimal point can be moved two places to the left.

17. $0.6 = 60\%$

0 is attached so the decimal point can be moved two places to the right. Attach a percent sign. (The decimal point is not written with whole numbers.)

19. $0.58 = 58\%$

Move the decimal point two places to the right and attach a percent sign.

21. $0.01 = 1\%$

Move the decimal point two places to the right and attach a percent sign.

23. $0.125 = 12.5\%$

Move the decimal point two places to the right and attach a percent sign.

25. $0.375 = 37.5\%$

Move the decimal point two places to the right and attach a percent sign.

27. $2 = 200\%$

Two zeros are attached so the decimal point can be moved two places to the right. Attach a percent sign.

29. $3.7 = 370\%$

One zero is attached so the decimal point can be moved two places to the right. Attach a percent sign.

31. $0.0312 = 3.12\%$

Move the decimal point two places to the right and attach a percent sign.

33. $4.162 = 416.2\%$

Move the decimal point two places to the right and attach a percent sign.

35. $0.0028 = 0.28\%$

Move the decimal point two places to the right and attach a percent sign.

37. Answers will vary. Some possibilities are: No common denominators are needed with percents. The denominator is always 100 with percent, which makes comparisons easier to understand.

39. Drop the percent sign and move the decimal point two places to the left.

45% of the refuse $= 0.45$

41. Drop the percent sign and move the decimal point two places to the left.

18% of the items $= 0.18$

43. Move the decimal point two places to the right and attach a percent sign.

0.035 property tax rate $= 3.5\%$

45. Attach two zeros and move the decimal point two places to the right. Attach a percent sign.

2 times the success rate of the last session $= 200\%$

47. Move the decimal point two places to the right and attach a percent sign.

0.005 of the population $= 0.5\%$

49. Drop the percent sign and move the decimal point two places to the left.

153.6% of normal $= 1.536$

51. 100% is all of the children.
So, 100% of 20 children is 20 children.
20 children are served both meals.

53. 200% is two times as many employees.
So, 200% of 210 employees is $2 \cdot 210 = 420$ employees. There are 420 employees this year.

55. 300% is three times as many chairs.
So, 300% of 90 chairs is $3 \cdot 90 = 270$ chairs.
We will need 270 chairs.

57. 50% is half of the tuition.
So, 50% of \$755 is $\frac{1}{2} \cdot \$755 = \$377.50$.
Financial aid will pay \$377.50.

59. 10% is found by moving the decimal point one place to the left.
So, 10% of 8200 commuters is 820 commuters.
820 commuters carpool.

61. 1% is found by moving the decimal point two places to the left.
So 1% of 2600 plants is 26 plants.
26 plants are poisonous.

63. **(a)** Since 100% means 100 parts out of 100 parts, 100% is all of the number.

(b) Answers will vary. For example, 100% of $72 is $72.

65. **(a)** Since 200% is two times a number, find 200% of the number by multiplying the number by 2 (double it).

(b) Answers will vary. For example, 200% of $20 is $2 \cdot \$20 = \$40$.

67. **(a)** Since 10% means 10 parts out of 100 parts or $\frac{1}{10}$, the shortcut for finding 10% of a number is to move the decimal point in the number one place to the left.

(b) Answers will vary. For example, 10% of $90 is $9.

69. 63% or 0.63 of the homeowners selected painting and wall papering as a project to do themselves.

71. **(a)** The second longest bar, landscaping and gardening, is the second most selected project.

(b) Landscaping and gardening was selected by 54% or 0.54 of the homeowners.

73. 26% or 0.26 of the women in the U.S. military are in the Air Force.

75. **(a)** The smallest portion of the circle graph corresponds to the Marine Corps.

(b) 5% or 0.05 of the women in the U.S. military are in the Marine Corps.

77. 16.8% or 0.168 of the population of Spain is 65 or older.

79. **(a)** The two countries with the lowest portion are Germany and Bulgaria.

(b) $16.5\% = 0.165$

81. Ninety-five parts of the one hundred parts are shaded.

$$\frac{95}{100} = 0.95 = 95\%$$

Five parts of the one hundred parts are unshaded.

$$\frac{5}{100} = 0.05 = 5\%$$

83. Three parts of the ten parts are shaded.

$$\frac{3}{10} = 0.3 = 30\%$$

Seven parts of the ten parts are unshaded.

$$\frac{7}{10} = 0.7 = 70\%$$

85. Fifty-five parts of the one hundred parts are shaded.

$$\frac{55}{100} = 0.55 = 55\%$$

Forty-five parts of the one hundred parts are unshaded.

$$\frac{45}{100} = 0.45 = 45\%$$

6.2 Percents and Fractions

6.2 Margin Exercises

1. **(a)** $50\% = \frac{50 \div 50}{100 \div 50} = \frac{1}{2}$

(b) $75\% = \frac{75 \div 25}{100 \div 25} = \frac{3}{4}$

(c) $48\% = \frac{48 \div 4}{100 \div 4} = \frac{12}{25}$

(d) $23\% = \frac{23}{100}$

(e) $125\% = \frac{125 \div 25}{100 \div 25} = \frac{5}{4} = 1\frac{1}{4}$

(f) $250\% = \frac{250 \div 50}{100 \div 50} = \frac{5}{2} = 2\frac{1}{2}$

2. **(a)** $37.5\% = \frac{37.5}{100} = \frac{37.5(10)}{100(10)} = \frac{375 \div 125}{1000 \div 125} = \frac{3}{8}$

(b) $62.5\% = \frac{62.5}{100} = \frac{62.5(10)}{100(10)} = \frac{625 \div 125}{1000 \div 125} = \frac{5}{8}$

(c) $4.5\% = \frac{4.5}{100} = \frac{4.5(10)}{100(10)} = \frac{45 \div 5}{1000 \div 5} = \frac{9}{200}$

(d) $66\frac{2}{3}\% = \frac{66\frac{2}{3}}{100} = \frac{\frac{200}{3}}{100} = \frac{200}{3} \div \frac{100}{1}$
$= \frac{200}{3} \cdot \frac{1}{100} = \frac{2}{3}$

(e) $10\frac{1}{3}\% = \frac{10\frac{1}{3}}{100} = \frac{\frac{31}{3}}{100} = \frac{31}{3} \div \frac{100}{1}$
$= \frac{31}{3} \cdot \frac{1}{100} = \frac{31}{300}$

(f) $87\frac{1}{2}\% = \frac{87\frac{1}{2}}{100} = \frac{\frac{175}{2}}{100} = \frac{175}{2} \div \frac{100}{1}$
$= \frac{\overset{7}{\cancel{175}}}{2} \cdot \frac{1}{\underset{4}{\cancel{100}}} = \frac{7}{8}$

3. **(a)**
$$\frac{1}{4} = \frac{p}{100}$$
$$4 \cdot p = 1 \cdot 100$$
$$\frac{4 \cdot p}{4} = \frac{100}{4}$$
$$p = 25$$

Thus, $\frac{1}{4} = 25\%$.

(b)
$$\frac{3}{10} = \frac{p}{100}$$
$$10 \cdot p = 3 \cdot 100$$
$$\frac{10 \cdot p}{10} = \frac{300}{10}$$
$$p = 30$$

Thus, $\frac{3}{10} = 30\%$.

(c)
$$\frac{6}{25} = \frac{p}{100}$$
$$25 \cdot p = 6 \cdot 100$$
$$\frac{25 \cdot p}{25} = \frac{600}{25}$$
$$p = 24$$

Thus, $\frac{6}{25} = 24\%$.

(d)
$$\frac{5}{8} = \frac{p}{100}$$
$$8 \cdot p = 5 \cdot 100$$
$$\frac{8 \cdot p}{8} = \frac{500}{8}$$
$$p = 62.5$$

Thus, $\frac{5}{8} = 62.5\%$.

(e)
$$\frac{1}{6} = \frac{p}{100}$$
$$6 \cdot p = 1 \cdot 100$$
$$\frac{6 \cdot p}{6} = \frac{100}{6}$$
$$p = 16.\overline{6} \approx 16.7$$

Thus, $\frac{1}{6} \approx 16.7\%$ or exactly $16\frac{2}{3}\%$.

(f)
$$\frac{2}{9} = \frac{p}{100}$$
$$9 \cdot p = 2 \cdot 100$$
$$\frac{9 \cdot p}{9} = \frac{200}{9}$$
$$p = 22.\overline{2} \approx 22.2$$

Thus, $\frac{2}{9} \approx 22.2\%$ or exactly $22\frac{2}{9}\%$.

4. **(a)** $\frac{3}{4}$ as a percent

$$\frac{3}{4} = 75\%$$

(b) 10% as a fraction

$$10\% = \frac{1}{10}$$

(c) $0.\overline{6} \approx 0.667$ as a fraction

$$0.\overline{6} \approx 0.667 = \frac{2}{3}$$

(d) $37\frac{1}{2}\%$ as a fraction

$$37\frac{1}{2}\% = \frac{3}{8}$$

(e) $\frac{7}{8}$ as a percent

$$\frac{7}{8} = 87\frac{1}{2}\% \text{ or } 87.5\%$$

(f) $\frac{1}{2}$ as a percent

$$\frac{1}{2} = 50\%$$

(g) $33\frac{1}{3}\%$ as a fraction

$$33\frac{1}{3}\% = \frac{1}{3}$$

(h) $1\frac{3}{4}$ as a percent

$$1\frac{3}{4} = 175\%$$

6.2 Section Exercises

1. $25\% = \frac{25}{100} = \frac{25 \div 25}{100 \div 25} = \frac{1}{4}$

3. $75\% = \frac{75}{100} = \frac{75 \div 25}{100 \div 25} = \frac{3}{4}$

5. $85\% = \frac{85}{100} = \frac{85 \div 5}{100 \div 5} = \frac{17}{20}$

7. $62.5\% = \frac{62.5}{100} = \frac{62.5(10)}{100(10)} = \frac{625 \div 125}{1000 \div 125} = \frac{5}{8}$

9. $6.25\% = \frac{6.25}{100} = \frac{6.25(100)}{100(100)} = \frac{625 \div 625}{10{,}000 \div 625} = \frac{1}{16}$

11. $16\frac{2}{3}\% = \frac{16\frac{2}{3}}{100} = 16\frac{2}{3} \div \frac{100}{1} = \frac{\overset{1}{\cancel{50}}}{3} \cdot \frac{1}{\underset{2}{\cancel{100}}} = \frac{1}{6}$

13. $6\frac{2}{3}\% = \frac{6\frac{2}{3}}{100} = \frac{\frac{20}{3}}{100} = \frac{20}{3} \div \frac{100}{1}$

$$= \frac{\overset{1}{\cancel{20}}}{3} \cdot \frac{1}{\underset{5}{\cancel{100}}} = \frac{1}{15}$$

15. $0.5\% = \frac{0.5}{100} = \frac{0.5(10)}{100(10)} = \frac{5 \div 5}{1000 \div 5} = \frac{1}{200}$

17. $180\% = \frac{180}{100} = \frac{180 \div 20}{100 \div 20} = \frac{9}{5} = 1\frac{4}{5}$

19. $375\% = \frac{375}{100} = \frac{375 \div 25}{100 \div 25} = \frac{15}{4} = 3\frac{3}{4}$

21. $\frac{1}{2}=\frac{p}{100}$

$2\cdot p=1\cdot 100$

$\frac{2\cdot p}{2}=\frac{100}{2}$

$p=50$

Thus, $\frac{1}{2}=50\%$.

23. $\frac{4}{5}=\frac{p}{100}$

$5\cdot p=4\cdot 100$

$\frac{5\cdot p}{5}=\frac{400}{5}$

$p=80$

Thus, $\frac{4}{5}=80\%$.

25. $\frac{7}{10}=\frac{p}{100}$

$10\cdot p=7\cdot 100$

$\frac{10\cdot p}{10}=\frac{700}{10}$

$p=70$

Thus, $\frac{7}{10}=70\%$.

27. The denominator is already 100.

$$\frac{37}{100}=37\%$$

29. $\frac{5}{8}=\frac{p}{100}$

$8\cdot p=5\cdot 100$

$\frac{8\cdot p}{8}=\frac{500}{8}=\frac{125}{2}$

$p=62.5$

Thus, $\frac{5}{8}=62.5\%$.

31. $\frac{7}{8}=\frac{p}{100}$

$8\cdot p=7\cdot 100$

$\frac{8\cdot p}{8}=\frac{700}{8}=\frac{175}{2}$

$p=87.5$

Thus, $\frac{7}{8}=87.5\%$.

33. $\frac{12}{25}=\frac{p}{100}$

$25\cdot p=12\cdot 100$

$\frac{25\cdot p}{25}=\frac{1200}{25}$

$p=48$

Thus, $\frac{12}{25}=48\%$.

35. As a different approach, we see that the denominator, 50, can be multiplied by 2 to get a denominator of 100. We'll multiply both the numerator and denominator by 2.

$$\frac{23}{50}=\frac{23\cdot 2}{50\cdot 2}=\frac{46}{100}=46\%$$

37. $\frac{7}{20}=\frac{p}{100}$

$20\cdot p=7\cdot 100$

$\frac{20\cdot p}{20}=\frac{700}{20}$

$p=35$

Thus, $\frac{7}{20}=35\%$.

39. $\frac{5}{6}=\frac{p}{100}$

$6\cdot p=5\cdot 100$

$\frac{6\cdot p}{6}=\frac{500}{6}=\frac{250}{3}$

$p\approx 83.3$

Thus, $\frac{5}{6}=83.3\%$ (rounded).

41. $\frac{5}{9}=\frac{p}{100}$

$9\cdot p=5\cdot 100$

$\frac{9\cdot p}{9}=\frac{500}{9}$

$p\approx 55.6$

Thus, $\frac{5}{9}=55.6\%$ (rounded).

43. $\frac{1}{7}=\frac{p}{100}$

$7\cdot p=1\cdot 100$

$\frac{7\cdot p}{7}=\frac{100}{7}$

$p\approx 14.3$

Thus, $\frac{1}{7}=14.3\%$ (rounded).

45. $0.5=\frac{5\div 5}{10\div 5}=\frac{1}{2}$ *fraction*

$0.5=0.50=50\%$ *percent*

47. $87.5\%=0.875$ *decimal*

$=\frac{875\div 125}{1000\div 125}=\frac{7}{8}$ *fraction*

49. $0.8=\frac{8\div 2}{10\div 2}=\frac{4}{5}$ *fraction*

$0.8=0.80=80\%$ *percent*

51. $\frac{1}{6}\approx 0.167$ *decimal*

$\approx 16.7\%$ *percent*

53. $0.7 = \frac{7}{10}$ *fraction*
$0.7 = 0.70 = 70\%$ *percent*

55. $12.5\% = 0.125$ *decimal*
$= \frac{125 \div 125}{1000 \div 125} = \frac{1}{8}$ *fraction*

57. $\frac{2}{3} \approx 0.667$ *decimal*
$\approx 66.7\%$ *percent*

59. $\frac{3}{50} = \frac{3 \cdot 2}{50 \cdot 2} = \frac{6}{100} = 0.06$ *decimal*
$= 6\%$ *percent*

61. $\frac{8}{100} = 8\%$ *percent*
$= 0.08$ *decimal*

63. $\frac{1}{200} = \frac{1 \cdot 5}{200 \cdot 5} = \frac{5}{1000} = 0.005$ *decimal*
$= 0.5\%$ *percent*

65. $2.5 = 2\frac{5}{10} = 2\frac{1}{2}$ *fraction*
$2.5 = 250\%$ *percent*

67. $3\frac{1}{4} = 3\frac{1 \cdot 25}{4 \cdot 25} = 3\frac{25}{100} = 3.25$ *decimal*
$= 325\%$ *percent*

69. There are many possible answers. Examples 2 and 3 show the steps that students should include in their answers.

71. 90 out of 500 people who used the Internet to find pet information, said they used it when buying a pet.
$\frac{90}{500} = \frac{90 \div 10}{500 \div 10} = \frac{9}{50}$ *fraction*
$\frac{9}{50} = \frac{9 \cdot 2}{50 \cdot 2} = \frac{18}{100} = 0.18$ *decimal*
$0.18 = 18\%$ *percent*

73. 13 adults out of 100 adults consumes the recommended 1000 mg of calcium daily.
$\frac{13}{100}$ *fraction*
$\frac{13}{100} = 0.13$ *decimal*
$0.13 = 13\%$ *percent*

75. 9 out of 15 are single parents.
$\frac{9}{15} = \frac{9 \div 3}{15 \div 3} = \frac{3}{5}$ *fraction*
$\frac{3}{5} = \frac{3 \cdot 2}{5 \cdot 2} = \frac{6}{10} = 0.6$ *decimal*
$0.6 = 0.60 = 60\%$ *percent*

77. 64 out of 80 employees have cellular phones. Therefore, $80 - 64 = 16$ employees do *not* have cellular phones.
$\frac{16}{80} = \frac{16 \div 16}{80 \div 16} = \frac{1}{5}$ *fraction*
$\frac{1}{5} = \frac{1 \cdot 2}{5 \cdot 2} = \frac{2}{10} = 0.2$ *decimal*
$0.2 = 0.20 = 20\%$ *percent*

79. 342 out of 380 do not have side effects.
$380 - 342 = 38$ do have side effects.
$\frac{38}{380} = \frac{38 \div 38}{380 \div 38} = \frac{1}{10}$ *fraction*
$\frac{1}{10} = 0.1$ *decimal*
$0.1 = 0.10 = 10\%$ *percent*

81. 2700 students out of 10,800 students expect to earn \$2500 to \$5000.
$\frac{2700 \div 2700}{10{,}800 \div 2700} = \frac{1}{4}$ *fraction*
$\frac{1}{4} = \frac{1 \cdot 25}{4 \cdot 25} = \frac{25}{100} = 0.25$ *decimal*
$0.25 = 25\%$ *percent*

83. 540 students out of 10,800 students expect to earn less than \$500.
$\frac{540 \div 540}{10{,}800 \div 540} = \frac{1}{20}$ *fraction*
$\frac{1}{20} = \frac{1 \cdot 5}{20 \cdot 5} = \frac{5}{100} = 0.05$ *decimal*
$0.05 = 5\%$ *percent*

85. 100% of a number means all of the parts or <u>100</u> parts out of <u>100</u> parts. 200% means two times as many parts and 300% means three times as many parts.

86. **(a)** 100% of 765 workers is <u>765 workers</u>.
(b) 200% of 48 letters is $2 \cdot 48 =$ <u>96 letters</u>.
(c) 300% of 7 videos is $3 \cdot 7 =$ <u>21 videos</u>.

87. 50% of a number is <u>50</u> parts out of <u>100</u> parts, which is <u>half</u> or $\frac{1}{2}$ of the parts.

88. 10% of a number is <u>10</u> parts out of <u>100</u> parts and is found by moving the decimal point <u>1</u> place to the <u>left</u>.

89. 1% of a number is <u>1</u> part out of <u>100</u> parts and is found by moving the decimal point <u>2</u> places to the <u>left</u>.

90. **(a)** 50% of 1050 homes is $\frac{1}{2} \cdot 1050 =$ <u>525 homes</u>.
(b) 10% of 370 printers is <u>37 printers</u>.
(c) 1% of \$8 sales is <u>\$0.08</u>.

91. Find 10% of 160, then add $\frac{1}{2}$ of 10%.

$$\begin{array}{ccccc} 10\% & + & 5\% & = & 15\% \\ \downarrow & & \downarrow & & \downarrow \\ 16 & + & 8 & = & 24 \end{array}$$

92. Find 100% of 160, then add 50% of 160.

$$\begin{array}{ccccc} 100\% & + & 50\% & = & 150\% \\ \downarrow & & \downarrow & & \downarrow \\ 160 & + & 80 & = & 240 \end{array}$$

93. From 100% of $450, subtract 10% of $450.

$$\begin{array}{ccccc} 100\% & - & 10\% & = & 90\% \\ \downarrow & & \downarrow & & \downarrow \\ \$450 & - & \$45 & = & \$405 \end{array}$$

94. To 100% of $800, add 100% of $800, and then add 10% of $800.

$$\begin{array}{ccccccc} 100\% & + & 100\% & + & 10\% & = & 210\% \\ \downarrow & & \downarrow & & \downarrow & & \downarrow \\ \$800 & + & \$800 & + & \$80 & = & \$1680 \end{array}$$

6.3 Using the Percent Proportion and Identifying the Components in a Percent Problem

6.3 Margin Exercises

1. **(a)** $\frac{1}{2} = \frac{25}{50}$

Cross product: $2 \cdot 25 = 50$; $1 \cdot 50 = 50$

The cross products are equivalent, so the proportion is *true*.

(b) $\frac{3}{4} = \frac{150}{200}$

Cross product: $3 \cdot 200 = 600$; $4 \cdot 150 = 600$

The cross products are equivalent, so the proportion is *true*.

(c) $\frac{7}{8} = \frac{180}{200}$

Cross product: $8 \cdot 180 = 1440$; $7 \cdot 200 = 1400$

The cross products are *not* equivalent, so the proportion is *false*.

(d) $\frac{32}{53} = \frac{160}{265}$

Cross product: $53 \cdot 160 = 8480$; $32 \cdot 265 = 8480$

The cross products are equivalent, so the proportion is *true*.

(e) $\frac{112}{41} = \frac{332}{123}$

Cross product:
$41 \cdot 332 = 13{,}612$; $112 \cdot 123 = 13{,}776$

The cross products are *not* equivalent, so the proportion is *false*.

2. **(a)** part = 12, percent = 16

$$\frac{\text{part}}{\text{whole}} = \frac{\text{percent}}{100}$$
$$\frac{12}{x} = \frac{16}{100}$$
$$x \cdot 16 = 12 \cdot 100$$
$$\frac{x \cdot 16}{16} = \frac{1200}{16}$$
$$x = 75$$

The whole is 75.

(b) part = 30, whole = 120

$$\frac{30}{120} = \frac{x}{100}$$
$$120 \cdot x = 30 \cdot 100$$
$$\frac{120 \cdot x}{120} = \frac{3000}{120}$$
$$x = 25$$

The percent is 25, written as 25%.

(c) whole = 210, percent = 20

$$\frac{x}{210} = \frac{20}{100}$$
$$x \cdot 100 = 210 \cdot 20$$
$$\frac{x \cdot 100}{100} = \frac{4200}{100}$$
$$x = 42$$

The part is 42.

(d) whole = 4000, percent = 32

$$\frac{x}{4000} = \frac{32}{100}$$
$$x \cdot 100 = 4000 \cdot 32$$
$$\frac{x \cdot 100}{100} = \frac{128{,}000}{100}$$
$$x = 1280$$

The part is 1280.

(e) part = 74, whole = 185

$$\frac{74}{185} = \frac{x}{100}$$
$$185 \cdot x = 74 \cdot 100$$
$$\frac{185 \cdot x}{185} = \frac{7400}{185}$$
$$x = 40$$

The percent is 40, written as 40%.

3. **(a)** The percent is 25. The number 25 appears with the symbol %.

(b) The percent is 65. The number 65 appears with the symbol %.

(c) The percent is $6\frac{1}{2}$ because $6\frac{1}{2}$ appears with the word *percent*.

(d) The percent is 3. The number 3 appears with the symbol %.

(e) The word *percent* has no number with it, so the percent is the unknown part of the problem.

4. **(a)** The whole is 900 since there are 900 total tests.

(b) The whole is 620 since there are 620 total students.

(c) The whole is \$590.
\$590 appears after the word *of.*

(d) For "\$105 is 3% of what number?," the whole is, "what number," the unknown part of the problem.

(e) The whole is 380.
380 appears after the word *of.*

5. **(a)** The percent is 25 and the whole is 900.
225 is the remaining number, so the part is 225.

$$\begin{array}{r}\text{Part} \rightarrow \\ \text{Whole} \rightarrow\end{array} \frac{225}{900} = \frac{25}{100} \begin{array}{l}\leftarrow \text{Percent} \\ \leftarrow \text{Always 100}\end{array}$$

(b) The percent is 65 and the whole is 620.
403 is the remaining number, so the part is 403.

$$\begin{array}{r}\text{Part} \rightarrow \\ \text{Whole} \rightarrow\end{array} \frac{403}{620} = \frac{65}{100} \begin{array}{l}\leftarrow \text{Percent} \\ \leftarrow \text{Always 100}\end{array}$$

(c) The percent is $6\frac{1}{2}$ and the whole is \$590.
The part is unknown.

$$\begin{array}{r}\text{Part} \rightarrow \\ \text{Whole} \rightarrow\end{array} \frac{\text{unknown}}{590} = \frac{6\frac{1}{2}}{100} \begin{array}{l}\leftarrow \text{Percent} \\ \leftarrow \text{Always 100}\end{array}$$

(d) The percent is 3 and the whole is unknown.
\$105 is the part.

$$\begin{array}{r}\text{Part} \rightarrow \\ \text{Whole} \rightarrow\end{array} \frac{105}{\text{unknown}} = \frac{3}{100} \begin{array}{l}\leftarrow \text{Percent} \\ \leftarrow \text{Always 100}\end{array}$$

(e) The percent is 80 and the whole is 380.
The part is unknown.

$$\begin{array}{r}\text{Part} \rightarrow \\ \text{Whole} \rightarrow\end{array} \frac{\text{unknown}}{380} = \frac{80}{100} \begin{array}{l}\leftarrow \text{Percent} \\ \leftarrow \text{Always 100}\end{array}$$

6.3 Section Exercises

1. part = 5, percent = 10

$$\frac{\text{part}}{\text{whole}} = \frac{\text{percent}}{100}$$

$$\frac{5}{x} = \frac{10}{100}$$
$$x \cdot 10 = 5 \cdot 100$$
$$\frac{x \cdot 10}{10} = \frac{500}{10}$$
$$x = 50$$

The whole is 50.

3. part = 30, percent = 20

$$\frac{30}{x} = \frac{20}{100}$$
$$x \cdot 20 = 30 \cdot 100$$
$$\frac{x \cdot 20}{20} = \frac{3000}{20}$$
$$x = 150$$

The whole is 150.

5. part = 28, percent = 40

$$\frac{28}{x} = \frac{40}{100}$$
$$x \cdot 40 = 28 \cdot 100$$
$$\frac{x \cdot 40}{40} = \frac{2800}{40}$$
$$x = 70$$

The whole is 70.

7. part = 15, whole = 60

$$\frac{15}{60} = \frac{x}{100} \quad \textbf{OR} \quad \frac{1}{4} = \frac{x}{100}$$
$$4 \cdot x = 1 \cdot 100$$
$$\frac{4 \cdot x}{4} = \frac{100}{4}$$
$$x = 25$$

The percent is 25, written as 25%.

9. part = 36, whole = 24

$$\frac{36}{24} = \frac{x}{100} \quad \textbf{OR} \quad \frac{3}{2} = \frac{x}{100}$$
$$2 \cdot x = 3 \cdot 100$$
$$\frac{2 \cdot x}{2} = \frac{300}{2}$$
$$x = 150$$

The percent is 150, written as 150%.

11. part = 9.25, whole = 27.75

$$\frac{9.25}{27.75} = \frac{x}{100} \quad \textbf{OR} \quad \frac{1}{3} = \frac{x}{100}$$
$$3 \cdot x = 1 \cdot 100$$
$$\frac{3 \cdot x}{3} = \frac{100}{3}$$
$$x = 33.3 \text{ (rounded)}$$

The percent is about 33.3, written as 33.3%.

13. whole = 52, percent = 50

$$\frac{x}{52} = \frac{50}{100}$$
$$x \cdot 100 = 52 \cdot 50$$
$$\frac{x \cdot 100}{100} = \frac{2600}{100}$$
$$x = 26$$

The part is 26.

15. whole = 72, percent = 30

$$\frac{x}{72} = \frac{30}{100}$$
$$x \cdot 100 = 72 \cdot 30$$
$$\frac{x \cdot 100}{100} = \frac{2160}{100}$$
$$x = 21.6$$

The part is 21.6.

17. whole = 94.4, part = 25

$$\frac{25}{94.4} = \frac{x}{100}$$
$$94.4 \cdot x = 25 \cdot 100$$
$$\frac{94.4 \cdot x}{94.4} = \frac{2500}{94.4}$$
$$x = 26.5 \text{ (rounded)}$$

The percent is about 26.5, written as 26.5%.

19. part = 46, percent = 40

$$\frac{46}{x} = \frac{40}{100}$$
$$x \cdot 40 = 46 \cdot 100$$
$$\frac{x \cdot 40}{40} = \frac{4600}{40}$$
$$x = 115$$

The whole is 115.

21. whole = 5000, part = 20

$$\frac{20}{5000} = \frac{x}{100}$$
$$5000 \cdot x = 20 \cdot 100$$
$$\frac{5000 \cdot x}{5000} = \frac{2000}{5000}$$
$$x = 0.4$$

The percent is 0.4, written as 0.4%.

23. whole = 4300, part = $107\frac{1}{2}$

$$\frac{107\frac{1}{2}}{4300} = \frac{x}{100}$$
$$4300 \cdot x = 107\tfrac{1}{2} \cdot 100$$
$$\frac{4300 \cdot x}{4300} = \frac{10{,}750}{4300}$$
$$x = 2.5$$

The percent is 2.5, written as 2.5%.

25. whole = 6480, part = 19.44

$$\frac{19.44}{6480} = \frac{x}{100}$$
$$6480 \cdot x = 19.44 \cdot 100$$
$$\frac{6480 \cdot x}{6480} = \frac{1944}{6480}$$
$$x = 0.3$$

The percent is 0.3, written as 0.3%.

27. 10% of how many bicycles is 60 bicycles?

10%	of how many bicycles	is 60 bicycles?
↑	↑	↑
percent	whole	part
10	unknown	60

$$\begin{matrix}\text{part} \rightarrow \\ \text{whole} \rightarrow\end{matrix} \frac{60}{\text{unknown}} = \frac{10}{100} \begin{matrix}\leftarrow \text{percent} \\ \leftarrow \text{always } 100\end{matrix}$$

29. 75% of \$800 is \$600.

75%	of \$800	is \$600.
↑	↑	↑
percent	whole	part
75	800	600

There is no "unknown."

$$\frac{600}{800} = \frac{75}{100}$$

31. What is 25% of \$970?

What is	25%	of \$970?
↑	↑	↑
part	percent	whole
unknown	25	970

$$\frac{\text{unknown}}{970} = \frac{25}{100}$$

33. 12 injections is 20% of what number of injections?

12 injections is	20%	of what number of injections?
↑	↑	↑
part	percent	whole
12	20	unknown

$$\frac{12}{\text{unknown}} = \frac{20}{100}$$

35. 34 trophies is 50% of 68 trophies.

34 trophies is	50%	of 68 trophies.
↑	↑	↑
part	percent	whole
34	50	68

There is no "unknown."

$$\frac{34}{68} = \frac{50}{100}$$

37. What percent of \$296 is \$177?

percent unknown ↑ — whole 296 ↑ — part 177 ↑

$$\frac{177}{296} = \frac{\text{unknown}}{100}$$

39. 54.34 is 3.25% of what number?

part 54.34 ↑ — percent 3.25 ↑ — whole unknown ↑

$$\frac{54.34}{\text{unknown}} = \frac{3.25}{100}$$

41. 0.68% of \$487 is what number?

percent 0.68 ↑ — whole 487 ↑ — part unknown ↑

$$\frac{\text{unknown}}{487} = \frac{0.68}{100}$$

43. Percent—the ratio of the part to the whole. It appears with the word *percent* or "%" after it. Whole—the entire quantity. Often appears after the word *of.* Part—the part being compared with the whole.

45. What percent of 810 trees is 640 trees?

percent unknown ↑ — whole 810 ↑ — part 640 ↑

$$\frac{640}{810} = \frac{\text{unknown}}{100}$$

47. 86 of 142 people is what percent?

part 86 ↑ — whole 142 ↑ — percent unknown ↑

$$\frac{86}{142} = \frac{\text{unknown}}{100}$$

49. 23% prefer fat-free dressing; 610 is the total number of customers; the number of customers who prefer fat-free dressing

percent 23 ↑ — whole 610 ↑ — part unknown ↑

$$\frac{\text{unknown}}{610} = \frac{23}{100}$$

51. 680 computer chips is what percent of 2000 computer chips?

part 680 ↑ — percent unknown ↑ — whole 2000 ↑

$$\frac{680}{2000} = \frac{\text{unknown}}{100}$$

53. 55% of 480 adults is how many adults?

percent 55 ↑ — whole 480 ↑ — part unknown ↑

$$\frac{\text{unknown}}{480} = \frac{55}{100}$$

55. 49.5% of 822 people is how many people?

percent 49.5 ↑ — whole 822 ↑ — part unknown ↑

$$\frac{\text{unknown}}{822} = \frac{49.5}{100}$$

57. 16.8% of total number of patients is 504 patients who were late.

percent 16.8 ↑ — whole unknown ↑ — part 504 ↑

$$\frac{504}{\text{unknown}} = \frac{16.8}{100}$$

59. 45% of 680 toys is what number?

percent 45 ↑ — whole 680 ↑ — part unknown ↑

$$\frac{\text{unknown}}{680} = \frac{45}{100}$$

6.4 Using Proportions to Solve Percent Problems

6.4 Margin Exercises

1. **(a)** percent is 10; whole is 1250

$$\frac{\text{part}}{\text{whole}} = \frac{\text{percent}}{100}$$

$$\frac{x}{1250} = \frac{10}{100} \quad \textbf{OR} \quad \frac{x}{1250} = \frac{1}{10}$$

$$x \cdot 10 = 1250 \cdot 1$$

$$\frac{x \cdot 10}{10} = \frac{1250}{10}$$

$$x = 125$$

10% of 1250 allergy tests is 125 allergy tests.

(b) percent is 15; whole is 3220

$$\frac{x}{3220} = \frac{15}{100} \quad \textbf{OR} \quad \frac{x}{3220} = \frac{3}{20}$$
$$x \cdot 20 = 3220 \cdot 3$$
$$\frac{x \cdot 20}{20} = \frac{9660}{20}$$
$$x = 483$$

15% of $3220 is $483.

(c) percent is 7; whole is 2700

$$\frac{x}{2700} = \frac{7}{100}$$
$$x \cdot 100 = 2700 \cdot 7$$
$$\frac{x \cdot 100}{100} = \frac{18{,}900}{100}$$
$$x = 189$$

7% of 2700 miles is 189 miles.

(d) percent is 39; whole is 1220

$$\frac{x}{1220} = \frac{39}{100}$$
$$x \cdot 100 = 1220 \cdot 39$$
$$\frac{x \cdot 100}{100} = \frac{47{,}580}{100}$$
$$x = 475.8$$

39% of 1220 meters is 475.8 meters.

2. **(a)** 55% = 0.55
part = (0.55)(10,000) = 5500
55% of 10,000 injections is 5500 injections.

(b) 16% = 0.16
part = (0.16)(120) = 19.2
16% of 120 miles is 19.2 miles.

(c) 135% = 1.35
part = (1.35)(60) = 81
135% of 60 dosages is 81 dosages.

(d) 0.5% = 0.005
part = (0.005)(238) = 1.19
0.5% of $238 is $1.19.

3. **(a)** percent is 45; whole is 2920
45% = 0.45
part = $0.45 \cdot 2920$ = 1314
There were 1314 advertising pieces of mail.

(b) percent is 23; whole is 14,100
23% = 0.23
part = (0.23)(14,100) = 3243
There are 3243 iced mocha customers.

4. **(a)** part is 750; percent is 25

$$\frac{750}{x} = \frac{25}{100} \quad \textbf{OR} \quad \frac{750}{x} = \frac{1}{4}$$
$$x \cdot 1 = 750 \cdot 4$$
$$x = 3000$$

750 tickets is 25% of 3000 tickets.

(b) part is 28; percent is 35

$$\frac{28}{x} = \frac{35}{100} \quad \textbf{OR} \quad \frac{28}{x} = \frac{7}{20}$$
$$x \cdot 7 = 28 \cdot 20$$
$$\frac{x \cdot 7}{7} = \frac{560}{7}$$
$$x = 80$$

28 antiques is 35% of 80 antiques.

(c) part is 387; percent is 36

$$\frac{387}{x} = \frac{36}{100} \quad \textbf{OR} \quad \frac{387}{x} = \frac{9}{25}$$
$$x \cdot 9 = 387 \cdot 25$$
$$\frac{x \cdot 9}{9} = \frac{9675}{9}$$
$$x = 1075$$

387 customers is 36% of 1075 customers.

(d) part is 292.5; percent is 37.5

$$\frac{292.5}{x} = \frac{37.5}{100}$$
$$x \cdot 37.5 = (292.5)(100)$$
$$\frac{x \cdot 37.5}{37.5} = \frac{29{,}250}{37.5}$$
$$x = 780$$

292.5 miles is 37.5% of 780 miles.

5. **(a)** part is 182; percent is 52

$$\frac{182}{x} = \frac{52}{100} \quad \textbf{OR} \quad \frac{182}{x} = \frac{13}{25}$$
$$x \cdot 13 = 182 \cdot 25$$
$$\frac{x \cdot 13}{13} = \frac{4550}{13}$$
$$x = 350$$

182 tons is 52% of 350 tons.

(b) part is 450; percent is 8

$$\frac{450}{x} = \frac{8}{100} \quad \textbf{OR} \quad \frac{450}{x} = \frac{2}{25}$$
$$x \cdot 2 = 450 \cdot 25$$
$$\frac{x \cdot 2}{2} = \frac{11{,}250}{2}$$
$$x = 5625$$

450 pounds of zinc is 8% of 5625 pounds of zinc.

6. **(a)** part is 21; whole is 105

$$\frac{21}{105}=\frac{x}{100} \textbf{ OR } \frac{1}{5}=\frac{x}{100}$$
$$5 \cdot x = 1 \cdot 100$$
$$\frac{5 \cdot x}{5}=\frac{100}{5}$$
$$x = 20$$

\$21 is 20% of \$105.

(b) part is 48; whole is 320

$$\frac{48}{320}=\frac{x}{100} \textbf{ OR } \frac{3}{20}=\frac{x}{100}$$
$$20 \cdot x = 3 \cdot 100$$
$$\frac{20 \cdot x}{20}=\frac{300}{20}$$
$$x = 15$$

48 Internet companies is 15% of 320 Internet companies.

(c) part is 1026; whole is 2280

$$\frac{1026}{2280}=\frac{x}{100} \textbf{ OR } \frac{513}{1140}=\frac{x}{100}$$
$$1140 \cdot x = 513 \cdot 100$$
$$\frac{1140 \cdot x}{1140}=\frac{51{,}300}{1140}$$
$$x = 45$$

1026 trials is 45% of 2280 trials.

(d) part is 432; whole is 108

$$\frac{432}{108}=\frac{x}{100} \textbf{ OR } \frac{4}{1}=\frac{x}{100}$$
$$1 \cdot x = 4 \cdot 100$$
$$x = 400$$

432 snowboarders is 400% of 108 snowboarders.

7. **(a)** part is 289; whole is 425

$$\frac{289}{425}=\frac{x}{100}$$
$$425 \cdot x = 289 \cdot 100$$
$$\frac{425 \cdot x}{425}=\frac{28{,}900}{425}$$
$$x = 68$$

\$289 is 68% of \$425.

(b) part is 32.3; whole is 38

$$\frac{32.3}{38}=\frac{x}{100}$$
$$38 \cdot x = (32.3)(100)$$
$$\frac{38 \cdot x}{38}=\frac{3230}{38}$$
$$x = 85$$

32.3 miles per gallon is 85% of 38 miles per gallon.

8. **(a)** part is 450; whole is 300

$$\frac{450}{300}=\frac{p}{100} \textbf{ OR } \frac{3}{2}=\frac{p}{100}$$
$$2 \cdot p = 3 \cdot 100$$
$$\frac{2 \cdot p}{2}=\frac{300}{2}$$
$$p = 150$$

450 students is 150% of 300 students.

(b) part is 504; whole is 360

$$\frac{504}{360}=\frac{x}{100} \textbf{ OR } \frac{7}{5}=\frac{x}{100}$$
$$5 \cdot x = 7 \cdot 100$$
$$\frac{5 \cdot x}{5}=\frac{700}{5}$$
$$x = 140$$

\$504 is 140% of \$360.

6.4 Section Exercises

1. 35% of 120 test tubes
$(0.35)(120) = 42$
part = 42 test tubes

3. 45% of 4080 military personnel
$(0.45)(4080) = 1836$
part = 1836 military personnel

5. 4% of 120 ft
$(0.04)(120) = 4.8$
part = 4.8 ft

7. 150% of 210 files
$(1.50)(210) = 315$
part = 315 files

9. 52.5% of 1560 trucks
$(0.525)(1560) = 819$
part = 819 trucks

11. 2% of \$164
$(0.02)(164) = 3.28$
part = \$3.28

13. 225% of 680 tables
$(2.25)(680) = 1530$
part = 1530 tables

15. 17.5% of 1040 homes
$(0.175)(1040) = 182$
part = 182 homes

17. 0.9% of \$2400
$(0.009)(2400) = 21.6$
part = \$21.60

19. part is 80; percent is 25

$$\frac{80}{x} = \frac{25}{100} \quad \textbf{OR} \quad \frac{80}{x} = \frac{1}{4}$$
$$x \cdot 1 = 80 \cdot 4$$
$$x = 320$$

80 e-mails is 25% of 320 e-mails.

21. part is 48; percent is 30

$$\frac{48}{x} = \frac{30}{100} \quad \textbf{OR} \quad \frac{48}{x} = \frac{3}{10}$$
$$x \cdot 3 = 48 \cdot 10$$
$$\frac{x \cdot 3}{3} = \frac{480}{3}$$
$$x = 160$$

30% of 160 hay bales is 48 hay bales.

23. part is 495; percent is 90

$$\frac{495}{x} = \frac{90}{100} \quad \textbf{OR} \quad \frac{495}{x} = \frac{9}{10}$$
$$x \cdot 9 = 495 \cdot 10$$
$$\frac{x \cdot 9}{9} = \frac{4950}{9}$$
$$x = 550$$

495 successful students is 90% of 550 students.

25. part is 462; percent is 140

$$\frac{462}{x} = \frac{140}{100} \quad \textbf{OR} \quad \frac{462}{x} = \frac{7}{5}$$
$$x \cdot 7 = 462 \cdot 5$$
$$\frac{x \cdot 7}{7} = \frac{2310}{7}$$
$$x = 330$$

462 mountain bikes is 140% of 330 mountain bikes.

27. part is 350; percent is 12.5

$$\frac{350}{x} = \frac{12.5}{100}$$
$$12.5 \cdot x = 35{,}000$$
$$\frac{12.5 \cdot x}{12.5} = \frac{35{,}000}{12.5}$$
$$x = 2800$$

$12\frac{1}{2}$% of 2800 is 350.

29. part is 18; whole is 36

$$\frac{18}{36} = \frac{x}{100} \quad \textbf{OR} \quad \frac{1}{2} = \frac{x}{100}$$
$$2 \cdot x = 1 \cdot 100$$
$$\frac{2 \cdot x}{2} = \frac{100}{2}$$
$$x = 50$$

18 bean burritos is 50% of 36 bean burritos.

31. part is 390; whole is 750

$$\frac{390}{750} = \frac{x}{100} \quad \textbf{OR} \quad \frac{13}{25} = \frac{x}{100}$$
$$25 \cdot x = 13 \cdot 100$$
$$\frac{25 \cdot x}{25} = \frac{1300}{25}$$
$$x = 52$$

390 SUVs is 52% of 750 SUVs.

33. part is 32; whole is 400

$$\frac{32}{400} = \frac{x}{100} \quad \textbf{OR} \quad \frac{2}{25} = \frac{x}{100}$$
$$25 \cdot x = 2 \cdot 100$$
$$\frac{25 \cdot x}{25} = \frac{200}{25}$$
$$x = 8$$

32 patients is 8% of 400 patients.

35. part is 54; whole is 3600

$$\frac{54}{3600} = \frac{x}{100} \quad \textbf{OR} \quad \frac{3}{200} = \frac{x}{100}$$
$$200 \cdot x = 3 \cdot 100$$
$$\frac{200 \cdot x}{200} = \frac{300}{200}$$
$$x = 1.5$$

54 CDs is 1.5% of 3600 CDs.

37. part is 64; whole is 344

$$\frac{64}{344} = \frac{x}{100} \quad \textbf{OR} \quad \frac{8}{43} = \frac{x}{100}$$
$$43 \cdot x = 8 \cdot 100$$
$$\frac{43 \cdot x}{43} = \frac{800}{43}$$
$$x \approx 18.6$$

$64 is 18.6% (rounded) of $344.

39. part is 23; whole is 250

$$\frac{23}{250} = \frac{x}{100}$$
$$250 \cdot x = 23 \cdot 100$$
$$\frac{250 \cdot x}{250} = \frac{2300}{250}$$
$$x = 9.2$$

23 tires is 9.2% of 250 tires.

41. 150% of $30 cannot be less than $30 because 150% is greater than 1 (100%). The answer must be greater than $30.
25% of $16 cannot be greater than $16 because 25% is less than 1 (100%). The answer must be less than $16.

43. part is unknown; whole is 240; percent is 22

$$\frac{x}{240}=\frac{22}{100} \quad \textbf{OR} \quad \frac{x}{240}=\frac{11}{50}$$
$$x \cdot 50 = 240 \cdot 11$$
$$\frac{x \cdot 50}{50}=\frac{2640}{50}$$
$$x = 52.8$$

The amount withheld is \$52.80.

45. part is unknown; whole is 335; percent is 13

$$\frac{x}{335}=\frac{13}{100}$$
$$x \cdot 100 = 335 \cdot 13$$
$$\frac{x \cdot 100}{100}=\frac{4355}{100}$$
$$x = 43.55 \approx 44$$

There are about 44 female crew members.

47. part is unknown; whole is 550; percent is 40

$$\frac{x}{550}=\frac{40}{100} \quad \textbf{OR} \quad \frac{x}{550}=\frac{2}{5}$$
$$5 \cdot x = 2 \cdot 550$$
$$\frac{5 \cdot x}{5}=\frac{1100}{5}$$
$$x = 220$$

The number of students who will receive scholarships is 220.

49. $\$95 + \$104 = \$199$
part is unknown; whole is 199; percent is 3

$$\frac{x}{199}=\frac{3}{100}$$
$$x \cdot 100 = 199 \cdot 3$$
$$\frac{x \cdot 100}{100}=\frac{597}{100}$$
$$x = 5.97$$

The total cost is $\$199 + \$5.97 = \$204.97$.

51. 67% of the children go to bed without brushing, so $100\% - 67\% = 33\%$ brush their teeth before going to bed.

53. 61% of $3400 = (0.61)(3400) = 2074$
2074 of the 3400 children brush less than one minute.

55. part is 960; whole is 48,000; percent is unknown

$$\frac{960}{48{,}000}=\frac{x}{100} \quad \textbf{OR} \quad \frac{1}{50}=\frac{x}{100}$$
$$50 \cdot x = 1 \cdot 100$$
$$\frac{50 \cdot x}{50}=\frac{100}{50}$$
$$x = 2$$

There are 2% of these jobs filled by women.

57. part is 240; whole is unknown; percent is 7.5%

$$\frac{240}{x}=\frac{7.5}{100}$$
$$7.5 \cdot x = 240 \cdot 100$$
$$\frac{7.5 \cdot x}{7.5}=\frac{24{,}000}{7.5}$$
$$x = 3200$$

She has monthly earnings of \$3200 and yearly earnings of $12(\$3200) = \$38{,}400$.

59. The graph show that *March* had sales of 7%, which is the lowest for the cold-and-flu season months.

61. The highest sales month was January when 15% of the total cans of chicken noodle soup were sold.

part is unknown; whole is 350; percent is 15

$$\frac{x}{350}=\frac{15}{100}$$
$$x \cdot 100 = 350 \cdot 15$$
$$\frac{x \cdot 100}{100}=\frac{5250}{100}$$
$$x = 52.5$$

There were 52.5 million cans of chicken noodle soup sold in January.

63. The circle graph shows the smallest sector represents 6% for Hardee's.

65. part is unknown; whole is 42; percent is 43

$$\frac{x}{42}=\frac{43}{100}$$
$$x \cdot 100 = 42 \cdot 43$$
$$\frac{x \cdot 100}{100}=\frac{1806}{100}$$
$$x = 18.06$$

The total annual sales for McDonald's is \$18.06 billion.

67. percent is 20; whole is 3100; part is unknown

$$\frac{x}{3100}=\frac{20}{100} \quad \textbf{OR} \quad \frac{x}{3100}=\frac{1}{5}$$
$$\frac{5 \cdot x}{5}=\frac{3100}{5}$$
$$x = 620$$

Add the application fee.

$$\$620 + \$25 = \$645$$

The total charge is \$645.

69. percent is 86; whole is 15,401; part is unknown

$$\frac{x}{15{,}401} = \frac{86}{100} \quad \textbf{OR} \quad \frac{x}{15{,}401} = \frac{43}{50}$$

$$50 \cdot x = 15{,}401 \cdot 43$$

$$\frac{50 \cdot x}{50} = \frac{662{,}243}{50}$$

$$x \approx 13{,}245$$

$15{,}401 - 13{,}245 = 2156$
2156 products were successful.

71. In the percent proportion, part is to whole as percent is to 100.

72. All percent problems involve a comparison between a part of something and the whole.

73. Since there are 27 grams of total carbohydrates per serving, and there are 4 calories per gram of carbohydrates, there are

$$27 \cdot 4 = 108$$

calories per serving from total carbohydrates.

The correct label is as follows.

Nutrition Facts	Amount/serving	%DV*	Amount/serving	%DV*
Serving Size 1/2 Pkg. (15 oz/42.5 g)	**Total Fat** 8g	**12%**	**Total Carbohydrates** 27g	**9%**
Servings Per package 2	Saturated Fat 4g	**20%**	Dietary Fiber Less Than 1g	**3%**
Calories 190	**Cholesterol** 0mg	**0%**	Sugars Less Than 1g	
Calories from Fat 70	**Sodium** 670mg	**28%**	**Protein** 4g	
	Vitamin A 0% • Vitamin C 0% • Calcium 2% • Iron 4%			
*Percent Daily Values (DV) are based on a 2,000 calorie diet.	Calories Per Gram Fat 9 • Carbohydrates 4 • Protein 4			

74. 9% of what number is 27 grams?
part is 27; whole is unknown; percent is 9

$$\frac{27}{x} = \frac{9}{100}$$

$$x \cdot 9 = 27 \cdot 100$$

$$\frac{x \cdot 9}{9} = \frac{2700}{9}$$

$$x = 300$$

300 grams of total carbohydrates are needed to meet the "recommended daily value."

75. 12% of what number is 8 grams?
part is 8; whole is unknown; percent is 12

$$\frac{8}{x} = \frac{12}{100}$$

$$x \cdot 12 = 8 \cdot 100$$

$$\frac{x \cdot 12}{12} = \frac{800}{12}$$

$$x \approx 67$$

67 grams of fat are needed to meet the "percent daily value."

76. Yes, since they would eat
$28\% \times 4$ servings $= 112\%$ of the daily value.

77. Each package has 2 servings and each serving has 3% of the recommended daily value of fiber.

$$\frac{100\%}{2 \cdot 3\%} = \frac{100\%}{6\%} \approx 16.7$$

17 packages (rounded) would have to be eaten. It may be possible but would result in a diet that is high in total fat, saturated fat, sodium, and total carbohydrates.

6.5 Using the Percent Equation

6.5 Margin Exercises

1. **(a)** $15\% = 0.15$

part = percent • whole

$x = (0.15)(880)$
$x = 132$

132 policy holders is 15% of 880 policy holders.

(b) $23\% = 0.23$

part = percent • whole

$x = (0.23)(840)$
$x = 193.2$

193.2 gallons is 23% of 840 gallons.

(c) $120\% = 1.2$

part = percent • whole

$x = (1.2)(220)$
$x = 264$

\$264 is 120% of \$220.

(d) $135\% = 1.35$

part = percent • whole

$x = (1.35)(1080)$
$x = 1458$

\$1458 is 135% of \$1080.

(e) $0.5\% = 0.005$

part = percent • whole

$x = (0.005)(1200)$
$x = 6$

6 test tubes is 0.5% of 1200 test tubes.

(f) $0.25\% = 0.0025$

part = percent • whole

$x = (0.0025)(1600)$
$x = 4$

4 lab tests is 0.25% of 1600 lab tests.

2. **(a)** 18 is 45% of what number?

part = percent • whole

$18 = (0.45)(x) \qquad 45\% = 0.45$

$\frac{18}{0.45} = \frac{(0.45)(x)}{0.45}$

$40 = x$

18 supervisors is 45% of 40 supervisors.

(b) 67.5 is 27% of what number?

part = percent • whole

$67.5 = (0.27)(x) \qquad 27\% = 0.27$

$\frac{67.5}{0.27} = \frac{(0.27)(x)}{0.27}$

$250 = x$

67.5 containers is 27% of 250 containers.

(c) 666 is 45% of what number?

part = percent • whole

$666 = (0.45)(x) \qquad 45\% = 0.45$

$\frac{666}{0.45} = \frac{(0.45)(x)}{0.45}$

$1480 = x$

666 inoculations is 45% of 1480 inoculations.

(d) $5\frac{1}{2}\%$ of what number is 66?

part = percent • whole

$66 = (0.055)(x) \qquad 5\frac{1}{2}\% = 5.5\% = 0.055$

$\frac{66}{0.055} = \frac{(0.055)(x)}{0.055}$

$1200 = x$

$5\frac{1}{2}\%$ of 1200 policies is 66 policies.

3. **(a)** The whole is 35 and the part is 7.

part = percent • whole

$7 = x \cdot 35$

$\frac{7}{35} = \frac{x \cdot 35}{35}$

$0.2 = x$

0.2 is 20%.
20% of 35 monitors is 7 monitors.

(b) The whole is 85 and the part is 34.

part = percent • whole

$34 = x \cdot 85$

$\frac{34}{85} = \frac{x \cdot 85}{85}$

$0.4 = x$

0.4 is 40%.
34 post office boxes is 40% of 85 post office boxes.

(c) The whole is 920 and the part is 1288.

part = percent • whole

$1288 = x \cdot 920$

$\frac{1288}{920} = \frac{x \cdot 920}{920}$

$1.4 = x$

1.4 is 140%.
140% of 920 invitations is 1288 invitations.

(d) The whole is 1125 and the part is 9.

part = percent • whole

$9 = x \cdot 1125$

$\frac{9}{1125} = \frac{x \cdot 1125}{1125}$

$0.008 = x$

0.008 is 0.8%.
9 world class runners is 0.8% of 1125 runners.

6.5 Section Exercises

1. part = percent • whole

$x = (0.25)(1080) \qquad 25\% = 0.25$

$x = 270$

25% of 1080 blood donors is 270 blood donors.

3. part = percent • whole

$x = (0.45)(3000) \qquad 45\% = 0.45$

$x = 1350$

45% of 3000 bath towels is 1350 bath towels.

5. part = percent • whole

$x = (0.32)(260) \qquad 32\% = 0.32$

$x = 83.2$

32% of 260 quarts is 83.2 quarts.

7. part = percent • whole

$x = (1.4)(2500) \qquad 140\% = 1.4$

$x = 3500$

3500 air bags is 140% of 2500 air bags.

9. part = percent • whole

$x = (0.124)(8300) \qquad 12.4\% = 0.124$

$x = 1029.2$

12.4% of 8300 meters is 1029.2 meters.

11. part = percent • whole

$x = (0.008)(520) \qquad 0.8\% = 0.008$

$x = 4.16$

$4.16 is 0.8% of $520.

13. part = percent • whole
$24 = (0.15)(x)$ $\quad 15\% = 0.15$
$\frac{24}{0.15} = \frac{(0.15)(x)}{0.15}$
$160 = x$

24 patients is 15% of 160 patients.

15. part = percent • whole
$130 = (0.40)(x)$ $\quad 40\% = 0.40$
$\frac{130}{0.40} = \frac{(0.40)(x)}{0.40}$
$325 = x$

40% of 325 salads is 130 salads.

17. part = percent • whole
$476 = (0.70)(x)$ $\quad 70\% = 0.70$
$\frac{476}{0.70} = \frac{(0.70)(x)}{0.70}$
$680 = x$

70% of 680 circuits is 476 circuits.

19. part = percent • whole
$135 = (0.125)(x)$ $\quad 12\frac{1}{2}\% = 12.5\% = 0.125$
$\frac{135}{0.125} = \frac{(0.125)(x)}{0.125}$
$1080 = x$

$12\frac{1}{2}\%$ of 1080 people is 135 people.

21. part = percent • whole
$3.75 = (0.0125)(x)$ $\quad 1\frac{1}{4}\% = 1.25\% = 0.0125$
$\frac{3.75}{0.0125} = \frac{(0.0125)(x)}{0.0125}$
$300 = x$

$1\frac{1}{4}\%$ of 300 gallons is 3.75 gallons.

23. part = percent • whole
$70 = x \cdot 140$
$\frac{70}{140} = \frac{x \cdot 140}{140}$
$0.5 = x$
0.5 is 50%

70 shipments is 50% of 140 shipments.

25. part = percent • whole
$114 = x \cdot 150$
$\frac{114}{150} = \frac{x \cdot 150}{150}$
$0.76 = x$
0.76 is 76%

114 tuxedos is 76% of 150 tuxedos.

27. part = percent • whole
$330 = x \cdot 264$
$\frac{330}{264} = \frac{x \cdot 264}{264}$
$1.25 = x$
1.25 is 125%

125% of $264 is $330.

29. part = percent • whole
$2.4 = x \cdot 160$
$\frac{2.4}{160} = \frac{x \cdot 160}{160}$
$0.015 = x$
0.015 is 1.5%

1.5% of 160 liters is 2.4 liters.

31. part = percent • whole
$170 = x \cdot 68$
$\frac{170}{68} = \frac{x \cdot 68}{68}$
$2.5 = x$
2.5 is 250%

170 cartons is 250% of 68 cartons.

33. You must first change the fraction in the percent to a decimal, then divide the percent by 100 to change it to a decimal.

$2\frac{1}{2}\% = 2.5\% = 0.025$
Change $2\frac{1}{2}$ to 2.5 — 2.5% as a decimal

35. 27% of 14 million
part = percent • whole
$x = (0.27)(14)$
$x = 3.78$
3.78 million or 3,780,000 office workers want more storage space.

37. 79% of 104.2 million homes
part = percent • whole
$x = (0.79)(104.2)$
$x \approx 82.3$
The number of homes that have WD-40 in them is 82.3 million homes (rounded).

39. part = percent • whole
$x = (0.84)(1100)$
$x = 924$
924 employers offer only one health plan.

41. **(a)** 5 of the 50 states

part = percent • whole

$5 = x \cdot 50$

$\frac{5}{50} = \frac{x \cdot 50}{50}$

$0.1 = x$

0.1 is 10%

10% of the states do not have a sales tax.

(b) 100% − 10% = 90% of the states do have a sales tax.

43. part = percent • whole

$461 = x \cdot 1250$

$\frac{461}{1250} = \frac{x \cdot 1250}{1250}$

$0.369 \approx x$

0.369 is 36.9%

36.9% (rounded) of these Americans rate their health as excellent.

45. 39% of 5400

part = percent • whole

$x = (0.39)(5400)$

$x = 2106$

2106 people said they spend "16–30 minutes" preparing weekday dinners.

47. 31–61+ minutes: 35% + 15% + 4% = 54%

54% of 5400

part = percent • whole

$x = (0.54)(5400)$

$x = 2916$

2916 people said they spend over 30 minutes preparing weekday dinners.

49. $338 million is 18.6% of what number?

part = percent • whole

$338 = (0.186)(x)$

$\frac{338}{0.186} = \frac{(0.186)(x)}{0.186}$

$1817.2 \approx x$

The total value of all mortgages made last year was $1817.2 million (rounded).

51. 220,917 is 46.2% of what number?

part = percent • whole

$220{,}917 = (0.462)(x)$

$\frac{220{,}917}{0.462} = \frac{(0.462)(x)}{0.462}$

$478{,}175 \approx x$

In 1967, 478,175 Mustangs (rounded) were sold.

53. part = percent • whole

$x = (0.325)(385{,}200)$ $\quad 32\frac{1}{2}\% = 0.325$

$x = 125{,}190$

There was an increase in sales of $125,190. Therefore, the volume of sales this year is $385,200 + $125,190 = $510,390.

55. Find the sales tax.

part = percent • whole

$x = (0.0775)(524)$ $\quad 7\frac{3}{4}\% = 0.0775$

$x = 40.61$

Add the sales tax to the purchase price.

$524 + $40.61 = $564.61

Subtract the trade-in.

$564.61 − $125 = $439.61

The total cost to the customer is $439.61.

6.6 Solving Application Problems with Percent

6.6 Margin Exercises

1. Use the sales tax formula:

amount of sales tax = *rate of tax* • *cost of item*

The sales tax rate is 6%.

(a) The cost of the bat is $29.

amount of sales tax = (6%)($29)

$a = (0.06)(\$29)$

$a = \$1.74$

The total cost is $29 + $1.74 = $30.74.

(b) The cost of the Game Boy game console is $119.

amount of sales tax = (6%)($119)

$a = (0.06)(\$119)$

$a = \$7.14$

The total cost is $119 + $7.14 = $126.14.

(c) The cost of the leather chair and ottoman is $1287.

amount of sales tax = (6%)($1287)

$a = (0.06)(\$1287)$

$a = \$77.22$

The total cost is $1287 + $77.22 = $1364.22.

(d) The cost of the pickup truck is $19,300.

amount of sales tax = (6%)($19,300)

$a = (0.06)(\$19{,}300)$

$a = \$1158$

The total cost is $19,300 + $1158 = $20,458.

2. Use the sales tax formula:

sales tax = rate of tax • cost of item

(a) The tax on a $320 patio set is $25.60.

$$\$25.60 = r \cdot \$320$$
$$\frac{25.60}{320} = \frac{r \cdot 320}{320}$$
$$0.08 = r$$

0.08 is 8%
The sales tax rate is 8%.

(b) The tax on a $24 sweatshirt is $1.56.

$$\$1.56 = r \cdot \$24$$
$$\frac{1.56}{24} = \frac{r \cdot 24}{24}$$
$$0.065 = r$$

0.065 is 6.5% or $6\frac{1}{2}$%.
The sales tax rate is 6.5% or $6\frac{1}{2}$%.

(c) The tax on a $13,520 Chrysler PT Cruiser is $540.80.

$$\$540.80 = r \cdot \$13{,}520$$
$$\frac{540.80}{13{,}520} = \frac{r \cdot 13{,}520}{13{,}520}$$
$$0.04 = r$$

0.04 is 4%
The sales tax rate is 4%.

3. Use the commission formula:

amount of commission
= rate of commission • amount of sales

(a) Rate of commission is 12%; sales are $28,750.

$$\begin{aligned} \textit{amount} &= (12\%)(\$28{,}750) \\ &= (0.12)(\$28{,}750) \\ &= \$3450 \end{aligned}$$

Jill earned a commission of $3450 for selling dental equipment.

(b) Rate of commission is 6%; sales are $175,500.

$$\begin{aligned} \textit{amount} &= (6\%)(\$175{,}500) \\ &= (0.06)(\$175{,}500) \\ &= \$10{,}530 \end{aligned}$$

Pam earned a commission of $10,530 for selling a home.

4. **(a)** A commission of $450 and sales worth $22,500:

$$\frac{\text{part}}{\text{whole}} = \frac{x}{100}$$
$$\frac{450}{22{,}500} = \frac{x}{100}$$
$$22{,}500 \cdot x = 450 \cdot 100$$
$$\frac{22{,}500 \cdot x}{22{,}500} = \frac{45{,}000}{22{,}500}$$
$$x = 2$$

The rate of commission is 2%.

(b) A commission of $2898 and sales worth $32,200:

$$\frac{\text{part}}{\text{whole}} = \frac{x}{100}$$
$$\frac{2898}{32{,}200} = \frac{x}{100}$$
$$32{,}200 \cdot x = 2898 \cdot 100$$
$$\frac{32{,}200 \cdot x}{32{,}200} = \frac{289{,}800}{32{,}200}$$
$$x = 9$$

The rate of commission is 9%.

5. **(a)** 42% discount on a leather recliner priced at $950:

amount of discount
= rate of discount • original price
= (0.42)($950)
= $399

sale price
= original price − amount of discount
= $950 − $399
= $551

The sale price for the Easy-Boy leather recliner is $551.

(b) 35% discount on a sweater set priced at $30.

amount of discount = (0.35)($30)
= $10.50

sale price = $30 − $10.50
= $19.50

6. **(a)** Production increased from 14,100 units to 19,035 units.

increase = 19,035 − 14,100 = 4935

$$\frac{\text{amount of increase (part)}}{\text{original value (whole)}} = \frac{\text{percent}}{100}$$
$$\frac{4935}{14{,}100} = \frac{x}{100}$$
$$14{,}100 \cdot x = 4935 \cdot 100$$
$$\frac{14{,}100 \cdot x}{14{,}100} = \frac{493{,}500}{14{,}100}$$
$$x = 35$$

The percent of increase is 35%.

(b) Flu cases rose from 496 cases to 620 cases.

increase = 620 – 496 = 124

$$\frac{124}{496} = \frac{x}{100}$$
$$496 \cdot x = 124 \cdot 100$$
$$\frac{496 \cdot x}{496} = \frac{12{,}400}{496}$$
$$x = 25$$

The percent of increase is 25%.

7. **(a)** The number of service calls decreased from 380 to 285.

decrease = 380 – 285 = 95

$$\frac{\text{amount of decrease (part)}}{\text{original value (whole)}} = \frac{x}{100}$$
$$\frac{95}{380} = \frac{x}{100}$$
$$380 \cdot x = 95 \cdot 100$$
$$\frac{380 \cdot x}{380} = \frac{9500}{380}$$
$$x = 25$$

The percent of decrease is 25%.

(b) The number of workers decreased from 4850 to 3977.

decrease = 4850 – 3977 = 873

$$\frac{873}{4850} = \frac{p}{100}$$
$$4850 \cdot p = 873 \cdot 100$$
$$\frac{4850 \cdot p}{4850} = \frac{87{,}300}{4850}$$
$$p = 18$$

The percent of decrease is 18%.

6.6 Section Exercises

1. *sales tax* = *rate of tax* • *cost of item*
= (4%)($6)
= (0.04)($6)
= $0.24

The amount of sales tax is $0.24 and the total cost is $6 + $0.24 = $6.24.

3. *sales tax* = *rate of tax* • *cost of item*

$$\$12.75 = r \cdot \$425$$
$$\frac{12.75}{425} = \frac{r \cdot 425}{425}$$
$$0.03 = r$$

0.03 is 3%

The tax rate is 3% and the total cost is $425 + $12.75 = $437.75.

5. *sales tax* = *rate of tax* • *cost of item*

$$\$14.20 = r \cdot \$284$$
$$\frac{14.20}{284} = \frac{r \cdot 284}{284}$$
$$0.05 = r$$

0.05 is 5%

The tax rate is 5% and the total cost is $284 + $14.20 = $298.20.

7. *sales tax* = *rate of tax* • *cost of item*
= ($5\frac{1}{2}$%)($12,229)
= (0.055)($12,229)
≈ $672.60

The amount of sales tax is $672.60 (rounded) and the total cost is $12,229 + $672.60 = $12,901.60.

9. *commission* = *rate of commission* • *sales*
= (8%)($280)
= (0.08)($280)
= $22.40

The amount of commission is $22.40.

11.
$$\frac{\text{part}}{\text{whole}} = \frac{\text{percent}}{100}$$
$$\frac{600}{3000} = \frac{x}{100} \quad \textbf{OR} \quad \frac{1}{5} = \frac{x}{100}$$
$$5 \cdot x = 1 \cdot 100$$
$$\frac{5 \cdot x}{5} = \frac{100}{5}$$
$$x = 20$$

The rate of commission is 20%.

13. *commission* = *rate of commission* • *sales*
= (3%)($6183.50)
= (0.03)($6183.50)
= $185.51 (rounded)

The amount of commission is $185.51.

15. *commission* = *rate of commission* • *sales*
= (9%)($73,500)
= (0.09)($73,500)
= $6615

The amount of commission is $6615.

17. *discount* = *rate of discount* • *original price*
= (10%)($199.99)
= (0.10)($199.99)
= $20.00 (rounded)

The amount of discount is $20.00 and the sale price is $199.99 – $20 = $179.99.

19. *discount* = *rate of discount* • *original price*

$\$54 = r \cdot \$180$

$\frac{54}{180} = \frac{r \cdot 180}{180}$

$0.3 = r$

0.3 is 30%

The rate of discount is 30% and the sale price is $180 – $54 = $126.

21. *discount* = *rate of discount* • *original price*

= (25%)($17.50)

= (0.25)($17.50)

= $4.38 (rounded)

The amount of discount is $4.38 and the sale price is $17.50 – $4.38 = $13.12.

23. *sales tax* = *rate of tax* • *cost of item*

= (15%)($58.40)

= (0.15)($58.40)

= $8.76

The amount of discount is $8.76 and the sale price is $58.40 – $8.76 = $49.64.

25. Answers will vary. A sample answer follows: On the basis of commission alone you would choose Company A. Other considerations might be: reputation of the company; expense allowances; other fringe benefits; travel; promotion and training, to name a few.

27. The cost of 6 skillets is

6($29.99) = $179.94.

Since the cost of the skillets is "$100.00 or more," we need to add $10.95 for shipping and insurance. The sales tax is

(5%)($179.94) = (0.05)($179.94) ≈ $9.00.

Thus, the total cost is

$179.94 + $10.95 + $9.00 = $199.89.

29. The cost of 3 pop-up hampers is

3($9.99) = $29.97.

The cost of 4 nonstick mini donut pans is

4($10.99) = $43.96.

The cost of the hampers and pans is $29.97 + $43.96 = $73.93. Since this cost is "$70.01 to $99.99," we need to add $9.95 for shipping and insurance.
The sales tax is

(5%)($73.93) = (0.05)($73.93) ≈ $3.70.

Thus, the total cost is

$73.93 + $9.95 + $3.70 = $87.58.

31. *sales tax* = *rate of tax* • *cost of item*

$\$99 = r \cdot \$1980$

$\frac{99}{1980} = \frac{r \cdot 1980}{1980}$

$0.05 = r$

0.05 is 5%

The rate of sales tax is 5%.

33. increase = 267,285 – 234,461 = 32,824

$\frac{\text{increase}}{\text{original}} = \frac{x}{100}$

$\frac{32{,}824}{234{,}461} = \frac{x}{100}$

$234{,}461 \cdot x = 32{,}824 \cdot 100$

$\frac{234{,}461 \cdot x}{234{,}461} = \frac{3{,}282{,}400}{234{,}461}$

$x \approx 14.0$

The percent of increase is 14.0% (rounded).

35. The decrease in accidents is 1276 – 989 = 287 accidents.

$\frac{\text{decrease}}{\text{original}} = \frac{x}{100}$

$\frac{287}{1276} = \frac{x}{100}$

$1276 \cdot x = 287 \cdot 100$

$\frac{1276 \cdot x}{1276} = \frac{28{,}700}{1276}$

$x \approx 22.5$

The percent of decrease is 22.5% (rounded).

37. *discount* = *rate of discount* • *original price*

= (60%)($335)

= (0.60)($335)

= $201

The sale price is $335 – $201 = $134.

39. *commission* = *rate of commission* • *sales*

= (3%)($9500)

= (0.03)($9500)

= $285.00

McKee's commission is $285.

41. *commission* = *rate of commission* • *sales*

$\$355 = r \cdot \$8875$

$\frac{355}{8875} = \frac{r \cdot 8875}{8875}$

$0.04 = r$

0.04 is 4%

The rate of commission for Poznick is 4%.

43. $\textit{discount} = \textit{rate of discount} \cdot \textit{original price}$
$= (18\%)(\$590)$
$= (0.18)(\$590)$
$= \$106.20$

The discount is \$106.20, and the sale price is \$590 − \$106.20 = \$483.80.

45. The stock decreased \$35.50 − \$33.50 = \$2.00 per share.

$$\frac{\text{decrease}}{\text{original}} = \frac{x}{100}$$
$$\frac{2}{35.50} = \frac{x}{100}$$
$$35.50 \cdot x = 2 \cdot 100$$
$$\frac{35.50 \cdot x}{35.50} = \frac{200}{35.50}$$
$$x \approx 5.6$$

The percent of decrease in price is 5.6% (rounded).

47. $\textit{discount} = \textit{rate of discount} \cdot \textit{original price}$
$= (6\%)(\$18.50)$
$= (0.06)(\$18.50)$
$= \$1.11$

The sale price is \$18.50 − \$1.11 = \$17.39.

$\textit{sales tax} = \textit{rate of tax} \cdot \textit{cost of item}$
$= (6\%)(\$17.39)$
$= (0.06)(\$17.39)$
$\approx \$1.04$

The cost of the dictionary is \$17.39 + \$1.04 = \$18.43.

49. $\textit{sales commission} = \textit{rate of commission} \cdot \textit{sales}$
$= (6\%)(\$129{,}605)$
$= (0.06)(\$129{,}605)$
$= \$7776.30$

$\textit{commission for agent} = (55\%)(\$7776.30)$
$= (0.55)(\$7776.30)$
$\approx \$4276.97$

The agent receives \$4276.97 (rounded).

51. $\textit{discount} = \textit{rate of discount} \cdot \textit{original price}$
$= (15\%)(\$10{,}214)$
$= (0.15)(\$10{,}214)$
$= \$1532.10$

The sale price is \$10,214 − \$1532.10 = \$8681.90.

$\textit{sales tax} = \textit{rate of tax} \cdot \textit{cost of item}$
$= (3\tfrac{3}{4}\%)(\$8681.90)$
$= (0.0375)(\$8681.90)$
$\approx \$325.57$

The total cost of the boat is \$8681.90 + \$325.57 = \$9007.47 (rounded).

53. The percent equation is

$$\text{part} = \underline{\text{percent}} \cdot \underline{\text{whole}}$$

54. The sales tax formula is

$$\text{sales tax} = \underline{\text{rate (percent) of tax}} \cdot \underline{\text{cost of item}}$$

55. $\textit{excise tax} = \textit{rate of tax} \cdot \textit{cost of item}$
$= (12.4\%)((\$123)$
$= (0.124)(\$123)$
$= \$15.25$ (rounded)

$\textit{sales tax} = \textit{rate of tax} \cdot \textit{cost of item}$
$= (6\tfrac{1}{2}\%)(\$123)$
$= (0.065)(\$123)$
$= \$8.00$ (rounded)

The total cost is \$123 + \$15.25 + \$8.00 = \$146.25.

56. Combined tax rate is 12.4% + 6.5% = 18.9%.

$\textit{sales tax} = \textit{rate of tax} \cdot \textit{cost of item}$
$= (18.9\%)(\$123.00)$
$= (0.189)(\$123.00)$
$\approx \$23.25$

The total cost is \$123 + \$23.25 = \$146.25. Yes, the answers are the same since both taxes are calculated on the same selling price.

57. $\textit{sales tax} = \textit{rate of tax} \cdot \textit{cost of item}$
$= (7\tfrac{3}{4}\%)(\$1248)$
$= (0.0775)(\$1248)$
$= \$96.72$

The total cost is

$$\$1248 + \$96.72 + \$13.40 = \$1358.12.$$

58. No. In Exercise 57 the excise tax of \$13.40 cannot be added to the sales tax rate of $7\frac{3}{4}\%$ because the excise tax is an amount, not a percent.

6.7 Simple Interest

6.7 Margin Exercises

1. **(a)** \$1000 at 5% for 1 year

$I = p \cdot r \cdot t$
$= (1000)(0.05)(1)$
$= 50$

The interest is \$50.

(b) \$3650 at 2% for 1 year

$I = p \cdot r \cdot t$
$= (3650)(0.02)(1)$
$= 73$

The interest is \$73.

2. **(a)** $820 at 6% for $3\frac{1}{2}$ years

$$\begin{aligned} I &= p \cdot r \cdot t \\ &= (820)(0.06)(3.5) \\ &= 172.2 \end{aligned}$$

The interest is $172.20.

(b) $4850 at 8% for $2\frac{1}{2}$ years

$$\begin{aligned} I &= p \cdot r \cdot t \\ &= (4850)(0.08)(2.5) \\ &= 970 \end{aligned}$$

The interest is $970.

(c) $16,800 at 3% for $2\frac{3}{4}$ years

$$\begin{aligned} I &= p \cdot r \cdot t \\ &= (16{,}800)(0.03)(2.75) \\ &= 1386 \end{aligned}$$

The interest is $1386.

3. **(a)** $1800 at 3% for 4 months

$$\begin{aligned} I &= p \cdot r \cdot t \\ &= (1800)(0.03)\left(\tfrac{4}{12}\right) \quad & 4 \text{ months} \\ &= 54\left(\tfrac{1}{3}\right) & = \left(\tfrac{4}{12}\right) \text{year} \\ &= \frac{54}{3} = 18 \end{aligned}$$

The interest is $18.

(b) $28,000 at $9\frac{1}{2}$% for 3 months

$$\begin{aligned} I &= p \cdot r \cdot t \\ &= (28{,}000)(0.095)\left(\tfrac{3}{12}\right) \\ &= (2660)\left(\tfrac{1}{4}\right) \\ &= \frac{2660}{4} = 665 \end{aligned}$$

The interest is $665.

4. **(a)** $3800 at $6\frac{1}{2}$% for 6 months

$$\begin{aligned} I &= p \cdot r \cdot t \\ &= (3800)(0.065)\left(\tfrac{6}{12}\right) \\ &= 247\left(\tfrac{1}{2}\right) \\ &= \frac{247}{2} = 123.5 \end{aligned}$$

The interest is $123.50.

amount due = *principal* + *interest*
= $3800 + $123.50
= $3923.50

The total amount due is $3923.50.

(b) $12,400 at 5% for 5 years

$$\begin{aligned} I &= p \cdot r \cdot t \\ &= (12{,}400)(0.05)(5) \\ &= 3100 \end{aligned}$$

The interest is $3100.

amount due = *principal* + *interest*
= $12,400 + $3100
= $15,500

The total amount due is $15,500.

(c) $2400 at 11% for $2\frac{3}{4}$ years

$$\begin{aligned} I &= p \cdot r \cdot t \\ &= (2400)(0.11)(2.75) \\ &= 726 \end{aligned}$$

The interest is $726.

amount due = *principal* + *interest*
= $2400 + $726
= $3126

The total amount due is $3126.

6.7 Section Exercises

1. $100 at 6% for 1 year

$$\begin{aligned} I &= p \cdot r \cdot t \\ &= (100)(0.06)(1) \\ &= 6 \end{aligned}$$

The interest is $6.

3. $700 at 5% for 3 years

$$\begin{aligned} I &= p \cdot r \cdot t \\ &= (700)(0.05)(3) \\ &= 105 \end{aligned}$$

The interest is $105.

5. $240 at 4% for 3 years

$$\begin{aligned} I &= p \cdot r \cdot t \\ &= (240)(0.04)(3) \\ &= 28.8 \end{aligned}$$

The interest is $28.80.

7. $2300 at $8\frac{1}{2}$% for $2\frac{1}{2}$ years

$$\begin{aligned} I &= p \cdot r \cdot t \\ &= (2300)(0.085)(2.5) \\ &= 488.75 \end{aligned}$$

The interest is $488.75.

9. $10,800 at $7\frac{1}{2}$% for $2\frac{3}{4}$ years

$$I = p \cdot r \cdot t$$
$$= (10{,}800)(0.075)(2.75)$$
$$= 2227.5$$

The interest is $2227.50.

11. $400 at 5% for 6 months

$$I = p \cdot r \cdot t$$
$$= (400)(0.05)\left(\tfrac{6}{12}\right)$$
$$= (20)\left(\tfrac{1}{2}\right)$$
$$= \frac{20}{2} = 10$$

The interest is $10.

13. $820 at 6% for 12 months

$$I = p \cdot r \cdot t$$
$$= (820)(0.06)\left(\tfrac{12}{12}\right)$$
$$= (49.2)(1)$$
$$= 49.2$$

The interest is $49.20.

15. $940 at 3% for 18 months

$$I = p \cdot r \cdot t$$
$$= (940)(0.03)\left(\tfrac{18}{12}\right)$$
$$= (28.2)(1.5)$$
$$= 42.3$$

The interest is $42.30.

17. $1225 at $5\frac{1}{2}$% for 3 months

$$I = p \cdot r \cdot t$$
$$= (1225)(0.055)\left(\tfrac{3}{12}\right)$$
$$= (67.375)\left(\tfrac{1}{4}\right)$$
$$= 16.84 \text{ (rounded)}$$

The interest is $16.84.

19. $15,300 at $7\frac{1}{4}$% for 7 months

$$I = p \cdot r \cdot t$$
$$= (15{,}300)(0.0725)\left(\tfrac{7}{12}\right)$$
$$= (1109.25)\left(\tfrac{7}{12}\right)$$
$$= 647.06 \text{ (rounded)}$$

The interest is $647.06.

21. $200 at 5% for 1 year

$$I = p \cdot r \cdot t$$
$$= (200)(0.05)(1)$$
$$= 10$$

The interest is $10.

$$\textit{amount due} = \textit{principal} + \textit{interest}$$
$$= \$200 + \$10$$
$$= \$210$$

The total amount due is $210.

23. $740 at 6% for 9 months

$$I = p \cdot r \cdot t$$
$$= (740)(0.06)\left(\tfrac{9}{12}\right)$$
$$= (44.4)\left(\tfrac{3}{4}\right)$$
$$= \frac{133.2}{4} = 33.3$$

The interest is $33.30.

$$\textit{amount due} = \textit{principal} + \textit{interest}$$
$$= \$740 + \$33.30$$
$$= \$773.30$$

The total amount due is $773.30.

25. $1800 at 9% for 18 months

$$I = p \cdot r \cdot t$$
$$= (1800)(0.09)\left(\tfrac{18}{12}\right)$$
$$= (162)(1.5)$$
$$= 243$$

The interest is $243.

$$\textit{amount due} = \textit{principal} + \textit{interest}$$
$$= \$1800 + \$243$$
$$= \$2043$$

The total amount due is $2043.

27. $3250 at 10% for 6 months

$$I = p \cdot r \cdot t$$
$$= (3250)(0.10)\left(\tfrac{6}{12}\right)$$
$$= (325)(0.5)$$
$$= 162.5$$

The interest is $162.50.

$$\textit{amount due} = \textit{principal} + \textit{interest}$$
$$= \$3250 + \$162.50$$
$$= \$3412.50$$

The total amount due is $3412.50.

29. $16,850 at $7\frac{1}{2}$% for 9 months

$$I = p \cdot r \cdot t$$
$$= (16{,}850)(0.075)\left(\tfrac{9}{12}\right)$$
$$= (1263.75)\left(\tfrac{3}{4}\right)$$
$$= 947.81 \text{ (rounded)}$$

The interest is $947.81.

amount due = *principal* + *interest*
= \$16,850 + \$947.81
= \$17,797.81

The total amount due is \$17,797.81.

31. The answer should include:
Amount of principal—This is the amount of money borrowed or loaned.
Interest rate—This is the percent used to calculate the interest.
Time of loan—The length of time that money is loaned or borrowed is an important factor in determining interest.

33. \$825 at 5% for 1 year

$I = p \cdot r \cdot t$
$= (825)(0.05)(1)$
$= 41.25$

She will earn \$41.25 in interest.

35. \$380,000 at 6% for 18 months

$I = p \cdot r \cdot t$
$= (380{,}000)(0.06)\left(\frac{18}{12}\right)$
$= (22{,}800)(1.5)$
$= 34{,}200$

The bank will earn \$34,200 in interest.

37. \$1200 at 8% for 5 months

$I = p \cdot r \cdot t$
$= (1200)(0.08)\left(\frac{5}{12}\right)$
$= (96)\left(\frac{5}{12}\right)$
$= 40$

The interest is \$40.

amount due = *principal* + *interest*
= \$1200 + \$40
= \$1240

The total amount due is \$1240.

39. \$14,800 at $2\frac{1}{4}$% for 10 months

$I = p \cdot r \cdot t$
$= (14{,}800)(0.0225)\left(\frac{10}{12}\right)$
$= (333)\left(\frac{5}{6}\right)$
$= 277.50$

She will earn \$277.50 in interest.

41. \$8800 at $7\frac{1}{4}$% for $\frac{1}{4}$ year

$I = p \cdot r \cdot t$
$= (8800)(0.0725)\left(\frac{1}{4}\right)$
$= (638)\left(\frac{1}{4}\right)$
$= 159.50$

Beverly will earn \$159.50 in interest.

43. \$11,500 at $8\frac{3}{4}$% for $\frac{3}{4}$ year

$I = p \cdot r \cdot t$
$= (11{,}500)(0.0875)(0.75)$
$= (1006.25)(0.75)$
$= 754.69$ (rounded)

The total amount in the account is \$11,500 + \$754.69 = \$12,254.69.

45. $6 \cdot \$550 = \$3300$
60% of \$3300
$x = (0.6)(3300)$
$x = 1980$
\$1980 at $12\frac{1}{2}$% for 9 months

$I = p \cdot r \cdot t$
$= (1980)(0.125)\left(\frac{9}{12}\right)$
$= (247.5)\left(\frac{3}{4}\right)$
$= 185.63$ (rounded)

The interest is \$185.63.

amount due = *principal* + *interest*
= \$1980 + \$185.63
= \$2165.63

The amount due is \$2165.63.

Summary Exercises on Percent

1. 6.25% = 0.0625
Drop the percent sign and move the decimal point two places to the left.

3. 0.375 = 37.5%
Move the decimal point two places to the right and attach a percent sign.

5. $87.5\% = \frac{87.5}{100} = \frac{87.5 \cdot 10}{100 \cdot 10} = \frac{875}{1000} = \frac{875 \div 125}{1000 \div 125} = \frac{7}{8}$

7. $\frac{5}{8} = \frac{p}{100}$
$8 \cdot p = 5 \cdot 100$
$\frac{8 \cdot p}{8} = \frac{500}{8}$
$p = 62.5$
Thus, $\frac{5}{8} = 62.5\%$.

9. percent is 6; whole is 780

$\frac{\text{part}}{780} = \frac{6}{100}$ **OR** $\frac{x}{780} = \frac{3}{50}$
$x \cdot 50 = 780 \cdot 3$
$\frac{x \cdot 50}{50} = \frac{2340}{50}$
$x = 46.8$

6% of \$780 is \$46.80.

11. part is 176; whole is 320

$$\frac{176}{320} = \frac{x}{100} \quad \textbf{OR} \quad \frac{11}{20} = \frac{x}{100}$$
$$20 \cdot x = 11 \cdot 100$$
$$\frac{20 \cdot x}{20} = \frac{1100}{20}$$
$$x = 55$$

55% of 320 policies is 176 policies.

13. part is 1016.4; percent is 280

$$\frac{1016.4}{x} = \frac{280}{100} \quad \textbf{OR} \quad \frac{1016.4}{x} = \frac{14}{5}$$
$$x \cdot 14 = (1016.4)(5)$$
$$\frac{x \cdot 14}{14} = \frac{5082}{14}$$
$$x = 363$$

1016.4 acres is 280% of 363 acres.

15. $\textit{sales tax} = \textit{rate of tax} \cdot \textit{cost of item}$
$= (5\frac{1}{2}\%)(\$176)$
$= (0.055)(\$176)$
$= \$9.68$

The amount of sales tax is \$9.68 and the total cost is \$176 + \$9.68 = \$185.68.

17. $\textit{commission} = \textit{rate of commission} \cdot \textit{sales}$
$= (8\%)(\$3765)$
$= (0.08)(\$3765)$
$= \$301.20$

The amount of commission is \$301.20.

19. $\textit{discount} = \textit{rate of discount} \cdot \textit{original price}$
$= (18\%)(\$49.95)$
$= (0.18)(\$49.95)$
$= \$8.99$ (rounded)

The amount of discount is \$8.99, and the sale price is \$49.95 − \$8.99 = \$40.96.

21. \$1000 at 3% for 1 year

$I = p \cdot r \cdot t$
$= (1000)(0.03)(1)$
$= 30$

The interest is \$30 and the total amount due is \$1000 + \$30 = \$1030.

23. \$1470 at 4% for 9 months

$I = p \cdot r \cdot t$
$= (1470)(0.04)(\frac{9}{12})$
$= (58.8)(\frac{3}{4})$
$= 44.1$

The interest is \$44.10 and the total amount due is \$1470 + \$44.10 = \$1514.10.

25. increase = 267,285 − 111,800 = 155,485

$$\frac{155{,}485}{111{,}800} = \frac{x}{100}$$
$$111{,}800 \cdot x = 155{,}485 \cdot 100$$
$$\frac{111{,}800 \cdot x}{111{,}800} = \frac{15{,}548{,}500}{111{,}800}$$
$$x \approx 139.1$$

The percent of increase for the five-year period is 139.1% (rounded).

6.8 Compound Interest

6.8 Margin Exercises

1. **(a)** \$500 at 4% for 2 years

Year	*Interest*	*Compound Amount*
1	(\$500)(0.04)(1) = \$20	\$500 + \$20 = \$520
2	(\$520)(0.04)(1) = \$20.80	\$520 + \$20.80 = \$540.80

The compound amount is \$540.80.

(b) \$2000 at 7% for 3 years

Year	*Interest*	*Compound Amount*
1	(\$2000)(0.07)(1) = \$140	\$2000 + \$140 = \$2140
2	(\$2140)(0.07)(1) = \$149.80	\$2140 + \$149.80 = \$2289.80
3	(\$2289.80)(0.07)(1) ≈ \$160.29	\$2289.80 + \$160.29 = \$2450.09

The compound amount is \$2450.09.

2. **(a)** \$1800 at 2% for 3 years

100% + 2% = 102% = 1.02
(\$1800)(1.02)(1.02)(1.02) ≈ \$1910.17

The compound amount is \$1910.17.

(b) \$900 at 3% for 2 years

100% + 3% = 103% = 1.03
(\$900)(1.03)(1.03) = \$954.81

The compound amount is \$954.81.

(c) \$2500 at 5% for 4 years

100% + 5% = 105% = 1.05
(\$2500)(1.05)(1.05)(1.05)(1.05) ≈ \$3038.77

The compound amount is \$3038.77.

3. Read from the compound interest table.

(a) \$1 at 3% for 6 years: \$1.1941 ≈ \$1.19

(b) \$1 at 6% for 8 years: \$1.5938 ≈ \$1.59

(c) \$1 at $4\frac{1}{2}$% for 12 years: \$1.6959 ≈ \$1.70

4. Use the table and multiply to find the compound amount. Then subtract the original deposit from the compound amount to find the interest.

(a) \$4000 at 3% for 10 years

compound amount = (\$4000)(1.3439)
= \$5375.60
interest = \$5375.60 − \$4000
= \$1375.60

(b) \$12,600 at $5\frac{1}{2}$% for 8 years

compound amount = (\$12,600)(1.5347)
= \$19,337.22
interest = \$19,337.22 − \$12,600
= \$6737.22

(c) \$32,700 at 6% for 12 years

compound amount = (\$32,700)(2.0122)
= \$65,798.94
interest = \$65,798.94 − \$32,700
= \$33,098.94

6.8 Section Exercises

1. \$500 at 4% for 2 years

Year	*Interest*	*Compound Amount*
1	(\$500)(0.04)(1) = \$20	\$500 + \$20 = \$520
2	(\$520)(0.04)(1) = \$20.80	\$520 + \$20.80 = \$540.80

The compound amount is \$540.80.

3. \$1800 at 3% for 3 years

Year	*Interest*	*Compound Amount*
1	(\$1800)(0.03)(1) = \$54	\$1800 + \$54 = \$1854
2	(\$1854)(0.03)(1) = \$55.62	\$1854 + \$55.62 = \$1909.62
3	(\$1909.62)(0.03)(1) ≈ \$57.29	\$1909.62 + \$57.29 = \$1966.91

The compound amount is \$1966.91.

5. \$3500 at 7% for 4 years

Year	*Interest*	*Compound Amount*
1	(\$3500)(0.07)(1) = \$245	\$3500 + \$245 = \$3745
2	(\$3745)(0.07)(1) = \$262.15	\$3745 + \$262.15 = \$4007.15
3	(\$4007.15)(0.07)(1) ≈ \$280.50	\$4007.15 + \$280.50 = \$4287.65
4	(\$4287.65)(0.07)(1) ≈ \$300.14	\$4287.65 + \$300.14 = \$4587.79

The compound amount is \$4587.79.

7. \$1000 at 5% for 2 years

100% + 5% = 105% = 1.05
(\$1000)(1.05)(1.05) = \$1102.50

The compound amount is \$1102.50.

9. \$1400 at 6% for 5 years

100% + 6% = 106% = 1.06
(\$1400)(1.06)(1.06)(1.06)(1.06)(1.06) ≈ \$1873.52

The compound amount is \$1873.52.

11. \$1180 at 7% for 8 years

100% + 7% = 107% = 1.07
(\$1180)(1.07)(1.07)(1.07)(1.07)
(1.07)(1.07)(1.07)(1.07) ≈ \$2027.46

The compound amount is \$2027.46.

13. \$10,940 at 8% for 6 years

100% + 8% = 108% = 1.08
(\$10,940)(1.08)(1.08)(1.08)(1.08)(1.08)(1.08)
≈ \$17,360.41

The compound amount is \$17,360.41.

15. \$1000 at 4% for 5 years

4% column, row 5 of the table gives 1.2167.

compound amount = (\$1000)(1.2167)
= \$1216.70
interest = \$1216.70 − \$1000
= \$216.70

17. \$8000 at 6% for 10 years

6% column, row 10 of the table gives 1.7908.

compound amount = (\$8000)(1.7908)
= \$14,326.40
interest = \$14,326.40 − \$8000
= \$6326.40

19. \$8428.17 at $4\frac{1}{2}\%$ for 6 years

4.50% column, row 6 of the table gives 1.3023.

compound amount = (\$8428.17)(1.3023)
≈ \$10,976.01
interest = \$10,976.01 – \$8428.17
= \$2547.84

21. Compound interest is interest paid on past interest as well as on the principal. Many people describe compound interest as "interest on interest."

23. \$8450 at 6% compounded annually for 8 years

6% column, row 8 of the table gives 1.5938.

compound amount = (\$8450)(1.5938)
= \$13,467.61

She will have \$13,467.61 at the end of 8 years.

25. **(a)** 6% column, row 9 from the table is 1.6895.

compound amount = (\$76,000)(1.6895)
= \$128,402

The total amount that should be repaid is \$128,402.

(b) The amount of interest earned is

\$128,402 – \$76,000 = \$52,402

27. **(a)** \$30,000 at 6% compounded annually for 2 years

6% column, row 2 from the table is 1.1236.
compound amount = (\$30,000)(1.1236)
= \$33,708

After 2 years she will have \$33,708. After the \$40,000 deposit, she will have \$73,708.

6% column, row 3 from the table is 1.1910.
compound amount = (\$73,708)(1.1910)
≈ \$87,786.23

The total amount will be \$87,786.23.

(b) The amount of interest earned is
\$87,786.23 – \$30,000 – \$40,000 = \$17,786.23.

29. The formula for simple interest is

Interest = <u>principal</u> · <u>rate</u> · <u>time</u>
or *I* = <u>*p*</u> · <u>*r*</u> · <u>*t*</u>.

30. The formula for amount due is

amount due = <u>principal</u> + <u>interest</u>.

31. Compound interest is interest calculated on

<u>principal</u> plus past <u>interest</u>.

32. Compound amount is the original

<u>principal</u> + compound <u>interest</u>.

33. **(a)** \$4350 at 6% simple interest for 6 years

(\$4350)(0.06)(6) = \$1566

\$4350 at 6% compounded annually for 6 years

(\$4350)(1.4185) ≈ \$6170.48

\$6170.48 – \$4350 = \$1820.48

The difference in the amount of interest is \$1820.48 – \$1566 = \$254.48.

(b) \$4350 at 6% simple interest for 12 years

(\$4350)(0.06)(12) = \$3132

\$4350 at 6% compounded annually for 12 years

(\$4350)(2.0122) = \$8753.07

\$8753.07 – \$4350 = \$4403.07

The difference is \$4403.07 – \$3132 = \$1271.07.

No, it has more than doubled. It is about five times as much (\$1271.07 ÷ \$254.48 ≈ 5).

(c) Examples will vary.

34. **(a)** \$20, 000 at 8% simple interest for 12 years
(\$20,000)(0.08)(12) = \$19,200

\$20,000 + \$19,200 = \$39,200

\$16,000 at 8% compounded annually for 12 years.
(\$16,000)(2.5182) = \$40,291.20

The compound interest account has the higher balance. It has \$40,291.20 – \$39,200 = \$1091.20 more.

(b) Greater amounts of interest are earned with compound interest than simple interest because interest is earned on both principal and past interest.

Chapter 6 Review Exercises

1. 35% = 0.35
Drop the percent sign and move the decimal point 2 places to the left.

2. 150% = 1.5
Drop the percent sign and move the decimal point 2 places to the left.

3. 99.44% = 0.9944
Drop the percent sign and move the decimal point 2 places to the left.

4. 0.085% = 0.00085
Drop the percent sign and move the decimal point 2 places to the left.

5. $3.15 = 315\%$
Move the decimal point 2 places to the right and attach a percent sign.

6. $0.02 = 2\%$
Move the decimal point 2 places to the right and attach a percent sign.

7. $0.875 = 87.5\%$
Move the decimal point 2 places to the right and attach a percent sign.

8. $0.002 = 0.2\%$
Move the decimal point 2 places to the right and attach a percent sign.

9. $15\% = \frac{15}{100} = \frac{15 \div 5}{100 \div 5} = \frac{3}{20}$

10. $37.5\% = \frac{37.5}{100} = \frac{37.5 \cdot 10}{100 \cdot 10} = \frac{375 \div 125}{1000 \div 125} = \frac{3}{8}$

11. $175\% = \frac{175}{100} = \frac{175 \div 25}{100 \div 25} = \frac{7}{4} = 1\frac{3}{4}$

12. $0.25\% = \frac{0.25}{100} = \frac{(0.25)(4)}{(100)(4)} = \frac{1}{400}$

13.
$$\frac{3}{4} = \frac{p}{100}$$
$$4 \cdot p = 3 \cdot 100$$
$$\frac{4 \cdot p}{4} = \frac{300}{4}$$
$$p = 75$$
$$\frac{3}{4} = 75\%$$

14.
$$\frac{5}{8} = \frac{p}{100}$$
$$8 \cdot p = 5 \cdot 100$$
$$\frac{8 \cdot p}{8} = \frac{500}{8}$$
$$p = 62.5$$
$$\frac{5}{8} = 62.5\% \text{ or } 62\frac{1}{2}\%$$

15. $3\frac{1}{4} = 3\frac{25}{100} = 3.25 = 325\%$

16.
$$\frac{1}{200} = \frac{p}{100}$$
$$200 \cdot p = 1 \cdot 100$$
$$\frac{200 \cdot p}{200} = \frac{100}{200}$$
$$p = 0.5$$
$$\frac{1}{200} = 0.5\%$$

17. $\frac{1}{8}$

$$\begin{array}{r} 0.125 \\ 8\overline{)1.000} \\ \underline{8} \\ 20 \\ \underline{16} \\ 40 \\ \underline{40} \\ 0 \end{array}$$

$\frac{1}{8} = 0.125$

18. From Exercise 17, $0.125 = 12.5\%$

19. $0.25 = \frac{25}{100} = \frac{25 \div 25}{100 \div 25} = \frac{1}{4}$

20. $0.25 = 25\%$

21. $180\% = \frac{180}{100} = \frac{180 \div 20}{100 \div 20} = \frac{9}{5} = 1\frac{4}{5}$

22. $180\% = 1.80 = 1.8$

23. part = 25, percent = 10

$$\frac{25}{x} = \frac{10}{100} \quad \textbf{OR} \quad \frac{25}{x} = \frac{1}{10}$$
$$x \cdot 1 = 25 \cdot 10$$
$$x = 250$$

The whole is 250.

24. whole = 480, percent = 5

$$\frac{x}{480} = \frac{5}{100} \quad \textbf{OR} \quad \frac{x}{480} = \frac{1}{20}$$
$$x \cdot 20 = 480 \cdot 1$$
$$\frac{x \cdot 20}{20} = \frac{480}{20}$$
$$x = 24$$

The part is 24.

25. 35% of 820 mailboxes is 287 mailboxes.

$$\begin{array}{l} \text{part} \rightarrow \\ \text{whole} \rightarrow \end{array} \frac{287}{820} = \frac{35}{100} \begin{array}{l} \leftarrow \text{percent} \\ \leftarrow \text{always } 100 \end{array}$$

26. 73 brooms is what percent of 90 brooms?

$$\frac{73}{90} = \frac{\text{unknown}}{100}$$

27. Find 14% of 160 bicycles.

$$\frac{\text{unknown}}{160} = \frac{14}{100}$$

28. 418 curtains is 16% of what number of curtains?

$$\frac{418}{\text{unknown}} = \frac{16}{100}$$

29. A golfer lost 3 of his 8 golf balls. What percent were lost?

$$\frac{3}{8} = \frac{\text{unknown}}{100}$$

30. What number is 88% of 1280 keys?

$$\frac{\text{unknown}}{1280} = \frac{88}{100}$$

31. whole is 950; percent is 18; part is unknown.
part = $(0.18)(950) = 171$
18% of 950 programs is 171 programs.

32. whole is 1450; percent is 60; part is unknown.

$$\frac{x}{1450} = \frac{60}{100} \quad \textbf{OR} \quad \frac{x}{1450} = \frac{3}{5}$$
$$5 \cdot x = 1450 \cdot 3$$
$$\frac{5 \cdot x}{5} = \frac{4350}{5}$$
$$x = 870$$

60% of 1450 reference books is 870 reference books.

33. whole is 5200; percent is 0.6; part is unknown.
part = $(0.006)(5200) = 31.2$
0.6% of 5200 acres is 31.2 acres.

34. whole is 1400; percent is 0.2; part is unknown.

$$\frac{x}{1400} = \frac{0.2}{100}$$
$$100 \cdot x = (0.2)(1400)$$
$$\frac{100 \cdot x}{100} = \frac{280}{100}$$
$$x = 2.8$$

0.2% of 1400 kilograms is 2.8 kilograms.

35. part is 105; percent is 14; whole is unknown.

$$\frac{105}{x} = \frac{14}{100} \quad \textbf{OR} \quad \frac{105}{x} = \frac{7}{50}$$
$$x \cdot 7 = 105 \cdot 50$$
$$\frac{x \cdot 7}{7} = \frac{5250}{7}$$
$$x = 750$$

105 crates is 14% of 750 crates.

36. part is 348; percent is 15; whole is unknown.

$$\frac{348}{x} = \frac{15}{100} \quad \textbf{OR} \quad \frac{348}{x} = \frac{3}{20}$$
$$x \cdot 3 = 348 \cdot 20$$
$$\frac{x \cdot 3}{3} = \frac{6960}{3}$$
$$x = 2320$$

348 test tubes is 15% of 2320 test tubes.

37. part is 677.6; percent is 140; whole is unknown.

$$\frac{677.6}{x} = \frac{140}{100} \quad \textbf{OR} \quad \frac{677.6}{x} = \frac{7}{5}$$
$$x \cdot 7 = 677.6 \cdot 5$$
$$\frac{x \cdot 7}{7} = \frac{3388}{7}$$
$$x = 484$$

677.6 miles is 140% of 484 miles.

38. part is 425; percent is 2.5; whole is unknown.

$$\frac{425}{x} = \frac{2.5}{100}$$
$$(2.5)(x) = 425 \cdot 100$$
$$\frac{(2.5)(x)}{2.5} = \frac{42{,}500}{2.5}$$
$$x = 17{,}000$$

2.5% of 17,000 cases is 425 cases.

39. part is 649; whole is 1180; percent is unknown.

$$\frac{649}{1180} = \frac{x}{100}$$
$$1180 \cdot x = 649 \cdot 100$$
$$\frac{1180 \cdot x}{1180} = \frac{64{,}900}{1180}$$
$$x = 55$$

649 tulip bulbs is 55% of 1180 tulip bulbs.

40. part is 85; whole is 1620; percent is unknown.

$$\frac{85}{1620} = \frac{x}{100} \quad \textbf{OR} \quad \frac{17}{324} = \frac{x}{100}$$
$$324 \cdot x = 17 \cdot 100$$
$$\frac{324 \cdot x}{324} = \frac{1700}{324}$$
$$x \approx 5.2$$

85 dinner rolls is 5.2% (rounded) of 1620 dinner rolls.

41. part is 36; whole is 380; percent is unknown.

$$\frac{36}{380} = \frac{x}{100} \quad \textbf{OR} \quad \frac{9}{95} = \frac{x}{100}$$
$$95 \cdot x = 9 \cdot 100$$
$$\frac{95 \cdot x}{95} = \frac{900}{95}$$
$$x \approx 9.5$$

36 pairs is 9.5% (rounded) of 380 pairs.

42. part is 200; whole is 650; percent is unknown.

$$\frac{200}{650} = \frac{x}{100} \quad \textbf{OR} \quad \frac{4}{13} = \frac{x}{100}$$
$$13 \cdot x = 4 \cdot 100$$
$$\frac{13 \cdot x}{13} = \frac{400}{13}$$
$$x \approx 30.8$$

200 cans is 30.8% (rounded) of 650 cans.

43. Add 95% of 1.2 million to 1.2 million.
part is unknown; whole is 1.2 million;
percent is 95

$$\frac{x}{1{,}200{,}000}=\frac{95}{100} \quad \textbf{OR} \quad \frac{x}{1{,}200{,}000}=\frac{19}{20}$$
$$20\cdot x = 19\cdot 1{,}200{,}000$$
$$\frac{20\cdot x}{20}=\frac{22{,}800{,}000}{20}$$
$$x = 1{,}140{,}000$$

$1{,}200{,}000 + 1{,}140{,}000 = 2{,}340{,}000$
The average cost of 30 seconds of advertising during the Super Bowl was $2.3 million (rounded).

44. part = 1000; whole = 9600; percent is unknown.

$$\frac{1000}{9600}=\frac{x}{100} \quad \textbf{OR} \quad \frac{5}{48}=\frac{x}{100}$$
$$48\cdot x = 5\cdot 100$$
$$\frac{48\cdot x}{48}=\frac{500}{48}$$
$$x \approx 10.4$$

10.4% (rounded) of the birds are in danger of extinction.

45. *part* = *percent* • *whole*
$x = (0.32)(454) \qquad 32\% = 0.32$
$x = 145.28$

32% of $454 is $145.28.

46. *part* = *percent* • *whole*
$x = (1.55)(120) \qquad 155\% = 1.55$
$x = 186$

155% of 120 trucks is 186 trucks.

47. *part* = *percent* • *whole*
$0.128 = x\cdot 32$
$$\frac{0.128}{32}=\frac{x\cdot 32}{32}$$
$0.004 = x$
0.004 is 0.4%

0.128 ounces is 0.4% of 32 ounces.

48. *part* = *percent* • *whole*
$304.5 = x\cdot 174$
$$\frac{304.5}{174}=\frac{x\cdot 174}{174}$$
$1.75 = x$
1.75 is 175%

304.5 meters is 175% of 174 meters.

49. *part* = *percent* • *whole*
$33.6 = (0.28)(x)$
$$\frac{33.6}{0.28}=\frac{(0.28)(x)}{0.28}$$
$120 = x$

33.6 miles is 28% of 120 miles.

50. *part* = *percent* • *whole*
$92 = (0.16)(x)$
$$\frac{92}{0.16}=\frac{(0.16)(x)}{0.16}$$
$575 = x$

$92 is 16% of $575.

51. *sales tax* = *rate of tax* • *cost of item*
= (5%)($630)
= (0.05)($630)
= $31.50

The sales tax is $31.50 and the total cost is $630 + $31.50 = $661.50.

52. *sales tax* = *rate of tax* • *cost of item*
$\$58.50 = r\cdot \$780$
$$\frac{58.50}{780}=\frac{r\cdot 780}{780}$$
$0.075 = r$
0.075 is $7\frac{1}{2}\%$

The tax rate is $7\frac{1}{2}\%$ and the total cost is $780 + $58.50 = $838.50.

53. *commission* = *rate of commission* • *sales*
= (8%)($3450)
= (0.08)($3450)
= $276

The amount of the commission is $276.

54. *commission* = *rate of commission* • *sales*
$\$3265 = r\cdot \$65{,}300$
$$\frac{3265}{65{,}300}=\frac{r\cdot 65{,}300}{65{,}300}$$
$0.05 = r$
0.05 is 5%

The rate of commission is 5%.

55. *discount* = *rate of discount* • *original price*
= (30%)($112.50)
= (0.30)($112.50)
= $33.75

The amount of discount is $33.75 and the sale price is $78.75 ($112.50 – $33.75)

56. *discount* = *rate of discount* • *original price*

$$\$63 = r \cdot \$252$$
$$\frac{63}{252} = \frac{r \cdot 252}{252}$$
$$0.25 = r$$

0.25 is 25%

The rate of discount is 25% and the sale price is

$252 − $63 = $189.

57. $200 at 4% for 1 year

$$\begin{aligned} I &= p \cdot r \cdot t \\ &= (200)(0.04)(1) \\ &= 8 \end{aligned}$$

The interest is $8.

58. $1080 at 5% for $1\frac{1}{4}$ years

$$\begin{aligned} I &= p \cdot r \cdot t \\ &= (1080)(0.05)(1.25) \quad 1\tfrac{1}{4} = 1.25 \\ &= (54)(1.25) \\ &= 67.5 \end{aligned}$$

The interest is $67.50.

59. $400 at 7% for 3 months

$$\begin{aligned} I &= p \cdot r \cdot t \\ &= (400)(0.07)\left(\tfrac{3}{12}\right) \\ &= (28)\left(\tfrac{3}{12}\right) \\ &= 7 \end{aligned}$$

The interest is $7.

60. $1560 at $6\frac{1}{2}$% for 18 months

$$\begin{aligned} I &= p \cdot r \cdot t \\ &= (1560)(0.065)\left(\tfrac{18}{12}\right) \\ &= (101.4)\left(\tfrac{18}{12}\right) \\ &= 152.1 \end{aligned}$$

The interest is $152.10.

61. $750 at $5\frac{1}{2}$% for 2 years

$$\begin{aligned} I &= p \cdot r \cdot t \\ &= (750)(0.055)(2) \\ &= 82.5 \end{aligned}$$

The total amount due is
$750 + $82.50 = $832.50.

62. $1530 at 6% for 9 months

$$\begin{aligned} I &= p \cdot r \cdot t \\ &= (1530)(0.06)\left(\tfrac{9}{12}\right) \\ &= (91.8)\left(\tfrac{9}{12}\right) \\ &= 68.85 \end{aligned}$$

The total amount due is
$1530 + $68.85 = $1598.85.

63. $4000 at 3% for 10 years

$$\begin{aligned} \textit{compound amount} &= (\$4000)(1.3439) \\ &= \$5375.60 \\ \textit{interest} &= \$5375.60 - \$4000 \\ &= \$1375.60 \end{aligned}$$

64. $1870 at 4% for 4 years

4% column, row 4 of the table gives 1.1699.

$$\begin{aligned} \textit{compound amount} &= (\$1870)(1.1699) \\ &\approx \$2187.71 \\ \textit{interest} &= \$2187.71 - \$1870 \\ &= \$317.71 \end{aligned}$$

65. $3600 at 8% for 3 years

8% column, row 3 of the table gives 1.2597.

$$\begin{aligned} \textit{compound amount} &= (\$3600)(1.2597) \\ &= \$4534.92 \\ \textit{interest} &= \$4534.92 - \$3600 \\ &= \$934.92 \end{aligned}$$

66. $12,500 at $5\frac{1}{2}$% for 5 years

$5\frac{1}{2}$% column, row 5 of the table gives 1.3070.

$$\begin{aligned} \textit{compound amount} &= (\$12{,}500)(1.3070) \\ &= \$16{,}337.50 \\ \textit{interest} &= \$16{,}337.50 - \$12{,}500 \\ &= \$3837.50 \end{aligned}$$

67. **[6.3]** whole = 80; percent = 15

$$\frac{x}{80} = \frac{15}{100} \quad \textbf{OR} \quad \frac{x}{80} = \frac{3}{20}$$
$$x \cdot 20 = 80 \cdot 3$$
$$\frac{x \cdot 20}{20} = \frac{240}{20}$$
$$x = 12$$

The part is 12.

68. **[6.3]** part = 738; percent = 45

$$\frac{738}{x} = \frac{45}{100} \quad \textbf{OR} \quad \frac{738}{x} = \frac{9}{20}$$
$$x \cdot 9 = 738 \cdot 20$$
$$\frac{x \cdot 9}{9} = \frac{14{,}760}{9}$$
$$x = 1640$$

The whole is 1640.

69. **[6.5]** 12% of 194 meters

$$x = (0.12)(194)$$
$$x = 23.28$$

12% of 194 meters is 23.28 meters.

70. **[6.4]** part is 327, whole is 218; percent is unknown.

$$\frac{327}{218}=\frac{x}{100}$$
$$218\cdot x=327\cdot 100$$
$$\frac{218\cdot x}{218}=\frac{32{,}700}{218}$$
$$x=150$$
150 is 150%

327 cars is 150% of 218 cars.

71. **[6.5]** 0.6% of \$85

$$x=(0.006)(85)$$
$$x=0.51$$

0.6% of \$85 is \$0.51.

72. **[6.4]** part is 396; percent is 20; whole is unknown.

$$\frac{396}{x}=\frac{20}{100}\quad \textbf{OR}\quad \frac{396}{x}=\frac{1}{5}$$
$$x\cdot 1=396\cdot 5$$
$$x=1980$$

396 employees is 20% of 1980 employees.

73. **[6.4]** part is 76; whole is 190; percent is unknown.

$$\frac{76}{190}=\frac{x}{100}\quad \textbf{OR}\quad \frac{38}{95}=\frac{x}{100}$$
$$95\cdot x=38\cdot 100$$
$$\frac{95\cdot x}{95}=\frac{3800}{95}$$
$$x=40$$
40 is 40%

76 chickens is 40% of 190 chickens.

74. **[6.4]** part is 214.484; percent is 43; whole is unknown.

$$\frac{214.484}{x}=\frac{43}{100}$$
$$x\cdot 43=(214.484)(100)$$
$$\frac{x\cdot 43}{43}=\frac{21{,}448.4}{43}$$
$$x=498.8$$

214.484 liters is 43% of 498.8 liters.

75. **[6.1]** $55\%=0.55$
Drop the percent sign and move the decimal point two places to the left.

76. **[6.1]** $300\%=3$
Drop the percent sign and move the decimal point two places to the left.

77. **[6.1]** $5=500\%$
Attach two zeros so the decimal point can be moved two places to the right. Attach a percent sign.

78. **[6.1]** $4.71=471\%$
Move the decimal point two places to the right and attach a percent sign.

79. **[6.1]** $8.6\%=0.086$
Drop the percent sign and move the decimal point two places to the left.

80. **[6.1]** $0.621=62.1\%$
Move the decimal point two places to the right and attach a percent sign.

81. **[6.1]** $0.375\%=0.00375$
Drop the percent sign and move the decimal point two places to the left.

82. **[6.1]** $0.0006=0.06\%$
Move the decimal point two places to the right and attach a percent sign.

83. **[6.2]** $\frac{3}{4}=\frac{3\cdot 25}{4\cdot 25}=\frac{75}{100}=0.75=75\%$

84. **[6.2]** $42\%=\frac{42}{100}=\frac{42\div 2}{100\div 2}=\frac{21}{50}$

85. **[6.2]**

$$87.5\%=\frac{87.5\cdot 10}{100\cdot 10}=\frac{875}{1000}=\frac{875\div 125}{1000\div 125}=\frac{7}{8}$$

86. **[6.2]** $\frac{3}{8}=\frac{p}{100}$
$$8\cdot p=3\cdot 100$$
$$8\cdot p=300$$
$$\frac{8\cdot p}{8}=\frac{300}{8}$$
$$p=37.5\text{ or }37\tfrac{1}{2}$$
$$\frac{3}{8}=37.5\%\text{ or }37\tfrac{1}{2}\%$$

87. **[6.2]**

$$32\tfrac{1}{2}\%=\frac{32.5\cdot 10}{100\cdot 10}=\frac{325}{1000}=\frac{325\div 25}{1000\div 25}=\frac{13}{40}$$

88. **[6.2]** $\frac{3}{5}=\frac{p}{100}$
$$5\cdot p=3\cdot 100$$
$$5\cdot p=300$$
$$\frac{5\cdot p}{5}=\frac{300}{5}$$
$$p=60$$
$$\frac{3}{5}=60\%$$

89. **[6.2]**

$$0.25\%=\frac{0.25\cdot 100}{100\cdot 100}=\frac{25\div 25}{10{,}000\div 25}=\frac{1}{400}$$

90. **[6.2]** $3\frac{3}{4}=3\frac{3\cdot 25}{4\cdot 25}=3\frac{75}{100}=3.75=375\%$

91. **[6.7]** \$16,850 at $10\frac{1}{2}\%$ for 9 months

$$\begin{aligned} I &= p \cdot r \cdot t \\ &= (16{,}850)(10\tfrac{1}{2}\%)(\tfrac{9}{12}) \\ &= (16{,}850)(0.105)(0.75) \quad \frac{9}{12} = 0.75 \\ &= 1326.94 \text{ (rounded)} \end{aligned}$$

The interest is \$1326.94.

92. **[6.7]** \$14,750 at 8% for 18 months

$$\begin{aligned} I &= p \cdot r \cdot t \\ &= (14{,}750)(8\%)(\tfrac{18}{12}) \\ &= (14{,}750)(0.08)(1.5) \\ &= 1770 \end{aligned}$$

The interest is \$1770 and the total amount due is \$14,750 + \$1770 = \$16,520.

93. **[6.4]** part is 5.5; whole is 11.5 + 5.5 = 17; percent is unknown.

$$\begin{aligned} \frac{5.5}{17} &= \frac{x}{100} \\ 17 \cdot x &= (5.5)(100) \\ \frac{17 \cdot x}{17} &= \frac{550}{17} \\ x &\approx 32.4 \end{aligned}$$

The capacity of the freezer is 32.4% (rounded) of the total capacity of the refrigerator.

94. **[6.8]** **(a)** \$12,500 at 6% compounded annually for 4 years
100% + 6% = 106% = 1.06

(\$12,500)(1.06)(1.06)(1.06)(1.06) ≈ \$15,780.96
The compound amount is \$15,780.96.

(b) The amount of interest is

\$15,780.96 − \$12,500 = \$3280.96.

95. **[6.6]** Add the two sales.

\$125,000 + \$290,000 = \$415,000
amount of commission
= *rate of commission* • *amount of sales*
= $(1\frac{1}{2}\%)$ (\$415,000)
= (0.015)(\$415,000)
= \$6225

The commission that he earned was \$6225.

96. **[6.4]** increase = 520 − 481 = 39
Use the percent proportion.

$$\begin{aligned} \frac{39}{481} &= \frac{x}{100} \\ 481 \cdot x &= 39 \cdot 100 \\ \frac{481 \cdot x}{481} &= \frac{3900}{481} \\ x &\approx 8.1 \end{aligned}$$

The increase is 8.1% (rounded).

97. **[6.6]** *amount of discount*
= *rate of discount* • *original price*
= (18%)(\$958)
= (0.18)(\$958)
= \$172.44

The price is \$958 − \$172.44 = \$785.56.

sales tax = *rate of tax* • *cost of item*
= (8%)(\$785.56)
= (0.08)(\$785.56)
= \$62.84 (rounded)

The cost of the washer/dryer set is

\$785.56 + \$62.84 = \$848.40.

98. **[6.4]** part = 749; percent = 71; whole is unknown

$$\begin{aligned} \frac{749}{x} &= \frac{71}{100} \\ 71 \cdot x &= 749 \cdot 100 \\ \frac{71 \cdot x}{71} &= \frac{74{,}900}{71} \\ x &\approx 1055 \end{aligned}$$

There were 1055 people (rounded) in the survey.

99. **[6.5]** 25% + 30% + 8% + 20% = 83% and 100% − 83% = 17%, so the couple saves 17% of their earnings.
Stephen's yearly income is

\$2075 • 12 = \$24,900.

Their total yearly income is

\$24,900 + \$32,500 = \$57,400.

The amount saved is

17% of \$57,400 = (0.17)(\$57,400)
= \$9758.

100. **[6.4]** decrease = 32.8 − 28.5 = 4.3
Use the percent proportion.

$$\begin{aligned} \frac{4.3}{32.8} &= \frac{x}{100} \\ 32.8 \cdot x &= (4.3)(100) \\ \frac{32.8 \cdot x}{32.8} &= \frac{430}{32.8} \\ x &\approx 13.1 \end{aligned}$$

The percent of decrease is 13.1% (rounded).

Chapter 6 Test

1. $65\% = 0.65$
Drop the percent sign and move the decimal point two places to the left.

2. $0.8 = 80\%$
0 is attached so the decimal point can be moved two places to the right. Attach a percent sign.

3. $1.75 = 175\%$
Move the decimal point two places to the right and attach a percent sign.

4. $0.875 = 87.5\%$ or $87\frac{1}{2}\%$
Move the decimal point two places to the right and attach a percent sign.

5. $300\% = 3.00$ or 3
Drop the percent sign and move the decimal point two places to the left.

6. $0.05\% = 0.0005$
Drop the percent sign, attach two zeros on the left, and move the decimal point two places to the left.

7. $12.5\% = \frac{12.5 \cdot 10}{100 \cdot 10} = \frac{125 \div 125}{1000 \div 125} = \frac{1}{8}$

8. $0.25\% = \frac{0.25}{100} = \frac{(0.25)(4)}{(100)(4)} = \frac{1}{400}$

9. $\frac{3}{5} = \frac{3 \cdot 20}{5 \cdot 20} = \frac{60}{100} = 60\%$

10.
$$\frac{5}{8} = \frac{p}{100}$$
$$8 \cdot p = 5 \cdot 100$$
$$\frac{8 \cdot p}{8} = \frac{500}{8}$$
$$p = 62.5$$
$$\frac{5}{8} = 62.5\% \text{ or } 62\frac{1}{2}\%$$

11. $2\frac{1}{2} = 2\frac{1 \cdot 50}{2 \cdot 50} = 2\frac{50}{100} = 2.50 = 250\%$

12. part is 32; percent is 4; whole is unknown
Use the percent proportion.

$$\frac{32}{x} = \frac{4}{100} \quad \textbf{OR} \quad \frac{32}{x} = \frac{1}{25}$$
$$x \cdot 1 = 32 \cdot 25$$
$$x = 800$$

32 sacks is 4% of 800 sacks.

13. part is 680; whole is 3400; percent is unknown.
Use the percent proportion.

$$\frac{680}{3400} = \frac{x}{100} \quad \textbf{OR} \quad \frac{1}{5} = \frac{x}{100}$$
$$5 \cdot x = 1 \cdot 100$$
$$\frac{5 \cdot x}{5} = \frac{100}{5}$$
$$x = 20$$

\$680 is 20% of \$3400.

14. part is 12,025; percent is 65; whole is unknown.
Use the percent proportion.

$$\frac{12{,}025}{x} = \frac{65}{100} \quad \textbf{OR} \quad \frac{12{,}025}{x} = \frac{13}{20}$$
$$x \cdot 13 = 12{,}025 \cdot 20$$
$$\frac{x \cdot 13}{13} = \frac{240{,}500}{13}$$
$$x = 18{,}500$$

The total down payment needed is \$18,500.

15. *sales tax* = *rate of tax* • *cost of item*
$$= (6\tfrac{1}{2}\%)(\$3240)$$
$$= (0.065)(\$3240)$$
$$= \$210.60$$

The total cost is

\$3240 + \$210.60 = \$3450.60.

16. *commission* = *rate of commission* • *sales*
$$\$628 = r \cdot \$7850$$
$$\frac{628}{7850} = \frac{r \cdot 7850}{7850}$$
$$0.08 = r$$
0.08 is 8%

The rate of commission paid is 8%.

17. decrease = 5760 − 4320 = 1440

$$\frac{1440}{5760} = \frac{x}{100} \quad \textbf{OR} \quad \frac{1}{4} = \frac{x}{100}$$
$$4 \cdot x = 1 \cdot 100$$
$$\frac{4 \cdot x}{4} = \frac{100}{4}$$
$$x = 25$$

The percent decrease is 25%.

18. A possible answer is:
Part is the increase in salary.
Whole is the last year's salary.
Percent of increase is unknown.

$$\frac{\text{amount of increase}}{\text{last year's salary}} = \frac{p}{100}$$

19. The interest formula is $I = p \cdot r \cdot t$.
If time is in months, it is expressed as a fraction with 12 as the denominator.

$$I = p \cdot r \cdot t$$
$$I = (\$1000)(0.05)\left(\tfrac{9}{12}\right) = \$37.50$$
$$I = (\$1000)(0.05)(2.5) = \$125$$

20. *amount of discount*
$= \textit{rate of discount} \cdot \textit{original price}$
$= (12\%)(\$96)$
$= (0.12)(\$96)$
$= \$11.52$

The amount of discount is \$11.52. The sale price is

$$\$96 - \$11.52 = \$84.48.$$

21. *amount of discount* $= (32.5\%)(\$280)$
$= (0.325)(\$280)$
$= \$91$

The amount of discount is \$91. The sale price is \$280 − \$91 = \$189.

22. \$4200 at 6% for $1\frac{1}{2}$ years

$$I = p \cdot r \cdot t$$
$$= (4200)(0.06)(1.5)$$
$$= 378$$

The interest is \$378.

23. \$6400 at 5% for 9 months

$$I = p \cdot r \cdot t$$
$$= (6400)(0.05)\left(\tfrac{9}{12}\right)$$
$$= (320)(0.75)$$
$$= 240$$

The interest is \$240.

24. \$5300 at 9% for 9 months

$$I = p \cdot r \cdot t$$
$$= (5300)(0.09)\left(\tfrac{9}{12}\right) \qquad \frac{9}{12} = \frac{3}{4} = 0.75$$
$$= (477)(0.75)$$
$$= 357.75$$

The interest is \$357.75 and the total amount due is \$5300 + \$357.75 = \$5657.75.

25. **(a)** \$4000 at 6% for 2 years
6% column, row 2 of the table gives 1.1236.

$$(\$4000)(1.1236) = \$4494.40$$

\$9494.40 (\$5000 + \$4494.40) at 6% for 2 years
6% column, row 2 of the table gives 1.1236.

$$(\$9494.40)(1.1236) \approx \$10{,}667.91$$

Rounded to the nearest dollar, the total amount after four years is \$10,668.

(b) The amount of interest earned is

$$\$10{,}668 - \$4000 - \$5000 = \$1668.$$

Cumulative Review Exercises (Chapters 1–6)

1. *Estimate:*
$$\begin{array}{r} 6000 \\ 90 \\ +\,700 \\ \hline 6790 \end{array}$$
Exact:
$$\begin{array}{r} 5608 \\ 94 \\ +\,739 \\ \hline 6441 \end{array}$$

2. *Estimate:*
$$\begin{array}{r} 1 \\ 50 \\ +\,8 \\ \hline 59 \end{array}$$
Exact:
$$\begin{array}{r} 0.560 \\ 49.614 \\ +\,8.400 \\ \hline 58.574 \end{array}$$

3. *Estimate:*
$$\begin{array}{r} 80{,}000 \\ -\,50{,}000 \\ \hline 30{,}000 \end{array}$$
Exact:
$$\begin{array}{r} 75{,}078 \\ -\,46{,}090 \\ \hline 28{,}988 \end{array}$$

4. *Estimate:*
$$\begin{array}{r} 8 \\ -\,4 \\ \hline 4 \end{array}$$
Exact:
$$\begin{array}{r} 7.8000 \\ -\,3.5029 \\ \hline 4.2971 \end{array}$$

5. *Estimate:*
$$\begin{array}{r} 7000 \\ \times\,700 \\ \hline 4{,}900{,}000 \end{array}$$
Exact:
$$\begin{array}{r} 6538 \\ \times\,708 \\ \hline 52\,304 \\ 4\,576\,60 \\ \hline 4{,}628{,}904 \end{array}$$

6. *Estimate:*
$$\begin{array}{r} 70 \\ \times\,9 \\ \hline 630 \end{array}$$
Exact:
$$\begin{array}{rl} 65.3 & \leftarrow \textit{1 decimal place} \\ \times\,8.7 & \leftarrow \textit{1 decimal place} \\ \hline 45\,71 & \\ 522\,4 & \\ \hline 568.11 & \leftarrow \textit{2 decimal places} \end{array}$$

7. *Estimate:*

$$40\overline{)40{,}000} = 1\,000$$

Exact:

$$43\overline{)38{,}786} = 902$$

387, 8, 0, 86, 86, 0

8. *Estimate:*

$$8\overline{)2000} = 250$$

Exact:

$$7.6_\wedge\overline{)2432.0_\wedge} = 320.$$

228, 152, 152, 00, 00, 0

9. *Estimate:*

$$1\overline{)7} = 7$$

Exact:

$$0.8_\wedge\overline{)6.7_\wedge 60} = 8.45$$

64, 36, 32, 40, 40, 0

10. $6^2 - 3(6)$
$= 36 - 3(6)$ *Exponent*
$= 36 - 18$ *Multiply*
$= 18$ *Subtract*

11. $\sqrt{49} + 5 \cdot 4 - 8$
$= 7 + 5 \cdot 4 - 8$ *Square root*
$= 7 + 20 - 8$ *Multiply*
$= 27 - 8$ *Add*
$= 19$ *Subtract*

12. $9 + 6 \div 3 + 7(4)$
$= 9 + 2 + 7(4)$ *Divide*
$= 9 + 2 + 28$ *Multiply*
$= 11 + 28 = 39$ *Add*

13. 4677 rounded to the nearest ten: 46$\underline{7}$7
Next digit is 5 or more. Tens place changes $(7 + 1 = 8)$. All digits to the right of the underlined place change to 0. **4680**

14. 7,583,281 rounded to the nearest hundred thousand: 7,$\underline{5}$83,281
Next digit is 5 or more. Hundred thousands place changes $(5 + 1 = 6)$. All digits to the right of the underlined place change to 0. **7,600,000**

15. $513.499 rounded to the nearest dollar:
Draw a cut-off line: $513|.499
First digit cut is less than 5 so the part you keep stays the same. **$513**

16. $362.735 rounded to the nearest cent:
Draw a cut-off line: $362.73|5
First digit cut is 5 or more, so the cents place changes $(3 + 1 = 4)$. **$362.74**

17. $\frac{3}{4} + \frac{5}{8} = \frac{6}{8} + \frac{5}{8} = \frac{6+5}{8} = \frac{11}{8} = 1\frac{3}{8}$

18. $\frac{1}{3} + \frac{3}{4} = \frac{4}{12} + \frac{9}{12} = \frac{4+9}{12} = \frac{13}{12} = 1\frac{1}{12}$

19.
$$\begin{array}{rcr} 5\frac{3}{4} & = & 5\frac{6}{8} \\ +\,7\frac{5}{8} & = & 7\frac{5}{8} \\ \hline & & 12\frac{11}{8} \end{array}$$

$12\frac{11}{8} = 12 + \frac{8}{8} + \frac{3}{8} = 13\frac{3}{8}$

20. $\frac{7}{8} - \frac{3}{4} = \frac{7}{8} - \frac{6}{8}$
$= \frac{7-6}{8}$
$= \frac{1}{8}$

21.
$$\begin{array}{rcrcr} 8\frac{3}{8} & = & 8\frac{3}{8} & = & 7\frac{11}{8} \\ -\,4\frac{1}{2} & = & 4\frac{4}{8} & = & 4\frac{4}{8} \\ \hline & & & & 3\frac{7}{8} \end{array}$$

22.
$$\begin{array}{rcrcr} 26\frac{1}{3} & = & 26\frac{5}{15} & = & 25\frac{20}{15} \\ -\,17\frac{4}{5} & = & 17\frac{12}{15} & = & 17\frac{12}{15} \\ \hline & & & & 8\frac{8}{15} \end{array}$$

23. $\left(\frac{7}{8}\right)\left(\frac{2}{3}\right) = \frac{7 \cdot \overset{1}{\cancel{2}}}{\underset{4}{\cancel{8}} \cdot 3} = \frac{7 \cdot 1}{4 \cdot 3} = \frac{7}{12}$

24. $7\frac{3}{4} \cdot 3\frac{3}{8} = \frac{31}{4} \cdot \frac{27}{8} = \frac{31 \cdot 27}{4 \cdot 8} = \frac{837}{32} = 26\frac{5}{32}$

25. $36 \cdot \frac{4}{5} = \frac{36}{1} \cdot \frac{4}{5} = \frac{36 \cdot 4}{1 \cdot 5} = \frac{144}{5} = 28\frac{4}{5}$

26. $\frac{5}{8} \div \frac{5}{7} = \frac{\overset{1}{\cancel{5}}}{8} \cdot \frac{7}{\underset{1}{\cancel{5}}} = \frac{7}{8}$

27. $12 \div \frac{3}{4} = \frac{12}{1} \div \frac{3}{4} = \frac{\overset{4}{\cancel{12}}}{1} \cdot \frac{4}{\underset{1}{\cancel{3}}} = \frac{16}{1} = 16$

28. $2\frac{3}{4} \div 7\frac{1}{2} = \frac{11}{4} \div \frac{15}{2} = \frac{11}{\underset{2}{\cancel{4}}} \cdot \frac{\overset{1}{\cancel{2}}}{15} = \frac{11 \cdot 1}{2 \cdot 15} = \frac{11}{30}$

29. $\frac{5}{8} \underline{\quad} \frac{2}{3}$

$\frac{5}{8} = \frac{15}{24}$ and $\frac{2}{3} = \frac{16}{24}$

$\frac{15}{24} < \frac{16}{24}$, so $\frac{5}{8} \boxed{<} \frac{2}{3}$.

30. $\frac{8}{15} \underline{\quad} \frac{11}{20}$

$\frac{8}{15} = \frac{32}{60}$ and $\frac{11}{20} = \frac{33}{60}$

$\frac{32}{60} < \frac{33}{60}$, so $\frac{8}{15} \boxed{<} \frac{11}{20}$.

31. $\frac{2}{3} \underline{\quad} \frac{7}{12}$

$\frac{2}{3} = \frac{8}{12}$

$\frac{8}{12} > \frac{7}{12}$, so $\frac{2}{3} \boxed{>} \frac{7}{12}$.

32. $\frac{2}{3}\left(\frac{7}{8} - \frac{1}{2}\right) = \frac{2}{3}\left(\frac{7}{8} - \frac{4}{8}\right)$

$= \frac{\overset{1}{\cancel{2}}}{\underset{1}{\cancel{3}}} \cdot \frac{\overset{1}{\cancel{3}}}{\underset{4}{\cancel{8}}}$

$= \frac{1}{4}$

33. $\frac{7}{8} \div \left(\frac{3}{4} + \frac{1}{8}\right) = \frac{7}{8} \div \left(\frac{6}{8} + \frac{1}{8}\right)$

$= \frac{7}{8} \div \frac{7}{8}$

$= \frac{\overset{1}{\cancel{7}}}{\underset{1}{\cancel{8}}} \cdot \frac{\overset{1}{\cancel{8}}}{\underset{1}{\cancel{7}}}$

$= 1$

34. $\left(\frac{5}{6} - \frac{5}{12}\right) - \left(\frac{1}{2}\right)^2 \cdot \frac{2}{3}$

$= \left(\frac{10}{12} - \frac{5}{12}\right) - \left(\frac{1}{2}\right)^2 \cdot \frac{2}{3}$

$= \frac{5}{12} - \left(\frac{1}{2}\right)^2 \cdot \frac{2}{3}$

$= \frac{5}{12} - \frac{1}{4} \cdot \frac{2}{3}$

$= \frac{5}{12} - \frac{2}{12} = \frac{3}{12} = \frac{1}{4}$

35. $\frac{3}{4} = 0.75$

$$\begin{array}{r} 0.75 \\ 4\overline{)3.00} \\ \underline{2\,8} \\ 20 \\ \underline{20} \\ 0 \end{array}$$

36. $\frac{3}{8} = 0.375$

$$\begin{array}{r} 0.375 \\ 8\overline{)3.000} \\ \underline{2\,4} \\ 60 \\ \underline{56} \\ 40 \\ \underline{40} \\ 0 \end{array}$$

37. $\frac{7}{12} = 0.583$ (rounded)

$$\begin{array}{r} 0.5833 \\ 12\overline{)7.0000} \\ \underline{6\,0} \\ 1\,00 \\ \underline{96} \\ 40 \\ \underline{36} \\ 40 \\ \underline{36} \\ 4 \end{array}$$

38. $\frac{11}{20} = 0.55$

$$\begin{array}{r} 0.55 \\ 20\overline{)11.00} \\ \underline{10\,0} \\ 1\,00 \\ \underline{1\,00} \\ 0 \end{array}$$

39. 2 hours to 40 minutes

$\frac{\text{2 hours}}{\text{40 minutes}} = \frac{\text{120 minutes}}{\text{40 minutes}}$

$= \frac{120}{40} = \frac{3}{1}$

40. 36 life jackets to 27 people

$\frac{36}{27} = \frac{36 \div 9}{27 \div 9} = \frac{4}{3}$

41. $1\frac{5}{8}$ to 13

$$\frac{1\frac{5}{8}}{13} = \frac{\frac{13}{8}}{\frac{13}{1}} = \frac{13}{8} \div \frac{13}{1} = \frac{\overset{1}{\cancel{13}}}{8} \cdot \frac{1}{\underset{1}{\cancel{13}}} = \frac{1}{8}$$

42. $\frac{5}{8} = \frac{45}{72}$

Cross products: $5 \cdot 72 = 360$; $8 \cdot 45 = 360$

The cross products are *equal,* so the proportion is *true.*

43. $\frac{64}{144} = \frac{48}{108}$

Cross products: $64 \cdot 108 = 6912$; $144 \cdot 48 = 6912$

The cross products are *equal,* so the proportion is *true.*

44.
$$\frac{1}{5} = \frac{x}{30}$$
$$5 \cdot x = 1 \cdot 30$$
$$\frac{5 \cdot x}{5} = \frac{30}{5}$$
$$x = 6$$

45.
$$\frac{224}{32} = \frac{28}{x}$$
$$\frac{28}{4} = \frac{28}{x} \quad \textit{Lowest terms}$$

Since the numerators are equal, the denominators must be equal, so $x = 4$.

46.
$$\frac{8}{x} = \frac{72}{144} \quad \textbf{OR} \quad \frac{8}{x} = \frac{1}{2}$$
$$8 \cdot 2 = x \cdot 1$$
$$16 = x$$

47.
$$\frac{x}{120} = \frac{7.5}{30}$$
$$30 \cdot x = 120(7.5)$$
$$\frac{30 \cdot x}{30} = \frac{900}{30}$$
$$x = 30$$

48. $78\% = 0.78$
Drop the percent sign and move the decimal point two places to the left.

49. $3\% = 0.03$
Drop the percent sign, attach one zero on the left, and move the decimal point two places to the left.

50. $200\% = 2.00$ or 2
Drop the percent sign and move the decimal point two places to the left.

51. $0.5\% = 0.005$
Drop the percent sign, attach two zeros on the left, and move the decimal point two places to the left.

52. $0.87 = 87\%$
Move the decimal point two places to the right and attach a percent sign.

53. $3.8 = 380\%$
0 is attached so the decimal point can be moved two places to the right. Attach a percent sign.

54. $0.023 = 2.3\%$
Move the decimal point two places to the right and attach a percent sign.

55. $8\% = 0.08 = \frac{8}{100} = \frac{8 \div 4}{100 \div 4} = \frac{2}{25}$

56. $62.5\% = 0.625 = \frac{625}{1000} = \frac{625 \div 125}{1000 \div 125} = \frac{5}{8}$

57. $175\% = 1.75 = 1\frac{75}{100} = 1\frac{75 \div 25}{100 \div 25} = 1\frac{3}{4}$

58.
$$\frac{7}{8} = \frac{x}{100}$$
$$8 \cdot x = 7 \cdot 100$$
$$\frac{8 \cdot x}{8} = \frac{700}{8}$$
$$x = 87.5$$
$$\frac{7}{8} = 87.5\% \text{ or } 87\frac{1}{2}\%$$

59. $\frac{3}{10} = \frac{3 \cdot 10}{10 \cdot 10} = \frac{30}{100} = 30\%$

60. $4\frac{1}{5} = 4\frac{1 \cdot 2}{5 \cdot 2} = 4\frac{2}{10} = 4.2 = 420\%$

61. 65% of 3280 DVDs

$x = (0.65)(3280)$
$x = 2132$ DVDs

62. whole is 720; percent is $5\frac{1}{2}$; part is unknown
Use the percent proportion.

$$\frac{x}{720} = \frac{5\frac{1}{2}}{100}$$
$$x \cdot 100 = (720)(5.5)$$
$$\frac{x \cdot 100}{100} = \frac{3960}{100}$$
$$x = 39.60$$

$5\frac{1}{2}\%$ of \$720 is \$39.60.

63. part is 72; percent is 40; whole is unknown

$$\frac{72}{x} = \frac{40}{100} \quad \textbf{OR} \quad \frac{72}{x} = \frac{2}{5}$$
$$x \cdot 2 = 72 \cdot 5$$
$$\frac{x \cdot 2}{2} = \frac{360}{2}$$
$$x = 180$$

72 tires is 40% of 180 tires.

64. part is 76.5; percent is $4\frac{1}{2}$; whole is unknown

$$\frac{76.5}{x} = \frac{4\frac{1}{2}}{100}$$
$$(4.5)(x) = 76.5 \cdot 100$$
$$\frac{(4.5)(x)}{4.5} = \frac{7650}{4.5}$$
$$x = 1700$$

$4\frac{1}{2}$% of 1700 miles is 76.5 miles.

65. part is 328; whole is 656; percent is unknown

$$\frac{328}{656} = \frac{x}{100} \text{ OR } \frac{1}{2} = \frac{x}{100}$$
$$2 \cdot x = 1 \cdot 100$$
$$\frac{2 \cdot x}{2} = \frac{100}{2}$$
$$x = 50$$

50% of 656 books is 328 books.

66. part is 72; whole is 180; percent is unknown

$$\frac{72}{180} = \frac{x}{100} \text{ OR } \frac{2}{5} = \frac{x}{100}$$
$$5 \cdot x = 2 \cdot 100$$
$$\frac{5 \cdot x}{5} = \frac{200}{5}$$
$$x = 40$$

72 hours is 40% of 180 hours.

67. *sales tax = rate of tax • cost of item*

$$= (4\%)(\$8.50)$$
$$= (0.04)(\$8.50)$$
$$= \$0.34$$

The sales tax is \$0.34 and the total cost is

$$\$8.50 + \$0.34 = \$8.84.$$

68. *sales tax = rate of tax • cost of item*

$$\$29.90 = r \cdot \$460$$
$$\frac{29.90}{460} = \frac{r \cdot 460}{460}$$
$$0.065 = r$$

0.065 is 6.5% or $6\frac{1}{2}$%

The tax rate is 6.5% or $6\frac{1}{2}$% and the total cost is

$$\$460 + \$29.90 = \$489.90.$$

69. *commission = rate of commission • sales*

$$= (8\%)(\$12{,}538)$$
$$= (0.08)(\$12{,}538)$$
$$= \$1003.04$$

The amount of commission is \$1003.04.

70. *commission = rate of commission • sales*

$$\$5632.50 = r \cdot \$225{,}300$$
$$\frac{5632.5}{225{,}300} = \frac{r \cdot 225{,}300}{225{,}300}$$
$$0.025 = r$$

0.025 is 2.5%

The rate of commission is 2.5% or $2\frac{1}{2}$%.

71. *discount = rate of discount • original price*

$$= (45\%)(\$456)$$
$$= (0.45)(\$456)$$
$$= \$205.20$$

The amount of discount is \$205.20 and the sale price is \$456 – \$205.20 = \$250.80.

72. *discount = rate of discount • original price*

$$\$162.75 = r \cdot \$1085$$
$$\frac{162.75}{1085} = \frac{r \cdot 1085}{1085}$$
$$0.15 = r$$

0.15 is 15%

The rate of discount is 15% and the sale price is

$$\$1085 - \$162.75 = \$922.25.$$

73. \$970 at 10% for $1\frac{1}{2}$ years

$$I = p \cdot r \cdot t$$
$$= (970)(10\%)(1\tfrac{1}{2})$$
$$= (970)(0.10)(1.5)$$
$$= 145.50$$

The total amount to be repaid is

$$\$970 + \$145.50 = \$1115.50.$$

74. \$22,250 at 7% for 6 months

$$I = p \cdot r \cdot t$$
$$= (22{,}250)(7\%)(\tfrac{6}{12})$$
$$= (22{,}250)(0.07)(0.5)$$
$$= 778.75$$

The total amount to be repaid is

$$\$22{,}250 + \$778.75 = \$23{,}028.75.$$

75. Set up a proportion.

$$\frac{11 \text{ patients}}{4 \text{ hours}} = \frac{x}{12 \text{ hours}}$$
$$4 \cdot x = 11 \cdot 12$$
$$\frac{4 \cdot x}{4} = \frac{11 \cdot \overset{3}{\cancel{12}}}{\underset{1}{\cancel{4}}}$$
$$x = 33$$

33 patients can be tested in 12 hours.

76. Set up a proportion.

$$\frac{12.5 \text{ ounces}}{5 \text{ gallons}} = \frac{x}{102 \text{ gallons}}$$
$$5 \cdot x = (12.5)(102)$$
$$\frac{5 \cdot x}{5} = \frac{1275}{5}$$
$$x = 255$$

255 ounces of Roundup is needed.

77. The increase in sales is

$$36{,}000 - 32{,}000 = 4000.$$

Use the percent proportion.

$$\frac{\text{increase}}{\text{original}} = \frac{x}{100}$$
$$\frac{4000}{32{,}000} = \frac{x}{100} \quad \textbf{OR} \quad \frac{1}{8} = \frac{x}{100}$$
$$8 \cdot x = 1 \cdot 100$$
$$\frac{8 \cdot x}{8} = \frac{100}{8}$$
$$x = 12.5$$

The percent of increase in sales in the northeast region is 12.5%.

78. The increase in sales is

$$66{,}300 - 65{,}000 = 1300.$$

$$\frac{\text{increase}}{\text{original}} = \frac{x}{100}$$
$$\frac{1300}{65{,}000} = \frac{x}{100} \quad \textbf{OR} \quad \frac{1}{50} = \frac{x}{100}$$
$$50 \cdot x = 1 \cdot 100$$
$$\frac{50 \cdot x}{50} = \frac{100}{50}$$
$$x = 2$$

The percent of increase in sales in the midwestern region is 2%.

79. The decrease in sales is

$$82{,}000 - 77{,}500 = 4500.$$

$$\frac{\text{decrease}}{\text{original}} = \frac{x}{100}$$
$$\frac{4500}{82{,}000} = \frac{x}{100} \quad \textbf{OR} \quad \frac{9}{164} = \frac{x}{100}$$
$$164 \cdot x = 9 \cdot 100$$
$$\frac{164 \cdot x}{164} = \frac{900}{164}$$
$$x \approx 5.5$$

The percent of decrease in sales in the southern region is 5.5% (rounded).

80. The decrease in sales is

$$54{,}000 - 49{,}600 = 4400.$$

$$\frac{\text{decrease}}{\text{original}} = \frac{x}{100}$$
$$\frac{4400}{54{,}000} = \frac{x}{100} \quad \textbf{OR} \quad \frac{11}{135} = \frac{x}{100}$$
$$135 \cdot x = 11 \cdot 100$$
$$\frac{135 \cdot x}{135} = \frac{1100}{135}$$
$$x \approx 8.1$$

The percent of decrease in sales in the western region is 8.1% (rounded).

81. part is 26,880; percent is 25

$$\frac{26{,}880}{x} = \frac{25}{100} \quad \textbf{OR} \quad \frac{26{,}880}{x} = \frac{1}{4}$$
$$x \cdot 1 = 26{,}880 \cdot 4$$
$$x = 107{,}520$$

Her total assets are $107,520.

82. **(a)** $15,000 at 5.5% compounded annually for 5 years

$100\% + 5.5\% = 105.5\% = 1.055$

$(\$15{,}000)(1.055)(1.055)(1.055)$
$(1.055)(1.055) \approx \$19{,}604.40$

The amount in the account is $19,604.40 (rounded).

(b) The amount of interest earned is

$\$19{,}604.40 - \$15{,}000 = \$4604.40.$

CHAPTER 7 MEASUREMENT

7.1 Problem Solving with English Measurement

7.1 Margin Exercises

1. **(a)** 1 c = 8 fl oz

   (b) 4 qt = 1 gal

   (c) 1 wk = 7 days

   (d) 3 ft = 1 yd

   (e) 1 ft = 12 in.

   (f) 16 oz = 1 lb

   (g) 1 T = 2000 lb

   (h) 60 min = 1 hr

   (i) 1 pt = 2 c

   (j) 24 hr = 1 day

   (k) 1 min = 60 sec

   (l) 1 qt = 2 pt

   (m) 5280 ft = 1 mi

2. **(a)** $5\frac{1}{2}$ ft to inches

   You are converting from a *larger* unit to a *smaller* unit, so multiply.

   Because 1 ft = 12 in., multiply by 12.

   $$5\frac{1}{2}\text{ ft} = 5\frac{1}{2}\cdot 12 = \frac{11}{\cancel{2}_1}\cdot\frac{\cancel{12}^6}{1} = \frac{66}{1} = 66\text{ in.}$$

   (b) 64 oz to pounds

   You are converting from a *smaller* unit to a *larger* unit, so divide.

   Because 16 oz = 1 lb, divide by 16.

   $$64\text{ oz} = \frac{64}{16} = \frac{64\div 16}{16\div 16} = \frac{4}{1} = 4\text{ lb}$$

   (c) 6 yd to feet

   You are converting from a *larger* unit to a *smaller* unit, so multiply.

   Because 1 yd = 3 ft, multiply by 3.

   $$6\text{ yd} = 6\cdot 3 = 18\text{ ft}$$

   (d) 2 T to pounds

   You are converting from a *larger* unit to a *smaller* unit, so multiply.

   Because 1 T = 2000 lb, multiply by 2000.

   $$2\text{ T} = 2\cdot 2000 = 4000\text{ lb}$$

   (e) 35 pt to quarts

   You are converting from a *smaller* unit to a *larger* unit, so divide.

   Because 1 qt = 2 pt, divide by 2.

   $$35\text{ pt} = \frac{35}{2} = 17\frac{1}{2}\text{ qt}$$

   (f) 20 min to hours

   You are converting from a *smaller* unit to a *larger* unit, so divide.

   Because 60 min = 1 hr, divide by 60.

   $$20\text{ min} = \frac{20}{60} = \frac{20\div 20}{60\div 20} = \frac{1}{3}\text{ hr}$$

   (g) 4 wk to days

   You are converting from a *larger* unit to a *smaller* unit, so multiply.

   Because 1 wk = 7 days, multiply by 7.

   $$4\text{ wk} = 4\cdot 7 = 28\text{ days}$$

3. **(a)** 36 in. to feet

   unit fraction $\left\}\frac{1\text{ ft}}{12\text{ in.}}\right.$

   $$36\text{ in.} = \frac{\cancel{36}^3\ \cancel{\text{in.}}}{1}\cdot\frac{1\text{ ft}}{\cancel{12}_1\ \cancel{\text{in.}}} = \frac{3\cdot 1\text{ ft}}{1} = 3\text{ ft}$$

   (b) 14 ft to inches

   unit fraction $\left\}\frac{12\text{ in.}}{1\text{ ft}}\right.$

   $$14\text{ ft} = \frac{14\ \cancel{\text{ft}}}{1}\cdot\frac{12\text{ in.}}{1\ \cancel{\text{ft}}} = \frac{14\cdot 12\text{ in.}}{1} = 168\text{ in.}$$

   (c) 60 in. to feet

   unit fraction $\left\}\frac{1\text{ft}}{12\text{ in.}}\right.$

   $$60\text{ in.} = \frac{\cancel{60}^5\ \cancel{\text{in.}}}{1}\cdot\frac{1\text{ ft}}{\cancel{12}_1\ \cancel{\text{in.}}} = \frac{5\cdot 1\text{ ft}}{1} = 5\text{ ft}$$

(d) 4 yd to feet

$$\left.\begin{array}{l}\text{unit}\\ \text{fraction}\end{array}\right\}\frac{3\text{ ft}}{1\text{ yd}}$$

$$4\text{ yd} = \frac{4\,\cancel{\text{yd}}}{1}\cdot\frac{3\text{ ft}}{1\,\cancel{\text{yd}}} = \frac{4\cdot 3\text{ ft}}{1} = 12\text{ ft}$$

(e) 39 ft to yards

$$\left.\begin{array}{l}\text{unit}\\ \text{fraction}\end{array}\right\}\frac{1\text{ yd}}{3\text{ ft}}$$

$$39\text{ ft} = \frac{\overset{13}{\cancel{39}}\,\cancel{\text{ft}}}{1}\cdot\frac{1\text{ yd}}{\underset{1}{\cancel{3}}\,\cancel{\text{ft}}} = \frac{13\cdot 1\text{ yd}}{1} = 13\text{ yd}$$

(f) 2 mi to feet

$$\left.\begin{array}{l}\text{unit}\\ \text{fraction}\end{array}\right\}\frac{5280\text{ ft}}{1\text{ mi}}$$

$$2\text{ mi} = \frac{2\,\cancel{\text{mi}}}{1}\cdot\frac{5280\text{ ft}}{1\,\cancel{\text{mi}}} = \frac{2\cdot 5280\text{ ft}}{1} = 10{,}560\text{ ft}$$

4. **(a)** 16 qt to gallons

$$\left.\begin{array}{l}\text{unit}\\ \text{fraction}\end{array}\right\}\frac{1\text{ gallon}}{4\text{ quarts}}$$

$$16\text{ qt} = \frac{\overset{4}{\cancel{16}}\,\cancel{\text{qt}}}{1}\cdot\frac{1\text{ gal}}{\underset{1}{\cancel{4}}\,\cancel{\text{qt}}} = \frac{4\cdot 1\text{ gal}}{1} = 4\text{ gal}$$

(b) 3 c to pints

$$\left.\begin{array}{l}\text{unit}\\ \text{fraction}\end{array}\right\}\frac{1\text{ pt}}{2\text{ c}}$$

$$3\text{ c} = \frac{3\,\cancel{\text{c}}}{1}\cdot\frac{1\text{ pt}}{2\,\cancel{\text{c}}} = \frac{3\cdot 1\text{ pt}}{2} = 1\frac{1}{2}\text{ pt or 1.5 pt}$$

(c) $3\frac{1}{2}$ T to pounds

$$\left.\begin{array}{l}\text{unit}\\ \text{fraction}\end{array}\right\}\frac{2000\text{ lb}}{1\text{ T}}$$

$$3\frac{1}{2}\text{ T} = \frac{3\frac{1}{2}\,\cancel{\text{T}}}{1}\cdot\frac{2000\text{ lb}}{1\,\cancel{\text{T}}} = \frac{7}{\underset{1}{\cancel{2}}}\cdot\frac{\overset{1000}{\cancel{2000}}}{1} = 7000\text{ lb}$$

(d) $1\frac{3}{4}$ lb to ounces

$$\left.\begin{array}{l}\text{unit}\\ \text{fraction}\end{array}\right\}\frac{16\text{ oz}}{1\text{ lb}}$$

$$1\frac{3}{4}\text{ lb} = \frac{1\frac{3}{4}\,\cancel{\text{lb}}}{1}\cdot\frac{16\text{ oz}}{1\,\cancel{\text{lb}}} = \frac{7}{\underset{1}{\cancel{4}}}\cdot\frac{\overset{4}{\cancel{16}}}{1} = 28\text{ oz}$$

(e) 4 oz to pounds

$$\left.\begin{array}{l}\text{unit}\\ \text{fraction}\end{array}\right\}\frac{1\text{ lb}}{16\text{ oz}}$$

$$4\text{ oz} = \frac{\overset{1}{\cancel{4}}\,\cancel{\text{oz}}}{1}\cdot\frac{1\text{ lb}}{\underset{4}{\cancel{16}}\,\cancel{\text{oz}}} = \frac{1}{4}\text{ lb or 0.25 lb}$$

5. **(a)** 4 T to ounces

$$\left.\begin{array}{l}\text{unit}\\ \text{fractions}\end{array}\right\}\frac{2000\text{ lb}}{1\text{ T}},\ \frac{16\text{ oz}}{1\text{ lb}}$$

$$\frac{4\,\cancel{\text{T}}}{1}\cdot\frac{2000\,\cancel{\text{lb}}}{1\,\cancel{\text{T}}}\cdot\frac{16\text{ oz}}{1\,\cancel{\text{lb}}} = 4\cdot 2000\cdot 16\text{ oz}$$

$$= 128{,}000\text{ oz}$$

(b) 3 mi to inches

$$\left.\begin{array}{l}\text{unit}\\ \text{fractions}\end{array}\right\}\frac{5280\text{ ft}}{1\text{ mi}},\ \frac{12\text{ in.}}{1\text{ ft}}$$

$$\frac{3\,\cancel{\text{mi}}}{1}\cdot\frac{5280\,\cancel{\text{ft}}}{1\,\cancel{\text{mi}}}\cdot\frac{12\text{ in.}}{1\,\cancel{\text{ft}}} = 3\cdot 5280\cdot 12\text{ in.}$$

$$= 190{,}080\text{ in.}$$

(c) 36 pt to gallons

$$\left.\begin{array}{l}\text{unit}\\ \text{fractions}\end{array}\right\}\frac{1\text{ qt}}{2\text{ pt}},\ \frac{1\text{ gal}}{4\text{ qt}}$$

$$\frac{36\,\cancel{\text{pt}}}{1}\cdot\frac{1\,\cancel{\text{qt}}}{2\,\cancel{\text{pt}}}\cdot\frac{1\text{ gal}}{4\,\cancel{\text{qt}}} = \frac{\overset{9}{\cancel{36}}}{1}\cdot\frac{1}{2}\cdot\frac{1}{\underset{1}{\cancel{4}}}\text{ gal} = \frac{9}{2}\text{ gal}$$

$$= 4\frac{1}{2}\text{ gal or 4.5 gal}$$

(d) 2 wk to minutes

$$\left.\begin{array}{l}\text{unit}\\ \text{fractions}\end{array}\right\}\frac{7\text{ days}}{1\text{ wk}},\ \frac{24\text{ hr}}{1\text{ day}},\ \frac{60\text{ min}}{1\text{ hr}}$$

$$\frac{2\,\cancel{\text{wk}}}{1}\cdot\frac{7\,\cancel{\text{days}}}{1\,\cancel{\text{wk}}}\cdot\frac{24\,\cancel{\text{hr}}}{1\,\cancel{\text{day}}}\cdot\frac{60\text{ min}}{1\,\cancel{\text{hr}}} = 2\cdot 7\cdot 24\cdot 60\text{ min}$$

$$= 20{,}160\text{ min}$$

6. **(a)** *Step 1*
The problem asks for the price per pound.

Step 2
Convert ounces to pounds. Then divide the cost by the pounds.

Step 3
To estimate, round \$3.29 to \$3. Then, there are 16 oz in a pound, so 12 oz is about 1 lb. Thus, \$3 ÷ 1 = \$3 per pound is our estimate.

Step 4

$$\frac{\overset{3}{\cancel{12}}\,\cancel{oz}}{1} \cdot \frac{1 \text{ lb}}{\underset{4}{\cancel{16}}\,\cancel{oz}} = \frac{3}{4} \text{ lb} = 0.75 \text{ lb}$$

Then divide: $\frac{\$3.29}{0.75 \text{ lb}} = 4.38\overline{6} \approx 4.39$

Step 5
The cheese costs $4.39 per pound (to the nearest cent).

Step 6
The answer, $4.39, is close to our estimate of $3.

(b) *Step 1*
The problem asks for the number of tons of furnishings they moved.

Step 2
First convert pounds to tons. Then multiply to find the tons for 5 houses.

Step 3
To estimate, round 11,000 lb to 10,000 lb. There are 2000 lb in a ton, so $10{,}000 \div 2000 = 5$ tons. Multiply by 5 to get 25 T as our estimate.

Step 4

$$\frac{\overset{11}{\cancel{11{,}000}}\,\cancel{lb}}{1} \cdot \frac{1 \text{ T}}{\underset{2}{\cancel{2000}}\,\cancel{lb}} = \frac{11}{2} \text{ T} = 5.5 \text{ T} = 5\frac{1}{2} \text{ T}$$

$5(5.5 \text{ T}) = 27.5 \text{ T}$

Step 5
The company moved 27.5 T or $27\frac{1}{2}$ T of furnishings.

Step 6
The answer, $27\frac{1}{2}$ T, is close to our estimate of 25 T.

7.1 Section Exercises

1. 1 yd = <u>3</u> ft

3. <u>8</u> fl oz = 1 c

5. 1 mi = <u>5280</u> ft

7. <u>2000</u> lb = 1 T

9. 1 min = <u>60</u> sec

11. 120 sec to minutes

$$120 \text{ sec} = \frac{\overset{2}{\cancel{120}}\,\cancel{sec}}{1} \cdot \frac{1 \text{ min}}{\underset{1}{\cancel{60}}\,\cancel{sec}} = 2 \text{ min}$$

13. 8 qt to gallons

$$8 \text{ qt} = \frac{\overset{2}{\cancel{8}}\,\cancel{qt}}{1} \cdot \frac{1 \text{ gal}}{\underset{1}{\cancel{4}}\,\cancel{qt}} = 2 \text{ gal}$$

15. 7 to 8 tons to pounds

$$\frac{7\,\cancel{T}}{1} \cdot \frac{2000 \text{ lb}}{1\,\cancel{T}} = 14{,}000 \text{ lb}$$

$$\frac{8\,\cancel{T}}{1} \cdot \frac{2000 \text{ lb}}{1\,\cancel{T}} = 16{,}000 \text{ lb}$$

An adult African elephant may weigh 14,000 to 16,000 lb.

17. 9 yd to feet

$$\frac{9\,\cancel{yd}}{1} \cdot \frac{3 \text{ ft}}{1\,\cancel{yd}} = 9 \cdot 3 \text{ ft} = 27 \text{ ft}$$

19. 7 lb to ounces

$$\frac{7\,\cancel{lb}}{1} \cdot \frac{16 \text{ oz}}{1\,\cancel{lb}} = 7 \cdot 16 \text{ oz} = 112 \text{ oz}$$

21. 5 qt to pints

$$\frac{5\,\cancel{qt}}{1} \cdot \frac{2 \text{ pt}}{1\,\cancel{qt}} = 5 \cdot 2 \text{ pt} = 10 \text{ pt}$$

23. 90 min to hours

$$\frac{90\,\cancel{min}}{1} \cdot \frac{1 \text{ hr}}{60\,\cancel{min}} = \frac{90}{60} \text{ hr} = 1\frac{1}{2} \text{ or } 1.5 \text{ hr}$$

25. 3 in. to feet

$$\frac{\overset{1}{\cancel{3}}\,\cancel{in.}}{1} \cdot \frac{1 \text{ ft}}{\underset{4}{\cancel{12}}\,\cancel{in.}} = \frac{1}{4} \text{ or } 0.25 \text{ ft}$$

27. 24 oz to pounds

$$\frac{24\,\cancel{oz}}{1} \cdot \frac{1 \text{ lb}}{16\,\cancel{oz}} = \frac{24}{16} \text{ lb} = 1\frac{1}{2} \text{ or } 1.5 \text{ lb}$$

29. 5 c to pints

$$\frac{5\,\cancel{c}}{1} \cdot \frac{1 \text{ pt}}{2\,\cancel{c}} = \frac{5}{2} \text{ pt} = 2\frac{1}{2} \text{ or } 2.5 \text{ pt}$$

31. $\frac{1}{2} \text{ ft} = \frac{\frac{1}{2}\,\cancel{ft}}{1} \cdot \frac{12 \text{ in.}}{1\,\cancel{ft}} = \frac{1}{2} \cdot 12 \text{ in.} = 6 \text{ in.}$

The ice will safely support a snowmobile or ATV or a person walking.

33. $2\frac{1}{2}$ T to lb

$$\frac{2\frac{1}{2}\,\cancel{T}}{1} \cdot \frac{2000 \text{ lb}}{1\,\cancel{T}} = \frac{5}{\underset{1}{\cancel{2}}} \cdot \frac{\overset{1000}{\cancel{2000}}}{1} \text{ lb} = 5000 \text{ lb}$$

35. $4\frac{1}{4}$ gal to quarts

$$\frac{4\frac{1}{4}\,\cancel{\text{gal}}}{1}\cdot\frac{4\text{ qt}}{1\,\cancel{\text{gal}}}=\frac{17}{\underset{1}{\cancel{4}}}\cdot\frac{\overset{1}{\cancel{4}}}{1}\text{ qt}=17\text{ qt}$$

37. $$\frac{1}{3}\text{ ft}=\frac{\frac{1}{3}\,\cancel{\text{ft}}}{1}\cdot\frac{12\text{ in.}}{1\,\cancel{\text{ft}}}=\frac{1}{\underset{1}{\cancel{3}}}\cdot\frac{\overset{4}{\cancel{12}}}{1}\text{ in.}=4\text{ in.}$$

Two-thirds of a foot would be twice as high; that is, 2(4 in.) = 8 in. The cactus could be 4 to 8 in. tall.

39. 6 yd to inches

$$\frac{6\,\cancel{\text{yd}}}{1}\cdot\frac{3\,\cancel{\text{ft}}}{1\,\cancel{\text{yd}}}\cdot\frac{12\text{ in.}}{1\,\cancel{\text{ft}}}=6\cdot 3\cdot 12\text{ in.}=216\text{ in.}$$

41. 112 c to quarts

$$\frac{\overset{\overset{28}{\cancel{56}}}{\cancel{112}}\,\cancel{\text{c}}}{1}\cdot\frac{1\,\cancel{\text{pt}}}{\underset{1}{\cancel{2}}\,\cancel{\text{c}}}\cdot\frac{1\text{ qt}}{\underset{1}{\cancel{2}}\,\cancel{\text{pt}}}=28\text{ qt}$$

43. 6 days to seconds

$$\frac{6\,\cancel{\text{days}}}{1}\cdot\frac{24\,\cancel{\text{hr}}}{1\,\cancel{\text{day}}}\cdot\frac{60\,\cancel{\text{min}}}{1\,\cancel{\text{hr}}}\cdot\frac{60\text{ sec}}{1\,\cancel{\text{min}}}$$
$$=6\cdot 24\cdot 60\cdot 60\text{ sec}$$
$$=518{,}400\text{ sec}$$

45. $1\frac{1}{2}$ T to ounces

$$\frac{1\frac{1}{2}\,\cancel{\text{T}}}{1}\cdot\frac{2000\,\cancel{\text{lb}}}{1\,\cancel{\text{T}}}\cdot\frac{16\text{ oz}}{1\,\cancel{\text{lb}}}=\frac{3}{\underset{1}{\cancel{2}}}\cdot\frac{\overset{1000}{\cancel{2000}}}{1}\cdot\frac{16}{1}\text{ oz}$$
$$=48{,}000\text{ oz}$$

47. **(a)** There is only one relationship that uses 1 to 16: 1 pound = 16 ounces

(b) 10 to 20 is the same as 1 to 2. There are 2 relationships like this: 1 pint = 2 cups and 1 quart = 2 pints.
10 quarts = 20 pints or 10 pints = 20 cups

(c) 120 to 2 is the same as 60 to 1. There are 2 relationships like this: 60 minutes = 1 hour and 60 seconds = 1 minute.
120 minutes = 2 hours or
120 seconds = 2 minutes

(d) 2 to 24 is the same as 1 to 12. Use 1 foot = 12 inches.
2 feet = 24 inches

(e) 6000 to 3 is the same as 2000 to 1. Use 2000 pounds = 1 ton.
6000 pounds = 3 tons

(f) 35 to 5 is the same as 7 to 1. Use 7 days = 1 week.
35 days = 5 weeks

49. $2\frac{3}{4}$ miles to inches

$$\frac{2\frac{3}{4}\,\cancel{\text{mi}}}{1}\cdot\frac{5280\,\cancel{\text{ft}}}{1\,\cancel{\text{mi}}}\cdot\frac{12\text{ in.}}{1\,\cancel{\text{ft}}}=\frac{11}{\underset{1}{\cancel{4}}}\cdot\frac{5280}{1}\cdot\frac{\overset{3}{\cancel{12}}}{1}\text{ in.}$$
$$=174{,}240\text{ in.}$$

51. $6\frac{1}{4}$ gal to fluid ounces

$$\frac{6\frac{1}{4}\,\cancel{\text{gal}}}{1}\cdot\frac{4\,\cancel{\text{qt}}}{1\,\cancel{\text{gal}}}\cdot\frac{32\text{ oz}}{1\,\cancel{\text{qt}}}=\frac{25}{4}\cdot 4\cdot 32\text{ oz}$$
$$=800\text{ fl oz}$$

53. 24,000 oz to tons

$$\frac{24{,}000\,\cancel{\text{oz}}}{1}\cdot\frac{1\,\cancel{\text{lb}}}{16\,\cancel{\text{oz}}}\cdot\frac{1\text{ T}}{2000\,\cancel{\text{lb}}}$$
$$=\frac{\overset{\overset{3}{\cancel{12}}}{\cancel{24{,}000}}}{1}\cdot\frac{1}{\underset{4}{\cancel{16}}}\cdot\frac{1}{\underset{1}{\cancel{2000}}}\text{ T}$$
$$=\frac{3}{4}\text{ or }0.75\text{ T}$$

For Exercises 55–61, the six problem-solving steps should be used, but are only shown for Exercise 55.

55. *Step 1*
The problem asks for the price per pound of strawberries.

Step 2
Convert ounces to pounds. Then divide the cost by the pounds.

Step 3
To estimate, round \$2.29 to \$2. Then, there are 16 oz in a pound, so 20 oz is a little more than 1 lb. Thus, $\$2 \div 1 = \$2$ per pound is our estimate.

Step 4

$$\frac{\overset{5}{\cancel{20}}\,\cancel{\text{oz}}}{1}\cdot\frac{1\text{ lb}}{\underset{4}{\cancel{16}}\,\cancel{\text{oz}}}=\frac{5}{4}\text{ lb}=1.25\text{ lb}$$

$$\frac{\$2.29}{1.25\text{ lb}}=1.832\approx 1.83$$

Step 5
The strawberries are \$1.83 per pound (to the nearest cent).

Step 6
The answer, \$1.83, is close to our estimate of \$2.

57. Find the total number of feet needed. Then convert feet to yards. Then multiply to find the cost.

$24 \cdot 2 = 48$ ft

$$\frac{\overset{16}{\cancel{48}}\ \cancel{ft}}{1} \cdot \frac{1 \text{ yd}}{\underset{1}{\cancel{3}}\ \cancel{ft}} = 16 \text{ yd}$$

$(16 \text{ yd})(\$8.75) = \$140$

It will cost \$140 to equip all the stations.

59. **(a)** Convert seconds per foot to seconds per mile.

$$\frac{1 \text{ sec}}{5 \text{ ft}} \cdot \frac{5280 \text{ ft}}{1 \text{ mi}} = \frac{\overset{1056}{\cancel{5280}}}{\underset{1}{\cancel{5}}} \text{ sec/mi} = 1056 \text{ sec/mi}$$

It would take the cockroach 1056 seconds to travel 1 mile.

(b) $$\frac{1056 \text{ sec}}{1 \text{ mi}} \cdot \frac{1 \text{ min}}{60 \text{ sec}} = \frac{\overset{88}{\cancel{1056}}}{\underset{5}{\cancel{60}}} \text{ min/mi}$$
$$= 17.6 \text{ min/mi}$$

It would take the cockroach 17.6 minutes to travel 1 mile.

61. **(a)** Find the total number of cups per week. Then convert cups to quarts.

$$\frac{2}{3} \cdot 15 \cdot 5 = \frac{2}{\underset{1}{\cancel{3}}} \cdot \frac{\overset{5}{\cancel{15}}}{1} \cdot \frac{5}{1} = 50 \text{ c}$$

$$\frac{\overset{25}{\cancel{50}}\ \cancel{c}}{1} \cdot \frac{1 \ \cancel{pt}}{\underset{1}{\cancel{2}}\ \cancel{c}} \cdot \frac{1 \text{ qt}}{2 \ \cancel{pt}} = \frac{25}{2} \text{ qt} = 12\frac{1}{2} \text{ qt}$$

The center needs $12\frac{1}{2}$ qt of milk per week.

(b) Convert quarts to gallons.

$$\frac{12\frac{1}{2}\ \cancel{qt}}{1} \cdot \frac{1 \text{ gal}}{4\ \cancel{qt}} = \frac{25}{2} \cdot \frac{1}{4} \text{ gal} = 3.125 \text{ gal}$$

The center should order 4 jugs, because you can't buy part of a jug.

63. **(a)** $$\frac{34{,}646{,}437}{24{,}907} \approx 1391$$

You would need to travel about 1391 times around Earth.

(b) $$\frac{34{,}646{,}437}{2786} \approx 12{,}436$$

You would need to drive about 12,436 times from Los Angeles to New York.

64. **(a)** $$\frac{34{,}646{,}437}{24{,}907} \approx \frac{35{,}000{,}000}{25{,}000}$$
$$= \frac{35{,}000}{25} = \frac{7000}{5} = 1400$$

The estimate is 1400 times around Earth.

(b) $$\frac{34{,}646{,}437}{2786} \approx \frac{35{,}000{,}000}{2800}$$
$$= \frac{350{,}000}{28} = \frac{50{,}000}{4} = 12{,}500$$

The estimate is 12,500 trips.

(c) They are very close and just as useful for making comparisons of such large numbers.

7.2 The Metric System—Length

7.2 Margin Exercises

1. The length of a baseball bat, the height of a doorknob from the floor, and a basketball player's arm length are each about 1 meter in length.

2. **(a)** The woman's height is 168 cm.

(b) The man's waist is 90 cm around.

(c) Louise ran the 100 m dash in the track meet.

(d) A postage stamp is 22 mm wide.

(e) Michael paddled his canoe 2 km down the river.

(f) The pencil lead is 1 mm thick.

(g) A stick of gum is 7 cm long.

(h) The highway speed limit is 90 km per hour.

(i) The classroom was 12 m long.

(j) A penny is about 18 mm across.

3. **(a)** 3.67 m to cm

unit fraction $\left.\right\}$ $\frac{100 \text{ cm}}{1 \text{ m}}$

$$\frac{3.67\ \cancel{m}}{1} \cdot \frac{100 \text{ cm}}{1\ \cancel{m}} = (3.67)(100) \text{ cm} = 367 \text{ cm}$$

(b) 92 cm to m

unit fraction $\left.\right\}$ $\frac{1 \text{ m}}{100 \text{ cm}}$

$$\frac{92\ \cancel{cm}}{1} \cdot \frac{1 \text{ m}}{100\ \cancel{cm}} = \frac{92}{100} \text{ m} = 0.92 \text{ m}$$

(c) 432.7 cm to m

unit fraction $\left.\right\}$ $\frac{1 \text{ m}}{100 \text{ cm}}$

$$\frac{432.7\ \cancel{cm}}{1} \cdot \frac{1 \text{ m}}{100\ \cancel{cm}} = \frac{432.7}{100} \text{ m} = 4.327 \text{ m}$$

(d) 65 mm to cm

unit fraction $\left.\right\}\dfrac{1\text{ cm}}{10\text{ mm}}$

$$\frac{65\text{ mm}}{1}\cdot\frac{1\text{ cm}}{10\text{ mm}}=\frac{65}{10}\text{ cm}=6.5\text{ cm}$$

(e) 0.9 m to mm

unit fraction $\left.\right\}\dfrac{1000\text{ mm}}{1\text{ m}}$

$$\frac{0.9\text{ m}}{1}\cdot\frac{1000\text{ mm}}{1\text{ m}}=(0.9)(1000)\text{ mm}=900\text{ mm}$$

(f) 2.5 cm to mm

unit fraction $\left.\right\}\dfrac{10\text{ mm}}{1\text{ cm}}$

$$\frac{2.5\text{ cm}}{1}\cdot\frac{10\text{ mm}}{1\text{ cm}}=(2.5)(10)\text{ mm}=25\text{ mm}$$

4. **(a)** $(43.5)(10)=435$
43.5∧ gives 435.

(b) $43.5\div 10=4.35$
4∧3.5 gives 4.35.

(c) $(28)(100)=2800$
28.00∧ gives 2800.

(d) $28\div 100=0.28$
∧28. gives 0.28.

(e) $(0.7)(1000)=700$
0.700∧ gives 700.

(f) $0.7\div 1000=0.0007$
∧000.7 gives 0.0007.

5. **(a)** 12.008 km to m
Count 3 places to the *right* on the conversion line.
12.008∧ km = 12,008 m

(b) 561.4 m to km
Count 3 places to the *left* on the conversion line.
∧561.4 m = 0.5614 km

(c) 20.7 cm to m
Count 2 places to the *left* on the conversion line.
∧20.7 cm = 0.207 m

(d) 20.7 cm to mm
Count 1 place to the *right* on the conversion line.
20.7∧ cm = 207 mm

(e) 4.66 m to cm
Count 2 places to the *right* on the conversion line.
4.66∧ m = 466 cm

(f) 85.6 mm to cm
Count 1 place to the *left* on the conversion line.
8∧5.6 mm = 8.56 cm

6. **(a)** 9 m to mm
Count 3 places to the *right* on the conversion line.
Three zeros are written in as placeholders.
9.000∧ m = 9000 mm

(b) 3 cm to m
Count 2 places to the *left* on the conversion line.
One zero is written in as a placeholder.
∧03. cm = 0.03 m

(c) 14.6 km to m
Count 3 places to the *right* on the conversion line.
Two zeros are written in as placeholders.
14.600∧ km = 14,600 m

(d) 5 mm to cm
Count 1 place to the *left* on the conversion line.
∧5. mm = 0.5 cm

(e) 70 m to km
Count 3 places to the *left* on the conversion line.
One zero is written in as a placeholder.
∧070. m = 0.07 km

(f) 0.8 m to cm
Count 2 places to the *right* on the conversion line.
One zero is written in as a placeholder.
0.80∧ m = 80 cm

7.2 Section Exercises

1. *Kilo* means 1000, so 1 km = 1000 m.

3. *Milli* means $\frac{1}{1000}$ or 0.001, so
1 mm = $\frac{1}{1000}$ or 0.001 m.

5. *Centi* means $\frac{1}{100}$ or 0.01, so 1 cm = $\frac{1}{100}$ or 0.01 m.

7. The width of your hand in centimeters •
Answers will vary; about 8 to 10 cm.

9. The width of your thumb in millimeters •
Answers will vary; about 20 to 25 mm.

11. The child was about 91 <u>cm</u> tall.

13. Ming-Na swam in the 200 <u>m</u> backstroke race.

15. Adriana drove 400 <u>km</u> on her vacation.

17. An aspirin tablet is 10 <u>mm</u> across.

19. A paper clip is about 3 <u>cm</u> long.

21. Dave's truck is 5 <u>m</u> long.

23. Some possible answers are: 35 mm film for cameras, track and field events, metric auto parts, and lead refills for mechanical pencils.

25. 7 m to cm

$$\frac{7\,\cancel{m}}{1} \cdot \frac{100\text{ cm}}{1\,\cancel{m}} = 7 \cdot 100\text{ cm} = 700\text{ cm}$$

27. 40 mm to m

$$\frac{40\,\cancel{mm}}{1} \cdot \frac{1\text{ m}}{1000\,\cancel{mm}} = \frac{40}{1000}\text{ m} = 0.040\text{ m or } 0.04\text{ m}$$

29. 9.4 km to m

$$\frac{9.4\,\cancel{km}}{1} \cdot \frac{1000\text{ m}}{1\,\cancel{km}} = (9.4)(1000)\text{ m} = 9400\text{ m}$$

31. 509 cm to m

$$\frac{509\,\cancel{cm}}{1} \cdot \frac{1\text{ m}}{100\,\cancel{cm}} = \frac{509}{100}\text{ m} = 5.09\text{ m}$$

33. 400 mm to cm
Count 1 place to the *left* on the conversion line.
40‸0. mm = 40.0 cm = 40 cm

35. 0.91 m to mm
Count 3 places to the *right* on the conversion line.
One zero is written in as a placeholder.
0.910‸ m = 910 mm

37. 82 cm to m
Count 2 places to the *left* on the conversion line.
‸82. cm = 0.82 m

0.82 m is less than 1 m, so 82 cm is **less than** 1 m.
The difference in length is 1 m − 0.82 m = 0.18 m or 100 cm − 82 cm = 18 cm.

39. 5 mm
1 mm
5 mm to centimeters
Count one place to the *left* on the conversion line.
‸5. mm = 0.5 cm
Similarly, 1 mm = 0.1 cm.

41. 18 m to km
Count 3 places to the *left* on the conversion line.
One zero is written in as a placeholder.
‸018. m = 0.018 km

The north fork of the Roe River is just under 0.018 km long.

43. 1.64 m to centimeters and millimeters
Count two (three) places to the *right* on the conversion line.
1.64‸ m = 164 cm;

1.640‸ m = 1640 mm

The median height for U.S. females who are 20 to 29 years old is about 164 cm, or equivalently, 1640 mm.

45. 5.6 mm to km

$$\frac{5.6\,\cancel{mm}}{1} \cdot \frac{1\,\cancel{m}}{1000\,\cancel{mm}} \cdot \frac{1\text{ km}}{1000\,\cancel{m}} = \frac{5.6}{1{,}000{,}000}\text{ km} = 0.0000056\text{ km}$$

7.3 The Metric System—Capacity and Weight (Mass)

7.3 Margin Exercises

1. The liter would be used to measure the amount of water in the bathtub, gasoline you buy for your car, and water in a pail.

2. **(a)** I bought 8 <u>L</u> of milk at the store.

(b) The nurse gave me 10 <u>mL</u> of cough syrup.

(c) This is a 100 <u>L</u> garbage can.

(d) It took 10 <u>L</u> of paint to cover the bedroom walls.

(e) My car's gas tank holds 50 <u>L</u>.

(f) I added 15 <u>mL</u> of oil to the pancake mix.

(g) The can of orange soda holds 350 <u>mL</u>.

(h) My friend gave me a 30 <u>mL</u> bottle of expensive perfume.

3. **(a)** 9 L to mL

$$\frac{9\,\cancel{L}}{1} \cdot \frac{1000\text{ mL}}{1\,\cancel{L}} = 9000\text{ mL}$$

(b) 0.75 L to mL
Count 3 places to the *right* on the (metric capacity) conversion line.
0.750‸ L = 750 mL

(c) 500 mL to L

$$\frac{500 \not{mL}}{1} \cdot \frac{1 \text{ L}}{1000 \not{mL}} = \frac{500}{1000} \text{ L} = 0.5 \text{ L}$$

(d) 5 mL to L
Count 3 places to the *left* on the (metric capacity) conversion line.
∧005. mL = 0.005 L

(e) 2.07 L to mL

$$\frac{2.07 \not{L}}{1} \cdot \frac{1000 \text{ mL}}{1 \not{L}} = 2070 \text{ mL}$$

(f) 3275 mL to L
Count 3 places to the *left* on the (metric capacity) conversion line.
3∧275. mL = 3.275 L

4. A small paperclip, one playing card from a deck of cards, and the check you wrote at the grocery store would each weigh about 1 gram.

5. **(a)** A thumbtack weights 800 mg.

(b) A teenager weighs 50 kg.

(c) This large cast-iron frying pan weighs 1 kg.

(d) Jerry's basketball weighed 600 g.

(e) Tamlyn takes a 500 mg calcium tablet every morning.

(f) On his diet, Greg can eat 90 g of meat for lunch.

(g) One strand of hair weighs 2 mg.

(h) One banana might weigh 150 g.

6. **(a)** 10 kg to g

$$\frac{10 \not{kg}}{1} \cdot \frac{1000 \text{ g}}{1 \not{kg}} = 10{,}000 \text{ g}$$

(b) 45 mg to g

$$\frac{45 \not{mg}}{1} \cdot \frac{1 \text{ g}}{1000 \not{mg}} = \frac{45}{1000} \text{ g} = 0.045 \text{ g}$$

(c) 6.3 kg to g
Count 3 places to the *right* on the [metric weight (mass)] conversion line.
6.300∧ kg = 6300 g

(d) 0.077 g to mg
Count 3 places to the *right* on the [metric weight (mass)] conversion line.
0.077∧ g = 77 mg

(e) 5630 g to kg
Count 3 places to the *left* on the [metric weight (mass)] conversion line.
5∧630. g = 5.63 kg

(f) 90 g to kg

$$\frac{90 \not{g}}{1} \cdot \frac{1 \text{ kg}}{1000 \not{g}} = \frac{9}{100} \text{ kg} = 0.09 \text{ kg}$$

7. **(a)** Gail bought a 4 L can of paint.
Use capacity units.

(b) The bag of chips weighed 450 g.
Use weight units.

(c) Give the child 5 mL of cough syrup.
Use capacity units.

(d) The width of the window is 55 cm.
Use length units.

(e) Akbar drives 18 km to work.
Use length units.

(f) Each computer weighs 5 kg.
Use weight units.

(g) A credit card is 55 mm wide.
Use length units.

7.3 Section Exercises

1. The glass held 250 mL of water. (Liquids are measured in mL or L.)

3. Dolores can make 10 L of soup in that pot. (Liquids are measured in mL or L.)

5. Our yellow Labrador dog grew up to weigh 40 kg. (Weight is measured in mg, g, or kg.)

7. Lori caught a small sunfish weighing 150 g. (Weight is measured in mg, g, or kg.)

9. Andre donated 500 mL of blood today. (Blood is a liquid, and liquids are measured in mL or L.)

11. The patient received a 250 mg tablet of medication each hour. (Weight is measured in mg, g, or kg.)

13. The gas can for the lawn mower holds 4 L. (Gasoline is a liquid, and liquids are measured in mL or L.)

15. Pam's backpack weighs 5 kg when it is full of books. (Weight is measured in mg, g, or kg.)

17. This is unreasonable (too much) since 4.1 liters would be about 4 quarts.

19. This is unreasonable (too much) since 5 kilograms of Epsom salts would be about 11 pounds with approximately a quart of water.

21. This is reasonable because 15 milliliters would be 3 teaspoons.

23. This is reasonable because 350 milligrams would be a little more than 1 tablet, which is about 325 milligrams.

25. Some capacity examples are 2 L bottles of soda and shampoo bottles marked in mL; weight examples are grams of fat listed on cereal boxes and vitamin doses in milligrams.

27. The unit for your answer (g) is in the numerator. The unit being changed (kg) is in denominator, so it will divide out. The unit fraction is $\frac{1000 \text{ g}}{1 \text{ kg}}$.

29. 15 L to mL
Count 3 places to the *right* on the conversion line.
$15.000_{\wedge}$ L = 15,000 mL

31. 3000 mL to L
Count 3 places to the *left* on the conversion line.
$3_{\wedge}000.$ mL = 3 L

33. 925 mL to L

$$\frac{925 \cancel{\text{mL}}}{1} \cdot \frac{1 \text{ L}}{1000 \cancel{\text{mL}}} = \frac{925}{1000} \text{ L} = 0.925 \text{ L}$$

35. 8 mL to L

$$\frac{8 \cancel{\text{mL}}}{1} \cdot \frac{1 \text{ L}}{1000 \cancel{\text{mL}}} = \frac{8}{1000} \text{ L} = 0.008 \text{ L}$$

37. 4.15 L to mL
Count 3 places to the *right* on the conversion line.
$4.150_{\wedge}$ L = 4150 mL

39. 8000 g to kg
Count 3 places to the *left* on the conversion line.
$8_{\wedge}000.$ g = 8 kg

41. 5.2 kg to g
Count 3 places to the *right* on the conversion line.
$5.200_{\wedge}$ kg = 5200 g

43. 0.85 g to mg
Count 3 places to the *right* on the conversion line.
$0.850_{\wedge}$ g = 850 mg

45. 30,000 mg to g
Count 3 places to the *left* on the conversion line.
$30_{\wedge}000.$ mg = 30 g

47. 598 mg to g
Count 3 places to the *left* on the conversion line.
$_{\wedge}598.$ mg = 0.598 g

49. 60 mL to L

$$\frac{60 \cancel{\text{mL}}}{1} \cdot \frac{1 \text{ L}}{1000 \cancel{\text{mL}}} = \frac{60}{1000} \text{ L} = 0.06 \text{ L}$$

51. 3 g to kg

$$\frac{3 \cancel{\text{g}}}{1} \cdot \frac{1 \text{ kg}}{1000 \cancel{\text{g}}} = \frac{3}{1000} \text{ kg} = 0.003 \text{ kg}$$

53. 0.99 L to mL

$$\frac{0.99 \cancel{\text{L}}}{1} \cdot \frac{1000 \text{ mL}}{1 \cancel{\text{L}}} = 990 \text{ mL}$$

55. The masking tape is 19 mm wide. (Length is measured in mm, cm, m.)

57. Buy a 60 mL jar of acrylic paint for art class. (Paint is a liquid and liquids are measured in mL or L.)

59. My waist measurement is 65 cm. (Length is measured in mm, cm, m.)

61. A single postage stamp weighs 90 mg. (Weight is measured in mg, g, or kg.)

63. Convert 300 mL to L.
Count 3 places to the *left* on the conversion line.
$_{\wedge}300.$ mL = 0.3 L

Each day 0.3 L of sweat is released.

65. Convert 1.34 kg to g.
Count 3 places to the *right* on the conversion line.
$1.340_{\wedge}$ kg = 1340 g

The average weight of a human brain is 1340 g.

67. Convert 900 mL to L.

$$\frac{900 \cancel{\text{mL}}}{1} \cdot \frac{1 \text{ L}}{1000 \cancel{\text{mL}}} = \frac{900}{1000} \text{ L} = 0.9 \text{ L}$$

On average, we breathe in and out roughly 0.9 L of air every 10 seconds.

69. Convert 3000 g to kg and 4000 g to kg.

$$\frac{3000 \cancel{\text{g}}}{1} \cdot \frac{1 \text{ kg}}{1000 \cancel{\text{g}}} = \frac{3000}{1000} \text{ kg} = 3 \text{ kg}$$

$$\frac{4000 \cancel{\text{g}}}{1} \cdot \frac{1 \text{ kg}}{1000 \cancel{\text{g}}} = \frac{4000}{1000} \text{ kg} = 4 \text{ kg}$$

A small adult cat weighs from 3 kg to 4 kg.

71. There are 1000 milligrams in a gram so 1005 mg is *greater* than 1 g. The difference in weight is 1005 mg − 1000 mg = 5 mg or 1.005 g − 1 g = 0.005 g.

73. Convert 1 kg to g.

$$\frac{1\ \cancel{kg}}{1} \cdot \frac{1000\ g}{1\ \cancel{kg}} = 1000\ g$$

Divide the 1000 g by the weight of 1 nickel.

$$\frac{1000\ \cancel{g}}{5\ \cancel{g}} = 200$$

There are 200 nickels in 1 kg of nickels.

75. **(a)** 1 Mm = 1,000,000 m

(b) 3.5 Mm to m

$$\frac{3.5\ \cancel{Mm}}{1} \cdot \frac{1{,}000{,}000\ m}{1\ \cancel{Mm}} = 3{,}500{,}000\ m$$

76. **(a)** 1 Gm = 1,000,000,000 m

(b) 2500 m to Gm

$$\frac{2500\ \cancel{m}}{1} \cdot \frac{1\ Gm}{1{,}000{,}000{,}000\ \cancel{m}} = 0.0000025\ Gm$$

77. **(a)** 1 Tm = 1,000,000,000,000 m

(b) $$\frac{1\ \cancel{Tm}}{1} \cdot \frac{1{,}000{,}000{,}000{,}000\ \cancel{m}}{1\ \cancel{Tm}} \cdot \frac{1\ Gm}{1{,}000{,}000{,}000\ \cancel{m}}$$

$$= \frac{1{,}000{,}000{,}000{,}000}{1{,}000{,}000{,}000}\ Gm = 1000\ Gm$$

So 1 Tm = 1000 Gm.

$$\frac{1\ \cancel{Tm}}{1} \cdot \frac{1{,}000{,}000{,}000{,}000\ \cancel{m}}{1\ \cancel{Tm}} \cdot \frac{1\ Mm}{1{,}000{,}000\ \cancel{m}}$$

$$= \frac{1{,}000{,}000{,}000{,}000}{1{,}000{,}000}\ Mm = 1{,}000{,}000\ Mm$$

So 1 Tm = 1,000,000 Mm.

78. 1 MB = 1,000,000 bytes
1 GB = 1,000,000,000 bytes
$2^{20} = 1{,}048{,}576$
$2^{30} = 1{,}073{,}741{,}824$

7.4 Problem Solving with Metric Measurement

7.4 Margin Exercises

1. *Step 1*
The problem asks for the cost of 75 cm of ribbon.

Step 2
The price is $0.89 per meter, but the length wanted is cm. Convert cm to m.

Step 3
There are 100 cm in a meter, so 75 cm is $\frac{3}{4}$ of a meter. Round the cost of 1 m from $0.89 to $0.88 so that it is divisible by 4.

An estimate is: $\frac{3}{4} \cdot \$0.88 = \$0.66$

Step 4
Convert 75 cm to m: 75 cm = 0.75 m

Find the cost.

$$\frac{0.75\ \cancel{m}}{1} \cdot \frac{\$0.89}{1\ \cancel{m}} = \$0.6675 \approx \$0.67$$

Step 5
The cost for 75 cm is $0.67 (rounded).

Step 6
The rounded answer, $0.67, is close to our estimate of $0.66.

2. Convert 1.2 grams to milligrams.
1.200∧ g = 1200 mg

Divide by the number of doses.

$$\frac{1200\ mg}{3\ doses} = 400\ mg/dose$$

Each dose should be 400 mg.

3. Convert centimeters to meters.
2 m 35 cm = 2.35 m
1 m 85 cm = 1.85 m

Add the measurements of the fabric.

one piece	2.35	m
other piece	+ 1.85	m
	4.20	m

Andrea has 4.2 m of fabric.

7.4 Section Exercises

1. Write 2 kg 50 g in terms of kilograms (the unit in the price).

2 kg	→	2.00 kg
50 g	→	+ 0.05 kg
		2.05 kg

$$\frac{\$0.65}{1\ \cancel{kg}} \cdot \frac{2.05\ \cancel{kg}}{1} = \$1.3325 \approx \$1.33$$

She will pay $1.33 (rounded) for the rice.

3. Convert 500 g to kg.

$$\frac{500\ \cancel{g}}{1} \cdot \frac{1\ kg}{1000\ \cancel{g}} = 0.5\ kg$$

Find the difference.
90 kg − 0.5 kg = 89.5 kg

The difference in the weights of the two dogs is 89.5 kg.

5. Write 5 L in terms of mL.

$$5\text{ L} = \frac{5\not{\text{L}}}{1} \cdot \frac{1000\text{ mL}}{1\not{\text{L}}} = 5000\text{ mL}$$

$$\frac{5000\not{\text{mL}}}{1} \cdot \frac{1\text{ beat}}{70\not{\text{mL}}} \approx 71.42857\text{ beats} \approx 71\text{ beats}$$

It takes 71 beats (rounded) to pass all the blood through the heart.

7. Multiply to find the total length in mm, then convert mm to cm.
$60\text{ mm} \cdot 30 = 1800\text{ mm}$

$$\frac{\overset{180}{\not{1800}}\not{\text{mm}}}{1} \cdot \frac{1\text{ cm}}{\underset{1}{\not{10}}\not{\text{mm}}} = 180\text{ cm}$$

The total length is 180 cm.
Divide to find the cost per cm.

$$\frac{\$3.29}{180\text{ cm}} \approx \$0.0183/\text{cm} \approx \$0.02/\text{cm}$$

The cost is $0.02/cm (rounded).

9. Convert centimeters to meters, then add the lengths of the two pieces.

$$\begin{array}{lcr} 2\text{ m }8\text{ cm} & = & 2.08\text{ m} \\ 2\text{ m }95\text{ cm} & = & +\ 2.95\text{ m} \\ \hline & & 5.03\text{ m} \end{array}$$

Together the boards are 5.03 m.

11. Convert milliliters to liters.
$85\text{ mL} = 0.085\text{ L}$
Multiply by the number of students.
$(0.085\text{ L})(45) = 3.825\text{ L}$
Four one-liter bottles should be ordered.

13. Three cups a day for one week is $3 \cdot 7 = 21$ cups in one week.

$$\frac{21\not{\text{cups}}}{1} \cdot \frac{90\text{ mg}}{1\not{\text{cup}}} = 1890\text{ mg}$$

Convert mg to g.

$$\frac{1890\not{\text{mg}}}{1} \cdot \frac{1\text{ g}}{1000\not{\text{mg}}} = 1.89\text{ g}$$

15. (a) $\frac{10{,}000\not{\text{m}}}{1\text{ sec}} \cdot \frac{1\text{ km}}{1000\not{\text{m}}} = 10\text{ km/sec}$

(b) $\frac{60\not{\text{sec}}}{1\text{ min}} \cdot \frac{10\text{ km}}{1\not{\text{sec}}} = 600\text{ km/min}$

(c) $\frac{60\not{\text{min}}}{1\text{ hr}} \cdot \frac{600\text{ km}}{1\not{\text{min}}} = 36{,}000\text{ km/hr}$

17. A box of 50 envelopes weighs 255 g. Subtract the packaging weight to find the net weight.

$$\begin{array}{r} 255\text{ g} \\ -\ 40\text{ g} \\ \hline 215\text{ g} \end{array}$$

There are 50 envelopes. Divide to find the weight of one envelope.

$$\frac{215\text{ g}}{50\text{ envelopes}} = \frac{4.3\text{ g}}{\text{envelope}}$$

Convert grams to milligrams.

$$\frac{4.3\not{\text{g}}}{1} \cdot \frac{1000\text{ mg}}{1\not{\text{g}}} = 4300\text{ mg}$$

18. Box of 1000 staples
Subtract the weight of the packaging to find the net weight.

$$\begin{array}{r} 350\text{ g} \\ -\ 20\text{ g} \\ \hline 330\text{ g} \end{array}$$

There are 1000 staples, so divide the net weight by 1000 or move the decimal point 3 places to the left to find the weight of one staple.

$330\text{ g} \div 1000 = 0.33\text{ g}$

The weight of one staple is 0.33 g or 330 mg.

19. Given that the weight of one sheet of paper is 3000 mg, find the weight of one sheet in grams.

$3000\text{ mg} = 3_\wedge000.\text{ g} = 3\text{ g}$

To find the net weight of the ream of paper, multiply 500 sheets by 3 g per sheet.

$$\frac{500\text{ sheets}}{1} \cdot \frac{3\text{ g}}{1\text{ sheet}} = 1500\text{ g}$$

The net weight of the paper is 1500 g.
Since the weight of the packaging is 50 g, the total weight is $1500\text{ g} + 50\text{ g} = 1550\text{ g}$.
Then express the total weight in kilograms.

$1550\text{ g} = 1_\wedge550.\text{ kg} = 1.55\text{ kg}$

20. Box of 100 small paper clips
Use the metric conversion line to find the weight of one paper clip in grams.

$500\text{ mg} = {}_\wedge500.\text{ g} = 0.5\text{ g}$

The net weight is
$(0.5\text{ g})(100) = 50\text{ g}$
The total weight is
$50\text{ g} + 5\text{ g} = 55\text{ g}.$

7.5 Metric–English Conversions and Temperature

7.5 Margin Exercises

1. (a) 23 m to yards

$$\frac{23\ \cancel{m}}{1}\cdot\frac{1.09\text{ yd}}{1\ \cancel{m}}\approx 25.1\text{ yd}$$

(b) 40 cm to inches

$$\frac{40\ \cancel{cm}}{1}\cdot\frac{0.39\text{ in.}}{1\ \cancel{cm}}\approx 15.6\text{ in.}$$

(c) 5 mi to kilometers

$$\frac{5\ \cancel{mi}}{1}\cdot\frac{1.61\text{ km}}{1\ \cancel{mi}}\approx 8.1\text{ km}$$

(d) 12 in. to centimeters

$$\frac{12\ \cancel{in.}}{1}\cdot\frac{2.54\text{ cm}}{1\ \cancel{in.}}\approx 30.5\text{ cm}$$

2. (a) 17 kg to pounds

$$\frac{17\ \cancel{kg}}{1}\cdot\frac{2.20\text{ lb}}{1\ \cancel{kg}}\approx 37.4\text{ lb}$$

(b) 5 L to quarts

$$\frac{5\ \cancel{L}}{1}\cdot\frac{1.06\text{ qt}}{1\ \cancel{L}}\approx 5.3\text{ qt}$$

(c) 90 g to ounces

$$\frac{90\ \cancel{g}}{1}\cdot\frac{0.035\text{ oz}}{1\ \cancel{g}}\approx 3.2\text{ oz}$$

(d) 3.5 gal to liters

$$\frac{3.5\ \cancel{gal}}{1}\cdot\frac{3.78\text{ L}}{1\ \cancel{gal}}\approx 13.2\text{ L}$$

(e) 145 lb to kilograms

$$\frac{145\ \cancel{lb}}{1}\cdot\frac{0.45\text{ kg}}{1\ \cancel{lb}}\approx 65.3\text{ kg}$$

(f) 8 oz to grams

$$\frac{8\ \cancel{oz}}{1}\cdot\frac{28.35\text{ g}}{1\ \cancel{oz}}\approx 226.8\text{ g}$$

3. (a) Set the living room thermostat at:
21 °C

(b) The baby has a fever of:
39 °C

(c) Wear a sweater outside because it's:
15 °C

(d) My iced tea is:
5 °C

(e) Time to go swimming! It's:
35 °C

(f) Inside a refrigerator (not the freezer) it's:
3 °C

(g) There is a blizzard outside. It's:
−20 °C

(h) I need hot water to get these clothes clean. It should be:
55 °C

4. Use $C=\frac{5(F-32)}{9}$

(a) 59 °F

$$C=\frac{5(59-32)}{9}=\frac{5\cdot\overset{3}{\cancel{27}}}{\underset{1}{\cancel{9}}}=15$$

59 °F = 15 °C

(b) 41 °F

$$C=\frac{5(41-32)}{9}=\frac{5\cdot\overset{1}{\cancel{9}}}{\underset{1}{\cancel{9}}}=5$$

41 °F = 5 °C

(c) 212 °F

$$C=\frac{5(212-32)}{9}=\frac{5\cdot\overset{20}{\cancel{180}}}{\underset{1}{\cancel{9}}}=100$$

212 °F = 100 °C

(d) 98.6 °F

$$C=\frac{5(98.6-32)}{9}=\frac{5\cdot\overset{7.4}{\cancel{66.6}}}{\underset{1}{\cancel{9}}}=37$$

98.6 °F = 37 °C

5. Use $F=\frac{9\cdot C}{5}+32.$

(a) 100 °C

$$F=\frac{9\cdot\overset{20}{\cancel{100}}}{\underset{1}{\cancel{5}}}+32=180+32=212$$

100 °C = 212 °F

(b) 25 °C

$$F=\frac{9\cdot\overset{5}{\cancel{25}}}{\underset{1}{\cancel{5}}}+32=45+32=77$$

25 °C = 77 °F

(c) 80 °C

$$F = \frac{9 \cdot \overset{16}{\cancel{80}}}{\underset{1}{\cancel{5}}} + 32 = 144 + 32 = 176$$

80 °C = 176 °F

(d) 5 °C

$$F = \frac{9 \cdot \overset{1}{\cancel{5}}}{\underset{1}{\cancel{5}}} + 32 = 9 + 32 = 41$$

5 °C = 41 °F

7.5 Section Exercises

1. 20 m to yards

$$\frac{20\ \cancel{m}}{1} \cdot \frac{1.09\ \text{yd}}{1\ \cancel{m}} \approx 21.8\ \text{yd}$$

3. 80 m to feet

$$\frac{80\ \cancel{m}}{1} \cdot \frac{3.28\ \text{ft}}{1\ \cancel{m}} \approx 262.4\ \text{ft}$$

5. 16 ft to meters

$$\frac{16\ \cancel{ft}}{1} \cdot \frac{0.30\ \text{m}}{1\ \cancel{ft}} \approx 4.8\ \text{m}$$

7. 150 g to ounces

$$\frac{150\ \cancel{g}}{1} \cdot \frac{0.035\ \text{oz}}{1\ \cancel{g}} \approx 5.3\ \text{oz}$$

9. 248 lb to kilograms

$$\frac{248\ \cancel{lb}}{1} \cdot \frac{0.45\ \text{kg}}{1\ \cancel{lb}} \approx 111.6\ \text{kg}$$

11. 28.6 L to quarts

$$\frac{28.6\ \cancel{L}}{1} \cdot \frac{1.06\ \text{qt}}{1\ \cancel{L}} \approx 30.3\ \text{qt}$$

13. **(a)** Convert 5 g to ounces.

$$\frac{5\ \cancel{g}}{1} \cdot \frac{0.035\ \text{oz}}{1\ \cancel{g}} \approx 0.2\ \text{oz}$$

About 0.2 oz of gold was used.

(b) The extra weight probably did not slow him down.

15. Convert 8.4 gal to liters.

$$\frac{8.4\ \cancel{gal}}{1} \cdot \frac{3.78\ \text{L}}{1\ \cancel{gal}} \approx 31.8\ \text{L}$$

The dishwasher uses about 31.8 L.

17. Convert 0.5 in. to centimeters.

$$\frac{0.5\ \cancel{in.}}{1} \cdot \frac{2.54\ \text{cm}}{1\ \cancel{in.}} \approx 1.3$$

The dwarf gobie is about 1.3 cm long.

19. A snowy day
28 °F is reasonable. 28 °C is a temperature for a summer day.

21. A high fever
40 °C is the more reasonable temperature because normal body temperature is about 37 °C.

23. Oven temperature
150 °C is equivalent to 302 °F so this is a reasonable oven temperature.

25. A drop of 20 Celsius degrees is more than a drop of 20 Fahrenheit degrees . There are 180 degrees between freezing and boiling on the Fahrenheit scale, but only 100 degrees on the Celsius scale, so each Celsius degree is a greater change in temperature.

27. 60 °F

$$C = \frac{5(60 - 32)}{9} = \frac{5 \cdot 28}{9} = \frac{140}{9} \approx 15.6 \approx 16$$

60 °F ≈ 16 °C

29. 104 °F

$$C = \frac{5(104 - 32)}{9} = \frac{5 \cdot 72}{9} = \frac{5 \cdot \overset{8}{\cancel{72}}}{\underset{1}{\cancel{9}}} = 40$$

104 °F = 40 °C

31. 8 °C

$$F = \frac{9 \cdot 8}{5} + 32 = \frac{72}{5} + 32 = 14.4 + 32 = 46.4$$

8 °C ≈ 46 °F

33. 35 °C

$$F = \frac{9 \cdot 35}{5} + 32 = \frac{9 \cdot \overset{7}{\cancel{35}}}{\underset{1}{\cancel{5}}} + 32 = 63 + 32 = 95$$

35 °C = 95 °F

35. 136 °F

$$C = \frac{5(136 - 32)}{9} = \frac{5 \cdot 104}{9} = \frac{520}{9} \approx 57.8$$

136 °F ≈ 58 °C

37. **(a)** Since the comfort range of the boots is from 24 °C to 4 °C, you would wear these boots in pleasant weather—above freezing, but not hot.

(b) Change 24 °C to Fahrenheit.

$$F = \frac{9 \cdot C}{5} + 32 = \frac{9 \cdot 24}{5} + 32 = \frac{216}{5} + 32 = 43.2 + 32 = 75.2$$

Thus, 24 °C ≈ 75 °F.

Change 4 °C to Fahrenheit.

$$F = \frac{9 \cdot C}{5} + 32 = \frac{9 \cdot 4}{5} + 32 = \frac{36}{5} + 32 = 7.2 + 32 = 39.2$$

Thus, 4 °C ≈ 39 °F.

The boots are designed for Fahrenheit temperatures of about 75 °F to about 39 °F.

(c) The range of metric temperatures in January would depend on where you live. In Minnesota, it's 0°C to −40°C, and in California it's 24°C to 0°C.

39. Length of model plane

$$\frac{6\ \cancel{ft}}{1} \cdot \frac{0.30\ m}{1\ \cancel{ft}} \approx 6(0.30\ m) = 1.8\ m$$

40. Weight of plane

$$\frac{11\ \cancel{lb}}{1} \cdot \frac{0.45\ kg}{1\ \cancel{lb}} \approx 11(0.45\ kg) = 4.95\ kg \approx 5.0\ kg$$

41. Length of flight path

$$\frac{1888.3\ \cancel{mi}}{1} \cdot \frac{1.61\ km}{1\ \cancel{mi}} \approx 1888.3(1.61\ km) \approx 3040.2\ km$$

42. Time of flight = 38 hr 23 min

43. Cruising altitude

$$\frac{1000\ \cancel{ft}}{1} \cdot \frac{0.30\ m}{1\ \cancel{ft}} \approx 1000(0.30\ m) = 300\ m$$

44. Fuel at the start

$$\frac{1\ \cancel{gal}}{1} \cdot \frac{3.78\ \cancel{L}}{1\ \cancel{gal}} \cdot \frac{1000\ mL}{1\ \cancel{L}} \approx (3.78)(1000\ mL) = 3780\ mL$$

Since it was "less than a gallon of fuel," there would be less than 3780 mL.

45. Fuel left after landing

$$\frac{2\ \cancel{fl\ oz}}{1} \cdot \frac{1\ \cancel{qt}}{32\ \cancel{fl\ oz}} \cdot \frac{0.95\ \cancel{L}}{1\ \cancel{qt}} \cdot \frac{1000\ mL}{1\ \cancel{L}} = \frac{950}{16} = 59.375 \approx 59.4\ mL$$

46. $\frac{1\ \cancel{gal}}{1} \cdot \frac{4\ \cancel{qt}}{1\ \cancel{gal}} \cdot \frac{32\ fl\ oz}{1\ \cancel{qt}} = 128\ fl\ oz$

$$\frac{2\ fl\ oz}{128\ fl\ oz} = \frac{1}{64} = 0.015625 \approx 1.6\%$$

Chapter 7 Review Exercises

1. 1 lb = <u>16</u> oz

2. <u>3</u> ft = 1 yd

3. 1 T = <u>2000</u> lb

4. <u>4</u> qt = 1 gal

5. 1 hr = <u>60</u> min

6. 1 c = <u>8</u> fl oz

7. <u>60</u> sec = 1 min

8. <u>5280</u> ft = 1 mi

9. <u>12</u> in. = 1 ft

10. 4 ft to inches

$$\frac{4\ \cancel{ft}}{1} \cdot \frac{12\ in.}{1\ \cancel{ft}} = 4 \cdot 12\ in. = 48\ in.$$

11. 6000 lb to tons

$$\frac{\overset{3}{\cancel{6000}}\ \cancel{lb}}{1} \cdot \frac{1\ T}{\underset{1}{\cancel{2000}}\ \cancel{lb}} = 3\ T$$

12. 64 oz to pounds

$$\frac{\overset{4}{\cancel{64}}\ \cancel{oz}}{1} \cdot \frac{1\ lb}{\underset{1}{\cancel{16}}\ \cancel{oz}} = 4\ lb$$

13. 18 hr to days

$$\frac{\overset{3}{\cancel{18}}\ \cancel{hr}}{1} \cdot \frac{1\ day}{\underset{4}{\cancel{24}}\ \cancel{hr}} = \frac{3}{4}\ \text{day or 0.75 day}$$

14. 150 min to hours

$$\frac{\overset{5}{\cancel{150}}\ \cancel{min}}{1} \cdot \frac{1\ hr}{\underset{2}{\cancel{60}}\ \cancel{min}} = \frac{5}{2}\ hr = 2\frac{1}{2}\ \text{hr or 2.5 hr}$$

15. $1\frac{3}{4}$ lb to ounces

$$\frac{1\frac{3}{4}\ \cancel{lb}}{1} \cdot \frac{16\ oz}{1\ \cancel{lb}} = \frac{7}{\underset{1}{\cancel{4}}} \cdot \frac{\overset{4}{\cancel{16}}}{1}\ oz = 28\ oz$$

16. $6\frac{1}{2}$ ft to inches

$$\frac{6\frac{1}{2}\ \cancel{ft}}{1} \cdot \frac{12\ in.}{1\ \cancel{ft}} = \frac{13}{\underset{1}{\cancel{2}}} \cdot \frac{\overset{6}{\cancel{12}}}{1}\ in. = 78\ in.$$

17. 7 gal to cups

$$\frac{7\ \cancel{\text{gal}}}{1} \cdot \frac{4\ \cancel{\text{qt}}}{1\ \cancel{\text{gal}}} \cdot \frac{2\ \cancel{\text{pt}}}{1\ \cancel{\text{qt}}} \cdot \frac{2\text{ c}}{1\ \cancel{\text{pt}}} = 7 \cdot 4 \cdot 2 \cdot 2\text{ c} = 112\text{ c}$$

18. 4 days to seconds

$$\frac{4\ \cancel{\text{days}}}{1} \cdot \frac{24\ \cancel{\text{hr}}}{1\ \cancel{\text{day}}} \cdot \frac{60\ \cancel{\text{min}}}{1\ \cancel{\text{hr}}} \cdot \frac{60\text{ sec}}{1\ \cancel{\text{min}}}$$
$$= 4 \cdot 24 \cdot 60 \cdot 60\text{ sec}$$
$$= 345{,}600\text{ sec}$$

19. **(a)** 12,460 ft to yards

$$\frac{12{,}460\ \cancel{\text{ft}}}{1} \cdot \frac{1\text{ yd}}{3\ \cancel{\text{ft}}} = \frac{12{,}460}{3}\text{ yd}$$
$$= 4153\tfrac{1}{3}\text{ yd (exact)}$$
or 4153.3 yd (rounded).

The average depth is $4153\frac{1}{3}$ yd (exact) or 4153.3 yd (rounded).

(b) 12,460 ft to miles

$$\frac{12{,}460\ \cancel{\text{ft}}}{1} \cdot \frac{1\text{ mi}}{5280\ \cancel{\text{ft}}} \approx 2.36\text{ mi} \approx 2.4\text{ mi}$$

The average depth is 2.4 mi (rounded).

20. *Step 1*
The problem asks for the amount of money the company made.

Step 2
Convert pounds to tons. Then multiply to find the total amount.

Step 3
To estimate, round 123,260 lb to 100,000 lb. Then, there are 2000 lb in a ton, so 100,000 lb is 50 T. So $50 \cdot \$40 = \$2000$ is our estimate.

Step 4

$$\frac{123{,}260\ \cancel{\text{lb}}}{1} \cdot \frac{1\text{ T}}{2000\ \cancel{\text{lb}}} = 61.63\text{ T}$$

$61.63 \cdot \$40 = \$2465.20$

Step 5
The company made $2465.20.

Step 6
The answer, $2465.20, is close to our estimate of $2000.

21. My thumb is 20 mm wide.

22. Her waist measurement is 66 cm.

23. The two towns are 40 km apart.

24. A basketball court is 30 m long.

25. The height of the picnic bench is 45 cm.

26. The eraser on the end of my pencil is 5 mm long.

27. 5 m to cm
Count 2 places to the *right* on the metric conversion line.
5.00∧ m = 500 cm

28. 8.5 km to m

$$\frac{8.5\ \cancel{\text{km}}}{1} \cdot \frac{1000\text{ m}}{1\ \cancel{\text{km}}} = 8500\text{ m}$$

29. 85 mm to cm
Count 1 place to the *left* on the metric conversion line.
8∧5. mm = 8.5 cm

30. 370 cm to m

$$\frac{370\ \cancel{\text{cm}}}{1} \cdot \frac{1\text{ m}}{100\ \cancel{\text{cm}}} = 3.7\text{ m}$$

31. 70 m to km
Count 3 places to the *left* on the metric conversion line.
∧070. m = 0.07 km

32. 0.93 m to mm

$$\frac{0.93\ \cancel{\text{m}}}{1} \cdot \frac{1000\text{ mm}}{1\ \cancel{\text{m}}} = 930\text{ mm}$$

33. The eye dropper holds 1 mL.

34. I can heat 3 L of water in this pan.

35. Loretta's hammer weighed 650 g.

36. Yongshu's suitcase weighted 20 kg when it was packed.

37. My fish tank holds 80 L of water.

38. I'll buy the 500 mL bottle of mouthwash.

39. Mara took a 200 mg antibiotic pill.

40. This piece of chicken weighs 100 g.

41. 5000 mL to L
Count 3 places to the *left* on the metric conversion line.
5∧000. mL = 5 L

42. 8 L to mL
Count 3 places to the *right* on the metric conversion line.
8.000∧ L = 8000 mL

43. 4.58 g to mg

$$\frac{4.58\ \cancel{\text{g}}}{1} \cdot \frac{1000\text{ mg}}{1\ \cancel{\text{g}}} = 4580\text{ mg}$$

44. 0.7 kg to g

$$\frac{0.7\ \cancel{kg}}{1} \cdot \frac{1000\ g}{1\ \cancel{kg}} = 700\ g$$

45. 6 mg to g
Count 3 places to the *left* on the metric conversion line.
$_\wedge$006. mg = 0.006 g

46. 35 mL to L
Count 3 places to the *left* on the metric conversion line.
$_\wedge$035. mL = 0.035 L

47. Convert milliliters to liters.

$$\frac{180\ \cancel{mL}}{1} \cdot \frac{1\ L}{1000\ \cancel{mL}} = \frac{180}{1000}\ L = 0.18\ L$$

Multiply 0.18 L by the number of servings.
(0.18 L)(175) = 31.5 L

For 175 servings, 31.5 L of punch are needed.

48. Convert kilograms to grams
10 kg = 10,000 g
Divide by the number of people.

$$\frac{10{,}000\ g}{28\ \text{people}} \approx 357$$

Jason is allowing 357 g (rounded) of turkey for each person.

49. Convert grams to kilograms.
Using the metric conversion line,
4 kg 750 g = 4.75 kg.
Subtract the weight loss from his original weight.

$$\begin{array}{r l} 92.00 & kg \\ -\ 4.75 & kg \\ \hline 87.25 & kg \end{array}$$

Yerald weighs 87.25 kg.

50. Convert to kilograms.
Using the metric conversion line,
2 kg 20 g = 2.02 kg.
Multiply by the price per kilogram.

$$\begin{array}{r l} \$1.49 & \leftarrow \text{2 decimal places} \\ \times\ 2.02 & \leftarrow \text{2 decimal places} \\ \hline 298 & \\ 2\ 980 & \\ \hline \$3.0098 & \leftarrow \text{4 decimal places} \end{array}$$

Young-Mi paid $3.01 (rounded) for the onions.

51. 6 m to yards

$$\frac{6\ \cancel{m}}{1} \cdot \frac{1.09\ yd}{1\ \cancel{m}} \approx 6.5\ yd$$

52. 30 cm to inches

$$\frac{30\ \cancel{cm}}{1} \cdot \frac{0.39\ in.}{1\ \cancel{cm}} \approx 11.7\ in.$$

53. 108 km to miles

$$\frac{108\ \cancel{km}}{1} \cdot \frac{0.62\ mi}{1\ \cancel{km}} \approx 67.0\ mi$$

54. 800 miles to km

$$\frac{800\ \cancel{mi}}{1} \cdot \frac{1.61\ km}{1\ \cancel{mi}} \approx 1288\ km$$

55. 23 quarts to L

$$\frac{23\ \cancel{qt}}{1} \cdot \frac{0.95\ L}{1\ \cancel{qt}} \approx 21.9\ L$$

56. 41.5 L to quarts

$$\frac{41.5\ \cancel{L}}{1} \cdot \frac{1.06\ qt}{1\ \cancel{L}} \approx 44.0\ qt$$

57. Water freezes at 0 °C.

58. Water boils at 100 °C.

59. Normal body temperature is about 37 °C.

60. Comfortable room temperature is about 20 °C.

61. 77 °F

$$C = \frac{5(77-32)}{9} = \frac{5(\overset{5}{\cancel{45}})}{\underset{1}{\cancel{9}}} = 25$$

77 °F = 25 °C

62. 92 °F

$$C = \frac{5(92-32)}{9} = \frac{5(\overset{20}{\cancel{60}})}{\underset{3}{\cancel{9}}} \approx 33.3$$

92 °F ≈ 33 °C

63. 6 °C

$$F = \frac{9 \cdot 6}{5} + 32 = \frac{54}{5} + 32 = 10.8 + 32 = 42.8$$

6 °C ≈ 43 °F

64. 49 °C

$$F = \frac{9 \cdot 49}{5} + 32 = 88.2 + 32 = 120.2 \approx 120$$

49 °C ≈ 120 °F

65. **[7.3]** I added 1 L of oil to my car.

66. **[7.3]** The box of books weighed 15 kg.

67. **[7.2]** Larry's shoe is 30 cm long.

68. **[7.3]** Jan used 15 mL of shampoo on her hair.

69. **[7.2]** My fingernail is 10 <u>mm</u> wide.

70. **[7.2]** I walked 2 <u>km</u> to school.

71. **[7.3]** The tiny bird weighed 15 g.

72. **[7.2]** The new library building is 18 <u>m</u> wide.

73. **[7.3]** The cookie recipe uses 250 <u>mL</u> of milk.

74. **[7.3]** Renee's pet mouse weighs 30 g.

75. **[7.3]** One postage stamp weighs 90 <u>mg</u>.

76. **[7.3]** I bought 30 <u>L</u> of gas for my car.

77. **[7.2]** 10.5 cm to mm
Count 1 place to the *right* on the metric conversion line.
$10.5_{\wedge}\text{ cm} = 105\text{ mm}$

78. **[7.1]** 45 min to hours

$$\frac{\overset{3}{\cancel{45}}\ \cancel{\text{min}}}{1} \cdot \frac{1\text{ hr}}{\underset{4}{\cancel{60}}\ \cancel{\text{min}}} = \frac{3}{4}\text{ hr or } 0.75\text{ hr}$$

79. **[7.1]** 90 in. to feet

$$\frac{\overset{15}{\cancel{90}}\ \cancel{\text{in.}}}{1} \cdot \frac{1\text{ ft}}{\underset{2}{\cancel{12}}\ \cancel{\text{in.}}} = \frac{15}{2}\text{ ft} = 7\frac{1}{2}\text{ ft or } 7.5\text{ ft}$$

80. **[7.2]** 1.3 m to cm
Count 2 places to the *right* on the metric conversion line.
$1.30_{\wedge}\text{ m} = 130\text{ cm}$

81. **[7.5]** 25 °C to Fahrenheit

$$F = \frac{9 \cdot \overset{5}{\cancel{25}}}{\underset{1}{\cancel{5}}} + 32 = 45 + 32 = 77$$

25 °C = 77 °F

82. **[7.1]** $3\frac{1}{2}$ gal to quarts

$$\frac{3\frac{1}{2}\ \cancel{\text{gal}}}{1} \cdot \frac{4\text{ qt}}{1\ \cancel{\text{gal}}} = \frac{7}{\underset{1}{\cancel{2}}} \cdot \frac{\overset{2}{\cancel{4}}}{1}\text{ qt} = 14\text{ qt}$$

83. **[7.3]** 700 mg to g
Count 3 places to the *left* on the metric conversion line.
${}_{\wedge}700.\text{ mg} = 0.7\text{ g}$

84. **[7.3]** 0.81 L to mL
Count 3 places to the *right* on the metric conversion line.
$0.810_{\wedge}\text{ L} = 810\text{ mL}$

85. **[7.1]** 5 lb to ounces

$$\frac{5\ \cancel{\text{lb}}}{1} \cdot \frac{16\text{ oz}}{1\ \cancel{\text{lb}}} = 5 \cdot 16\text{ oz} = 80\text{ oz}$$

86. **[7.3]** 60 kg to g

$$\frac{60\ \cancel{\text{kg}}}{1} \cdot \frac{1000\text{ g}}{1\ \cancel{\text{kg}}} = 60{,}000\text{ g}$$

87. **[7.3]** 1.8 L to mL
Count 3 places to the *right* on the metric conversion line.
$1.800_{\wedge}\text{ L} = 1800\text{ mL}$

88. **[7.5]** 86 °F to Celsius

$$C = \frac{5(86 - 32)}{9} = \frac{5(\overset{6}{\cancel{54}})}{\underset{1}{\cancel{9}}} = 30$$

86 °F = 30 °C

89. **[7.2]** 0.36 m to cm
Count 2 places to the *right* on the metric conversion line.
$0.36_{\wedge}\text{ m} = 36\text{ cm}$

90. **[7.3]** 55 mL to L
Count 3 places to the *left* on the metric conversion line.
${}_{\wedge}055.\text{ mL} = 0.055\text{ L}$

91. **[7.4]** Convert centimeters to meters.

$$2\text{ m } 4\text{ cm} = 2 + 2.04\text{ m} = 2.04\text{ m}$$
$$78\text{ cm} = -\,0.78\text{ m}$$
$$1.26\text{ m}$$

The board is 1.26 m long.

92. **[7.1]** 3000 lb to tons

$$\frac{3000\ \cancel{\text{lb}}}{1} \cdot \frac{1\text{ T}}{2000\ \cancel{\text{lb}}} = \frac{3}{2}\text{ T} = 1.5\text{ T}$$

Multiply by 12 days.
(1.5 T)(12) = 18 T

18 T of cookies are sold in all.

93. **[7.5]** Convert ounces to grams.

$$\frac{4\ \cancel{\text{oz}}}{1} \cdot \frac{28.35\text{ g}}{1\ \cancel{\text{oz}}} \approx 113\text{ g}$$

350 °F to Celsius

$$C = \frac{5(350 - 32)}{9} = \frac{5(318)}{9} = \frac{1590}{9} \approx 177\text{ °C}$$

94. **[7.5]** Convert kilograms to pounds.

$$\frac{80.9\ \cancel{\text{kg}}}{1}\cdot\frac{2.20\text{ lb}}{1\ \cancel{\text{kg}}}\approx 178.0\text{ lb}$$

Convert meters to feet.

$$\frac{1.83\ \cancel{\text{m}}}{1}\cdot\frac{3.28\text{ ft}}{1\ \cancel{\text{m}}}\approx 6.0\text{ ft}$$

Jalo weighs about 178.0 lb and is about 6.0 ft.

95. **[7.5]** 44 ft to meters

$$\frac{44\ \cancel{\text{ft}}}{1}\cdot\frac{0.30\text{ m}}{1\ \cancel{\text{ft}}}\approx 13.2\text{ m}$$

96. **[7.5]** 6.7 m to feet

$$\frac{6.7\ \cancel{\text{m}}}{1}\cdot\frac{3.28\text{ ft}}{1\ \cancel{\text{m}}}\approx 22.0\text{ ft}$$

97. **[7.5]** 4 yd to meters

$$\frac{4\ \cancel{\text{yd}}}{1}\cdot\frac{0.91\text{ m}}{1\ \cancel{\text{yd}}}\approx 3.6\text{ m}$$

98. **[7.5]** 102 cm to inches

$$\frac{102\ \cancel{\text{cm}}}{1}\cdot\frac{0.39\text{ in.}}{1\ \cancel{\text{cm}}}\approx 39.8\text{ in.}$$

99. **[7.5]** 220,000 kg to pounds

$$\frac{220{,}000\ \cancel{\text{kg}}}{1}\cdot\frac{2.20\text{ lb}}{1\ \cancel{\text{kg}}}\approx 484{,}000\text{ lb}$$

100. **[7.5]** 5 gal to liters

$$\frac{5\ \cancel{\text{gal}}}{1}\cdot\frac{3.78\text{ L}}{1\ \cancel{\text{gal}}}\approx 18.9\text{ L}$$

6 gal to liters

$$\frac{6\ \cancel{\text{gal}}}{1}\cdot\frac{3.78\text{ L}}{1\ \cancel{\text{gal}}}\approx 22.7\text{ L}$$

101. **[7.1]** **(a)** From Exercise 99:

$$\frac{484{,}000\ \cancel{\text{lb}}}{1}\cdot\frac{1\text{ T}}{2000\ \cancel{\text{lb}}}=242\text{ T}$$

(b) From Exercise 97:

$$\frac{4\ \cancel{\text{yd}}}{1}\cdot\frac{3\ \cancel{\text{ft}}}{1\ \cancel{\text{yd}}}\cdot\frac{12\text{ in.}}{1\ \cancel{\text{ft}}}=144\text{ in.}$$

102. **[7.2]** **(a)** From Exercise 98:

$$\frac{102\ \cancel{\text{cm}}}{1}\cdot\frac{1\text{ m}}{100\ \cancel{\text{cm}}}=1.02\text{ m}$$

(b) From Exercise 96:

$$\frac{6.7\ \cancel{\text{m}}}{1}\cdot\frac{100\text{ cm}}{1\ \cancel{\text{m}}}=670\text{ cm}$$

Chapter 7 Test

1. 9 gal to quarts

$$\frac{9\ \cancel{\text{gal}}}{1}\cdot\frac{4\text{ qt}}{1\ \cancel{\text{gal}}}=36\text{ qt}$$

2. 45 ft to yards

$$\frac{45\ \cancel{\text{ft}}}{1}\cdot\frac{1\text{ yd}}{3\ \cancel{\text{ft}}}=\frac{45}{3}\text{ yd}=15\text{ yd}$$

3. 135 min to hours

$$\frac{\overset{27}{\cancel{135}}\ \cancel{\text{min}}}{1}\cdot\frac{1\text{ hr}}{\underset{12}{\cancel{60}}\ \cancel{\text{min}}}=\frac{27}{12}\text{ hr}=2.25\text{ hr or }2\frac{1}{4}\text{ hr}$$

4. 9 in. to feet

$$\frac{\overset{3}{\cancel{9}}\ \cancel{\text{in.}}}{1}\cdot\frac{1\text{ ft}}{\underset{4}{\cancel{12}}\ \cancel{\text{in.}}}=\frac{3}{4}\text{ ft or }0.75\text{ ft}$$

5. $3\frac{1}{2}$ lb to ounces

$$\frac{3\frac{1}{2}\ \cancel{\text{lb}}}{1}\cdot\frac{16\text{ oz}}{1\ \cancel{\text{lb}}}=\frac{7}{\underset{1}{\cancel{2}}}\cdot\frac{\overset{8}{\cancel{16}}}{1}\text{ oz}=56\text{ oz}$$

6. 5 days to minutes

$$\frac{5\ \cancel{\text{days}}}{1}\cdot\frac{24\ \cancel{\text{hr}}}{1\ \cancel{\text{day}}}\cdot\frac{60\text{ min}}{1\ \cancel{\text{hr}}}=\frac{5\cdot 24\cdot 60}{1}\text{ min}=7200\text{ min}$$

7. My husband weighs 75 <u>kg</u>.

8. I hiked 5 <u>km</u> this morning.

9. She bought 125 <u>mL</u> of cough syrup.

10. This apple weighs 180 g.

11. This page is 21 <u>cm</u> wide.

12. My watch band is 10 <u>mm</u> wide.

13. I bought 10 <u>L</u> of soda for the picnic.

14. The bracelet is 16 <u>cm</u> long.

15. 250 cm to meters
Count 2 places to the *left* on the metric conversion line.

$2_\wedge 50.$ cm = 2.5 m

16. 4.6 km to meters

$$\frac{4.6\ \cancel{\text{km}}}{1}\cdot\frac{1000\text{ m}}{1\ \cancel{\text{km}}}=4600\text{ m}$$

17. 5 mm to centimeters
Count 1 place to the *left* on the metric conversion line.

$_\wedge 5.$ mm = 0.5 cm

18. 325 mg to grams

$$\frac{325\ \cancel{mg}}{1} \cdot \frac{1\ g}{1000\ \cancel{mg}} = \frac{325}{1000}\ g = 0.325\ g$$

19. 16 L to milliliters
Count 3 places to the *right* on the metric conversion line.
16.000‸ L = 16,000 mL

20. 0.4 kg to grams

$$\frac{0.4\ \cancel{kg}}{1} \cdot \frac{1000\ g}{1\ \cancel{kg}} = 400\ g$$

21. 10.55 m to centimeters
Count 2 places to the *right* on the metric conversion line.
10.55‸ m = 1055 cm

22. 95 mL to liters
Count 3 places to the *left* on the metric conversion line.
‸095. mL = 0.095 L

23. Convert 460 in. to feet.

$$\frac{460\ \cancel{in.}}{1} \cdot \frac{1\ ft}{12\ \cancel{in.}} = \frac{460}{12}\ ft = \frac{115}{3}\ ft$$

Divide by 12 months

$$\frac{\frac{115}{3}\ ft}{12\ months} = \frac{115}{3} \cdot \frac{1}{12} \approx 3.2\ ft/month$$

It rains about 3.2 ft per month.

24. **(a)** Convert 2.9 g to mg.

$$\frac{2.9\ \cancel{g}}{1} = \frac{1000\ mg}{1\ \cancel{g}} = 2900\ mg$$

Find the difference.
2900 mg − 590 mg = 2310 mg
It has 2310 mg more sodium.

(b) 2900 mg is more than 2400 mg
2900 mg − 2400 mg = 500 mg

$$\frac{500\ \cancel{mg}}{1} \cdot \frac{1\ g}{1000\ \cancel{mg}} = 0.5\ g$$

The "Super Melt" has 500 mg or 0.5 g more sodium than the 2400 mg recommended daily amount.

25. The water is almost boiling.
On the Celsius scale, water boils at 100° so, 95 °C would be almost boiling.

26. The tomato plants may freeze tonight.
Water freezes at 0 °C.

27. 6 ft to meters

$$\frac{6\ \cancel{ft}}{1} \cdot \frac{0.30\ m}{1\ \cancel{ft}} \approx 1.8\ m$$

28. 125 lb to kilograms

$$\frac{125\ \cancel{lb}}{1} \cdot \frac{0.45\ kg}{1\ \cancel{lb}} \approx 56.3\ kg$$

29. 50 L to gallons

$$\frac{50\ \cancel{L}}{1} \cdot \frac{0.26\ gal}{1\ \cancel{L}} \approx 13\ gal$$

30. 8.1 km to miles

$$\frac{8.1\ \cancel{km}}{1} \cdot \frac{0.62\ mi}{1\ \cancel{km}} \approx 5.0\ mi$$

31. 74 °F to Celsius

$$C = \frac{5(74-32)}{9} = \frac{5(\overset{14}{\cancel{42}})}{\underset{3}{\cancel{9}}} = \frac{70}{3} \approx 23\ °C$$

74 °F ≈ 23 °C

32. 2 °C to Fahrenheit

$$F = \frac{9 \cdot 2}{5} + 32 = \frac{18}{5} + 32 = 3.6 + 32 \approx 36°$$

2 °C ≈ 36 °F

33. Convert 1 m 20 cm to meters.

$$1\ m + 20\ cm = 1\ m + \frac{20\ \cancel{cm}}{1} \cdot \frac{1\ m}{100\ \cancel{cm}}$$
$$= 1\ m + 0.2\ m = 1.2\ m$$

Multiply to find the number of meters for 5 pillows.
5(1.2 m) = 6 m

Convert 6 m to yards.

$$\frac{6\ \cancel{m}}{1} \cdot \frac{1.09\ yd}{1\ \cancel{m}} \approx 6.54\ yd$$

Multiply by \$3.98 to find the cost.

$$\frac{6.54\ yd}{1} \cdot \frac{\$3.98}{\cancel{yd}} \approx \$26.03$$

It will cost about \$26.03.

34. Possible answers: Use same system as rest of the world; easier system for children to learn; less use of fractional numbers; compete internationally.

Cumulative Review Exercises (Chapters 1–7)

1. *Estimate:* *Exact:*

$$\begin{array}{r} 100 \\ 3 \\ +\ 70 \\ \hline 173 \end{array} \qquad \begin{array}{r} 107.500 \\ 2.548 \\ +\ 68.790 \\ \hline 178.838 \end{array}$$

Attach zeros

2. *Estimate:* *Exact:*

$$\begin{array}{r} 30{,}000 \\ -\ 800 \\ \hline 29{,}200 \end{array} \qquad \begin{array}{r} 31{,}007 \\ -\ 829 \\ \hline 30{,}178 \end{array}$$

3. *Estimate:* *Exact:*

$$\begin{array}{r} 90{,}000 \\ \times\ 200 \\ \hline 18{,}000{,}000 \end{array} \qquad \begin{array}{r} 92{,}075 \\ \times\ 183 \\ \hline 276\ 225 \\ 7\ 366\ 00 \\ 9\ 207\ 5 \\ \hline 16{,}849{,}725 \end{array}$$

4. *Estimate:* *Exact:*

$$\begin{array}{r} 60 \\ \times\ 5 \\ \hline 300 \end{array} \qquad \begin{array}{rl} 56.52 & \leftarrow \textit{2 decimal places} \\ \times\ 4.7 & \leftarrow \textit{1 decimal place} \\ \hline 39\ 564 & \\ 226\ 08 & \\ \hline 265.644 & \leftarrow \textit{3 decimal places} \end{array}$$

5. *Estimate:* *Exact:*

$$40\overline{)20{,}000}\ \text{quotient } 500 \qquad 37\overline{)19{,}610}\ \text{quotient } 530$$

$$\begin{array}{r} 530 \\ 37\overline{)19{,}610} \\ 18\ 5 \\ \hline 1\ 11 \\ 1\ 11 \\ \hline 00 \\ 0 \\ \hline 0 \end{array}$$

6. *Estimate:* *Exact:*

$$\begin{array}{r} 5 \\ 8\overline{)40} \end{array} \qquad \begin{array}{r} 4.6 \\ 8.3_\wedge\overline{)38.1_\wedge 8} \\ 33\ 2 \\ \hline 4\ 98 \\ 4\ 98 \\ \hline 0 \end{array}$$

7. *Estimate:* *Exact:*

$$\begin{array}{r} 2 \\ +\ 4 \\ \hline 6 \end{array} \qquad \begin{array}{rcl} 1\frac{7}{10} & = & 1\frac{7}{10} \\ +\ 3\frac{4}{5} & = & 3\frac{8}{10} \\ \hline & & 4\frac{15}{10} \end{array}$$

$$4\frac{15}{10} = 4 + \frac{10}{10} + \frac{5}{10} = 5\frac{1}{2}$$

8. *Estimate:* *Exact:*

$$\begin{array}{r} 6 \\ -\ 1 \\ \hline 5 \end{array} \qquad \begin{array}{rcl} 5\frac{1}{2} & = & 5\frac{7}{14} \\ -\ 1\frac{2}{7} & = & 1\frac{4}{14} \\ \hline & & 4\frac{3}{14} \end{array}$$

9. *Exact:*

$$3\frac{1}{6} \cdot 4\frac{2}{3} = \frac{19}{\cancel{6}_3} \cdot \frac{\cancel{14}^7}{3} = \frac{133}{9} = 14\frac{7}{9}$$

Estimate: $3 \cdot 5 = 15$

10. *Exact:*

$$2\frac{1}{4} \div \frac{9}{10} = \frac{\cancel{9}^1}{\cancel{4}_2} \cdot \frac{\cancel{10}^5}{\cancel{9}_1} = \frac{5}{2} = 2\frac{1}{2}$$

Estimate: $2 \div 1 = 2$

11. $3 - 2\frac{5}{16} = 2\frac{16}{16} - 2\frac{5}{16} = \frac{11}{16}$

12. $7 + 484{,}099 + 3939$

$$\begin{array}{r} 7 \\ 484{,}099 \\ +\ 3\,939 \\ \hline 488{,}045 \end{array}$$

13. $12 \cdot 2\frac{2}{9} = \frac{\cancel{12}^4}{1} \cdot \frac{20}{\cancel{9}_3} = \frac{80}{3} = 26\frac{2}{3}$

14.

$$\begin{array}{r} 13.03 \\ 0.066_\wedge\overline{)0.860_\wedge 00} \\ 66 \\ \hline 200 \\ 198 \\ \hline 2\ 0 \\ 0 \\ \hline 2\ 00 \\ 1\ 98 \\ \hline 2 \end{array}$$

$0.86 \div 0.066 \approx 13.0$

15. $\frac{3}{8}+\frac{5}{6}=\frac{9}{24}+\frac{20}{24}=\frac{29}{24}=1\frac{5}{24}$

16. $8-0.9207$

$$\begin{array}{r} 8.0000 \\ -\ 0.9207 \\ \hline 7.0793 \end{array}$$

17. $3\frac{3}{4}\div 6$

$$3\frac{3}{4}\div 6=\frac{15}{4}\div\frac{6}{1}=\frac{\overset{5}{\cancel{15}}}{4}\cdot\frac{1}{\underset{2}{\cancel{6}}}=\frac{5}{8}$$

18.

$$\begin{array}{r} 3\,0\,7\ \mathbf{R}\,38 \\ 47\overline{)1\,4,\,4\,6\,7} \\ 1\,4\,1 \\ \hline 3\,6 \\ 0 \\ \hline 3\,6\,7 \\ 3\,2\,9 \\ \hline 3\,8 \end{array}$$

19. $(2.54)(0.003)$

$$\begin{array}{rl} 2.54 & \leftarrow\ \textit{2 decimal places} \\ \times\ 0.003 & \leftarrow\ \textit{3 decimal places} \\ \hline 0.00762 & \leftarrow\ \textit{5 decimal places} \end{array}$$

20.

$$\begin{array}{ll} 24-12\div 6(8)+(25-25) & \textit{Parentheses} \\ =24-12\div 6(8)+0 & \textit{Divide} \\ =24-2(8)+0 & \textit{Multiply} \\ =24-16+0 & \textit{Subtract} \\ =8+0=8 & \textit{Add} \end{array}$$

21.

$$\begin{array}{ll} 3^2+2^5\cdot\sqrt{64} & \textit{Exponents} \\ =9+32\cdot\sqrt{64} & \textit{Square root} \\ =9+32\cdot 8 & \textit{Multiply} \\ =9+256=265 & \textit{Add} \end{array}$$

22. In words, 307.19 is three hundred seven and nineteen hundredths.

23. Eighty-two ten-thousandths is 0.0082.

24.

$$\begin{array}{r} 0.67=0.6700 \\ 0.067=0.0670 \\ 0.6=0.6000 \\ 0.6007=0.6007 \end{array}$$

From smallest to largest:
0.067, 0.6, 0.6007, 0.67

25.

Size	Cost per Unit
package of 16	$\frac{\$3.87}{16}\approx\$0.242$
package of 22	$\frac{\$5.96}{22}\approx\$0.271$
package of 36	$\frac{\$11.69}{36}\approx\$0.325$

The best buy is 16 diapers for $3.87.

26. $\frac{7}{8}$

$$\begin{array}{r} 0.\,8\,7\,5 \\ 8\overline{)7.\,0\,0\,0} \\ 6\,4 \\ \hline 6\,0 \\ 5\,6 \\ \hline 4\,0 \\ 4\,0 \\ \hline 0 \end{array}$$

$\frac{7}{8}=0.875$ *decimal*

27. $\frac{7}{8}=0.875=87.5\%$ or $87\frac{1}{2}\%$

28. $0.05=\frac{5}{100}=\frac{5}{5\cdot 20}=\frac{1}{20}$

29. $0.05=5\%$

30. $350\%=\frac{350}{100}=\frac{35}{10}=3\frac{5}{10}=3\frac{1}{2}$

31. $350\%=3_\wedge50.=3.5$

32. 13 g to 18 g

$\frac{13}{18}$

33. 18 g to 3 g

$\frac{18}{3}=\frac{18\div 3}{3\div 3}=\frac{6}{1}$

34. Multiply the smallest numbers by 3. $1\times 3=3$, and 3 is in the table, so the ratio of light chicken to cod is $\frac{3}{1}$. $3\times 3=9$, and 9 is in the table, so the ratio of salmon to light chicken is $\frac{3}{1}$. $9\times 3=27$, and 27 is larger than any value in the table, so we have found all the possible 3 to 1 ratios.

35. $\frac{x}{16}=\frac{3}{4}$
$4\cdot x=3\cdot 16$
$\frac{4\cdot x}{4}=\frac{48}{4}$
$x=12$

36. $\frac{0.9}{0.75}=\frac{2}{x}$
$0.9\cdot x=2\cdot 0.75$
$0.9\cdot x=1.5$
$\frac{0.9\cdot x}{0.9}=\frac{1.5}{0.9}$
$x\approx 1.67$

37. Use the percent proportion.
part is 4; whole is 80; percent is unknown

$\frac{4}{80}=\frac{x}{100}$
$x\cdot 80=4\cdot 100$
$\frac{x\cdot 80}{80}=\frac{400}{80}$
$x=5$

$4 is 5% of $80.

38. Use the percent proportion.
part is 36; percent is 120; whole is unknown

$\frac{36}{x}=\frac{120}{100}$
$120\cdot x=36\cdot 100$
$120\cdot x=3600$
$\frac{120\cdot x}{120}=\frac{3600}{120}$
$x=30$

36 hours is 120% of 30 hours.

39. $2\frac{1}{2}$ ft to inches

$\frac{2\frac{1}{2}\text{ ft}}{1}\cdot\frac{12\text{ in.}}{1\text{ ft}}=\frac{5}{2}\cdot\frac{12}{1}\text{ in.}=30\text{ in.}$

40. 105 sec to minutes

$\frac{105\text{ sec}}{1}\cdot\frac{1\text{ min}}{60\text{ sec}}=\frac{21}{12}\text{ min}=1\frac{3}{4}\text{ min or }1.75\text{ min}$

41. 2.8 m to centimeters
Count 2 places to the *right* on the metric conversion line.
$2.80_\wedge$ m = 280 cm

42. 65 mg to grams

$\frac{65\text{ mg}}{1}\cdot\frac{1\text{ g}}{1000\text{ mg}}=\frac{65}{1000}\text{ g}=0.065\text{ g}$

43. 198 km to miles

$\frac{198\text{ km}}{1}\cdot\frac{0.62\text{ mi}}{1\text{ km}}\approx 122.76\text{ miles}$

44. 50 °F to Celsius

$C=\frac{5(F-32)}{9}=\frac{5(50-32)}{9}=\frac{5(18)}{9}=10$

50 °F = 10 °C

45. Ron bought the tube of toothpaste weighing 100 g.

46. The teacher's desk is 140 cm long.

47. The hallway is 3 m wide.

48. Joe's hammer weighed 1 kg.

49. Anne took a 500 mg tablet of Vitamin C.

50. Tia added 125 mL of milk to her cereal.

51. John has a slight fever.
38 °C is reasonable. Body temperature is about 37 °C or 98.6 °F.

52. You'll need a light jacket outside.
12 °C is reasonable. 45 °C would be too hot and 0 °C would be very cold.

53. *amount of discount* = *rate of discount* · *cost of item*
= (0.10)($189.94)
≈ $18.99

The amount of discount is $18.99.
The sale price is $189.94 − $18.99 = $170.95.

54. 24 prints for $10.35 plus $6\frac{1}{2}$% sales tax

The amount of tax is .065(10.35) ≈ .67, or 67 cents. The total cost is $10.35 + $0.67 = $11.02. The cost per unit print is then

$\frac{\$11.02}{24}\approx \$0.46.$

55. Find the total ounces, then convert oz to pounds.
12 oz + 48 · 4 oz = 12 oz + 192 oz = 204 oz

$\frac{204\text{ oz}}{1}\cdot\frac{1\text{ lb}}{16\text{ oz}}=12\frac{3}{4}\text{ lb or }12.75\text{ lb}$

The carton would weigh $12\frac{3}{4}$ lb or 12.75 lb.

56. **(a)** Set up a proportion.

$\frac{1\text{ cm}}{12\text{ km}}=\frac{7.8\text{ cm}}{x}$
$x\cdot 1=(7.8)(12)$
$x=93.6$

The distance is 93.6 kilometers.

(b) $\frac{93.6 \cancel{km}}{1} \cdot \frac{0.62 \text{ mi}}{1 \cancel{km}} \approx 58.0 \text{ mi}$

The distance is 58.0 mi (rounded).

57. \$8750 at 9% for $3\frac{1}{2}$ years

$$\begin{aligned} I &= p \cdot r \cdot t \\ &= (\$8750)(0.09)(3.5) \\ &= \$2756.25 \end{aligned}$$

The interest is \$2756.25 and the total amount due is \$8750 + \$2756.25 = \$11,506.25.

58. Use the percent proportion.
part is 31; whole is 35; percent is unknown

$$\frac{31}{35} = \frac{x}{100}$$
$$35 \cdot x = 31 \cdot 100$$
$$\frac{35 \cdot x}{35} = \frac{3100}{35}$$
$$x \approx 88.57$$

The percent of the problems that were correct is 88.6% (rounded).

59. Convert 650 g to kg.
$_{\wedge}650.\text{ g} = 0.650 \text{ kg}$

Multiply to find the cost.
(0.650)(\$14.98) ≈ \$9.74

Mark paid \$9.74 (rounded).

60. Add the amounts needed for the recipes.

$$\begin{aligned} 2\frac{1}{4} &= 2\frac{1}{4} \\ 1\frac{1}{2} &= 1\frac{2}{4} \\ +\frac{3}{4} &= \frac{3}{4} \\ \hline & 3\frac{6}{4} = 4\frac{2}{4} = 4\frac{1}{2} \text{ cups} \end{aligned}$$

Find the total amount of brown sugar.

$$\begin{aligned} & 2\frac{1}{3} \\ + & 2\frac{1}{3} \\ \hline & 4\frac{2}{3} \text{ cups} \end{aligned}$$

Subtract the amount to be used from the total amount.

$$\begin{aligned} 4\frac{2}{3} &= 4\frac{4}{6} \\ -4\frac{1}{2} &= 4\frac{3}{6} \\ \hline & \frac{1}{6} \end{aligned}$$

The Jackson family will have $\frac{1}{6}$ cup more than the amount needed.

61. Set up a proportion.

$$\frac{6 \text{ rows}}{5 \text{ cm}} = \frac{x}{100 \text{ cm}}$$
$$5 \cdot x = 6 \cdot 100$$
$$\frac{5 \cdot x}{5} = \frac{600}{5}$$
$$x = 120$$

Akuba will knit 120 rows.

62. $\frac{3}{8}$ of 5600 students

$$\frac{3}{8} \cdot 5600 = \frac{3}{\underset{1}{\cancel{8}}} \cdot \frac{\overset{700}{\cancel{5600}}}{1} = 2100$$

The survey showed that 2100 students work 20 hours or more per week.

63. $\frac{\$6}{40 \text{ applications}} = \$0.15/\text{application}$

64. decrease = 8 − 5 = 3 days

$$\frac{3 \text{ days}}{8 \text{ days}} = \frac{x}{100}$$
$$8 \cdot x = 3 \cdot 100$$
$$\frac{8 \cdot x}{8} = \frac{300}{8}$$
$$x = 37.5$$

The length of a hospital stay has decreased by 37.5%.

65. Total items = 13 + 3 + 1 = 17

(a) $\frac{13 \text{ advertisements}}{17 \text{ total items}} \approx 76.5\%$

(b) $\frac{1 \text{ financial statement}}{17 \text{ total items}} \approx 5.9\%$

66. Total e-mails = 12 + 12 = 24

(a) $\frac{12 \text{ spam}}{24 \text{ total}} = \frac{12 \div 12}{24 \div 12} = \frac{1}{2}$

(b) $\frac{1}{2} = \frac{1 \cdot 50}{2 \cdot 50} = \frac{50}{100} = 0.50 \text{ or } 0.5$

(c) $\frac{50}{100} = 50\%$

CHAPTER 8 GEOMETRY

8.1 Basic Geometric Terms

8.1 Margin Exercises

1. **(a)** The figure has two endpoints, so it is a *line segment*.

(b) The figure starts at point S and goes on forever in one direction, so it is a *ray*.

(c) The figure goes on forever in both directions, so it is a *line*.

(d) The figure has two endpoints, so it is a *line segment*.

2. **(a)** The lines cross at E, so they are *intersecting lines*.

(b) The lines never intersect (cross), so they appear to be *parallel lines*.

(c) The lines never intersect (cross), so they appear to be *parallel lines*.

3. **(a)** The angle is named $\angle 3$, $\angle CQD$, or $\angle DQC$.

(b) Darken rays $\overrightarrow{TW}$ and $\overrightarrow{TZ}$.

(c) The angle can be named $\angle 1$, $\angle R$, $\angle MRN$, and $\angle NRM$.

4. **(a)** The angle measures exactly 90° and there is a small square at the vertex, so it is a *right angle*.

(b) The angle measures exactly 180°, so it is a *straight angle*.

(c) The angle measures between 90° and 180°, so it is *obtuse*.

(d) The angle measures between 0° and 90°, so it is *acute*.

5. The lines in (b) show perpendicular lines, because they intersect at right angles. The lines in (a) show intersecting lines. They cross, but not at right angles.

8.1 Section Exercises

1. This is a *line* named $\overleftrightarrow{CD}$ or $\overleftrightarrow{DC}$. A line is a straight row of points that goes on forever in both directions.

3. The figure has two endpoints so it is a *line segment* named $\overline{GF}$ or $\overline{FG}$.

5. This is a *ray* named $\overrightarrow{PQ}$. A ray is a part of a line that has only one endpoint and goes on forever in one direction.

7. The lines are *perpendicular* because they intersect at right angles.

9. These lines appear to be *parallel* lines. Parallel lines are lines in the same plane that never intersect (cross).

11. The lines intersect so they are *not* parallel. At their intersection they *do not* form a right angle so they are not perpendicular. The lines are *intersecting*.

13. The angle can be named $\angle AOS$ or $\angle SOA$. The middle letter, O, identifies the vertex.

15. The angle can be named $\angle CRT$ or $\angle TRC$. The middle letter, R, identifies the vertex.

17. The angle can be named $\angle AQC$ or $\angle CQA$. The middle letter, Q, identifies the vertex.

19. The angle is a *right angle*, as indicated by the small square at the vertex. Right angles measure exactly 90°.

21. The measure of the angle is between 0° and 90°, so it is an *acute angle*.

23. Two rays in a straight line pointing opposite directions measure 180°. An angle that measures 180° is called a *straight angle*.

25. **(a)** If your car "did a 360," the car turned around in a complete circle.

(b) If the governor's view on taxes "took a 180° turn," he or she took the opposite view. For example, he or she may have opposed taxes but now supports them.

27. "$\angle UST$ is 90°" is *true* because $\overleftrightarrow{UQ}$ is perpendicular to $\overleftrightarrow{ST}$.

28. "$\overleftrightarrow{SQ}$ and $\overleftrightarrow{PQ}$ are perpendicular" is *true* because they form a 90° angle, as indicated by the small red square.

29. "The measure of $\angle USQ$ is less than the measure of $\angle PQR$" is *false*. $\angle USQ$ is a straight angle and so is $\angle PQR$, therefore each measures 180°. A true statement would be: $\angle USQ$ has the same measure as $\angle PQR$.

30. "$\overleftrightarrow{ST}$ and $\overleftrightarrow{PR}$ are intersecting" is *false*. $\overleftrightarrow{ST}$ and $\overleftrightarrow{PR}$ are parallel and will never intersect.

31. "$\overleftrightarrow{QU}$ and $\overleftrightarrow{TS}$ are parallel" is *false*. $\overleftrightarrow{QU}$ and $\overleftrightarrow{TS}$ are perpendicular because they intersect at right angles.

32. "$\angle UST$ and $\angle UQR$ measure the same number of degrees" is *true* because both angles are formed by perpendicular lines, so they both measure 90°.

8.2 Angles and Their Relationships

8.2 Margin Exercises

1. $\angle AOB$ and $\angle BOC$ are complementary angles because $45° + 45° = 90°$.
$\angle COD$ and $\angle DOE$ are complementary angles because $70° + 20° = 90°$.

2. **(a)** The complement of 35° is 55°, because $90° - 35° = 55°$.

(b) The complement of 80° is 10°, because $90° - 80° = 10°$.

3. $\angle CRF$ and $\angle BRF$ are supplementary angles because $127° + 53° = 180°$.
$\angle CRE$ and $\angle ERB$ are supplementary angles because $53° + 127° = 180°$.
$\angle BRF$ and $\angle BRE$ are supplementary angles because $53° + 127° = 180°$.
$\angle CRE$ and $\angle CRF$ are supplementary angles because $53° + 127° = 180°$.

4. **(a)** The supplement of 175° is 5°, because $180° - 175° = 5°$.

(b) The supplement of 30° is 150°, because $180° - 30° = 150°$.

5. $\angle BOC \cong \angle AOD$ because they each measure 150°.
$\angle AOB \cong \angle DOC$ because they each measure 30°.

6. $\angle SPB$ and $\angle MPD$ are vertical angles because they do *not* share a common side and the angles are formed by intersecting lines.
Similarly, $\angle BPD$ and $\angle SPM$ are vertical angles.

7. **(a)** $\angle TOS$
$\angle TOS$ is a right angle, so its measure is 90°.

(b) $\angle QOR$
$\angle QOR$ and $\angle SOT$ are vertical angles, so they are also congruent and have the same measure.
The measure of $\angle SOT$ is 90°, so the measure of $\angle QOR$ is 90°.

(c) $\angle POQ$
$\angle POQ$ and $\angle TOV$ are vertical angles, so they are congruent and have the same measure.
The measure of $\angle TOV$ is 52°, so the measure of $\angle POQ$ is 52°.

(d) $\angle VOR$
$\angle VOR$ and $\angle TOV$ are complementary angles, so the sum of their angle measures is 90°. Since the degree measure of $\angle TOV$ is 52°, the measure of $\angle VOR$ equals $90° - 52° = 38°$.

(e) $\angle POS$
$\angle POS$ and $\angle VOR$ are vertical angles, so they are congruent. From part (d), the measure of $\angle VOR$ is 38°, so the measure of $\angle POS$ is 38°.

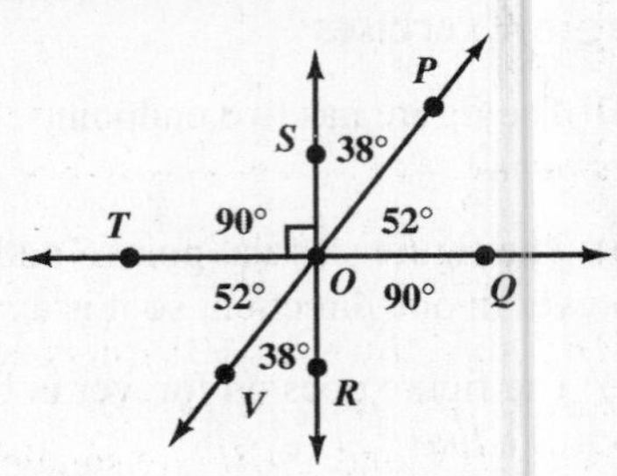

8.2 Section Exercises

1. The pairs of complementary angles are:
$\angle EOD$ and $\angle COD$ because $75° + 15° = 90°$;
$\angle AOB$ and $\angle BOC$ because $25° + 65° = 90°$.

3. The pairs of supplementary angles are:
$\angle HNE$ and $\angle ENF$ because $77° + 103° = 180°$;
$\angle HNG$ and $\angle GNF$ because $103° + 77° = 180°$;
$\angle HNE$ and $\angle HNG$ because $77° + 103° = 180°$;
$\angle ENF$ and $\angle GNF$ because $103° + 77° = 180°$.

5. The complement of 40° is 50°, because $90° - 40° = 50°$.

7. The complement of 86° is 4°, because $90° - 86° = 4°$.

9. The supplement of 130° is 50°, because $180° - 130° = 50°$.

11. The supplement of 90° is 90°, because $180° - 90° = 90°$.

13. $\angle SON \cong \angle TOM$ because the angles each measure 70°.
$\angle TOS \cong \angle MON$ because the angles each measure 110°.

15. Because $\angle COE$ and $\angle GOH$ are vertical angles, they are also congruent. This means they have the same measure. $\angle COE$ measures 63°, so $\angle GOH$ measures 63°.

The sum of the measures of $\angle COE$, $\angle AOC$, and $\angle AOH$ equals 180°. Therefore, $\angle AOC$ measures $180° - (63° + 37°) = 180° - 100° = 80°$. Since $\angle AOC$ and $\angle GOF$ are vertical, they are congruent, so $\angle GOF$ measures 80°. Since $\angle AOH$ and $\angle EOF$ are vertical, they are congruent, so $\angle EOF$ measures 37°.

17. Two angles are complementary if their sum is 90°. Two angles are supplementary if their sum is 180°. Drawings will vary; examples are:

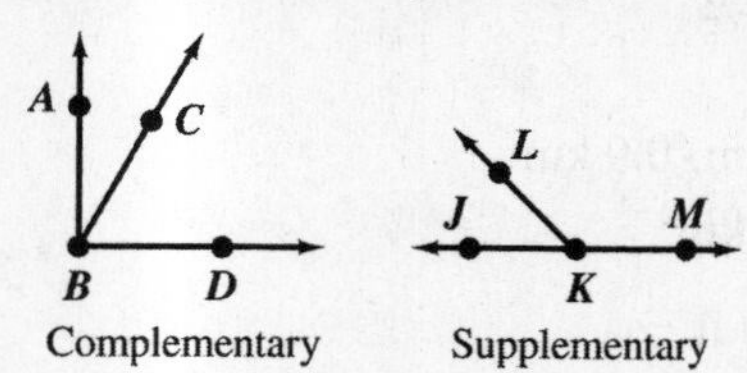

19. $\angle ABC$ and $\angle BCD$ are congruent because they both measure 42°, so $\angle ABC \cong \angle BCD$.
Since $\angle ABC$ and $\angle ABF$ are supplementary, $\angle ABF$ measures $180° - 42° = 138°$.
Since $\angle BCD$ and $\angle ECD$ are supplementary, $\angle ECD$ measures $180° - 42° = 138°$.
Since $\angle ABF$ and $\angle ECD$ each measure 138°, $\angle ABF \cong \angle ECD$.

21. No; obtuse angles are $> 90°$, so their sum would be $> 180°$.

8.3 Rectangles and Squares

8.3 Margin Exercises

1. (a), (b), and (e) are figures with four sides that meet to form 90° angles, so they are rectangles. (c) and (d) do not have four sides. (f) and (g) have four sides, but they do not meet to form right angles.

2. **(a)** The length of the rectangle is 17 cm and the width is 10 cm.

$$\begin{aligned} P &= 2 \cdot \text{length} + 2 \cdot \text{width} \\ &= 2 \cdot 17 \text{ cm} + 2 \cdot 10 \text{ cm} \\ &= 34 \text{ cm} + 20 \text{ cm} \\ &= 54 \text{ cm} \end{aligned}$$

(b) The length is 10.5 ft and the width is 7 ft.

$$\begin{aligned} P &= 2 \cdot \text{length} + 2 \cdot \text{width} \\ &= 2 \cdot 10.5 \text{ ft} + 2 \cdot 7 \text{ ft} \\ &= 21 \text{ ft} + 14 \text{ ft} \\ &= 35 \text{ ft} \end{aligned}$$

(c) 6 m wide and 11 m long

$$\begin{aligned} P &= 2 \cdot \text{length} + 2 \cdot \text{width} \\ &= 2 \cdot 11 \text{ m} + 2 \cdot 6 \text{ m} \\ &= 22 \text{ m} + 12 \text{ m} \\ &= 34 \text{ m} \end{aligned}$$

(d) 0.9 km by 2.8 km

$$\begin{aligned} P &= 2 \cdot \text{length} + 2 \cdot \text{width} \\ &= 2 \cdot 2.8 \text{ km} + 2 \cdot 0.9 \text{ km} \\ &= 5.6 \text{ km} + 1.8 \text{ km} \\ &= 7.4 \text{ km} \end{aligned}$$

3. **(a)** The length is 9 ft and the width is 4 ft.

$$\begin{aligned} A &= \text{length} \cdot \text{width} \\ &= 9 \text{ ft} \cdot 4 \text{ ft} \\ &= 36 \text{ ft}^2 \end{aligned}$$

(b) A rectangle is 6 m long and 0.5 m wide.

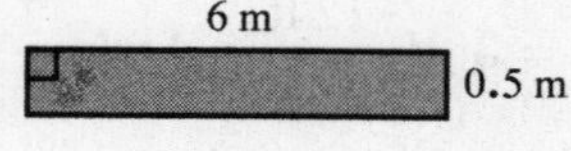

$$\begin{aligned} A &= \text{length} \cdot \text{width} \\ &= 6 \text{ m} \cdot 0.5 \text{ m} \\ &= 3 \text{ m}^2 \end{aligned}$$

(c) 3.5 yd by 2.5 yd

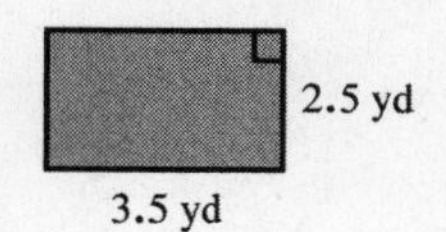

$$\begin{aligned} A &= \text{length} \cdot \text{width} \\ &= 3.5 \text{ yd} \cdot 2.5 \text{ yd} \\ &= 8.75 \text{ yd}^2 \end{aligned}$$

4. **(a)** Each side measures 2 ft.

$$\begin{aligned} P &= 4 \cdot s \\ &= 4 \cdot 2 \text{ ft} \\ &= 8 \text{ ft} \end{aligned} \qquad \begin{aligned} A &= s \cdot s \\ &= 2 \text{ ft} \cdot 2 \text{ ft} \\ &= 4 \text{ ft}^2 \end{aligned}$$

(b) 10.5 cm on each side

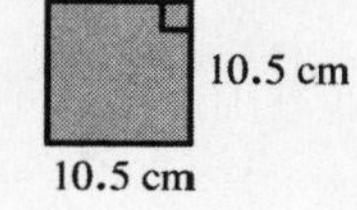

$$\begin{aligned} P &= 4 \cdot s \\ &= 4 \cdot 10.5 \text{ cm} \\ &= 42 \text{ cm} \end{aligned} \qquad \begin{aligned} A &= s \cdot s \\ &= 10.5 \text{ cm} \cdot 10.5 \text{ cm} \\ &= 110.25 \text{ cm}^2 \end{aligned}$$

(c) 2.1 miles on a side

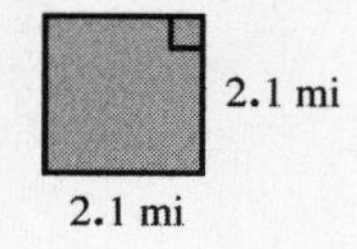

$$\begin{aligned} P &= 4 \cdot s \\ &= 4 \cdot 2.1 \text{ mi} \\ &= 8.4 \text{ mi} \end{aligned} \qquad \begin{aligned} A &= s \cdot s \\ &= 2.1 \text{ mi} \cdot 2.1 \text{ mi} \\ &= 4.41 \text{ mi}^2 \end{aligned}$$

5. The cost of the carpet is \$19.95 per square yard.

(a) The length is 6.5 yd and the width is 5 yd.

$$\begin{aligned} A &= \text{length} \cdot \text{width} \\ &= 6.5 \text{ yd} \cdot 5 \text{ yd} \\ &= 32.5 \text{ yd}^2 \end{aligned}$$

continued

The cost of the carpet is

$$\frac{32.5 \text{ yd}^2}{1} \cdot \frac{\$19.95}{1 \text{ yd}^2} = \$648.375 \approx \$648.38.$$

(b) First change the measurements from feet to yards.

$$\frac{21 \text{ ft}}{3} = 7 \text{ yd} \quad \frac{9 \text{ ft}}{3} = 3 \text{ yd} \quad \frac{12 \text{ ft}}{3} = 4 \text{ yd}$$

Next, break up the room into two pieces.

Area of rectangle $=$ length $\cdot$ width
$= 7 \text{ yd} \cdot 4 \text{ yd}$
$= 28 \text{ yd}^2$

Area of square $= s \cdot s$
$= 3 \text{ yd} \cdot 3 \text{ yd}$
$= 9 \text{ yd}^2$

Total area $= 28 \text{ yd}^2 + 9 \text{ yd}^2$
$= 37 \text{ yd}^2$

The cost of the carpet is

$$\frac{37 \text{ yd}^2}{1} \cdot \frac{\$19.95}{1 \text{ yd}^2} = \$738.15.$$

(c) A rectangular classroom that is 24 ft long and 18 feet wide.

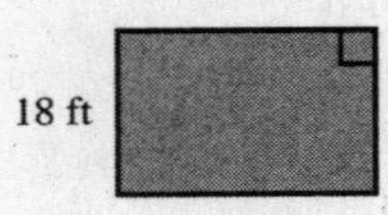

Change feet to yards.

$$\frac{24 \text{ ft}}{3} = 8 \text{ yd} \quad \frac{18 \text{ ft}}{3} = 6 \text{ yd}$$

$A =$ length $\cdot$ width
$= 8 \text{ yd} \cdot 6 \text{ yd}$
$= 48 \text{ yd}^2$

The cost of the carpet is

$$\frac{48 \text{ yd}^2}{1} \cdot \frac{\$19.95}{1 \text{ yd}^2} = \$957.60.$$

8.3 Section Exercises

1. $P = 2 \cdot \text{length} + 2 \cdot \text{width}$
$= 2 \cdot 8 \text{ yd} + 2 \cdot 6 \text{ yd}$
$= 16 \text{ yd} + 12 \text{ yd}$
$= 28 \text{ yd}$
$A = \text{length} \cdot \text{width}$
$= 8 \text{ yd} \cdot 6 \text{ yd}$
$= 48 \text{ yd}^2$

3. $P = 4 \cdot s$
$= 4 \cdot 0.9 \text{ km}$
$= 3.6 \text{ km}$

$A = s \cdot s$
$= 0.9 \text{ km} \cdot 0.9 \text{ km}$
$= 0.81 \text{ km}^2$

5. 10 ft by 10 ft

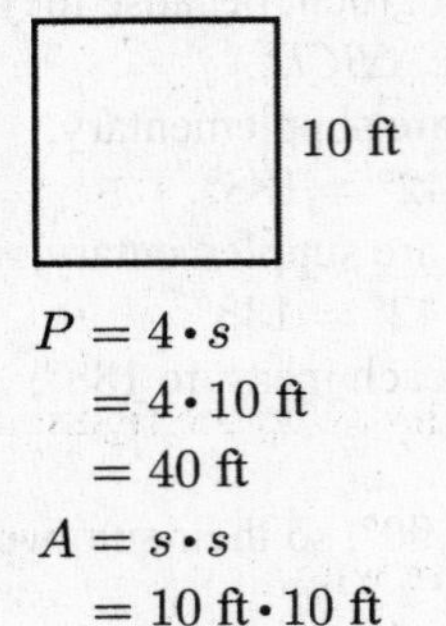

$P = 4 \cdot s$
$= 4 \cdot 10 \text{ ft}$
$= 40 \text{ ft}$

$A = s \cdot s$
$= 10 \text{ ft} \cdot 10 \text{ ft}$
$= 100 \text{ ft}^2$

7. 14 m by 0.5 m

14 m
0.5 m

$P = 2 \cdot \text{length} + 2 \cdot \text{width}$
$= 2 \cdot 14 \text{ m} + 2 \cdot 0.5 \text{ m}$
$= 28 \text{ m} + 1 \text{ m}$
$= 29 \text{ m}$

$A = \text{length} \cdot \text{width}$
$= 14 \text{ m} \cdot 0.5 \text{ m}$
$= 7 \text{ m}^2$

9. A storage building that is 76.1 ft by 22 ft

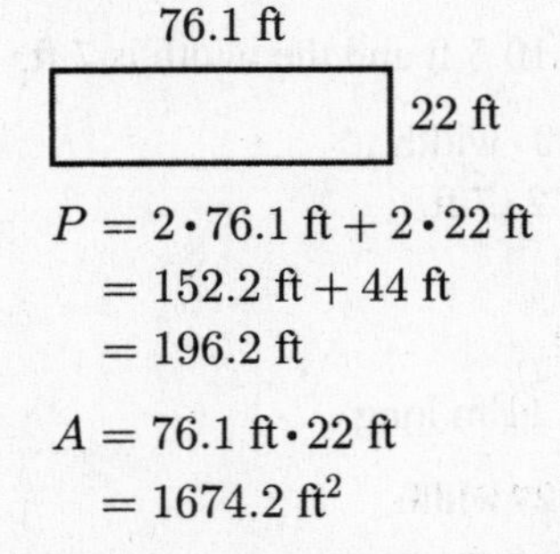

$P = 2 \cdot 76.1 \text{ ft} + 2 \cdot 22 \text{ ft}$
$= 152.2 \text{ ft} + 44 \text{ ft}$
$= 196.2 \text{ ft}$

$A = 76.1 \text{ ft} \cdot 22 \text{ ft}$
$= 1674.2 \text{ ft}^2$

11. A square nature preserve 3 mi wide

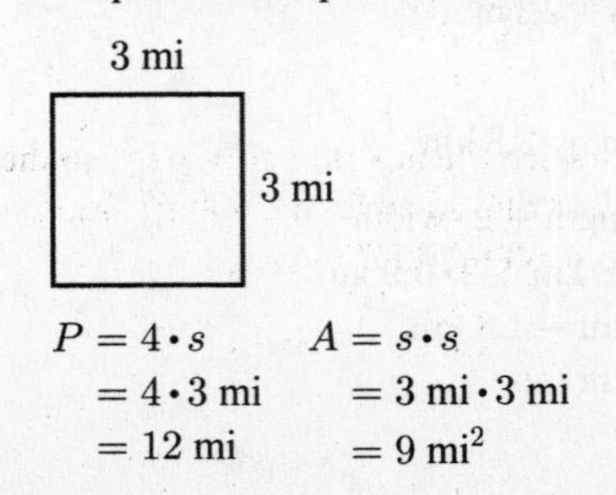

$P = 4 \cdot s$ $\quad$ $A = s \cdot s$
$= 4 \cdot 3 \text{ mi}$ $\quad$ $= 3 \text{ mi} \cdot 3 \text{ mi}$
$= 12 \text{ mi}$ $\quad$ $= 9 \text{ mi}^2$

13. $P = 12\text{ m} + 7\text{ m} + 3\text{ m} + 5\text{ m} + 9\text{ m} + 2\text{ m}$
$= 38\text{ m}$

Draw a horizontal line and break up the figure into two rectangles.

rectangle with	rectangle with
length: 7 m	length: 9 m
width: 3 m	width: 2 m
$A = 7\text{ m} \cdot 3\text{ m}$	$A = 9\text{ m} \cdot 2\text{ m}$
$= 21\text{ m}^2$	$= 18\text{ m}^2$

Total area $= 21\text{ m}^2 + 18\text{ m}^2 = 39\text{ m}^2$

15. $P = 28\text{ m} + 17\text{ m} + 12\text{ m} + 4\text{ m} + 4\text{ m} + 4\text{ m} + 12\text{ m} + 17\text{ m}$
$= 98\text{ m}$

Draw a vertical line and break up the figure into a rectangle and a square.

rectangle with	square with
length: 28 m	side: 4 m
width: 17 m	
$A = 28\text{ m} \cdot 17\text{ m}$	$A = 4\text{ m} \cdot 4\text{ m}$
$= 476\text{ m}^2$	$= 16\text{ m}^2$

Total area $= 476\text{ m}^2 + 16\text{ m}^2 = 492\text{ m}^2$

17. The unlabeled side is 12 in. (6 in. + 6 in.).

$P = 15\text{ in.} + 6\text{ in.} + 6\text{ in.} + 6\text{ in.} + 6\text{ in.} + 6\text{ in.} + 9\text{ in.} + 12\text{ in.} + 6\text{ in.} + 6\text{ in.}$
$= 78\text{ in.}$

Draw two horizontal lines to break up the figure.

$A = 6\text{ in.} \cdot 6\text{ in.} + 18\text{ in.} \cdot 9\text{ in.} + 6\text{ in.} \cdot 6\text{ in.}$
$= 36\text{ in.}^2 + 162\text{ in.}^2 + 36\text{ in.}^2$
$= 234\text{ in.}^2$

19. $P = 4 \cdot s$
$= 4 \cdot 12\text{ m}$
$= 48\text{ m}$

$A = s \cdot s$
$= 12\text{ m} \cdot 12\text{ m}$
$= 144\text{ m}^2$

21. Find the area of the room.

$A = \text{length} \cdot \text{width}$
$= 25\text{ ft} \cdot 20\text{ ft}$
$= 500\text{ ft}^2$

The area of each tile is $1\text{ ft} \cdot 1\text{ ft} = 1\text{ ft}^2$, so they will need 500 tiles to cover the 500 ft^2 floor.
Total cost $= 500 \cdot \$0.92 = \$460$

23. $P = 2 \cdot 4.4\text{ m} + 2 \cdot 5.1\text{ m}$
$= 8.8\text{ m} + 10.2\text{ m} = 19\text{ m}$

$\text{Cost} = \dfrac{19\text{ m}}{1} \cdot \dfrac{\$4.99}{1\text{ m}} = \$94.81$

Tyra will have to spend \$94.81 for the strip.

25. **(1)** Panoramic:

$P = 2 \cdot 14\text{ in.} + 2 \cdot 4\text{ in.}$
$= 28\text{ in.} + 8\text{ in.}$
$= 36\text{ in.}$
$A = 14\text{ in.} \cdot 4\text{ in.}$
$= 56\text{ in.}^2$

(2) 4 in. × 6 in.:

$P = 2 \cdot 6\text{ in.} + 2 \cdot 4\text{ in.}$
$= 12\text{ in.} + 8\text{ in.}$
$= 20\text{ in.}$
$A = 6\text{ in.} \cdot 4\text{ in.}$
$= 24\text{ in.}^2$

(3) 4 in. × 7 in.:

$P = 2 \cdot 7\text{ in.} + 2 \cdot 4\text{ in.}$
$= 14\text{ in.} + 8\text{ in.}$
$= 22\text{ in.}$
$A = 7\text{ in.} \cdot 4\text{ in.}$
$= 28\text{ in.}^2$

27. $A = \text{length} \cdot \text{width}$
$5300\text{ yd}^2 = 100\text{ yd} \cdot \text{width}$
$\dfrac{5300\text{ yd}^2}{100\text{ yd}} = \text{width}$
$53\text{ yd} = \text{width}$

The width is 53 yd.

29. Length of inner rectangle $= 172\text{ ft} - 12\text{ ft} - 12\text{ ft}$
$= 148\text{ ft}$
Width of inner rectangle $= 124\text{ ft} - 12\text{ ft} - 12\text{ ft}$
$= 100\text{ ft}$

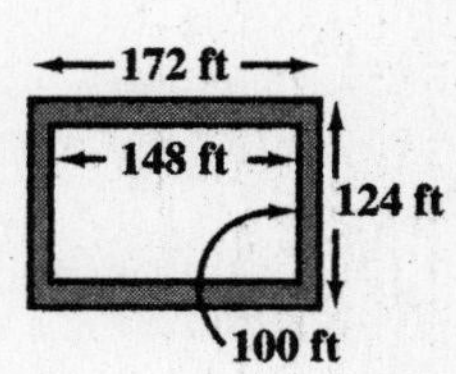

Shaded area $= 172\text{ ft} \cdot 124\text{ ft} - 148\text{ ft} \cdot 100\text{ ft}$
$= 21{,}328\text{ ft}^2 - 14{,}800\text{ ft}^2$
$= 6528\text{ ft}^2$

31. Divide 12 by 2 to get 6, which is what the length plus the width must equal. The only whole number possibilities are 1 and 5, 2 and 4, and 3 and 3.

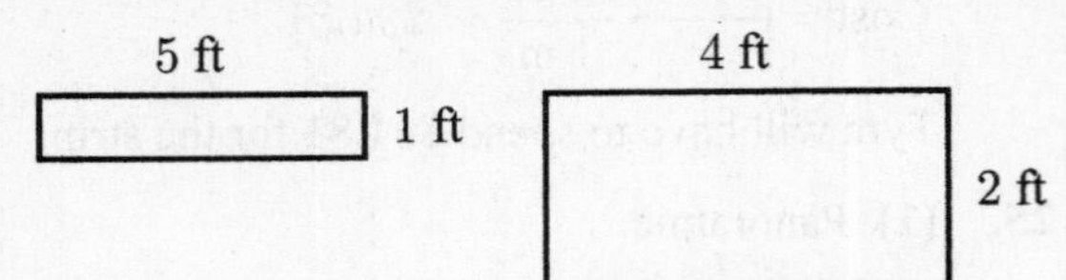

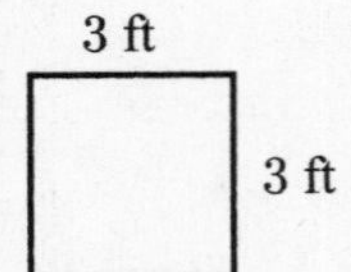

32. **(a)** 5 ft by 1 ft: $A = 5 \text{ ft} \cdot 1 \text{ ft} = 5 \text{ ft}^2$
4 ft by 2 ft: $A = 4 \text{ ft} \cdot 2 \text{ ft} = 8 \text{ ft}^2$
3 ft by 3 ft: $A = 3 \text{ ft} \cdot 3 \text{ ft} = 9 \text{ ft}^2$

(b) The square plot 3 ft by 3 ft has the greatest area.

33. The four possibilities for 16 ft of fencing are:

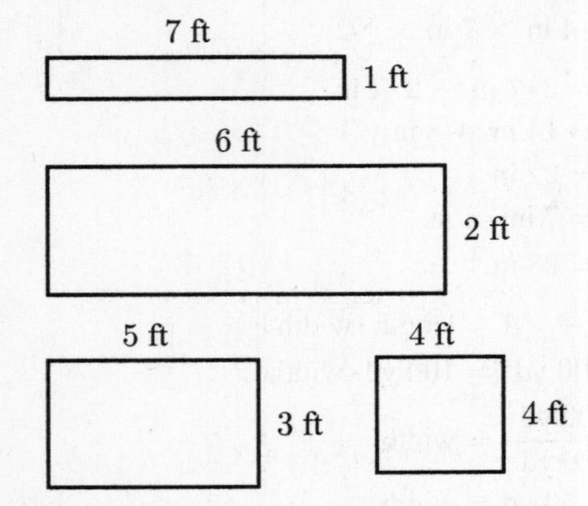

5 ft
3 ft
4 ft
4 ft

34. **(a)** 7 ft × 1 ft: $A = 7 \text{ ft} \cdot 1 \text{ ft} = 7 \text{ ft}^2$
6 ft × 2 ft: $A = 6 \text{ ft} \cdot 2 \text{ ft} = 12 \text{ ft}^2$
5 ft × 3 ft: $A = 5 \text{ ft} \cdot 3 \text{ ft} = 15 \text{ ft}^2$
4 ft × 4 ft: $A = 4 \text{ ft} \cdot 4 \text{ ft} = 16 \text{ ft}^2$

(b) The square plots have the greatest area.

35. **(a)**

3 ft
2 ft

$$P = 2 \cdot 3 \text{ ft} + 2 \cdot 2 \text{ ft} = 6 \text{ ft} + 4 \text{ ft} = 10 \text{ ft}$$
$$A = 3 \text{ ft} \cdot 2 \text{ ft} = 6 \text{ ft}^2$$

(b)

6 ft
4 ft

$$P = 2 \cdot 6 \text{ ft} + 2 \cdot 4 \text{ ft} = 12 \text{ ft} + 8 \text{ ft} = 20 \text{ ft}$$
$$A = 6 \text{ ft} \cdot 4 \text{ ft} = 24 \text{ ft}^2$$

(c) The perimeter of the enlarged plot is twice the perimeter of the original plot. The area of the enlarged plot is four times ($2^2 = 4$) the area of the original plot.

36. **(a)**

9 ft
6 ft

$$P = 2 \cdot 9 \text{ ft} + 2 \cdot 6 \text{ ft} = 18 \text{ ft} + 12 \text{ ft} = 30 \text{ ft}$$
$$A = 9 \text{ ft} \cdot 6 \text{ ft} = 54 \text{ ft}^2$$

(b) The perimeter is three times the original and the area is nine times ($3^2 = 9$) the original.

(c) The perimeter will be four times the original and the area will be 16 times ($4^2 = 16$) the original.

8.4 Parallelograms and Trapezoids

8.4 Margin Exercises

1. **(a)** $P = 15 \text{ m} + 27 \text{ m} + 15 \text{ m} + 27 \text{ m} = 84 \text{ m}$

(b) $P = 6.91 \text{ km} + 10.3 \text{ km} + 6.91 \text{ km} + 10.3 \text{ km} = 34.42 \text{ km}$

2. **(a)** base: 50 ft, height: 42 ft

$$A = b \cdot h = 50 \text{ ft} \cdot 42 \text{ ft} = 2100 \text{ ft}^2$$

(b) base: 3.8 cm, height: 2.3 cm

$$A = b \cdot h = 3.8 \text{ cm} \cdot 2.3 \text{ cm} = 8.74 \text{ cm}^2$$

(c) base: $12\frac{1}{2}$ mi, height: $4\frac{3}{4}$ mi

$$A = b \cdot h = 12\tfrac{1}{2} \text{ mi} \cdot 4\tfrac{3}{4} \text{ mi} = 12.5 \text{ mi} \cdot 4.75 \text{ mi} = 59.375 \text{ mi}^2 \text{ or } 59\tfrac{3}{8} \text{ mi}^2$$

3. **(a)** $P = 9.8 \text{ in.} + 6.3 \text{ in.} + 6.9 \text{ in.} + 5.6 \text{ in.} = 28.6 \text{ in.}$

(b) $P = 1.28 \text{ km} + 0.85 \text{ km} + 1.7 \text{ km} + 2 \text{ km}$
$= 5.83 \text{ km}$

(c) $P = 39.7 \text{ cm} + 29.2 \text{ cm} + 74.9 \text{ cm} + 16.4 \text{ cm}$
$= 160.2 \text{ cm}$

4. **(a)** height: 30 ft, short base: 40 ft, long base: 60 ft

$$\text{Area} = \frac{1}{2} \cdot \text{height} \cdot (\text{short base} + \text{long base})$$
$$A = \frac{1}{2} \cdot h \cdot (b + B)$$
$$A = \frac{1}{2} \cdot 30 \text{ ft} \cdot (40 \text{ ft} + 60 \text{ ft})$$
$$= 15 \text{ ft} \cdot (100 \text{ ft})$$
$$= 1500 \text{ ft}^2$$

(b) height: 11 cm, short base: 14 cm, long base: 19 cm

$$A = \frac{1}{2} \cdot h \cdot (b + B)$$
$$A = \frac{1}{2} \cdot 11 \text{ cm} \cdot (14 \text{ cm} + 19 \text{ cm})$$
$$= \frac{1}{2} \cdot 11 \text{ cm} \cdot (33 \text{ cm})$$
$$= 0.5 \cdot 363 \text{ cm}$$
$$= 181.5 \text{ cm}^2$$

(c) height: 4.7 m, short base: 9 m, long base: 10.5 m

$$A = 0.5 \cdot h \cdot (b + B)$$
$$= 0.5 \cdot 4.7 \text{ m} \cdot (9 \text{ m} + 10.5 \text{ m})$$
$$= 0.5 \cdot 4.7 \text{ m} \cdot 19.5 \text{ m}$$
$$= 45.825 \text{ m}^2$$

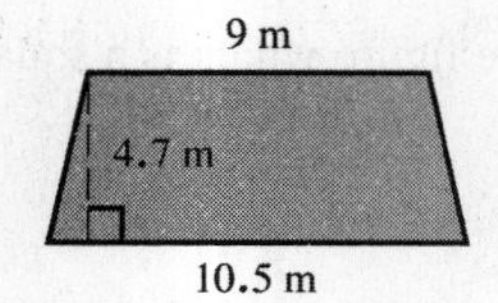

5. **(a)** Break up the figure into two pieces, a parallelogram and a trapezoid.
Area of a parallelogram:

$$A = b \cdot h$$
$$= 8 \text{ m} \cdot 5 \text{ m}$$
$$= 40 \text{ m}^2$$

Area of trapezoid:

$$A = \frac{1}{2} \cdot h \cdot (b + B)$$
$$= \frac{1}{2} \cdot 4 \text{ m} \cdot (8 \text{ m} + 14 \text{ m})$$
$$= \frac{1}{2} \cdot 4\text{m} \cdot (22 \text{ m})$$
$$= 44 \text{ m}^2$$

Total area $= 40 \text{ m}^2 + 44 \text{ m}^2 = 84 \text{ m}^2$

(b) Break up the figure into two pieces.
Area of square:

$$A = s \cdot s$$
$$= 5 \text{ yd} \cdot 5 \text{ yd}$$
$$= 25 \text{ yd}^2$$

Area of trapezoid:

$$A = \frac{1}{2} \cdot h \cdot (b + B)$$
$$= \frac{1}{2} \cdot 5 \text{ yd} \cdot (5 \text{ yd} + 10 \text{ yd})$$
$$= 0.5 \cdot 5 \text{ yd} \cdot 15 \text{ yd}$$
$$= 37.5 \text{ yd}^2$$

Total area $= 25 \text{ yd}^2 + 37.5 \text{ yd}^2 = 62.5 \text{ yd}^2$

6. **(a)** Floor (a) area is 84 m^2:

$$\text{Cost} = \frac{84 \text{ m}^2}{1} \cdot \frac{\$18.50}{1 \text{ m}^2} = \$1554$$

(b) Floor (b) area is 62.5 yd^2:

$$\text{Cost} = \frac{62.5 \text{ yd}^2}{1} \cdot \frac{\$28}{1 \text{ yd}^2} = \$1750$$

8.4 Section Exercises

1. $P = 46 \text{ m} + 58 \text{ m} + 46 \text{ m} + 58 \text{ m}$
$= 208 \text{ m}$

Do not include the height (43 m).

3. $P = 4 \cdot 51.8 \text{ m}$
$= 207.2 \text{ m}$

Do not include the height (48.3 m).

5. $P = 3 \text{ km} + 0.8 \text{ km} + 0.95 \text{ km} + 1.03 \text{ km}$
$= 5.78 \text{ km}$

Do not include the height (0.4 km).

7. $A = b \cdot h$
$= 31 \text{ mm} \cdot 25 \text{ mm}$
$= 775 \text{ mm}^2$

9. $A = b \cdot h$
$= 5.5 \text{ ft} \cdot 3.5 \text{ ft}$
$= 19.25 \text{ ft}^2$

11. $A = \frac{1}{2} \cdot h \cdot (b + B)$
$= 0.5 \cdot 42 \text{ cm} \cdot (61.4 \text{ cm} + 86.2 \text{ cm})$
$= 0.5 \cdot 42 \text{ cm} \cdot (147.6 \text{ cm})$
$= 3099.6 \text{ cm}^2$

13. $A = \frac{1}{2} \cdot h \cdot (b + B)$

$= \frac{1}{2} \cdot 45 \text{ ft} \cdot (80 \text{ ft} + 110 \text{ ft})$

$= \frac{1}{\cancel{2}_1} \cdot 45 \text{ ft} \cdot (\overset{95}{\cancel{190}} \text{ ft}) = 4275 \text{ ft}^2$

$\text{Cost} = \frac{4275 \text{ ft}^2}{1} \cdot \frac{\$0.33}{1 \text{ ft}^2} = \$1410.75$

It will cost $1410.75 to put sod on the yard.

15. Area of one parallelogram:

$A = b \cdot h$
$= 5 \text{ in.} \cdot 3.5 \text{ in.}$
$= 17.5 \text{ in.}^2$

$\text{Total area} = 25 \cdot 17.5 \text{ in.}^2 = 437.5 \text{ in.}^2$

The total area is 437.5 in.^2.

17. Perimeter is the distance around the edges of a figure. Height is not part of the perimeter. Square units are used for area, not perimeter.
$P = 2.5 \text{ cm} + 2.5 \text{ cm} + 2.5 \text{ cm} + 2.5 \text{ cm}$
$= 10 \text{ cm}$

19. Area of rectangle:

$A = \text{length} \cdot \text{width}$
$= 1.3 \text{ m} \cdot 0.8 \text{ m}$
$= 1.04 \text{ m}^2$

Area of parallelogram:

$A = \text{base} \cdot \text{height}$
$= 1.8 \text{ m} \cdot 1.1 \text{ m}$
$= 1.98 \text{ m}^2$

$\text{Total area} = 1.04 \text{ m}^2 + 1.98 \text{ m}^2$
$= 3.02 \text{ m}^2$

21. **(1)** Top figure (square):

$A = s \cdot s$
$= 96 \text{ ft} \cdot 96 \text{ ft}$
$= 9216 \text{ ft}^2$

(2) Middle figure (parallelogram):

$A = b \cdot h$
$= 96 \text{ ft} \cdot 72 \text{ ft}$
$= 6912 \text{ ft}^2$

(3) Bottom figure (square), same as top figure:

$A = 9216 \text{ ft}^2$

$\text{Total area} = 9216 \text{ ft}^2 + 6912 \text{ ft}^2 + 9216 \text{ ft}^2$
$= 25{,}344 \text{ ft}^2$

8.5 Triangles

8.5 Margin Exercises

1. **(a)** $P = 31 \text{ mm} + 16 \text{ mm} + 25 \text{ mm} = 72 \text{ mm}$

(b) $P = 25.9 \text{ m} + 16.2 \text{ m} + 11.7 \text{ m} = 53.8 \text{ m}$

(c) $P = 6\frac{1}{2} \text{ yd} + 9\frac{3}{4} \text{ yd} + 11\frac{1}{4} \text{ yd}$
$= 6\frac{2}{4} \text{ yd} + 9\frac{3}{4} \text{ yd} + 11\frac{1}{4} \text{ yd}$
$= 26\frac{6}{4} \text{ yd} = 27\frac{1}{2} \text{ yd or } 27.5 \text{ yd}$

2. **(a)** base: 26 m, height: 20 m

$A = \frac{1}{2} \cdot \text{base} \cdot \text{height} = \frac{1}{2} \cdot b \cdot h$
$= \frac{1}{2} \cdot 26 \text{ m} \cdot 20 \text{ m}$
$= 260 \text{ m}^2$

(b) base: 2.1 cm, height: 1.7 cm

$A = 0.5 \cdot b \cdot h$
$= 0.5 \cdot 2.1 \text{ cm} \cdot 1.7 \text{ cm}$
$= 1.785 \text{ cm}^2$

(c) base: 6 in., height: 4.5 in.

$A = 0.5 \cdot b \cdot h$
$= 0.5 \cdot 6 \text{ in.} \cdot 4.5 \text{ in.}$
$= 13.5 \text{ in.}^2$

(d) base: $9\frac{1}{2}$ ft, height: 7 ft

$A = 0.5 \cdot b \cdot h$
$= 0.5 \cdot 9.5 \text{ ft} \cdot 7 \text{ ft}$
$= 33.25 \text{ ft}^2 \text{ or } 33\frac{1}{4} \text{ ft}^2$

3. Find the area of the entire figure which is a square.

$A = s \cdot s$
$= 25 \text{ m} \cdot 25 \text{ m}$
$= 625 \text{ m}^2$

Find the area of the unshaded parts which are triangles.

$A = \frac{1}{2} \cdot b \cdot h$
$= \frac{1}{2} \cdot 25 \text{ m} \cdot 10 \text{ m}$
$= 125 \text{ m}^2$

Subtract to find the area of the shaded part.

entire area		*unshaded triangle*		*unshaded triangle*
625 m^2	$-$	125 m^2	$-$	125 m^2

$= 375 \text{ m}^2$ *shaded part*

4. The shaded part is 375 m^2. The cost of the carpet is $27 per square meter.

$$\frac{375 \text{ m}^2}{1} \cdot \frac{\$27}{1 \text{ m}^2} = \$10{,}125$$

The cost for carpeting is $10,125.
The unshaded part is $2(125 \text{ m}^2) = 250 \text{ m}^2$.
The cost for the vinyl floor covering is $18 per square meter.

$$\frac{250 \text{ m}^2}{1} \cdot \frac{\$18}{1 \text{ m}^2} = \$4500$$

The cost for the vinyl floor covering is $4500.
The total cost of covering the floor is $\$10{,}125 + \$4500 = \$14{,}625$.

5. **(a)** *Step 1* $48° + 59° = 107°$
Step 2 $180° - 107° = 73°$
$\angle G$ measures 73°.

(b) *Step 1* $90° + 55° = 145°$
Step 2 $180° - 145° = 35°$
$\angle C$ measures 35°.

(c) *Step 1* $60° + 60° = 120°$
Step 2 $180° - 120° = 60°$
$\angle X$ measures 60°.

8.5 Section Exercises

1. $P = 72 \text{ m} + 58 \text{ m} + 72 \text{ m} = 202 \text{ m}$

$$A = \frac{1}{2} \cdot b \cdot h = \frac{1}{2} \cdot 58 \text{ m} \cdot 66 \text{ m} = 1914 \text{ m}^2$$

3. $P = 15.6 \text{ cm} + 25.3 \text{ cm} + 18 \text{ cm} = 58.9 \text{ cm}$

$$A = \frac{1}{2} \cdot b \cdot h = \frac{1}{2} \cdot 25.3 \text{ cm} \cdot 11 \text{ cm} = 139.15 \text{ cm}^2$$

5. $P = 9 \text{ yd} + 7 \text{ yd} + 10\frac{1}{4} \text{ yd} = 26\frac{1}{4} \text{ yd}$ or 26.25 yd

$$A = \frac{1}{2} \cdot b \cdot h = \frac{1}{2} \cdot 10\tfrac{1}{4} \text{ yd} \cdot 6 \text{ yd} = 30\tfrac{3}{4} \text{ yd}^2 \text{ or } 30.75 \text{ yd}^2$$

7. $P = 35.5 \text{ cm} + 21.3 \text{ cm} + 28.4 \text{ cm} = 85.2 \text{ cm}$

$$A = \frac{1}{2} \cdot b \cdot h = \frac{1}{2} \cdot 28.4 \text{ cm} \cdot 21.3 \text{ cm} = 302.46 \text{ cm}^2$$

9. $A = 12 \text{ m} \cdot 12 \text{ m} + \frac{1}{2} \cdot 12 \text{ m} \cdot 9 \text{ m}$
$= 144 \text{ m}^2 + 54 \text{ m}^2$
$= 198 \text{ m}^2$

11. First, find the area of the entire figure.

$$A = l \cdot w = 52 \text{ m} \cdot 37 \text{ m} = 1924 \text{ m}^2$$

Find the area of the unshaded triangle.

$$A = \frac{1}{2} \cdot b \cdot h = \frac{1}{2} \cdot 52 \text{ m} \cdot 10 \text{ m} = 260 \text{ m}^2$$

Shaded area $= 1924 \text{ m}^2 - 260 \text{ m}^2 = 1664 \text{ m}^2$

13. *Step 1* Add the two angles given.
$90° + 58° = 148°$
Step 2 Subtract the sum from 180°.
$180° - 148° = 32°$
The third angle measures 32°.

15. *Step 1* $72° + 60° = 132°$
Step 2 $180° - 132° = 48°$
The third angle measures 48°.

17. A triangle *cannot* have two right angles. Right angles are 90°, so two right angles are 180°, and the sum of all *three* angles in a triangle equals 180°.

19. $A = \frac{1}{2} \cdot b \cdot h$
$= \frac{1}{2} \cdot 3\frac{1}{2} \text{ ft} \cdot 4\frac{1}{2} \text{ ft}$
$= \frac{1}{2} \cdot \frac{7}{2} \text{ ft} \cdot \frac{9}{2} \text{ ft}$
$= \frac{63}{8} \text{ ft}^2$
$= 7\frac{7}{8} \text{ ft}^2$ or 7.875 ft^2

To make the flap, $7\frac{7}{8} \text{ ft}^2$ or 7.875 ft^2 of canvas is needed.

21. **(a)** To find the amount of curbing needed to go around the space, find the perimeter of the triangle.
$P = 42 \text{ m} + 32 \text{ m} + 52.8 \text{ m} = 126.8 \text{ m}$
126.8 m of curbing will be needed.

(b) To find the amount of sod needed to cover the space, find the area. It is a right triangle, so the perpendicular sides are the base and height.

$$A = \frac{1}{2} \cdot 32 \text{ m} \cdot 42 \text{ m} = 672 \text{ m}^2$$

672 m^2 of sod will be needed.

23. **(a)** Area of one side of the house:

$$A = \frac{1}{2} \cdot b \cdot h$$
$$= \frac{1}{2} \cdot 8 \text{ m} \cdot 8 \text{ m}$$
$$= 32 \text{ m}^2$$

(b) Area of one roof section:

$$A = 0.5 \cdot b \cdot h$$
$$= 0.5 \cdot 3 \text{ m} \cdot 9 \text{ m}$$
$$= 13.5 \text{ m}^2$$

25. **(a)** The frontage of the lot is

$$100 \text{ yd} + 75 \text{ yd} = 175 \text{ yd.}$$

(b) Draw a vertical line to break the region into a triangle and a rectangle. For the triangular region, the base is 75 yd and the height is 100 yd.

$$A = 0.5 \cdot b \cdot h$$
$$= 0.5 \cdot 75 \text{ yd} \cdot 100 \text{ yd}$$
$$= 3750 \text{ yd}^2$$

Rectangular area:

$$A = l \cdot w$$
$$= 100 \text{ yd} \cdot 50 \text{ yd}$$
$$= 5000 \text{ yd}^2$$

Total area $= 3750 \text{ yd}^2 + 5000 \text{ yd}^2 = 8750 \text{ yd}^2$

8.6 Circles

8.6 Margin Exercises

1. **(a)** diameter: 40 ft

$$r = \frac{d}{2} = \frac{40 \text{ ft}}{2} = 20 \text{ ft}$$

(b) diameter: 11 cm

$$r = \frac{d}{2} = \frac{11 \text{ cm}}{2} = 5.5 \text{ cm}$$

(c) radius: 32 yd

$$d = 2 \cdot r = 2 \cdot 32 \text{ yd} = 64 \text{ yd}$$

(d) radius: 9.5 m

$$d = 2 \cdot r = 2 \cdot 9.5 \text{ m} = 19 \text{ m}$$

2. **(a)** diameter: 150 ft

$$C = \pi \cdot d$$
$$\approx 3.14 \cdot 150 \text{ ft}$$
$$= 471 \text{ ft}$$

(b) radius: 7 in.

$$C = 2 \cdot \pi \cdot r$$
$$\approx 2 \cdot 3.14 \cdot 7 \text{ in.}$$
$$\approx 44.0 \text{ in.}$$

(c) diameter: 0.9 km

$$C = \pi \cdot d$$
$$\approx 3.14 \cdot 0.9 \text{ km}$$
$$\approx 2.8 \text{ km}$$

(d) radius: 4.6 m

$$C = 2 \cdot \pi \cdot r$$
$$\approx 2 \cdot 3.14 \cdot 4.6 \text{ m}$$
$$\approx 28.9 \text{ m}$$

3. **(a)** radius: 1 cm

$$A = \pi \cdot r \cdot r$$
$$\approx 3.14 \cdot 1 \text{ cm} \cdot 1 \text{ cm}$$
$$\approx 3.1 \text{ cm}^2$$

(b) diameter: 12 m, so $r = 6$ m

$$A = \pi \cdot r \cdot r$$
$$\approx 3.14 \cdot 6 \text{ m} \cdot 6 \text{ m}$$
$$\approx 113.0 \text{ m}^2$$

(c) radius: 1.8 km

$$A = \pi \cdot r \cdot r$$
$$\approx 3.14 \cdot 1.8 \text{ km} \cdot 1.8 \text{ km}$$
$$\approx 10.2 \text{ km}^2$$

(d) diameter: 8.4 cm, so $r = 4.2$ cm

$$A = \pi \cdot r \cdot r$$
$$\approx 3.14 \cdot 4.2 \text{ cm} \cdot 4.2 \text{ cm}$$
$$\approx 55.4 \text{ cm}^2$$

4. **(a)** radius: 24 m
Area of circle

$$A = \pi \cdot r \cdot r$$
$$\approx 3.14 \cdot 24 \text{ m} \cdot 24 \text{ m}$$
$$= 1808.64 \text{ m}^2$$

Area of semicircle

$$\frac{1808.64 \text{ m}^2}{2} \approx 904.3 \text{ m}^2$$

(b) diameter: 35.4 ft, so $r = 17.7$ ft
Area of circle

$$A = \pi \cdot r \cdot r$$
$$\approx 3.14 \cdot 17.7 \text{ ft} \cdot 17.7 \text{ ft}$$
$$= 983.7306 \text{ ft}^2$$

Area of semicircle

$$\frac{983.7306 \text{ ft}^2}{2} \approx 491.9 \text{ ft}^2$$

(c) radius: 9.8 m
Area of circle

$$A = \pi \cdot r \cdot r$$
$$\approx 3.14 \cdot 9.8 \text{ m} \cdot 9.8 \text{ m}$$
$$= 301.5656 \text{ m}^2$$

Area of semicircle

$$\frac{301.5656 \text{ m}^2}{2} \approx 150.8 \text{ m}^2$$

5. $C = \pi \cdot d$
$\approx 3.14 \cdot 3$ m
$= 9.42$ m

$$\text{Cost of binding} = \frac{9.42 \text{ m}}{1} \cdot \frac{\$4.50}{1 \text{ m}} = \$42.39$$

6. diameter: 3 m, so $r = 1\frac{1}{2}$ m

$$A = \pi \cdot r \cdot r$$
$$\approx 3.14 \cdot 1\tfrac{1}{2} \text{ m} \cdot 1\tfrac{1}{2} \text{ m}$$
$$= 7.065 \text{ m}^2$$

$$\text{Total cost} = \frac{7.065}{1} \cdot \frac{\$2}{\text{m}^2} = \$14.13$$

7. **(a)** Some possibilities are:

*dia*meter; *dia*gonal
*fract*ion; *fract*ure
*para*llel; *para*medic
*per*cent; *per* capita
*peri*meter; *peri*scope
*rad*ius; *rad*iate
*rect*angle; *rect*ify
*sub*tract; *sub*marine

(b) *Peri* in perimeter means "around," so perimeter is the distance around the edges of a shape.

8.6 Section Exercises

1. The radius is 9 mm so the diameter is

$$d = 2 \cdot r$$
$$= 2 \cdot 9 \text{ mm}$$
$$= 18 \text{ mm.}$$

3. The diameter is 0.7 km, so the radius is

$$\frac{d}{2} = \frac{0.7 \text{ km}}{2} = 0.35 \text{ km.}$$

5. radius: 11 ft

$$C = 2 \cdot \pi \cdot r$$
$$\approx 2 \cdot 3.14 \cdot 11 \text{ ft}$$
$$\approx 69.1 \text{ ft}$$
$$A = \pi \cdot r \cdot r$$
$$\approx 3.14 \cdot 11 \text{ ft} \cdot 11 \text{ ft}$$
$$\approx 379.9 \text{ ft}^2$$

7. diameter: 2.6 m

$$C = \pi \cdot d$$
$$\approx 3.14 \cdot 2.6 \text{ m}$$
$$\approx 8.2 \text{ m}$$

$$\text{radius: } \frac{2.6 \text{ m}}{2} = 1.3 \text{ m}$$
$$A = \pi \cdot r \cdot r$$
$$\approx 3.14 \cdot 1.3 \text{ m} \cdot 1.3 \text{ m}$$
$$\approx 5.3 \text{ m}^2$$

9. $d = 15$ cm

$$C = \pi \cdot d$$
$$\approx 3.14 \cdot 15 \text{ cm}$$
$$= 47.1 \text{ cm}$$
$$r = \frac{15 \text{ cm}}{2} = 7.5 \text{ cm}$$
$$A = \pi \cdot r \cdot r$$
$$\approx 3.14 \cdot 7.5 \text{ cm} \cdot 7.5 \text{ cm}$$
$$\approx 176.6 \text{ cm}^2$$

11. $d = 7\frac{1}{2}$ ft

$$C = \pi \cdot d$$
$$\approx 3.14 \cdot 7.5 \text{ ft}$$
$$\approx 23.6 \text{ ft}$$
$$r = \frac{7\frac{1}{2} \text{ ft}}{2} = 3.75 \text{ ft}$$
$$A = \pi \cdot r \cdot r$$
$$\approx 3.14 \cdot 3.75 \text{ ft} \cdot 3.75 \text{ ft}$$
$$\approx 44.2 \text{ ft}^2$$

13. $d = 8.65$ km

$$C = \pi \cdot d$$
$$\approx 3.14 \cdot 8.65 \text{ km}$$
$$\approx 27.2 \text{ km}$$
$$r = \frac{8.65 \text{ km}}{2} = 4.325 \text{ km}$$
$$A = \pi \cdot r \cdot r$$
$$\approx 3.14 \cdot 4.325 \text{ km} \cdot 4.325 \text{ km}$$
$$\approx 58.7 \text{ km}^2$$

15. Area of the circle:

$$A = \pi \cdot r \cdot r$$
$$\approx 3.14 \cdot 10 \text{ cm} \cdot 10 \text{ cm}$$
$$= 314 \text{ cm}^2$$

Area of the semicircle:

$$\frac{314 \text{ cm}^2}{2} = 157 \text{ cm}^2$$

continued

Area of the triangle:

$$A = \frac{1}{2} \cdot b \cdot h$$
$$= \frac{1}{2} \cdot 20 \text{ cm} \cdot 10 \text{ cm}$$
$$= 100 \text{ cm}^2$$

The shaded area is about
$157 \text{ cm}^2 - 100 \text{ cm}^2 = 57 \text{ cm}^2$.

17. Area of large circle $\approx 3.14 \cdot 12 \text{ cm} \cdot 12 \text{ cm}$
$= 452.16 \text{ cm}^2$
Area of small circle $\approx 3.14 \cdot 9 \text{ cm} \cdot 9 \text{ cm}$
$= 254.34 \text{ cm}^2$
Shaded area $\approx 452.16 \text{ cm}^2 - 254.34 \text{ cm}^2$
$= 197.8 \text{ cm}^2$

19. Answers will vary. A sample answer follows: π is the ratio of circumference of a circle to its diameter. If you divide the circumference of any circle by its diameter, the answer is always a little more than 3. The approximate value is 3.14, which we call π (pi). Your test question could involve finding the circumference or the area of a circle.

21. $A = \pi \cdot r \cdot r$
$\approx 3.14 \cdot 50 \text{ yd} \cdot 50 \text{ yd}$
$= 7850 \text{ yd}^2$

The watered area is about 7850 yd^2.

23. A point on the tire tread moves the length of the circumference in one complete turn.

$C = \pi \cdot d$
$\approx 3.14 \cdot 29.10 \text{ in.}$
$\approx 91.4 \text{ in.}$

Bonus question:

$$\frac{1 \text{ revolution}}{91.4 \text{ inches}} \cdot \frac{12 \text{ inches}}{1 \text{ foot}} \cdot \frac{5280 \text{ feet}}{1 \text{ mile}} \approx 693 \text{ revolutions/mile}$$

25.

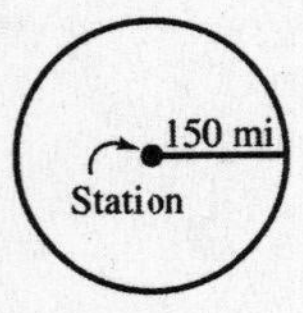

$A = \pi \cdot r \cdot r$
$\approx 3.14 \cdot 150 \text{ mi} \cdot 150 \text{ mi}$
$= 70{,}650 \text{ mi}^2$

There are about $70{,}650 \text{ mi}^2$ in the broadcast area.

27.

Watch

1 inch

Watch: $d = 1 \text{ in.}, r = \frac{1}{2} \text{ in.}$

$C = \pi \cdot 1 \text{ in.}$
$\approx 3.14 \cdot 1 \text{ in.}$
$\approx 3.1 \text{ in.}$
$A = \pi \cdot r \cdot r$
$\approx 3.14 \cdot \frac{1}{2} \text{ in.} \cdot \frac{1}{2} \text{ in.}$
$\approx 0.8 \text{ in.}^2$

Wall Clock

Wall clock: $r = 3$ in.

$C = 2 \cdot \pi \cdot 3 \text{ in.}$
$\approx 2 \cdot 3.14 \cdot 3 \text{ in.}$
$\approx 18.8 \text{ in.}$
$A = \pi \cdot r \cdot r$
$\approx 3.14 \cdot 3 \text{ in.} \cdot 3 \text{ in.}$
$\approx 28.3 \text{ in.}^2$

29.

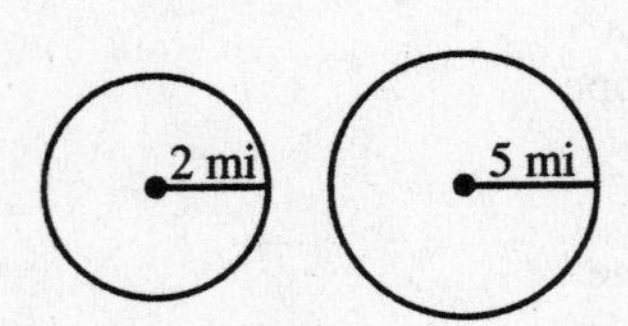

Area of larger circle:
$A = \pi \cdot r \cdot r$
$\approx 3.14 \cdot 5 \text{ mi} \cdot 5 \text{ mi}$
$= 78.5 \text{ mi}^2$

Area of smaller circle:
$A = \pi \cdot r \cdot r$
$\approx 3.14 \cdot 2 \text{ mi} \cdot 2 \text{ mi}$
$\approx 12.6 \text{ mi}^2$

The difference in the area covered is about

$$78.5 \text{ mi}^2 - 12.6 \text{ mi}^2 = 65.9 \text{ mi}^2.$$

31. (a) $C = 144 \text{ cm}$
$C = \pi \cdot d$
$144 \text{ cm} = \pi \cdot d$
$144 \text{ cm} \approx 3.14 \cdot d$
$\frac{144 \text{ cm}}{3.14} \approx d$
$d \approx 45.9 \text{ cm}$

The diameter is about 45.9 cm.

(b) Divide the circumference by π (3.14).

33. Find the area of the rectangle with length 29 ft and width 16 ft.

$$A = \text{length} \cdot \text{width}$$
$$= 29 \text{ ft} \cdot 16 \text{ ft}$$
$$= 464 \text{ ft}^2$$

The semicircles are of equal area, and the sum of their areas equals that of a circle. The radius is 8 ft.

$$A = \pi \cdot r \cdot r$$
$$\approx 3.14 \cdot 8 \text{ ft} \cdot 8 \text{ ft}$$
$$= 200.96 \text{ ft}^2$$

Total area $\approx 464 \text{ ft}^2 + 200.96 \text{ ft}^2 = 664.96 \text{ ft}^2$

Cost of the sod $= \dfrac{664.96 \text{ ft}^2}{1} \cdot \dfrac{\$1.76}{1 \text{ ft}^2} \approx \$1170.33$

35. The prefix *rad* tells you that radius is a ray from the center of the circle. The prefix *dia* means the diameter goes through the circle, and the prefix *circum* means the circumference is the distance around.

37. The diameter is $7\frac{1}{2}$ in., so the radius is

$$\frac{7\frac{1}{2} \text{ in.}}{2} = 3\tfrac{3}{4} \text{ in. or } 3.75 \text{ in.}$$

$$A = \pi \cdot r \cdot r$$
$$\approx 3.14 \cdot 3.75 \text{ in.} \cdot 3.75 \text{ in.}$$
$$= 44.15625 \text{ in.}^2$$

The area of a small pizza is about 44.2 in.^2.

38. The diameter is 13 in., so the radius is

$$\frac{13 \text{ in.}}{2} = 6\tfrac{1}{2} \text{ in. or } 6.5 \text{ in.}$$

$$A = \pi \cdot r \cdot r$$
$$\approx 3.14 \cdot 6.5 \text{ in.} \cdot 6.5 \text{ in.}$$
$$= 132.665 \text{ in.}^2$$

The area of a medium pizza is about 132.7 in.^2.

39. The diameter is 16 in., so the radius is

$$\frac{16 \text{ in.}}{2} = 8 \text{ in.}$$

$$A = \pi \cdot r \cdot r$$
$$\approx 3.14 \cdot 8 \text{ in.} \cdot 8 \text{ in.}$$
$$= 200.96 \text{ in.}^2$$

The area of a large pizza is about 201.0 in.^2.

40.

Size	Cost per square inch
Small	$\dfrac{\$2.80}{44.2 \text{ in.}^2} \approx \$0.063$
Medium	$\dfrac{\$6.50}{132.7 \text{ in.}^2} \approx \$0.049$
Large	$\dfrac{\$9.30}{201.0 \text{ in.}^2} \approx \$0.046$ (*)

The best buy is the large pizza.

41.

Size	Cost per square inch
Small	$\dfrac{\$3.70}{44.2 \text{ in.}^2} \approx \$0.084$
Medium	$\dfrac{\$8.95}{132.7 \text{ in.}^2} \approx \$0.067$ (*)
Large	$\dfrac{\$14.30}{201.0 \text{ in.}^2} \approx \$0.071$

The best buy is the medium pizza.

42. Cost of small pizza $= \$4.35 - \$0.95 = \$3.40$

Size	Cost per square inch
Small	$\dfrac{\$3.40}{44.2 \text{ in.}^2} \approx \$0.077$ (*)
Medium	$\dfrac{\$10.95}{132.7 \text{ in.}^2} \approx \$0.083$
Large	$\dfrac{\$15.65}{201.0 \text{ in.}^2} \approx \$0.078$

The best buy is the small pizza.

Summary Exercises on Perimeter, Circumference, and Area

1. (a)

All sides have the same length.

(b)

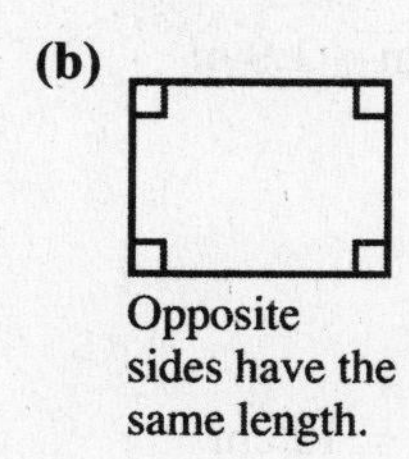

Opposite sides have the same length.

(c)

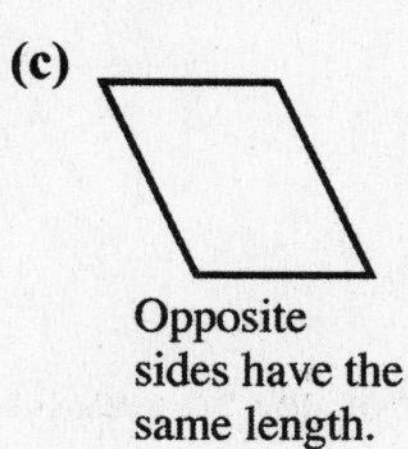

Opposite sides have the same length.

3. Add up the lengths of all the sides to find the perimeter.

5. parallelogram: (f) $A = b \cdot h$

square: (e) $A = s^2$

trapezoid: (d) $A = \frac{1}{2} \cdot h \cdot (b + B)$

circle: (b) $A = \pi \cdot r^2$

rectangle: (c) $A = l \cdot w$

triangle: (a) $A = \frac{1}{2} \cdot b \cdot h$

7. 1st triangle:
$P = 18 \text{ ft} + 21 \text{ ft} + 24 \text{ ft} = 63 \text{ ft}$

$$A = \frac{1}{2} \cdot b \cdot h$$
$$= \frac{1}{2} \cdot 24 \text{ ft} \cdot 15 \text{ ft}$$
$$= 180 \text{ ft}^2$$

2nd triangle:
$P = 9.5 \text{ cm} + 9.5 \text{ cm} + 13.4 \text{ cm} = 32.4 \text{ cm}$

$$A = \frac{1}{2} \cdot b \cdot h$$
$$= \frac{1}{2} \cdot 9.5 \text{ cm} \cdot 9.5 \text{ cm}$$
$$= 45.125 \text{ cm}^2 \approx 45.1 \text{ cm}^2$$

9. The figure is a rectangle.

$$P = 2 \cdot l + 2 \cdot w$$
$$= 2 \cdot 10\tfrac{1}{2} \text{ in.} + 2 \cdot 3 \text{ in.}$$
$$= 21 \text{ in.} + 6 \text{ in.}$$
$$= 27 \text{ in.}$$
$$A = l \cdot w$$
$$= 10\tfrac{1}{2} \text{ in.} \cdot 3 \text{ in.}$$
$$= 31\tfrac{1}{2} \text{ in.}^2 \text{ or } 31.5 \text{ in.}^2$$

11. The figure is a parallelogram.

$$P = 1.3 \text{ m} + 1.9 \text{ m} + 1.3 \text{ m} + 1.9 \text{ m}$$
$$= 6.4 \text{ m}$$
$$A = b \cdot h$$
$$= 1.3 \text{ m} \cdot 1.7 \text{ m}$$
$$= 2.21 \text{ m}^2 \approx 2.2 \text{ m}^2$$

13. **(a)** $r = 6 \text{ cm}$; $d = 2 \cdot 6 \text{ cm} = 12 \text{ cm}$

(b) $C = 2 \cdot \pi \cdot r$
$$\approx 2 \cdot \pi \cdot 6 \text{ cm}$$
$$\approx 37.7 \text{ cm}$$

(c) $A = \pi \cdot r \cdot r$
$$\approx 3.14 \cdot 6 \text{ cm} \cdot 6 \text{ cm}$$
$$\approx 113.0 \text{ cm}^2$$

15. **(a)** $d = 9 \text{ ft}$; $r = \frac{9 \text{ ft}}{2} = 4.5 \text{ ft}$

(b) $C = \pi \cdot d$
$$\approx 3.14 \cdot 9 \text{ ft}$$
$$\approx 28.3 \text{ ft}$$

(c) $A = \pi \cdot r \cdot r$
$$\approx 3.14 \cdot 4.5 \text{ ft} \cdot 4.5 \text{ ft}$$
$$\approx 63.6 \text{ ft}^2$$

17. Subtract the area of the triangle from the area of the rectangle.

Triangle:
$A = \frac{1}{2} \cdot b \cdot h = \frac{1}{2} \cdot 15 \text{ yd} \cdot 13 \text{ yd} = 97.5 \text{ yd}^2$

Rectangle: $A = l \cdot w = 21 \text{ yd} \cdot 15 \text{ yd} = 315 \text{ yd}^2$

Shaded area $= 315 \text{ yd}^2 - 97.5 \text{ yd}^2 = 217.5 \text{ yd}^2$

The shaded area is 217.5 yd^2.

19. $C = 2 \cdot \pi \cdot r \approx 2 \cdot 3.14 \cdot 2.33 \text{ ft} \approx 14.6 \text{ ft}$
The rag traveled about 14.6 ft with each revolution of the wheel.

Bonus question:

$$\frac{1 \text{ revolution}}{14.6 \text{ ft}} \cdot \frac{5280 \text{ ft}}{1 \text{ mi}} \approx 361.6 \text{ revolutions/mile}$$

The wheel makes about 361.6 revolutions in one mile. The Mormons used 360, which is the answer you get using $2\frac{1}{3}$ ft instead of 2.33 ft as the radius.

8.7 Volume

8.7 Margin Exercises

1. **(a)** $V = \text{length} \cdot \text{width} \cdot \text{height}$
$$= 8 \text{ m} \cdot 3 \text{ m} \cdot 3 \text{ m}$$
$$= 72 \text{ m}^3$$

(b) $V = \text{length} \cdot \text{width} \cdot \text{height}$
$$= 23.4 \text{ cm} \cdot 15.2 \text{ cm} \cdot 52.3 \text{ cm}$$
$$= 18{,}602.064 \text{ cm}^3$$
$$\approx 18{,}602.1 \text{ cm}^3$$

(c) $V = \text{length} \cdot \text{width} \cdot \text{height}$
$$= 6\tfrac{1}{4} \text{ ft} \cdot 3\tfrac{1}{2} \text{ ft} \cdot 2 \text{ ft}$$
$$= 6.25 \text{ ft} \cdot 3.5 \text{ ft} \cdot 2 \text{ ft}$$
$$= 43\tfrac{3}{4} \text{ ft}^3 \text{ or } 43.75 \text{ ft}^3$$
$$\approx 43.8 \text{ ft}^3$$

2. **(a)** $V = \frac{4}{3} \cdot \pi \cdot r^3$
$$\approx \frac{4 \cdot 3.14 \cdot 12 \text{ in.} \cdot 12 \text{ in.} \cdot 12 \text{ in.}}{3}$$
$$\approx 7234.6 \text{ in.}^3$$

(b) $V = \frac{4}{3} \cdot \pi \cdot r^3$

$\approx \frac{4 \cdot 3.14 \cdot 3.5 \text{ m} \cdot 3.5 \text{ m} \cdot 3.5 \text{ m}}{3}$

$\approx 179.5 \text{ m}^3$

(c) $V = \frac{4}{3} \cdot \pi \cdot r^3$

$\approx \frac{4 \cdot 3.14 \cdot 2.7 \text{ cm} \cdot 2.7 \text{ cm} \cdot 2.7 \text{ cm}}{3}$

$\approx 82.4 \text{ cm}^3$

3. **(a)** $V = \frac{2}{3} \cdot \pi \cdot r^3$

$\approx \frac{2 \cdot 3.14 \cdot 15 \text{ ft} \cdot 15 \text{ ft} \cdot 15 \text{ ft}}{3}$

$\approx 7065 \text{ ft}^3$

(b) $V = \frac{2}{3} \cdot \pi \cdot r^3$

$\approx \frac{2 \cdot 3.14 \cdot 6 \text{ cm} \cdot 6 \text{ cm} \cdot 6 \text{ cm}}{3}$

$\approx 452.2 \text{ cm}^3$

4. **(a)** $V = \pi \cdot r \cdot r \cdot h$

$\approx 3.14 \cdot 4 \text{ ft} \cdot 4 \text{ ft} \cdot 12 \text{ ft}$

$\approx 602.9 \text{ ft}^3$

(b) $r = \frac{7 \text{ cm}}{2} = 3.5 \text{ cm}$

$V = \pi \cdot r \cdot r \cdot h$

$\approx 3.14 \cdot 3.5 \text{ cm} \cdot 3.5 \text{ cm} \cdot 6 \text{ cm}$

$\approx 230.8 \text{ cm}^3$

5. First find the area of the circular base.

$B = \pi \cdot r \cdot r$

$\approx 3.14 \cdot 2 \text{ ft} \cdot 2 \text{ ft}$

$= 12.56 \text{ ft}^2$

Now find the volume of the cone.

$V = \frac{B \cdot h}{3}$

$\approx \frac{12.56 \text{ ft}^2 \cdot 11 \text{ ft}}{3}$

$\approx 46.1 \text{ ft}^3$

6. First find the area of the square base.

$B = s \cdot s$

$= 10 \text{ m} \cdot 10 \text{ m}$

$= 100 \text{ m}^2$

Now find the volume of the pyramid.

$V = \frac{B \cdot h}{3}$

$= \frac{100 \text{ m}^2 \cdot 8 \text{ m}}{3}$

$\approx 266.7 \text{ m}^3$

8.7 Section Exercises

1. The figure is a rectangular solid.

$V = l \cdot w \cdot h$

$= 12.5 \text{ cm} \cdot 4 \text{ cm} \cdot 11 \text{ cm}$

$= 550 \text{ cm}^3$

3. The figure is a sphere.

$V = \frac{4}{3} \cdot \pi \cdot r^3$

$\approx \frac{4 \cdot 3.14 \cdot 22 \text{ m} \cdot 22 \text{ m} \cdot 22 \text{ m}}{3}$

$\approx 44{,}579.6 \text{ m}^3$

5. The figure is a hemisphere.

$V = \frac{2}{3} \cdot \pi \cdot r^3$

$\approx \frac{2 \cdot 3.14 \cdot 12 \text{ in.} \cdot 12 \text{ in.} \cdot 12 \text{ in.}}{3}$

$\approx 3617.3 \text{ in.}^3$

7. The figure is a cylinder.

$V = \pi \cdot r^2 \cdot h$

$\approx 3.14 \cdot 5 \text{ ft} \cdot 5 \text{ ft} \cdot 6 \text{ ft}$

$= 471 \text{ ft}^3$

9. The figure is a cone.

First find the area of the circular base.

$B = \pi \cdot r \cdot r$

$\approx 3.14 \cdot 5 \text{ m} \cdot 5 \text{ m}$

$= 78.5 \text{ m}^2$

Now find the volume of the cone.

$V = \frac{B \cdot h}{3}$

$\approx \frac{78.5 \text{ m}^2 \cdot 16 \text{ m}}{3}$

$\approx 418.7 \text{ m}^3$

11. The figure is a pyramid.

First find the area of the rectangular base.

$B = l \cdot w$

$= 8 \text{ cm} \cdot 15 \text{ cm}$

$= 120 \text{ cm}^2$

Now find the volume of the pyramid.

$V = \frac{B \cdot h}{3}$

$= \frac{120 \text{ cm}^2 \cdot 20 \text{ cm}}{3}$

$= 800 \text{ cm}^3$

13. Use the formula for the volume of a rectangular solid.

$$V = l \cdot w \cdot h$$
$$= 3\text{ in.} \cdot 8\text{ in.} \cdot \frac{3}{4}\text{ in.}$$
$$= \frac{\overset{6}{\cancel{24}}}{1}\text{ in.}^2 \cdot \frac{3}{\underset{1}{\cancel{4}}}\text{ in.}$$
$$= 18\text{ in.}^3$$

The volume of the box is 18 in.^{3}.

15. Use the formula for the volume of a sphere.

$$\text{radius} = \frac{16.8\text{ cm}}{2} = 8.4\text{ cm}$$
$$V = \frac{4}{3} \cdot \pi \cdot r^3$$
$$\approx \frac{4 \cdot 3.14 \cdot 8.4\text{ cm} \cdot 8.4\text{ cm} \cdot 8.4\text{ cm}}{3}$$
$$\approx 2481.5\text{ cm}^3$$

17. Use the formula for the volume of a pyramid.

First find the area of the square base.

$$B = s \cdot s$$
$$= 145\text{ m} \cdot 145\text{ m}$$
$$= 21{,}025\text{ m}^2$$

Now find the volume of the pyramid.

$$V = \frac{B \cdot h}{3}$$
$$= \frac{21{,}025\text{ m}^2 \cdot 93\text{ m}}{3}$$
$$= 651{,}775\text{ m}^3$$

The volume of the ancient stone pyramid is 651,775 m^3.

19. Use the formula for the volume of a cylinder.

$$\text{radius} = \frac{5\text{ ft}}{2} = 2.5\text{ ft}$$
$$V = \pi \cdot r^2 \cdot h$$
$$\approx 3.14 \cdot 2.5\text{ ft} \cdot 2.5\text{ ft} \cdot 200\text{ ft}$$
$$= 3925\text{ ft}^3$$

The volume of the city sewer pipe is about 3925 ft^3.

21. First, the student used the diameter of 7 cm, but should have used the radius of 3.5 cm in the formula.
Secondly, the units for volume are cm^3, not cm^2.
The correct answer is $V \approx 192.3\text{ cm}^3$.

23. The volume of the base is

$$V = l \cdot w \cdot h$$
$$= 11\text{ cm} \cdot 9\text{ cm} \cdot 3\text{ cm}$$
$$= 297\text{ cm}^3$$

The volume of the column is

$$V = l \cdot w \cdot h$$
$$= 9\text{ cm} \cdot 2\text{ cm} \cdot 12\text{ cm}$$
$$= 216\text{ cm}^3$$

Thus, the volume of shape is

$$297\text{ cm}^3 + 216\text{ cm}^3 = 513\text{ cm}^3.$$

8.8 Pythagorean Theorem

8.8 Margin Exercises

1. **(a)** $\sqrt{36} = 6$ because $6 \cdot 6 = 36$.

(b) $\sqrt{25} = 5$ because $5 \cdot 5 = 25$.

(c) $\sqrt{9} = 3$ because $3 \cdot 3 = 9$.

(d) $\sqrt{100} = 10$ because $10 \cdot 10 = 100$.

(e) $\sqrt{121} = 11$ because $11 \cdot 11 = 121$.

2. **(a)** $\sqrt{11}$
Calculator shows 3.31662479; round to 3.317.

(b) $\sqrt{40}$
Calculator shows 6.32455532; round to 6.325.

(c) $\sqrt{56}$
Calculator shows 7.48331477; round to 7.483.

(d) $\sqrt{196}$
Calculator shows 14; $\sqrt{196} = 14$ because $14 \cdot 14 = 196$.

(e) $\sqrt{147}$
Calculator shows 12.12435565; round to 12.124.

3. **(a)** legs: 5 in. and 12 in.

$$\text{hypotenuse} = \sqrt{(\text{leg})^2 + (\text{leg})^2}$$
$$= \sqrt{(5)^2 + (12)^2}$$
$$= \sqrt{25 + 144}$$
$$= \sqrt{169}$$
$$= 13\text{ in.}$$

(b) hypotenuse: 25 cm, leg: 7 cm

$$\text{leg} = \sqrt{(\text{hypotenuse})^2 - (\text{leg})^2}$$
$$= \sqrt{(25)^2 - (7)^2}$$
$$= \sqrt{625 - 49}$$
$$= \sqrt{576}$$
$$= 24\text{ cm}$$

(c) legs: 17 m and 13 m

$$\begin{aligned}\text{hypotenuse} &= \sqrt{(\text{leg})^2 + (\text{leg})^2}\\ &= \sqrt{(17)^2 + (13)^2}\\ &= \sqrt{289 + 169}\\ &= \sqrt{458}\\ &\approx 21.4 \text{ m}\end{aligned}$$

(d) hypotenuse: 20 ft, leg: 18 ft

$$\begin{aligned}\text{leg} &= \sqrt{(\text{hypotenuse})^2 - (\text{leg})^2}\\ &= \sqrt{(20)^2 - (18)^2}\\ &= \sqrt{400 - 324}\\ &= \sqrt{76}\\ &\approx 8.7 \text{ ft}\end{aligned}$$

(e) legs: 8 mm and 5 mm

$$\begin{aligned}\text{hypotenuse} &= \sqrt{(\text{leg})^2 + (\text{leg})^2}\\ &= \sqrt{(8)^2 + (5)^2}\\ &= \sqrt{64 + 25}\\ &= \sqrt{89}\\ &\approx 9.4 \text{ mm}\end{aligned}$$

4. **(a)** hypotenuse: 25 ft, leg: 20 ft

$$\begin{aligned}\text{leg} &= \sqrt{(\text{hypotenuse})^2 - (\text{leg})^2}\\ &= \sqrt{(25)^2 - (20)^2}\\ &= \sqrt{625 - 400}\\ &= \sqrt{225}\\ &= 15\end{aligned}$$

The bottom of the ladder is 15 ft from the building.

(b) legs: 11 ft and 8 ft

$$\begin{aligned}\text{hypotenuse} &= \sqrt{(\text{leg})^2 + (\text{leg})^2}\\ &= \sqrt{(11)^2 + (8)^2}\\ &= \sqrt{121 + 64}\\ &= \sqrt{185}\\ &\approx 13.6 \text{ ft}\end{aligned}$$

The ladder is about 13.6 ft long.

(c) hypotenuse: 17 ft, leg: 10 ft

17 ft

10 ft

$$\begin{aligned}\text{leg} &= \sqrt{(\text{hypotenuse})^2 - (\text{leg})^2}\\ &= \sqrt{(17)^2 - (10)^2}\\ &= \sqrt{289 - 100}\\ &= \sqrt{189}\\ &\approx 13.7 \text{ ft}\end{aligned}$$

The ladder will reach about 13.7 ft high up on the building.

8.8 Section Exercises

1. $\sqrt{16} = 4$ because $4 \cdot 4 = 16$.

3. $\sqrt{64} = 8$ because $8 \cdot 8 = 64$.

5. $\sqrt{11}$
Calculator shows 3.31662479; round to 3.317.

7. $\sqrt{5}$
Calculator shows 2.236067977; rounds to 2.236.

9. $\sqrt{73}$
Calculator shows 8.544003745; rounds to 8.544.

11. $\sqrt{101}$
Calculator shows 10.04987562; rounds to 10.050.

13. $\sqrt{190}$
Calculator shows 13.78404875; rounds to 13.784.

15. $\sqrt{1000}$
Calculator shows 31.6227766; rounds to 31.623.

17. 30 is about halfway between 25 and 36, so $\sqrt{30}$ should be about halfway between 5 and 6, or about 5.5. Using a calculator, $\sqrt{30} \approx 5.477$. Similarly, $\sqrt{26}$ should be a little more than $\sqrt{25}$. By calculator, $\sqrt{26} \approx 5.099$. Also, $\sqrt{35}$ should be a little less than $\sqrt{36}$. By calculator, $\sqrt{35} \approx 5.916$.

19. legs: 15 ft and 36 ft

$$\begin{aligned}\text{hypotenuse} &= \sqrt{(\text{leg})^2 + (\text{leg})^2}\\ &= \sqrt{(15)^2 + (36)^2}\\ &= \sqrt{225 + 1296}\\ &= \sqrt{1521}\\ &= 39 \text{ ft}\end{aligned}$$

21. legs: 8 in. and 15 in.

$$\begin{aligned}\text{hypotenuse} &= \sqrt{(\text{leg})^2 + (\text{leg})^2}\\ &= \sqrt{(8)^2 + (15)^2}\\ &= \sqrt{64 + 225}\\ &= \sqrt{289}\\ &= 17 \text{ in.}\end{aligned}$$

23. hypotenuse: 20 mm, leg: 16 mm

$$\begin{aligned} \text{leg} &= \sqrt{(\text{hypotenuse})^2 - (\text{leg})^2} \\ &= \sqrt{(20)^2 - (16)^2} \\ &= \sqrt{400 - 256} \\ &= \sqrt{144} \\ &= 12 \text{ mm} \end{aligned}$$

25. legs: 8 in. and 3 in.

$$\begin{aligned} \text{hypotenuse} &= \sqrt{(\text{leg})^2 + (\text{leg})^2} \\ &= \sqrt{(8)^2 + (3)^2} \\ &= \sqrt{64 + 9} \\ &= \sqrt{73} \\ &\approx 8.5 \text{ in.} \end{aligned}$$

27. legs: 7 yd and 4 yd

$$\begin{aligned} \text{hypotenuse} &= \sqrt{(\text{leg})^2 + (\text{leg})^2} \\ &= \sqrt{(7)^2 + (4)^2} \\ &= \sqrt{49 + 16} \\ &= \sqrt{65} \\ &\approx 8.1 \text{ yd} \end{aligned}$$

29. hypotenuse: 22 cm, leg: 17 cm

$$\begin{aligned} \text{leg} &= \sqrt{(\text{hypotenuse})^2 - (\text{leg})^2} \\ &= \sqrt{(22)^2 - (17)^2} \\ &= \sqrt{484 - 289} \\ &= \sqrt{195} \\ &\approx 14.0 \text{ cm} \end{aligned}$$

31. legs: 1.3 m and 2.5 m

$$\begin{aligned} \text{hypotenuse} &= \sqrt{(\text{leg})^2 + (\text{leg})^2} \\ &= \sqrt{(1.3)^2 + (2.5)^2} \\ &= \sqrt{1.69 + 6.25} \\ &= \sqrt{7.94} \\ &\approx 2.8 \text{ m} \end{aligned}$$

33. hypotenuse: 11.5 cm, leg: 8.2 cm

$$\begin{aligned} \text{leg} &= \sqrt{(\text{hypotenuse})^2 - (\text{leg})^2} \\ &= \sqrt{(11.5)^2 - (8.2)^2} \\ &= \sqrt{132.25 - 67.24} \\ &= \sqrt{65.01} \\ &\approx 8.1 \text{ cm} \end{aligned}$$

35. hypotenuse: 21.6 km, leg: 13.2 km

$$\begin{aligned} \text{leg} &= \sqrt{(\text{hypotenuse})^2 - (\text{leg})^2} \\ &= \sqrt{(21.6)^2 - (13.2)^2} \\ &= \sqrt{466.56 - 174.24} \\ &= \sqrt{292.32} \\ &\approx 17.1 \text{ km} \end{aligned}$$

37. legs: 4 ft and 7 ft

$$\begin{aligned} \text{hypotenuse} &= \sqrt{(\text{leg})^2 + (\text{leg})^2} \\ &= \sqrt{(4)^2 + (7)^2} \\ &= \sqrt{16 + 49} \\ &= \sqrt{65} \\ &\approx 8.1 \text{ ft} \end{aligned}$$

The length of the loading ramp is about 8.1 ft.

39. hypotenuse: 1000 m, leg: 800 m

$$\begin{aligned} \text{leg} &= \sqrt{(\text{hypotenuse})^2 - (\text{leg})^2} \\ &= \sqrt{(1000)^2 - (800)^2} \\ &= \sqrt{1{,}000{,}000 - 640{,}000} \\ &= \sqrt{360{,}000} \\ &= 600 \text{ m} \end{aligned}$$

41. legs: 4.5 ft and 3.5 ft

$$\begin{aligned} \text{hypotenuse} &= \sqrt{(\text{leg})^2 + (\text{leg})^2} \\ &= \sqrt{(4.5)^2 + (3.5)^2} \\ &= \sqrt{20.25 + 12.25} \\ &= \sqrt{32.5} \\ &\approx 5.7 \text{ ft} \end{aligned}$$

The diagonal brace is about 5.7 ft long.

43.

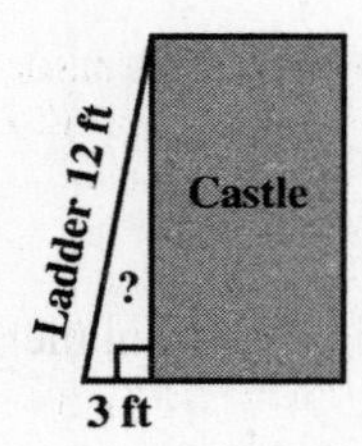

$$\begin{aligned} \text{leg} &= \sqrt{(\text{hypotenuse})^2 - (\text{leg})^2} \\ &= \sqrt{(12)^2 - (3)^2} \\ &= \sqrt{144 - 9} \\ &= \sqrt{135} \\ &\approx 11.6 \text{ ft} \end{aligned}$$

The ladder will reach about 11.6 ft high on the building.

45. The student did not square the numbers correctly: 9^2 is 81 and 7^2 is 49. Also, the final answer is rounded to thousandths instead of tenths. The correct answer is:

$$\begin{aligned} \sqrt{(9)^2 + (7)^2} &= \sqrt{81 + 49} \\ &= \sqrt{130} \\ &\approx 11.4 \text{ in.} \end{aligned}$$

47. legs: 90 ft and 90 ft

$$\begin{aligned}\text{hypotenuse} &= \sqrt{(\text{leg})^2 + (\text{leg})^2}\\ &= \sqrt{(90)^2 + (90)^2}\\ &= \sqrt{8100 + 8100}\\ &= \sqrt{16{,}200}\\ &\approx 127.3\end{aligned}$$

The distance from home plate to second base is about 127.3 ft.

48. (a)

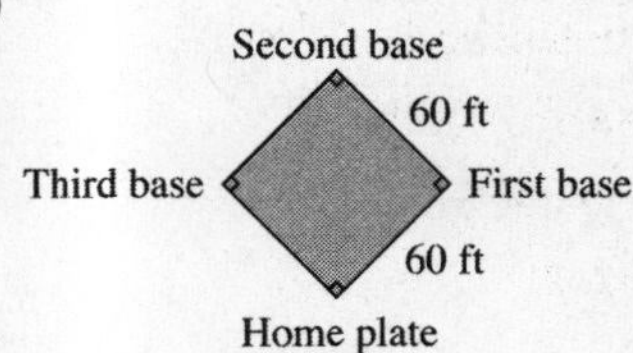

(b) legs: 60 ft and 60 ft

$$\begin{aligned}\text{hypotenuse} &= \sqrt{(\text{leg})^2 + (\text{leg})^2}\\ &= \sqrt{(60)^2 + (60)^2}\\ &= \sqrt{3600 + 3600}\\ &= \sqrt{7200}\\ &\approx 84.9\end{aligned}$$

The distance from home plate to second base is about 84.9 ft.

49. The distance from third to first is the same as the distance from home to second because the baseball diamond is a square.

50. One possibility is:

$$\text{major league}\left\{\frac{90\text{ ft}}{127.3\text{ ft}} = \frac{60\text{ ft}}{x}\right\}\text{softball}$$

$$\begin{aligned}90 \cdot x &= 127.3 \cdot 60\\ \frac{90 \cdot x}{90} &= \frac{7638}{90}\\ x &\approx 84.9\text{ ft}\end{aligned}$$

8.9 Similar Triangles

8.9 Margin Exercises

1. (a) Corresponding angles have the same measure.

The corresponding angles are 1 and 4, 2 and 5, 3 and 6.
$\overline{PN}$ and $\overline{ZX}$ are opposite corresponding angles 3 and 6.
$\overline{PM}$ and $\overline{ZY}$ are opposite corresponding angles 1 and 4.
$\overline{NM}$ and $\overline{XY}$ are opposite corresponding angles 2 and 5.
Thus, the corresponding sides are $\overline{PN}$ and $\overline{ZX}$, $\overline{PM}$ and $\overline{ZY}$, $\overline{NM}$ and $\overline{XY}$.

(b) The corresponding angles are 1 and 6, 2 and 4, 3 and 5.
The corresponding sides are $\overline{AB}$ and $\overline{EF}$, $\overline{BC}$ and $\overline{FG}$, $\overline{AC}$ and $\overline{EG}$.

2. Find the length of $\overline{EF}$.

$\frac{EF}{CB} = \frac{1}{3}$ *Ratio from Example 1*

Replace EF with x and CB with 33.

$$\begin{aligned}\frac{x}{33} &= \frac{1}{3}\\ 3 \cdot x &= 33 \cdot 1\\ \frac{3 \cdot x}{3} &= \frac{33}{3}\\ x &= 11\end{aligned}$$

$\overline{EF}$ has a length of 11 m.

3. (a) Find the length of $\overline{AB}$.
From Example 2,

$$\frac{PR}{AC} = \frac{7}{14} = \frac{1}{2}, \text{ so } \frac{PQ}{AB} = \frac{1}{2}.$$

Replace PQ with 3 and AB with y.

$$\begin{aligned}\frac{3}{y} &= \frac{1}{2}\\ y \cdot 1 &= 3 \cdot 2\\ y &= 6\end{aligned}$$

$\overline{AB}$ is 6 ft.
Find the perimeter.
Perimeter = 14 ft + 10 ft + 6 ft = 30 ft

(b) Set up ratios.

$$\frac{PQ}{AB} = \frac{10\text{ m}}{30\text{ m}} = \frac{1}{3}, \text{ so } \frac{QR}{BC} = \frac{1}{3} \text{ and } \frac{PR}{AC} = \frac{1}{3}.$$

Replace QR with x and BC with 18.

$$\begin{aligned}\frac{x}{18} &= \frac{1}{3}\\ 3 \cdot x &= 18 \cdot 1\\ \frac{3 \cdot x}{3} &= \frac{18}{3}\\ x &= 6\end{aligned}$$

$\overline{QR}$ is 6 m.
The perimeter of triangle PQR is

10 m + 8 m + 6 m = 24 m.

Next replace PR with 8 and AC with y.

$$\begin{aligned}\frac{8}{y} &= \frac{1}{3}\\ y \cdot 1 &= 8 \cdot 3\\ y &= 24\end{aligned}$$

$\overline{AC}$ is 24 m.
The perimeter of triangle ABC is

30 m + 18 m + 24 m = 72 m.

4. **(a)** Write a proportion.

$$\frac{h}{5} = \frac{48}{12}$$
$$12 \cdot h = 5 \cdot 48$$
$$\frac{12 \cdot h}{12} = \frac{240}{12}$$
$$h = 20$$

The flagpole is 20 ft high.

(b) Write a proportion.

$$\frac{h}{7.2} = \frac{12.5}{5}$$
$$5 \cdot h = 7.2 \cdot 12.5$$
$$\frac{5 \cdot h}{5} = \frac{90}{5}$$
$$h = 18$$

The flagpole is 18 m high.

8.9 Section Exercises

1. The triangles have the same shape, so they are similar.

3. The triangles do not have the same shape, so they are not similar.

5. The triangles have the same shape, so they are similar.

7. The corresponding angles are $\angle 1$ and $\angle 4$, $\angle 2$ and $\angle 5$, $\angle 3$ and $\angle 6$.
The corresponding sides are $\overline{AB}$ and $\overline{PQ}$, $\overline{BC}$ and $\overline{QR}$, $\overline{AC}$ and $\overline{PR}$.

9. The corresponding angles are $\angle 1$ and $\angle 6$, $\angle 2$ and $\angle 5$, $\angle 3$ and $\angle 4$.
The corresponding sides are $\overline{MP}$ and $\overline{QS}$, $\overline{MN}$ and $\overline{QR}$, $\overline{NP}$ and $\overline{RS}$.

11.
$$\frac{AB}{PQ} = \frac{9 \text{ m}}{6 \text{ m}} = \frac{9}{6} = \frac{3}{2}$$
$$\frac{AC}{PR} = \frac{12 \text{ m}}{8 \text{ m}} = \frac{12}{8} = \frac{3}{2}$$
$$\frac{BC}{QR} = \frac{15 \text{ m}}{10 \text{ m}} = \frac{15}{10} = \frac{3}{2}$$

13. Set up a ratio of corresponding sides.

$$\frac{6 \text{ mm}}{12 \text{ mm}} = \frac{6}{12} = \frac{1}{2}$$

Write a proportion to find a.

$$\frac{a}{10} = \frac{1}{2}$$
$$a \cdot 2 = 10 \cdot 1$$
$$\frac{a \cdot 2}{2} = \frac{10}{2}$$
$$a = 5 \text{ mm}$$

Write a proportion to find b.

$$\frac{b}{6} = \frac{1}{2}$$
$$b \cdot 2 = 6 \cdot 1$$
$$\frac{b \cdot 2}{2} = \frac{6}{2}$$
$$b = 3 \text{ mm}$$

15. Set up a ratio of corresponding sides.

$$\frac{6 \text{ cm}}{12 \text{ cm}} = \frac{6}{12} = \frac{1}{2}$$

Write a proportion to find a.

$$\frac{a}{12} = \frac{1}{2}$$
$$2 \cdot a = 12 \cdot 1$$
$$\frac{2 \cdot a}{2} = \frac{12}{2}$$
$$a = 6 \text{ cm}$$

Write a proportion to find b.

$$\frac{7.5}{b} = \frac{1}{2}$$
$$1 \cdot b = 2 \cdot 7.5$$
$$b = 15 \text{ cm}$$

17. Write a proportion to find x.

$$\frac{x}{18.6} = \frac{28}{21} \quad \textbf{OR} \quad \frac{x}{18.6} = \frac{4}{3}$$
$$3 \cdot x = 4 \cdot 18.6$$
$$\frac{3 \cdot x}{3} = \frac{74.4}{3}$$
$$x = 24.8 \text{ m}$$

$$P = 24.8 \text{ m} + 28 \text{ m} + 20 \text{ m} = 72.8 \text{ m}$$

Write a proportion to find y.

$$\frac{y}{20} = \frac{21}{28} \quad \textbf{OR} \quad \frac{y}{20} = \frac{3}{4}$$
$$4 \cdot y = 20 \cdot 3$$
$$\frac{4 \cdot y}{4} = \frac{60}{4}$$
$$y = 15 \text{ m}$$

$$P = 15 \text{ m} + 21 \text{ m} + 18.6 \text{ m} = 54.6 \text{ m}$$

19. Set up a ratio of corresponding sides to find the length of the missing sides in triangle FGH.

$$\frac{DE}{GH} = \frac{CE}{FH} \quad \text{Let } x = FH.$$
$$\frac{12}{8} = \frac{12}{x}$$
$$12 \cdot x = 8 \cdot 12$$
$$\frac{12 \cdot x}{12} = \frac{96}{12}$$
$$x = 8$$

Each missing side of triangle FGH is 8 cm.

$$\textit{Perimeter of triangle FGH} = 8 \text{ cm} + 8 \text{ cm} + 8 \text{ cm} = 24 \text{ cm}$$

Set up a ratio of corresponding sides to find the height h of triangle FGH.

$$\frac{10.4}{12} = \frac{h}{8}$$
$$12 \cdot h = 8 \cdot 10.4$$
$$\frac{12 \cdot h}{12} = \frac{83.2}{12}$$
$$h \approx 6.9 \text{ cm}$$

$$\textit{Area of triangle FGH} = 0.5 \cdot b \cdot h \approx 0.5 \cdot 8 \text{ cm} \cdot 6.9 \text{ cm} = 27.6 \text{ cm}^2$$

21. Write a proportion to find h.

$$\frac{2}{16} = \frac{3}{h}$$
$$2 \cdot h = 3 \cdot 16$$
$$\frac{2 \cdot h}{2} = \frac{48}{2}$$
$$h = 24 \text{ ft}$$

The height of the house is 24 ft.

23. Let s be the length of the stick's shadow. Write a proportion to find s.

$$\frac{h}{3} = \frac{6}{s}$$
$$\frac{24}{3} = \frac{6}{s} \quad (h = 24 \text{ from Exercise 21})$$
$$\frac{8}{1} = \frac{6}{s} \quad \text{(reduce)}$$
$$8 \cdot s = 1 \cdot 6$$
$$\frac{8 \cdot s}{8} = \frac{6}{8}$$
$$s = \frac{3}{4} \text{ ft or } 0.75 \text{ ft}$$

25. Answers will vary. A sample answer follows: One dictionary definition is "resembling, but not identical." Examples of similar objects are sets of different size pots or measuring cups; small and large size cans of beans; child's tennis shoe and adult's tennis shoe.

27. Using the hint, we can write a proportion to find x.

$$\frac{x}{120} = \frac{100}{100 + 140} \quad \textbf{OR} \quad \frac{x}{120} = \frac{5}{12}$$
$$x \cdot 12 = 120 \cdot 5$$
$$\frac{x \cdot 12}{12} = \frac{600}{12}$$
$$x = 50 \text{ m}$$

29. Write a proportion to find n.

$$\frac{50}{n} = \frac{100}{100 + 120} \quad \textbf{OR} \quad \frac{50}{n} = \frac{5}{11}$$
$$5 \cdot n = 50 \cdot 11$$
$$\frac{5 \cdot n}{5} = \frac{550}{5}$$
$$n = 110 \text{ m}$$

Chapter 8 Review Exercises

1. The figure has two endpoints so it is a *line segment* named $\overline{AB}$ or $\overline{BA}$.

2. This is a *line* named $\overleftrightarrow{CD}$ or $\overleftrightarrow{DC}$. A line is a straight row of points that goes on forever in both directions.

3. This is a *ray* named $\overrightarrow{OP}$. A ray is a part of a line that has only one endpoint and goes on forever in one direction.

4. These lines appear to be *parallel* lines. Parallel lines are lines in the same plane that never intersect (cross).

5. The lines are *perpendicular* because they intersect at right angles.

6. The lines intersect so they are *not* parallel. At their intersection they *do not* form a right angle so they are not perpendicular. The lines are *intersecting*.

7. The measure of the angle is between 0° and 90°, so it is an *acute angle.*

8. The measure of the angle is between 90° and 180°, so it is an *obtuse angle.*

9. Two rays in a straight line pointing opposite directions measure 180°. An angle that measures 180° is called a *straight angle.*

10. The angle is a *right angle*, as indicated by the small square at the vertex. Right angles measure exactly 90°.

11. $\angle 4$ measures 60° since $\angle 4$ and the 60° angle are vertical angles, which are congruent.
$\angle 4 + \angle 3 = 90°$, so $\angle 3 = 90° - 60° = 30°$.
$\angle 1$ also measures 30° since $\angle 1$ and $\angle 3$ are vertical angles.
$\angle 2$ measures 90° since $\angle 2$ and the 90° angle marked with the small square are vertical angles.

12. $\angle 1$ measures $100°$ since it is a vertical angle with the $100°$ angle and vertical angles are congruent.
$\angle 4 + 35° + 100° = 180°$ since they form a straight angle. Therefore, $\angle 4$ measures $180° - 35° - 100° = 45°$.
$\angle 2$ also measures $45°$ since $\angle 2$ and $\angle 4$ are vertical angles and vertical angles are congruent.
$\angle 3$ measures $35°$ since it is a vertical angle with the $35°$ angle.

13. $\angle AOB$ and $\angle BOC$,
$\angle BOC$ and $\angle COD$,
$\angle COD$ and $\angle DOA$, and
$\angle DOA$ and $\angle AOB$
are supplementary because the sum of each pair of angles is $180°$.

14. $\angle ERH$ and $\angle HRG$,
$\angle HRG$ and $\angle GRF$,
$\angle FRG$ and $\angle FRE$, and
$\angle FRE$ and $\angle ERH$
are supplementary because the sum of each pair of angles is $180°$.

15. **(a)** The complement of $80°$ is $10°$, because $90° - 80° = 10°$.

(b) The complement of $45°$ is $45°$, because $90° - 45° = 45°$.

(c) The complement of $7°$ is $83°$, because $90° - 7° = 83°$.

16. **(a)** The supplement of $155°$ is $25°$, because $180° - 155° = 25°$.

(b) The supplement of $90°$ is $90°$, because $180° - 90° = 90°$.

(c) The supplement of $33°$ is $147°$, because $180° - 33° = 147°$.

17. $P = 1.5\text{ m} + 0.92\text{ m} + 1.5\text{ m} + 0.92\text{ m}$
$= 4.84\text{ m}$

18. The figure is a square. So all 4 sides have the same measurement.
$P = 32\text{ in.} + 32\text{ in.} + 32\text{ in.} + 32\text{ in.}$
$= 128\text{ in.}$

19. $P = 4 \cdot s$
$= 4 \cdot 38\text{ cm}$
$= 152\text{ cm}$

To trim all the edges, 152 cm of trim is needed.

20. $P = 2 \cdot \text{length} + 2 \cdot \text{width}$
$= 2 \cdot 12\text{ ft} + 2 \cdot 8\tfrac{1}{2}\text{ ft}$
$= 24\text{ ft} + 17\text{ ft}$
$= 41\text{ ft}$

21. $A = \text{length} \cdot \text{width}$
$= 27\text{ mm} \cdot 18\text{ mm}$
$= 486\text{ mm}^2$

22. $A = \text{length} \cdot \text{width}$
$= 5\tfrac{1}{2}\text{ ft} \cdot 3\text{ ft}$
$= 5.5\text{ ft} \cdot 3\text{ ft}$
$= 16.5\text{ ft}^2 \text{ or } 16\tfrac{1}{2}\text{ ft}^2$

23. $A = s^2$
$= 6.3\text{ m} \cdot 6.3\text{ m}$
$= 39.69\text{ m}^2$
$\approx 39.7\text{ m}^2$

24. The figure is a parallelogram.

$P = 11\text{ cm} + 14\text{ cm} + 11\text{ cm} + 14\text{ cm}$
$= 50\text{ cm}$
$A = b \cdot h$
$= 14\text{ cm} \cdot 10\text{ cm}$
$= 140\text{ cm}^2$

25. The figure is a trapezoid.

$P = 26\text{ ft} + 21.1\text{ ft} + 37\text{ ft} + 18\text{ ft}$
$= 102.1\text{ ft}$
$A = \frac{1}{2} \cdot h \cdot (b + B)$
$= \frac{1}{2} \cdot 18\text{ ft} \cdot (26\text{ ft} + 37\text{ ft})$
$= \frac{1}{2} \cdot 18\text{ ft} \cdot 63\text{ ft}$
$= 567\text{ ft}^2$

26. The figure is a trapezoid.

$P = 33.9\text{ m} + 59.7\text{ m} + 34.2\text{ m} + 72.4\text{ m}$
$= 200.2\text{ m}$
$A = 0.5 \cdot h \cdot (b + B)$
$= 0.5 \cdot 31.4\text{ m} \cdot (59.7\text{ m} + 72.4\text{ m})$
$= 0.5 \cdot 31.4\text{ m} \cdot 132.1\text{ m}$
$= 2073.97\text{ m}^2 \approx 2074.0\text{ m}^2$

27. $P = 153\text{ cm} + 153\text{ cm} + 212\text{ cm}$
$= 518\text{ cm}$
$A = \frac{1}{2} \cdot b \cdot h$
$= \frac{1}{2} \cdot 212\text{ cm} \cdot 110\text{ cm}$
$= 11{,}660\text{ cm}^2$

28. $P = 5\text{ m} + 12.3\text{ m} + 9.8\text{ m}$
$= 27.1\text{ m}$
$A = \frac{1}{2} \cdot b \cdot h$
$= \frac{1}{2} \cdot 9.8\text{ m} \cdot 4.2\text{ m}$
$= 20.58\text{ m}^2$

29. $P = 8 \text{ ft} + 3\frac{1}{2} \text{ ft} + 8\frac{3}{4} \text{ ft}$
$= 8 \text{ ft} + 3.5 \text{ ft} + 8.75 \text{ ft}$
$= 20.25 \text{ ft or } 20\frac{1}{4} \text{ ft}$

$A = \frac{1}{2} \cdot b \cdot h$
$= \frac{1}{2} \cdot 3.5 \text{ ft} \cdot 8 \text{ ft}$
$= 14 \text{ ft}^2$

30. *Step 1* $70° + 40° = 110°$
Step 2 $180° - 110° = 70°$
The unlabeled angle measures 70°.

31. *Step 1* $90° + 66° = 156°$
Step 2 $180° - 156° = 24°$
The unlabeled angle measures 24°.

32. $d = 2 \cdot r = 2 \cdot 68.9 \text{ m} = 137.8 \text{ m}$
The diameter of the field is 137.8 m.

33. $r = \frac{d}{2} = \frac{3 \text{ in.}}{2} = 1\frac{1}{2} \text{ in. or } 1.5 \text{ in.}$

The radius of the juice can is $1\frac{1}{2}$ in. or 1.5 in.

34. radius: 1 cm, $d = 2$ cm

$C = \pi \cdot d$
$\approx 3.14 \cdot 2 \text{ cm}$
$\approx 6.3 \text{ cm}$
$A = \pi \cdot r \cdot r$
$\approx 3.14 \cdot 1 \text{ cm} \cdot 1 \text{ cm}$
$\approx 3.1 \text{ cm}^2$

35. radius: 17.4 m, $d = 34.8$ m

$C = \pi \cdot d$
$\approx 3.14 \cdot 34.8 \text{ m}$
$\approx 109.3 \text{ m}$
$A = \pi \cdot r \cdot r$
$\approx 3.14 \cdot 17.4 \text{ m} \cdot 17.4 \text{ m}$
$\approx 950.7 \text{ m}^2$

36. diameter: 12 in., radius: 6 in.

$C = \pi \cdot d$
$\approx 3.14 \cdot 12 \text{ in.}$
$\approx 37.7 \text{ in.}$
$A = \pi \cdot r \cdot r$
$\approx 3.14 \cdot 6 \text{ in.} \cdot 6 \text{ in.}$
$\approx 113.0 \text{ in.}^2$

37. Area of circle
radius: 3.6 m

$A = \pi \cdot r \cdot r$
$\approx 3.14 \cdot 3.6 \text{ m} \cdot 3.6 \text{ m}$
$= 40.6944 \text{ m}^2$

Area of semicircle

$\frac{40.6944 \text{ m}^2}{2} = 20.3472 \text{ m}^2 \approx 20.3 \text{ m}^2$

38. Draw a horizontal line to break up the shaded area into 2 rectangles that have a length of 8 inches and a width of 4 inches.

$A = \text{length} \cdot \text{width}$
$= 8 \text{ in.} \cdot 4 \text{ in.}$
$= 32 \text{ in.}^2$

Total area $= 2 \times 32 \text{ in.}^2 = 64 \text{ in.}^2$

39. Draw a vertical line to break up the figure into 3 parts.

(1) a rectangle with length 24 km and width 12 km
$A = 24 \text{ km} \cdot 12 \text{ km}$
$= 288 \text{ km}^2$

(2) a parallelogram with base 24 km and height 11 km

$A = b \cdot h$
$= 24 \text{ km} \cdot 11 \text{ km}$
$= 264 \text{ km}^2$

(3) a square with side 11 km

$A = s \cdot s$
$= 11 \text{ km} \cdot 11 \text{ km}$
$= 121 \text{ km}^2$

Total area $= 288 \text{ km}^2 + 264 \text{ km}^2 + 121 \text{ km}^2$
$= 673 \text{ km}^2$

40. Draw a vertical line to break up the figure into 2 parts.

(1) a rectangle with length 45 m and width 21 m
$A = 45 \text{ m} \cdot 21 \text{ m}$
$= 945 \text{ m}^2$

(2) a triangle with base 15 m and height 10 m

$A = \frac{1}{2} \cdot h \cdot b$
$= \frac{1}{2} \cdot 15 \text{ m} \cdot 10 \text{ m}$
$= 75 \text{ m}^2$

Total area $= 945 \text{ m}^2 + 75 \text{ m}^2$
$= 1020 \text{ m}^2$

41. Draw 2 horizontal lines to break up the figure into 3 parts.

(1) 2 rectangles with length 15 ft and width 6 ft

$$\begin{aligned} A &= \text{length} \cdot \text{width} \\ &= 15 \text{ ft} \cdot 6 \text{ ft} \\ &= 90 \text{ ft}^2 \end{aligned}$$

(2) a square 7 ft by 7 ft

$$\begin{aligned} A &= s \cdot s \\ &= 7 \text{ ft} \cdot 7 \text{ ft} \\ &= 49 \text{ ft}^2 \end{aligned}$$

$$\begin{aligned} \text{Total area} &= 2 \times 90 \text{ ft}^2 + 49 \text{ ft}^2 \\ &= 229 \text{ ft}^2 \end{aligned}$$

42. Find the area of the trapezoid and subtract the area of the triangle.

Area of trapezoid:

$$\begin{aligned} A &= \frac{1}{2} \cdot h \cdot (b + B) \\ &= \frac{1}{2} \cdot 12 \text{ ft} \cdot (11 \text{ ft} + 18 \text{ ft}) \\ &= \frac{1}{2} \cdot 12 \text{ ft} \cdot (29 \text{ ft}) \\ &= 174 \text{ ft}^2 \end{aligned}$$

Area of triangle:

$$\begin{aligned} A &= \frac{1}{2} \cdot b \cdot h \\ &= \frac{1}{2} \cdot 12 \text{ ft} \cdot 7 \text{ ft} \\ &= 42 \text{ ft}^2 \end{aligned}$$

$$\begin{aligned} \text{Total shaded area} &= 174 \text{ ft}^2 - 42 \text{ ft}^2 \\ &= 132 \text{ ft}^2 \end{aligned}$$

43. Find the area of the rectangle and subtract the area of the two unshaded triangles.

Area of rectangle:

$$\begin{aligned} A &= \text{length} \cdot \text{width} \\ &= (48 \text{ cm} + 48 \text{ cm}) \cdot 74 \text{ cm} \\ &= 96 \text{ cm} \cdot 74 \text{ cm} \\ &= 7104 \text{ cm}^2 \end{aligned}$$

Area of triangle:

$$\begin{aligned} A &= \frac{1}{2} \cdot b \cdot h \\ &= \frac{1}{2} \cdot 48 \text{ cm} \cdot 36 \text{ cm} \\ &= 864 \text{ cm}^2 \end{aligned}$$

Total shaded area

$$\begin{aligned} &= 7104 \text{ cm}^2 - (864 \text{ cm}^2 + 864 \text{ cm}^2) \\ &= 7104 \text{ cm}^2 - 1728 \text{ cm}^2 \\ &= 5376 \text{ cm}^2 \end{aligned}$$

44. Find the area of the rectangle and subtract the area of the semicircle.

Area of rectangle:

$$\begin{aligned} A &= \text{length} \cdot \text{width} \\ &= 32 \text{ ft} \cdot 21 \text{ ft} \\ &= 672 \text{ ft}^2 \end{aligned}$$

Area of circle:

$$\text{radius} = \frac{d}{2} = \frac{21}{2} = 10.5 \text{ ft}$$

$$\begin{aligned} A &= \pi \cdot r \cdot r \\ &\approx 3.14 \cdot 10.5 \text{ ft} \cdot 10.5 \text{ ft} \\ &\approx 346.2 \text{ ft}^2 \end{aligned}$$

Area of semicircle:

$$\frac{346.2 \text{ ft}^2}{2} = 173.1 \text{ ft}^2$$

Total shaded area is approximately $672 \text{ ft}^2 - 173.1 \text{ ft}^2 = 498.9 \text{ ft}^2$.

45. Find the area of the rectangle.

$$\begin{aligned} A &= \text{length} \cdot \text{width} \\ &= 21 \text{ yd} \cdot 14 \text{ yd} \\ &= 294 \text{ yd}^2 \end{aligned}$$

The two semicircles at the ends make one circle.

$$\begin{aligned} A &= \pi \cdot r \cdot r \\ &\approx 3.14 \cdot 7 \text{ yd} \cdot 7 \text{ yd} \\ &\approx 153.9 \text{ yd}^2 \end{aligned}$$

The total area is approximately $294 \text{ yd}^2 + 153.9 \text{ yd}^2 = 447.9 \text{ yd}^2$.

46. The figure is a rectangular solid.

$$\begin{aligned} V &= \text{length} \cdot \text{width} \cdot \text{height} \\ &= 4 \text{ in.} \cdot 3 \text{ in.} \cdot 2\frac{1}{2} \text{ in.} \\ &= 30 \text{ in.}^3 \end{aligned}$$

47. The figure is a rectangular solid.

$$\begin{aligned} V &= \text{length} \cdot \text{width} \cdot \text{height} \\ &= 6 \text{ cm} \cdot 4 \text{ cm} \cdot 4 \text{ cm} \\ &= 96 \text{ cm}^3 \end{aligned}$$

48. The figure is a rectangular solid.

$$\begin{aligned} V &= \text{length} \cdot \text{width} \cdot \text{height} \\ &= 30 \text{ mm} \cdot 20 \text{ mm} \cdot 75 \text{ mm} \\ &= 45{,}000 \text{ mm}^3 \end{aligned}$$

49. The figure is a sphere.

$$V = \frac{4}{3} \cdot \pi \cdot r^3$$
$$\approx \frac{4}{3} \cdot 3.14 \cdot (4 \text{ m})^3$$
$$= \frac{4}{3} \cdot 3.14 \cdot 64 \text{ m}^3$$
$$\approx 267.9 \text{ m}^3$$

50. The figure is a hemisphere.

$$V = \frac{2}{3} \cdot \pi \cdot r^3$$
$$\approx \frac{2}{3} \cdot 3.14 \cdot (6 \text{ ft})^3$$
$$= \frac{2}{3} \cdot 3.14 \cdot 216 \text{ ft}^3$$
$$\approx 452.2 \text{ ft}^3$$

51. The figure is a cylinder.

$$V = \pi \cdot r^2 \cdot h$$
$$\approx 3.14 \cdot 5 \text{ cm} \cdot 5 \text{ cm} \cdot 7 \text{ cm}$$
$$= 549.5 \text{ cm}^3$$

52. The figure is a cylinder.

$$V = \pi \cdot r^2 \cdot h \quad (d = 24 \text{ m}; r = 12 \text{ m})$$
$$\approx 3.14 \cdot 12 \text{ m} \cdot 12 \text{ m} \cdot 4 \text{ m}$$
$$\approx 1808.6 \text{ m}^3$$

53. The figure is a cone.

$$V = \frac{1}{3} \cdot \pi \cdot r^2 \cdot h$$
$$\approx \frac{1}{3} \cdot 3.14 \cdot 7 \text{ m} \cdot 7 \text{ m} \cdot 10 \text{ m}$$
$$\approx 512.9 \text{ m}^3$$

54. The figure is a pyramid.

$$V = \frac{1}{3} \cdot B \cdot h$$
$$= \frac{1}{3}(4 \text{ yd} \cdot 3 \text{ yd}) \cdot 4 \text{ yd}$$
$$= \frac{1}{3} \cdot 12 \text{ yd}^2 \cdot 4 \text{ yd}$$
$$= 16 \text{ yd}^3$$

55. $\sqrt{49} = 7$ because $7 \cdot 7 = 49$.

56. $\sqrt{8}$
Calculator shows 2.828427125; round to 2.828.

57. $\sqrt{3000}$
Calculator shows 54.77225575; round to 54.772.

58. $\sqrt{144} = 12$ because $12 \cdot 12 = 144$.

59. $\sqrt{58}$
Calculator shows 7.615773106; round to 7.616.

60. $\sqrt{625} = 25$ because $25 \cdot 25 = 625$.

61. $\sqrt{105}$
Calculator shows 10.24695077; round to 10.247.

62. $\sqrt{80}$
Calculator shows 8.94427191; round to 8.944.

63.
$$\text{hypotenuse} = \sqrt{(\text{leg})^2 + (\text{leg})^2}$$
$$= \sqrt{(15)^2 + (8)^2}$$
$$= \sqrt{225 + 64}$$
$$= \sqrt{289}$$
$$= 17 \text{ in.}$$

64.
$$\text{leg} = \sqrt{(\text{hypotenuse})^2 - (\text{leg})^2}$$
$$= \sqrt{(25)^2 - (24)^2}$$
$$= \sqrt{625 - 576}$$
$$= \sqrt{49}$$
$$= 7 \text{ cm}$$

65.
$$\text{leg} = \sqrt{(\text{hypotenuse})^2 - (\text{leg})^2}$$
$$= \sqrt{(15)^2 - (11)^2}$$
$$= \sqrt{225 - 121}$$
$$= \sqrt{104}$$
$$\approx 10.2 \text{ cm}$$

66.
$$\text{hypotenuse} = \sqrt{(\text{leg})^2 + (\text{leg})^2}$$
$$= \sqrt{(6)^2 + (4)^2}$$
$$= \sqrt{36 + 16}$$
$$= \sqrt{52}$$
$$\approx 7.2 \text{ in.}$$

67.
$$\text{hypotenuse} = \sqrt{(\text{leg})^2 + (\text{leg})^2}$$
$$= \sqrt{(2.2)^2 + (1.3)^2}$$
$$= \sqrt{4.84 + 1.69}$$
$$= \sqrt{6.53}$$
$$\approx 2.6 \text{ m}$$

68.
$$\text{leg} = \sqrt{(\text{hypotenuse})^2 - (\text{leg})^2}$$
$$= \sqrt{(12)^2 - (8.5)^2}$$
$$= \sqrt{144 - 72.25}$$
$$= \sqrt{71.75}$$
$$\approx 8.5 \text{ km}$$

69. Set up a ratio of corresponding sides.

$$\frac{40\text{ ft}}{20\text{ ft}} = \frac{40}{20} = \frac{2}{1}$$

Write a proportion to find y.

$$\frac{y}{15} = \frac{2}{1}$$
$$y \cdot 1 = 2 \cdot 15$$
$$y = 30\text{ ft}$$

Write a proportion to find x.

$$\frac{x}{17} = \frac{2}{1}$$
$$x \cdot 1 = 17 \cdot 2$$
$$x = 34\text{ ft}$$

$$P = 34\text{ ft} + 30\text{ ft} + 40\text{ ft}$$
$$= 104\text{ ft}$$

70. Set up a ratio of corresponding sides.

$$\frac{4\text{ m}}{6\text{ m}} = \frac{4}{6} = \frac{2}{3}$$

Write a proportion to find x.

$$\frac{6}{x} = \frac{2}{3}$$
$$2 \cdot x = 3 \cdot 6$$
$$\frac{2 \cdot x}{2} = \frac{18}{2}$$
$$x = 9\text{ m}$$

Write a proportion to find y.

$$\frac{5}{y} = \frac{2}{3}$$
$$2 \cdot y = 5 \cdot 3$$
$$\frac{2 \cdot y}{2} = \frac{15}{2}$$
$$y = 7.5\text{ m}$$

$$P = 9\text{ m} + 7.5\text{ m} + 6\text{ m}$$
$$= 22.5\text{ m}$$

71. Set up a ratio of corresponding sides.

$$\frac{16\text{ mm}}{12\text{ mm}} = \frac{16}{12} = \frac{4}{3}$$

Write a proportion to find x.

$$\frac{x}{9} = \frac{4}{3}$$
$$3 \cdot x = 9 \cdot 4$$
$$\frac{3 \cdot x}{3} = \frac{36}{3}$$
$$x = 12\text{ mm}$$

Write a proportion to find y.

$$\frac{10}{y} = \frac{4}{3}$$
$$y \cdot 4 = 10 \cdot 3$$
$$\frac{y \cdot 4}{4} = \frac{30}{4}$$
$$y = 7.5\text{ mm}$$

$$P = 12\text{ mm} + 10\text{ mm} + 16\text{ mm}$$
$$= 38\text{ mm}$$

72. **[8.3]** The figure is a square.

$$P = 4 \cdot s$$
$$= 4 \cdot 4\tfrac{1}{2}\text{ in.}$$
$$= 18\text{ in.}$$
$$A = s \cdot s$$
$$= 4\tfrac{1}{2}\text{ in.} \cdot 4\tfrac{1}{2}\text{ in.}$$
$$= \tfrac{9}{2}\text{ in.} \cdot \tfrac{9}{2}\text{ in.}$$
$$= \tfrac{81}{4}\text{ in.}^2$$
$$= 20\tfrac{1}{4}\text{ in.}^2 \text{ or } 20.3\text{ in.}^2 \text{ (rounded)}$$

73. **[8.4]** The figure is a trapezoid.

$$P = 2.3\text{ cm} + 2.5\text{ cm} + 2.1\text{ cm} + 3.4\text{ cm}$$
$$= 10.3\text{ cm}$$
$$A = 0.5 \cdot h \cdot (b + B)$$
$$= 0.5 \cdot 2.1\text{ cm} \cdot (2.5\text{ cm} + 3.4\text{ cm})$$
$$= 0.5 \cdot 2.1\text{ cm} \cdot 5.9\text{ cm}$$
$$\approx 6.2\text{ cm}^2$$

74. **[8.6]** The figure is a circle.

$$C = \pi \cdot d$$
$$\approx 3.14 \cdot 13\text{ m}$$
$$\approx 40.8\text{ m}$$
$$A = \pi \cdot r^2 \left(r = \frac{13\text{ m}}{2} = 6.5\text{ m}\right)$$
$$\approx 3.14 \cdot 6.5\text{ m} \cdot 6.5\text{ m}$$
$$\approx 132.7\text{ m}^2$$

75. **[8.4]** The figure is a parallelogram.

$$P = 17\text{ ft} + 10\text{ ft} + 17\text{ ft} + 10\text{ ft}$$
$$= 54\text{ ft}$$
$$A = b \cdot h$$
$$= 10\text{ ft} \cdot 14\text{ ft}$$
$$= 140\text{ ft}^2$$

76. **[8.5]** The figure is a triangle.

$$P = 6\tfrac{1}{4}\text{ yd} + 7\tfrac{1}{2}\text{ yd} + 6\tfrac{1}{4}\text{ yd}$$
$$= 20\text{ yd}$$
$$A = \frac{1}{2} \cdot b \cdot h = \frac{1}{2} \cdot 7\tfrac{1}{2}\text{ yd} \cdot 5\text{ yd}$$
$$= 18\tfrac{3}{4}\text{ yd}^2 \text{ or } 18.8\text{ yd}^2 \text{ (rounded)}$$

77. **[8.3]** The figure is a rectangle.

$P = 0.7 \text{ km} + 2.8 \text{ km} + 0.7 \text{ km} + 2.8 \text{ km}$
$= 7 \text{ km}$
$A = \text{length} \cdot \text{width}$
$= 2.8 \text{ km} \cdot 0.7 \text{ km}$
$= 1.96 \text{ km}^2$
$\approx 2.0 \text{ km}^2$

78. **[8.6]** The figure is a circle.

$C = 2 \cdot \pi \cdot r$
$\approx 2 \cdot 3.14 \cdot 8.5 \text{ m}$
$\approx 53.4 \text{ m}$
$A = \pi \cdot r^2$
$\approx 3.14 \cdot 8.5 \text{ m} \cdot 8.5 \text{ m}$
$\approx 226.9 \text{ m}^2$

79. **[8.4]** The figure is a parallelogram.

$P = 24 \text{ mm} + 15 \text{ mm} + 24 \text{ mm} + 15 \text{ mm}$
$= 78 \text{ mm}$
$A = b \cdot h$
$= 24 \text{ mm} \cdot 12 \text{ mm}$
$= 288 \text{ mm}^2$

80. **[8.5]** The figure is a triangle.

$P = 13 \text{ mi} + 9 \text{ mi} + 15.8 \text{ mi}$
$= 37.8 \text{ mi}$
$A = 0.5 \cdot b \cdot h$
$= 0.5 \cdot 9 \text{ mi} \cdot 13 \text{ mi}$
$= 58.5 \text{ mi}^2$

81. **[8.1]** $\overleftrightarrow{WX}$ and $\overleftrightarrow{YZ}$ are parallel lines.

82. **[8.1]** $\overline{QR}$ is a line segment.

83. **[8.1]** $\angle CQD$ is an acute angle.

84. **[8.1]** $\overleftrightarrow{PQ}$ and $\overleftrightarrow{NO}$ are intersecting lines.

85. **[8.1]** $\angle APB$ is a right angle measuring 90°.

86. **[8.1]** $\overrightarrow{AB}$ is a ray.

87. **[8.1]** $\overleftrightarrow{T}$ is a straight angle measuring 180°.

88. **[8.1]** $\angle FEG$ is an obtuse angle.

89. **[8.1]** $\overleftrightarrow{LM}$ and $\overleftrightarrow{JK}$ are perpendicular lines.

90. **[8.2]** The complement of an angle measuring 9° is $90° - 9° = 81°$.

91. **[8.2]** The supplement of an angle measuring 42° is $180° - 42° = 138°$.

92. **[8.3]** Find the unknown length of the horizontal line.

$17 \text{ m} - 1 \text{ m} - 1 \text{ m} = 15 \text{ m}$

$P = 17 \text{ m} + 16 \text{ m} + 1 \text{ m} + 12 \text{ m}$
$+ 15 \text{ m} + 12 \text{ m} + 1 \text{ m} + 16 \text{ m}$
$= 90 \text{ m}$

Draw two short horizontal lines to break up the figure into 3 rectangles to find the area.
Two rectangles with length 12 m and width 1 m:

$A = 12 \text{ m} \cdot 1 \text{ m}$
$= 12 \text{ m}^2$

A rectangle with length 17 m and width $16 \text{ m} - 12 \text{ m} = 4 \text{ m}$:

$A = 17 \text{ m} \cdot 4 \text{ m}$
$= 68 \text{ m}^2$

$\text{Total area} = 12 \text{ m}^2 + 12 \text{ m}^2 + 68 \text{ m}^2$
$= 92 \text{ m}^2$

93. **[8.4]** $P = 72 \text{ cm} + 19 \text{ cm} + 50 \text{ cm}$
$+ 19 \text{ cm} + 72 \text{ cm} + 50 \text{ cm}$
$= 282 \text{ cm}$

Break up the figure into a parallelogram and a rectangle.
A parallelogram with base 72 cm and height 45 cm:

$A = 72 \text{ cm} \cdot 45 \text{ cm}$
$= 3240 \text{ cm}^2$

A rectangle with base 19 cm and height 50 cm:

$A = 19 \text{ cm} \cdot 50 \text{ cm}$
$= 950 \text{ cm}^2$

$\text{Total area} = 3240 \text{ cm}^2 + 950 \text{ cm}^2$
$= 4190 \text{ cm}^2$

94. **[8.7]** The figure is a cylinder.

$V = \pi \cdot r^2 \cdot h$
$\approx 3.14 \cdot 2 \text{ ft} \cdot 2 \text{ ft} \cdot 8 \text{ ft}$
$\approx 100.5 \text{ ft}^3$

95. **[8.7]** The figure is a cube.

$V = \text{length} \cdot \text{width} \cdot \text{height}$
$= 1\frac{1}{2} \text{ in.} \cdot 1\frac{1}{2} \text{ in.} \cdot 1\frac{1}{2} \text{ in.}$
$= \frac{3}{2} \text{ in.} \cdot \frac{3}{2} \text{ in.} \cdot \frac{3}{2} \text{ in.}$
$= \frac{27}{8} \text{ in.}^3$
$\approx 3.4 \text{ in.}^3 \text{ or } 3\frac{3}{8} \text{ in.}^3$

96. **[8.7]** The figure is a rectangular solid.

$V = \text{length} \cdot \text{width} \cdot \text{height}$
$= 3.5 \text{ m} \cdot 3 \text{ m} \cdot 0.7 \text{ m}$
$= 7.35 \text{ m}^3$
$\approx 7.4 \text{ m}^3$

97. **[8.7]** The figure is a pyramid.

$$V = \frac{1}{3} \cdot B \cdot h$$
$$= \frac{1}{3} \cdot (11 \text{ cm} \cdot 9 \text{ cm}) \cdot 17 \text{ cm}$$
$$= \frac{1}{3} \cdot 99 \text{ cm}^2 \cdot 17 \text{ cm}$$
$$= 561 \text{ cm}^3$$

98. **[8.7]** The figure is a cone.

$$V = \frac{1}{3} \cdot \pi \cdot r^2 \cdot h$$
$$\approx \frac{1}{3} \cdot 3.14 \cdot 9 \text{ cm} \cdot 9 \text{ cm} \cdot 15 \text{ cm}$$
$$\approx 1271.7 \text{ cm}^3$$

99. **[8.7]** The figure is a sphere.

$$V = \frac{4}{3} \cdot \pi \cdot r^3$$
$$\approx \frac{4}{3} \cdot 3.14 \cdot 7 \text{ m} \cdot 7 \text{ m} \cdot 7 \text{ m}$$
$$\approx 1436.0 \text{ m}^3$$

100. **[8.8]** $\text{leg} = \sqrt{(\text{hypotenuse})^2 - (\text{leg})^2}$

$$= \sqrt{(14)^2 - (6)^2}$$
$$= \sqrt{196 - 36}$$
$$= \sqrt{160}$$
$$\approx 12.6 \text{ km}$$

101. **[8.2]** *Step 1* $60° + 48° = 108°$
Step 2 $180° - 108° = 72°$
$\angle D$ measures $72°$.

102. **[8.9]** Set up a ratio of corresponding sides.

$$\frac{18 \text{ m}}{15 \text{ m}} = \frac{18}{15} = \frac{6}{5}$$

Write a proportion to find x.

$$\frac{x}{10} = \frac{6}{5}$$
$$5 \cdot x = 10 \cdot 6$$
$$\frac{5 \cdot x}{5} = \frac{60}{5}$$
$$x = 12 \text{ mm}$$

Write a proportion to find y.

$$\frac{16.8}{y} = \frac{6}{5}$$
$$y \cdot 6 = 16.8 \cdot 5$$
$$\frac{y \cdot 6}{6} = \frac{84}{6}$$
$$y = 14 \text{ mm}$$

103. **[8.6]** The prefix *dec* in decade means 10 and the prefix *cent* in century means 100, so divide 200 (two centuries) by 10. The answer is 20 decades.

Chapter 8 Test

1. $\angle LOM$ is an acute angle, so the answer is (e).

2. $\angle YOX$ is a right angle, so the answer is (a). Its measure is 90°.

3. $\overrightarrow{GH}$ is a ray, so the answer is (d).

4. $\overleftrightarrow{W}$ is a straight angle, so the answer is (g). Its measure is 180°.

5. Parallel lines in the same plane never intersect. Perpendicular lines intersect to form a right angle.

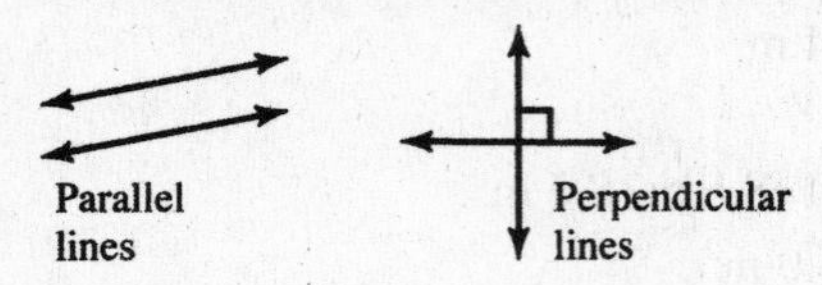

6. The complement of an angle measuring 81° is $90° - 81° = 9°$.

7. The supplement of an angle measuring 20° is $180° - 20° = 160°$.

8. $\angle 1$ measures 50° since $\angle 1$ and the 50° angle are vertical angles, so they are congruent.
$\angle 3$ measures 90° since it is a vertical angle with the 90° angle (the one marked with the square).
$\angle 4 + 50° + \angle 3 = 180°$
$\angle 4 + 50° + 90° = 180°$
$\angle 4$ measures 40° ($180° - 50° - 90°$).
$\angle 2$ also measures 40° since $\angle 2$ and $\angle 4$ are vertical angles.

9. The figure is a rectangle.

$$P = 4 \text{ ft} + 7\tfrac{1}{2} \text{ ft} + 4 \text{ ft} + 7\tfrac{1}{2} \text{ ft}$$
$$= 23 \text{ ft}$$
$$A = \text{length} \cdot \text{width}$$
$$= 7\tfrac{1}{2} \text{ ft} \cdot 4 \text{ ft}$$
$$= 30 \text{ ft}^2$$

10. The figure is a square.

$$P = 4 \cdot s$$
$$= 4 \cdot 18 \text{ mm}$$
$$= 72 \text{ mm}$$
$$A = s \cdot s$$
$$= 18 \text{ mm} \cdot 18 \text{ mm}$$
$$= 324 \text{ mm}^2$$

11. The figure is a parallelogram.

$$P = 5.9 \text{ m} + 7.2 \text{ m} + 5.9 \text{ m} + 7.2 \text{ m}$$
$$= 26.2 \text{ m}$$
$$A = b \cdot h$$
$$= 7.2 \text{ m} \cdot 4.6 \text{ m}$$
$$= 33.12 \text{ m}^2$$

12. The figure is a trapezoid.

$$P = 29 \text{ cm} + 46.4 \text{ cm} + 57 \text{ cm} + 37 \text{ cm}$$
$$= 169.4 \text{ cm}$$
$$A = \frac{1}{2} \cdot h \cdot (b + B)$$
$$= \frac{1}{2} \cdot 37 \text{ cm} \cdot (29 \text{ cm} + 57 \text{ cm})$$
$$= \frac{1}{2} \cdot 37 \text{ cm} \cdot 86 \text{ cm}$$
$$= 1591 \text{ cm}^2$$

13. $P = 11.8 \text{ m} + 8.65 \text{ m} + 12 \text{ m}$
$$= 32.45 \text{ m}$$
$$A = 0.5 \cdot b \cdot h$$
$$= 0.5 \cdot 12 \text{ m} \cdot 8 \text{ m}$$
$$= 48 \text{ m}^2$$

14. $P = 9 \text{ yd} + 15\frac{4}{5} \text{ yd} + 13 \text{ yd}$
$$= 37\tfrac{4}{5} \text{ yd or } 37.8 \text{ yd}$$
$$A = \frac{1}{2} \cdot b \cdot h$$
$$= \frac{1}{2} \cdot 13 \text{ yd} \cdot 9 \text{ yd}$$
$$= 58.5 \text{ yd}^2 \text{ or } 58\tfrac{1}{2} \text{ yd}^2$$

15. *Step 1* $90° + 35° = 125°$
Step 2 $180° - 125° = 55°$
The third angle measures 55°.

16. $\text{radius} = \frac{d}{2} = \frac{25 \text{ in.}}{2} = 12.5 \text{ in. or } 12\frac{1}{2} \text{ in.}$

17. $C = 2 \cdot \pi \cdot r$
$$\approx 2 \cdot 3.14 \cdot 0.9 \text{ km}$$
$$\approx 5.7 \text{ km}$$

18. $r = \frac{d}{2} = \frac{16.2 \text{ cm}}{2} = 8.1 \text{ cm}$
$$A = \pi \cdot r^2$$
$$\approx 3.14 \cdot 8.1 \text{ cm} \cdot 8.1 \text{ cm}$$
$$\approx 206.0 \text{ cm}^2$$

19. $A = \frac{\pi \cdot r^2}{2}$
$$\approx \frac{3.14 \cdot 5 \text{ m} \cdot 5 \text{ m}}{2}$$
$$= \frac{78.5}{2} \text{ m}^2$$
$$= 39.25 \text{ m}^2$$
$$\approx 39.3 \text{ m}^2$$

20. The figure is a rectangular solid.

$$V = \text{length} \cdot \text{width} \cdot \text{height}$$
$$= 30 \text{ m} \cdot 18 \text{ m} \cdot 12 \text{ m}$$
$$= 6480 \text{ m}^3$$

21. The figure is a sphere.

$$V = \frac{4}{3} \cdot \pi \cdot r^3$$
$$\approx \frac{4}{3} \cdot 3.14 \cdot 2 \text{ ft} \cdot 2 \text{ ft} \cdot 2 \text{ ft}$$
$$= \frac{100.48}{3} \text{ ft}^3$$
$$\approx 33.5 \text{ ft}^3$$

22. The figure is a cylinder.

$$V = \pi \cdot r^2 \cdot h$$
$$\approx 3.14 \cdot 18 \text{ ft} \cdot 18 \text{ ft} \cdot 5 \text{ ft}$$
$$= 5086.8 \text{ ft}^3$$

23. $\text{hypotenuse} = \sqrt{(\text{leg})^2 + (\text{leg})^2}$
$$= \sqrt{(7)^2 + (6)^2}$$
$$= \sqrt{49 + 36}$$
$$= \sqrt{85}$$
$$\approx 9.2 \text{ cm}$$

24. Set up a ratio of corresponding sides.

$$\frac{10 \text{ cm}}{15 \text{ cm}} = \frac{10}{15} = \frac{2}{3}$$

Write a proportion to find y.

$$\frac{y}{18} = \frac{2}{3}$$
$$y \cdot 3 = 18 \cdot 2$$
$$\frac{y \cdot 3}{3} = \frac{36}{3}$$
$$y = 12 \text{ cm}$$

Write a proportion to find z.

$$\frac{z}{9} = \frac{2}{3}$$
$$z \cdot 3 = 9 \cdot 2$$
$$\frac{z \cdot 3}{3} = \frac{18}{3}$$
$$z = 6 \text{ cm}$$

25. Linear units like cm are used to measure perimeter, radius, diameter, and circumference. Area is measured in square units like cm^2 (squares that measure 1 cm on each side). Volume is measured in cubic units like cm^3.

Cumulative Review Exercises (Chapters 1–8)

1. *Estimate:* / *Exact:*

Estimate	Exact
300	319
60,000	58,028
+ 6 000	+ 6 227
66,300	64,574

2. *Estimate:*

$$\begin{array}{r} 20 \\ -\ 10 \\ \hline 10 \end{array}$$

Exact:

$$\begin{array}{r} 20.070 \\ -\ 9.828 \\ \hline 10.242 \end{array}$$

3. *Estimate:*

$$\begin{array}{r} 4 \\ \times\ 7 \\ \hline 28 \end{array}$$

Exact:

$$\begin{array}{r} 3.664 \\ \times\ 7.3 \\ \hline 1\ 0992 \\ 25\ 648 \\ \hline 26.7472 \end{array}$$

4. *Estimate:*

$$\begin{array}{r} 30{,}000 \\ \times\ \ 70 \\ \hline 2{,}100{,}000 \end{array}$$

Exact:

$$\begin{array}{r} 28{,}419 \\ \times\ \ 73 \\ \hline 85\ 257 \\ 1\ 989\ 33 \\ \hline 2{,}074{,}587 \end{array}$$

5. *Estimate:*

$$3\overline{)600}\quad = 200$$

Exact:

$$2.8_\wedge\overline{)5\,6\,2.\,2_\wedge 4}\quad = 200.8$$

$$\begin{array}{r} 5\,6 \\ \hline 2 \\ 0 \\ \hline 2\,2 \\ 0 \\ \hline 2\,2\,4 \\ 2\,2\,4 \\ \hline 0 \end{array}$$

6. *Estimate:*

$$50\overline{)5000}\quad = 100$$

Exact:

$$52\overline{)4888}\quad = 94$$

$$\begin{array}{r} 4\,6\,8 \\ \hline 2\,0\,8 \\ 2\,0\,8 \\ \hline 0 \end{array}$$

7. *Estimate:*

$$\begin{array}{r} 5 \\ +\ 5 \\ \hline 10 \end{array}$$

Exact:

$$\begin{array}{rcl} 4\frac{1}{2} & = & 4\frac{5}{10} \\ +\ 4\frac{9}{10} & = & 4\frac{9}{10} \\ \hline & & 8\frac{14}{10} = 9\frac{2}{5} \end{array}$$

8. *Estimate:*

$$\begin{array}{r} 3 \\ -\ 2 \\ \hline 1 \end{array}$$

Exact:

$$\begin{array}{rcccl} 3\frac{1}{6} & = & \frac{19}{6} & = & \frac{76}{24} \\ -\ 1\frac{7}{8} & = & \frac{15}{8} & = & \frac{45}{24} \\ \hline & & & & \frac{31}{24} = 1\frac{7}{24} \end{array}$$

9. *Exact:*

$$3\frac{1}{9}\cdot 1\frac{5}{7} = \frac{28}{9}\cdot\frac{12}{7} = \frac{\overset{4}{\cancel{28}}}{\underset{3}{\cancel{9}}}\cdot\frac{\overset{4}{\cancel{12}}}{\underset{1}{\cancel{7}}} = \frac{16}{3} = 5\frac{1}{3}$$

Estimate: $3\cdot 2 = 6$

10. $3\frac{3}{5} \div 8 = \frac{18}{5} \div \frac{8}{1} = \frac{\overset{9}{\cancel{18}}}{5}\cdot\frac{1}{\underset{4}{\cancel{8}}} = \frac{9}{20}$

11. $1 - 0.0868$

$$\begin{array}{r} 1.0000 \\ -\ 0.0868 \\ \hline 0.9132 \end{array}$$

12. $81\overline{)5749}\quad = 70$ **R79**

$$\begin{array}{r} 5\,6\,7 \\ \hline 7\,9 \\ 0 \\ \hline 7\,9 \end{array}$$

13. $10 \div \frac{5}{16} = \frac{10}{1} \div \frac{5}{16} = \frac{\overset{2}{\cancel{10}}}{1}\cdot\frac{16}{\underset{1}{\cancel{5}}} = 32$

14. $(0.006)(0.013)$

$$\begin{array}{rl} 0.006 & \leftarrow \textit{3 decimal places} \\ \times\ 0.013 & \leftarrow \textit{3 decimal places} \\ \hline 18 & \\ 6 & \\ \hline 0.000078 & \leftarrow \textit{6 decimal places} \end{array}$$

15. $40{,}020 - 915$

$$\begin{array}{r} 40{,}020 \\ -\ 915 \\ \hline 39{,}105 \end{array}$$

16. $0.7 \div 0.036 \approx 19.44$

Move the decimal point three places to the right in the dividend and divisor.

$$36\overline{)700.000}\quad = 19.444$$

$$\begin{array}{r} 3\,6 \\ \hline 3\,4\,0 \\ 3\,2\,4 \\ \hline 1\,6\,0 \\ 1\,4\,4 \\ \hline 1\,6\,0 \\ 1\,4\,4 \\ \hline 1\,6\,0 \\ 1\,4\,4 \\ \hline 1\,6 \end{array}$$

17. $6\frac{1}{6} - 1\frac{3}{4}$

$$\begin{aligned} 6\frac{1}{6} &= 6\frac{2}{12} = 5\frac{14}{12} \\ -1\frac{3}{4} &= 1\frac{9}{12} = 1\frac{9}{12} \\ &\qquad\qquad\; 4\frac{5}{12} \end{aligned}$$

18. $752.6 + 83 + 0.485$

$$\begin{array}{r} 752.600 \\ 83.000 \\ +\,0.485 \\ \hline 836.085 \end{array}$$

19. $16 - (10 - 2) \div 2(3) + 5 = 16 - 8 \div 2(3) + 5$
$= 16 - 4(3) + 5$
$= 16 - 12 + 5$
$= 4 + 5 = 9$

20. $2^4 \div \sqrt{64} + 6^2 = 16 \div 8 + 36$
$= 2 + 36 = 38$

21. 0.0208 is two hundred eight ten-thousandths.

22. Six hundred sixty and five hundredths in numbers is 660.05.

23. From smallest to largest:

$2.55 = 2.5500$ *largest*
$2.505 = 2.5050$
$2.055 = 2.0550$ *smallest*
$2.5005 = 2.5005$

Answer: 2.055; 2.5005; 2.505; 2.55

24. *Per* means divide and *cent* means 100, so divide by 100 to change a percent to a decimal.

25. $0.02 = \frac{2}{100} = \frac{2 \div 2}{100 \div 2} = \frac{1}{50}$

26. $0.02 = 2\%$
Move the decimal point two places to the *right* and attach a percent sign.

27. $1\frac{3}{4} = 1\frac{75}{100} = 1.75$

28. $1.75 = 175\%$

29. $40\% = 0.4 = \frac{4}{10} = \frac{4 \div 2}{10 \div 2} = \frac{2}{5}$

30. $40\% = 0.4$

31. 4 feet to 6 inches
4 feet $= 4 \cdot 12$ inches $= 48$ inches

$$\frac{48 \text{ inches}}{6 \text{ inches}} = \frac{48 \div 6}{6 \div 6} = \frac{8}{1}$$

32. 21 sunny days to 9 cloudy days

$$\frac{21 \text{ days}}{9 \text{ days}} = \frac{21}{9} = \frac{7}{3}$$

33.
$$\begin{aligned} \frac{5}{13} &= \frac{x}{91} \\ 13 \cdot x &= 5 \cdot 91 \\ \frac{13 \cdot x}{13} &= \frac{455}{13} \\ x &= 35 \end{aligned}$$

34.
$$\begin{aligned} \frac{207}{69} &= \frac{300}{x} \\ 207 \cdot x &= 69 \cdot 300 \\ \frac{207 \cdot x}{207} &= \frac{20{,}700}{207} \\ x &= 100 \end{aligned}$$

35.
$$\begin{aligned} \frac{4.5}{x} &= \frac{6.7}{3} \\ 6.7 \cdot x &= 3 \cdot 4.5 \\ \frac{6.7 \cdot x}{6.7} &= \frac{13.5}{6.7} \\ x &\approx 2.01 \end{aligned}$$

36. part is 72; whole is 45; percent is unknown.
Use the percent proportion.

$$\begin{aligned} \frac{72}{45} &= \frac{x}{100} \\ 45 \cdot x &= 72 \cdot 100 \\ \frac{45 \cdot x}{45} &= \frac{7200}{45} \\ x &= 160 \end{aligned}$$

72 patients is 160% of 45 patients.

37. part is 18; percent is 3; whole is unknown.
Use the percent proportion.

$$\begin{aligned} \frac{18}{x} &= \frac{3}{100} \\ 3 \cdot x &= 18 \cdot 100 \\ \frac{3 \cdot x}{3} &= \frac{1800}{3} \\ x &= 600 \end{aligned}$$

\$18 is 3% of \$600.

38. $2\frac{1}{4}$ hours to minutes

$$\frac{2\frac{1}{4} \text{ hours}}{1} \cdot \frac{60 \text{ minutes}}{1 \text{ hour}} = \frac{9}{\cancel{4}_1} \cdot \frac{\cancel{60}^{15}}{1} \text{ minutes}$$
$$= 135 \text{ minutes}$$

39. 40 oz to pounds

$$\frac{40 \text{ ounces}}{1} \cdot \frac{1 \text{ pound}}{16 \text{ ounces}} = \frac{40}{16} \text{ pounds}$$
$$= 2\frac{1}{2} \text{ or } 2.5 \text{ pounds}$$

40. 8 cm to meters
Count 2 places to the *left* on the metric conversion line.
8 cm = 0.08 m

41. 1.8 L to mL
Count 3 places to the *right* on the metric conversion line.
1.8 L = 1800 mL

42. Her wristwatch strap is 15 mm wide.

43. Jon added 2 L of oil to his car.

44. The child weighs 15 kg.

45. The bookcase is 90 cm high.

46. Water freezes at 0°C and boils at 100°C.

47. The figure is a rectangle.

$$\begin{aligned} P &= 3\tfrac{1}{2}\text{ in.} + 2\text{ in.} + 3\tfrac{1}{2}\text{ in.} + 2\text{ in.} \\ &= 11\text{ in.} \\ A &= \text{length} \cdot \text{width} \\ &= 3\tfrac{1}{2}\text{ in.} \cdot 2\text{ in.} \\ &= 7\text{ in.}^2 \end{aligned}$$

48. The figure is a triangle.

$$\begin{aligned} P &= 2.1\text{ m} + 2\text{ m} + 1.7\text{ m} \\ &= 5.8\text{ m} \\ A &= 0.5 \cdot b \cdot h \\ &= 0.5 \cdot 2.1\text{ m} \cdot 1.5\text{ m} \\ &= 1.575\text{ m}^2 \end{aligned}$$

49. The figure is a circle.

$$\begin{aligned} C &= 2 \cdot \pi \cdot r \\ &\approx 2 \cdot 3.14 \cdot 5\text{ ft} \\ &= 31.4\text{ ft} \\ A &= \pi \cdot r \cdot r \\ &\approx 3.14 \cdot 5\text{ ft} \cdot 5\text{ ft} \\ &= 78.5\text{ ft}^2 \end{aligned}$$

50. The figure is a parallelogram.

$$\begin{aligned} P &= 24\text{ cm} + 14\text{ cm} + 24\text{ cm} + 14\text{ cm} \\ &= 76\text{ cm} \\ A &= \text{base} \cdot \text{height} \\ &= 24\text{ cm} \cdot 11\text{ cm} \\ &= 264\text{ cm}^2 \end{aligned}$$

51. The figure is a trapezoid.

$$\begin{aligned} P &= 13\text{ m} + 10\text{ m} + 7\text{ m} + 10.8\text{ m} \\ &= 40.8\text{ m} \\ A &= 0.5 \cdot h \cdot (b + B) \\ &= 0.5 \cdot 9.5\text{ m} \cdot (7\text{ m} + 13\text{ m}) \\ &= 0.5 \cdot 9.5\text{ m} \cdot 20\text{ m} \\ &= 95\text{ m}^2 \end{aligned}$$

52.

$$\begin{aligned} P &= 10\text{ yd} + 10\text{ yd} + 10\text{ yd} + 10\text{ yd} + 10\text{ yd} \\ &= 50\text{ yd} \end{aligned}$$

Break up the figure into 2 parts.

(1) a square

$$\begin{aligned} A &= s \cdot s \\ &= 10\text{ yd} \cdot 10\text{ yd} \\ &= 100\text{ yd}^2 \end{aligned}$$

(2) a triangle

$$\begin{aligned} A &= \frac{1}{2} \cdot b \cdot h \\ &= \frac{1}{2} \cdot 10\text{ yd} \cdot 8.4\text{ yd} \\ &= 42\text{ yd}^2 \end{aligned}$$

$$\begin{aligned} \text{Total area} &= 100\text{ yd}^2 + 42\text{ yd}^2 \\ &= 142\text{ yd}^2 \end{aligned}$$

53.

$$\begin{aligned} \text{hypotenuse} &= \sqrt{(\text{leg})^2 + (\text{leg})^2} \\ &= \sqrt{(19)^2 + (15)^2} \\ &= \sqrt{361 + 225} \\ &= \sqrt{586} \\ &\approx 24.2 \end{aligned}$$

$y \approx 24.2$ mm

54. Set up a ratio of corresponding sides.

$$\frac{5.5\text{ ft}}{15\text{ ft}} = \frac{5.5}{15} = \frac{1.1}{3}$$

Write a proportion to find x.

$$\begin{aligned} \frac{x}{22} &= \frac{1.1}{3} \\ 3 \cdot x &= 1.1 \cdot 22 \\ \frac{3 \cdot x}{3} &= \frac{24.2}{3} \\ x &\approx 8.1\text{ ft} \end{aligned}$$

$$\begin{aligned} P &\approx 8.1\text{ ft} + 5.5\text{ ft} + 5.5\text{ ft} \\ &= 19.1\text{ ft} \end{aligned}$$

55. part is 53; whole is 90; percent is unknown.

$$\begin{aligned} \frac{53}{90} &= \frac{x}{100} \\ 90 \cdot x &= 53 \cdot 100 \\ \frac{90 \cdot x}{90} &= \frac{5300}{90} \\ x &\approx 58.89 \approx 59 \end{aligned}$$

Mei Ling has 59% (rounded) of the necessary credits.

56. $15\frac{1}{2}$ ounces of Brand T for
$2.99 – $0.30 (coupon) = $2.69

$$\frac{\$2.69}{15\frac{1}{2}\text{ ounces}} \approx \$0.174 \text{ per ounce } (*)$$

14 ounces of Brand F for $2.49

$$\frac{\$2.49}{14\text{ ounces}} \approx \$0.178 \text{ per ounce}$$

18 ounces of Brand H for
$3.89 – $0.40 (coupon) = $3.49

$$\frac{\$3.49}{18\text{ ounces}} \approx \$0.194 \text{ per ounce}$$

The best buy is Brand T at $15\frac{1}{2}$ ounces for $2.99 with a $0.30 coupon.

57. Use the formula for the volume of a cylinder.

$$\begin{aligned} V &= \pi \cdot r^2 \cdot h\left(r = \frac{d}{2} = \frac{13\text{ cm}}{2} = 6.5\text{ cm}\right) \\ &\approx 3.14 \cdot 6.5\text{ cm} \cdot 6.5\text{ cm} \cdot 17\text{ cm} \\ &= 2255.305 \\ &\approx 2255.3\text{ cm}^3 \end{aligned}$$

58. Original area = 3 in. × 3 in. = 9 in.^{2}
XL Area = 4 in. × 4 in. = 16 in.^{2}

$$\frac{\text{XL}}{\text{Original}} = \frac{16}{9} = 1.\overline{7}$$

Thus, the XL area is about 178% of the Original area; that is, XL is 78% larger than Original (rounded).

59. First add the amount of canvas material that was used.

$$\begin{aligned} 1\frac{2}{3} &= 1\frac{8}{12} \\ +1\frac{3}{4} &= 1\frac{9}{12} \\ \hline 2\frac{17}{12} &= 3\frac{5}{12}\text{ yd} \end{aligned}$$

Subtract $3\frac{5}{12}$ from the amount of canvas material that Steven bought.

$$\begin{aligned} 4\frac{1}{2} &= 4\frac{6}{12} \\ -3\frac{5}{12} &= 3\frac{5}{12} \\ \hline & 1\frac{1}{12}\text{ yd} \end{aligned}$$

There will be $1\frac{1}{12}$ yd left.

60.
$$\begin{aligned} \frac{25\text{ pounds}}{140\text{ pounds}} &= \frac{x}{200\text{ people}} \\ 140 \cdot x &= 25 \cdot 200 \\ \frac{140 \cdot x}{140} &= \frac{5000}{140} \\ x &\approx 35.7 \end{aligned}$$

To feed 200 people, the cooks will need about 35.7 pounds of meat.

Similarly, for potatoes:

$$\begin{aligned} \frac{140 \cdot x}{140} &= \frac{35 \cdot 200}{140} \\ x &= 50\text{ pounds} \end{aligned}$$

And for carrots:

$$\begin{aligned} \frac{140 \cdot x}{140} &= \frac{15 \cdot 200}{140} \\ x &\approx 21.4\text{ pounds} \end{aligned}$$

61. Convert 85 cm to m.
85 cm = 0.85 m
She will need $4 \cdot 0.85$ m = 3.4 m.

62. Find the discount amount.

$$\begin{aligned} \text{discount amount} &= \text{discount rate} \cdot \text{cost} \\ &= 65\% \cdot 64 \\ &= 0.65 \cdot 64 \\ &= 41.6 \end{aligned}$$

Subtract the discount amount from the regular price.

$64 – $41.6 = $22.4

Lance will pay $22.40.

63. **(a)** (5 days/week)(6 weeks) = 30 days, so Britain averages six weeks of vacation.

(b) (5 days/week)(8 weeks) = 40 days, so Brazil averages eight weeks of vacation.

64. **(a)** $\frac{\text{Brazil}}{\text{U.S.}} = \frac{40\text{ days}}{20\text{ days}} = \frac{40 \div 20}{20 \div 20} = \frac{2}{1}$

(b) $\frac{\text{Canada}}{\text{France}} = \frac{24\text{ days}}{36\text{ days}} = \frac{24 \div 12}{36 \div 12} = \frac{2}{3}$

65. **(a)** $\frac{\text{U.S.}}{\text{Italy}} = \frac{20\text{ days}}{42\text{ days}} = \frac{10}{21} \approx 0.476 = 47.6\%$

(b) $\frac{\text{France}}{\text{U.S.}} = \frac{36\text{ days}}{20\text{ days}} = \frac{9}{5} = 1.8 = 180\%$

66. **(a)** Canada days – U.S. days = 24 – 20 = 4

4 is what percent of 20?

$$\frac{4}{20} = \frac{4 \div 4}{20 \div 4} = \frac{1}{5} = 0.2 = 20\%$$

(b) Britain days – U.S. days = 30 – 20 = 10

10 is what percent of 20?

$$\frac{10}{20} = \frac{10 \div 10}{20 \div 10} = \frac{1}{2} = 0.5 = 50\%$$

(c) Brazil days – U.S. days = 40 – 20 = 20

20 is what percent of 20%

$$\frac{20}{20} = 1 = 100\%$$

CHAPTER 9 BASIC ALGEBRA

9.1 Signed Numbers

9.1 Margin Exercises

1. **(a)** A temperature at the North Pole of 70 degrees below 0 is written $-70°$.

   (b) Your checking account is overdrawn by 15 dollars is written as $-\$15$.

   (c) The altitude of a place 284 feet below sea level is written -284 ft.

2. **(a)** -8 is negative.

   (b) $-\frac{3}{4}$ is negative.

   (c) 1 is positive.

   (d) 0 is neither positive nor negative.

3. **(a)** $-1, 1, -3, 3$

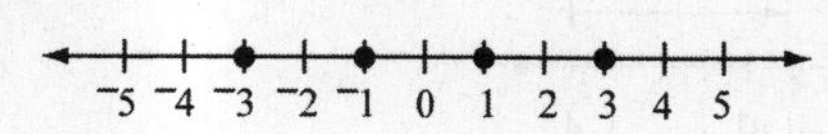

   (b) $-2, 4, 0, -1, -4$

   ⁻5 ⁻4 ⁻3 ⁻2 ⁻1 0 1 2 3 4 5

4. **(a)** $4 > 0$ because 4 is to the *right* of 0 on the number line.

   (b) $-1 < 0$ because -1 is to the *left* of 0 on the number line.

   (c) $-3 < -1$ because -3 is to the *left* of -1 on the number line.

   (d) $-8 > -9$ because -8 is to the *right* of -9 on the number line.

   (e) $0 > -3$ because 0 is to the *right* of -3 on the number line.

5. **(a)** The *distance* from 0 to 5 is 5, so $|5| = 5$.

   (b) The *distance* from 0 to -5 is also 5, so $|-5| = 5$.

   (c) The *distance* from 0 to -17 is 17, so $|-17| = 17$.

   (d) $-|-9|$
   First, $|-9| = 9$. But there is a negative sign outside the absolute value bars. So, -9 is the simplified expression.

   (e) $-|2| = -(2) = -2$
   Because the negative sign is outside the absolute value bars, the answer is negative.

6. **(a)** The opposite of 4 is $-(4) = -4$.

   (b) The opposite of 10 is $-(10) = -10$.

   (c) The opposite of 49 is $-(49) = -49$.

   (d) The opposite of $\frac{2}{5}$ is $-\left(\frac{2}{5}\right) = -\frac{2}{5}$.

   (e) The opposite of 0 is 0. Zero is neither positive nor negative.

7. **(a)** The opposite of -4 is $-(-4) = 4$.

   (b) The opposite of -10 is $-(-10) = 10$.

   (c) The opposite of -25 is $-(-25) = 25$.

   (d) The opposite of -1.9 is $-(-1.9) = 1.9$.

   (e) The opposite of -0.85 is $-(-0.85) = 0.85$.

   (f) The opposite of $-\frac{3}{4}$ is $-\left(-\frac{3}{4}\right) = \frac{3}{4}$.

9.1 Section Exercises

1. Water freezes at 32 degrees above 0 on the Fahrenheit temperature scale.

   $+32$ or 32 degrees

3. The price of the stock fell \$12.

   $-\$12$

5. Keith lost $6\frac{1}{2}$ lb while he was sick with the flu.

   $-6\frac{1}{2}$ lb

7. Mount McKinley, the tallest mountain in the United States, rises 20,320 ft above sea level.

   $+20{,}320$ or 20,320 ft

9. $4, -1, 2, 0, -5$

   -5 -4 -3 -2 -1 0 1 2 3 4 5

11. $-\frac{1}{2}, -3, \frac{7}{4}, -4\frac{1}{2}, 3\frac{1}{4}$

   -5 -4 -3 -2 -1 0 1 2 3 4 5

13. $3, 4.5, -1.5, 2.2, -0.5$

   -5 -4 -3 -2 -1 0 1 2 3 4 5

15. $9 < 14$ because 9 is to the *left* of 14 on a number line.

17. $0 > -2$ because 0 is to the *right* of -2 on a number line.

19. $-6 < 3$ because -6 is to the *left* of 3 on a number line.

21. $1 > -1$ because 1 is to the *right* of -1 on a number line.

23. $-11 < -2$ because -11 is to the *left* of -2 on a number line.

25. $-72 > -75$ because -72 is to the *right* of -75 on a number line.

27. $-1.2,\ 0.6,\ -0.5,\ 0$

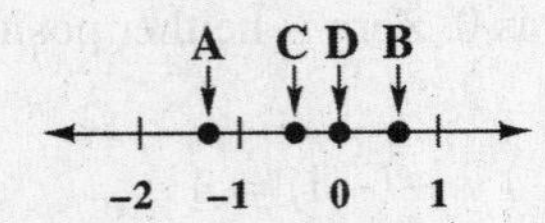

28. From lowest to highest, the patients' scores are:

$$-1.2, -0.5, 0, 0.6$$

29. Since $-1.2 < -1$, Patient A may be at risk.

Since $0.6 > 0$, Patient B is above normal.

Since $-1.0 \leq -0.5 \leq 0$, Patient C is normal.

Since $-1.0 \leq 0 \leq 0$, Patient D is normal.

30. **(a)** The patient would think the interpretation was "above normal" and wouldn't get treatment.

(b) The sign of the score makes no difference in Patient D's score of 0 because 0 is neither positive nor negative.

31. The *distance* from 0 to 3 is 3, so $|3| = 3$.

33. The *distance* from 0 to -10 is 10, so $|-10| = 10$.

35. The *distance* from 0 to 0 is 0, so $|0| = 0$.

37. $-|-18|$
First, $|-18| = 18$. But there is a negative sign outside the absolute value bars. So, -18 is the simplified expression.

39. $-|32| = -(32) = -32$
Because the negative sign is outside the absolute value bars, the answer is negative.

41. The opposite of 7 is -7.

43. The opposite of -14 is $-(-14) = 14$.

45. The opposite of $\frac{2}{3}$ is $-\frac{2}{3}$.

47. The opposite of -8.3 is $-(-8.3) = 8.3$.

49. The opposite of $-\frac{1}{6}$ is $-(-\frac{1}{6}) = \frac{1}{6}$.

51. $|-5| = 5$

5 is greater than 0, so $|-5| > 0$ is true.

53. $-(-6) = 6$

0 is less than 6, so $0 < -(-6)$ is true.

55. $-|-4| = -4$ and $-|-7| = -7$. Since -4 is to the right of -7 on the number line, $-4 > -7$, and $-|-4| < -|-7|$ is false.

9.2 Adding and Subtracting Signed Numbers

9.2 Margin Exercises

1. **(a)**

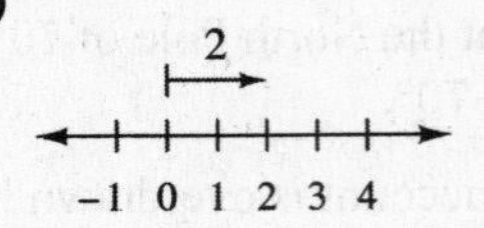

(b)

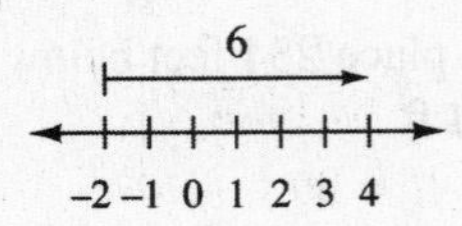

(c)

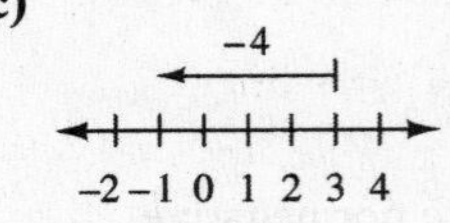

(d)

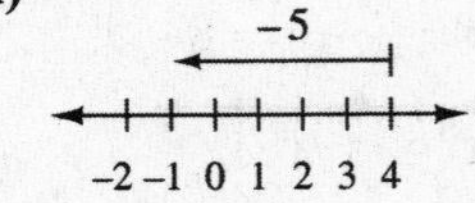

2. **(a)** $3 + (-2) = 1$

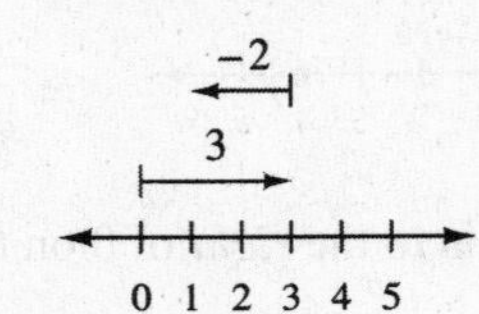

(b) $-4 + 1 = -3$

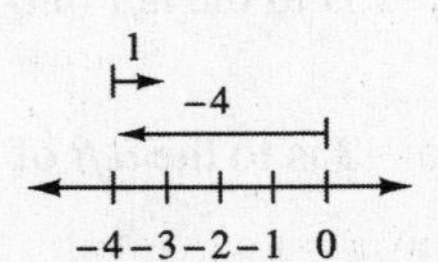

(c) $-3 + 7 = 4$

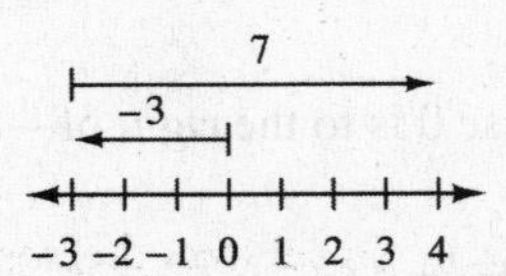

(d) $-1 + (-4) = -5$

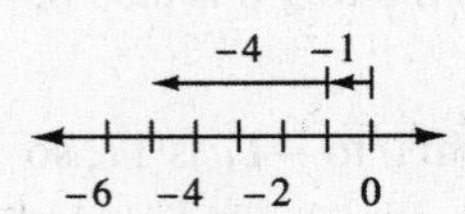

3. **(a)** $-4 + (-4)$
Add the absolute values.

$$|-4| = 4; \quad |-4| = 4$$

$$4 + 4 = 8$$

Write a negative sign in front of the sum.

$$-4 + (-4) = -8$$

(b) $-3+(-20)$
Add the absolute values.

$$|-3|=3; \quad |-20|=20$$

$$3+20=23$$

Write a negative sign in front of the sum.

$$-3+(-20)=-23$$

(c) $-31+(-5)$
Add the absolute values.

$$|-31|=31; \quad |-5|=5$$

$$31+5=36$$

Write a negative sign in front of the sum.

$$-31+(-5)=-36$$

(d) $-10+(-8)$
Add the absolute values.

$$|-10|=10; \quad |-8|=8$$

$$10+8=18$$

Write a negative sign in front of the sum.

$$-10+(-8)=-18$$

(e) $-\frac{9}{10}+\left(-\frac{3}{5}\right)$
Add the absolute values.

$$\left|-\frac{9}{10}\right|=\frac{9}{10}; \quad \left|-\frac{3}{5}\right|=\frac{3}{5}$$

$$\frac{9}{10}+\frac{3}{5}=\frac{9}{10}+\frac{6}{10}=\frac{15}{10}=\frac{3}{2}$$

Write a negative sign in front of the sum.

$$-\frac{9}{10}+\left(-\frac{3}{5}\right)=-\frac{3}{2}$$

4. **(a)** $10+(-2)$

The signs are different, so subtract the absolute values.

$$|10|=10; \quad |-2|=2$$

$$10-2=8$$

The positive number, 10, has the larger absolute value, so the answer is positive.

$$10+(-2)=8$$

(b) $-7+8$

The signs are different, so subtract the absolute values.

$$|-7|=7; \quad |8|=8$$

$$8-7=1$$

The positive number, 8, has the larger absolute value, so the answer is positive.

$$-7+8=1$$

(c) $-11+11$

The signs are different, so subtract the absolute values.

$$|-11|=11; \quad |11|=11$$

$$11-11=0$$

The numbers have equal absolute values, so the answer is neither positive nor negative.

(d) $23+(-32)$

The signs are different, so subtract the absolute values.

$$|23|=23; \quad |-32|=32$$

$$32-23=9$$

The negative number, -32, has the larger absolute value, so the answer is negative.

$$23+(-32)=-9$$

(e) $-\frac{7}{8}+\frac{1}{4}$

The signs are different, so subtract the absolute values.

$$\left|-\frac{7}{8}\right|=\frac{7}{8}; \quad \left|\frac{1}{4}\right|=\frac{1}{4}$$

$$\frac{7}{8}-\frac{1}{4}=\frac{7}{8}-\frac{2}{8}=\frac{5}{8}$$

The negative number, $-\frac{7}{8}$, has the larger absolute value, so the answer is negative.

$$-\frac{7}{8}+\frac{1}{4}=-\frac{5}{8}$$

5. **(a)** The additive inverse of 12 is -12. The sum of the number and its inverse is $12+(-12)=0$.

(b) The additive inverse of -9 is $-(-9)$ or 9. The sum of the number and its inverse is $(-9)+9=0$.

(c) The additive inverse of 3.5 is -3.5. The sum of the number and its inverse is $3.5+(-3.5)=0$.

(d) The additive inverse of $-\frac{7}{10}$ is $-\left(-\frac{7}{10}\right)$ or $\frac{7}{10}$. The sum of the number and its inverse is $\left(-\frac{7}{10}\right)+\frac{7}{10}=0$.

(e) The additive inverse of 0 is 0. The sum of the number and its inverse is $0+0=0$.

6. **(a)** $-7-2$ in words is negative seven minus positive two.

(b) -10 in words is negative ten.

(c) $3-(-5)$ in words is positive three minus negative five.

(d) 4 in words is positive four.

(e) $-8-(-6)$ in words is negative eight minus negative six.

(f) $2-9$ in words is positive two minus positive nine.

7. **(a)** $-6-5$
Change positive 5 to its opposite (-5) and add.

$$-6+(-5)=-11$$

(b) $3-(-10)$
Change negative 10 to its opposite $(+10)$ and add.

$$3-(-10)=3+(+10)=13$$

(c) $-8-(-2)$
Change negative 2 to its opposite $(+2)$ and add.

$$-8-(-2)=-8+(+2)=-6$$

(d) $4-9$
Change positive 9 to its opposite (-9) and add.

$$4-9=4+(-9)=-5$$

(e) $-7-(-15)$
Change negative 15 to its opposite $(+15)$ and add.

$$-7-(-15)=-7+(+15)=8$$

(f) $-\frac{2}{3}-\left(-\frac{5}{12}\right)$
Change negative $\frac{5}{12}$ to its opposite $\left(+\frac{5}{12}\right)$ and add.

$$-\frac{2}{3}+\left(+\frac{5}{12}\right)=-\frac{8}{12}+\left(+\frac{5}{12}\right)$$
$$=-\frac{3}{12}=-\frac{1}{4}$$

8. **(a)** $6-7+(-3)$
$=6+(-7)+(-3)$
$=-1+(-3)=-4$

(b) $-2+(-3)-(-5)$
$=-5-(-5)$
$=-5+(+5)=0$

(c) $-3-(-9)-(-5)$
$=-3+(+9)-(-5)$
$=6+(+5)=11$

(d) $8-(-2)+(-6)$
$=8+(+2)+(-6)$
$=10+(-6)=4$

9. **(a)** $-1-2+3-4$
$=-1+(-2)+3-4$
$=-3+3+(-4)$
$=0+(-4)=-4$

(b) $7-6-5+(-4)$
$=7+(-6)-5+(-4)$
$=1+(-5)+(-4)$
$=-4+(-4)=-8$

(c) $-6+(-15)-(-19)+(-25)$
$=-21+(+19)+(-25)$
$=-2+(-25)=-27$

(d)

-19.2			
-6.7 →	-25.9		
15.8	15.8 →	-10.1	
17.1	17.1	17.1 →	7
-5.4	-5.4	-5.4	-5.4
			1.6

9.2 Section Exercises

1. $-2+5=3$

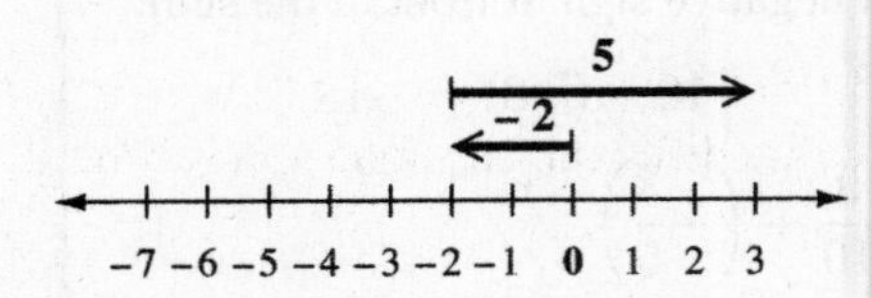

3. $-5+(-2)=-7$

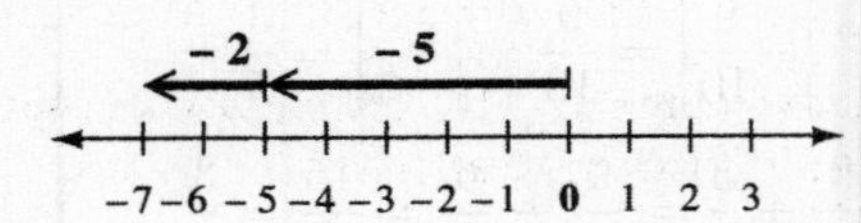

5. $3+(-4)=-1$

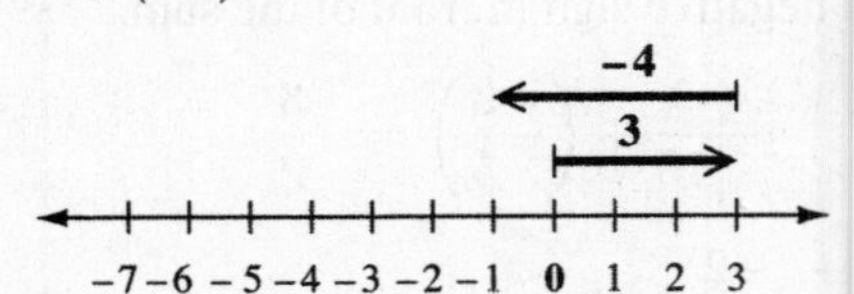

7. $-8+5$
The signs are different, so subtract the absolute values.

$$|-8|=8;\quad |5|=5$$
$$8-5=3$$

The negative number, -8, has the larger absolute value, so the answer is negative.

$$-8+5=-3$$

9. $-1+8$
The signs are different, so subtract the absolute values.

$$|-1|=1;\quad |8|=8$$
$$8-1=7$$

The positive number, 8, has the larger absolute value, so the answer is positive.

$$-1+8=7$$

11. $-2+(-5)$
Add the absolute values.

$$|-2|=2;\quad |-5|=5$$
$$2+5=7$$

Write a negative sign in front of the sum, because both numbers are negative.

$$-2+(-5)=-7$$

13. $6+(-5)=1$
The signs are different, so subtract the absolute values.

$$|6|=6;\quad |-5|=5$$
$$6-5=1$$

The positive number, 6, has the larger absolute value, so the answer is positive.

$$6+(-5)=1$$

15. $4+(-12)=-8$
The signs are different, so subtract the absolute values.

$$|4|=4;\quad |-12|=12$$
$$12-4=8$$

The negative number, -12, has the larger absolute value, so the answer is negative.

$$4+(-12)=-8$$

17. $-10+(-10)$
Add the absolute values.

$$|-10|=10;\quad |-10|=10$$
$$10+10=20$$

Write a negative sign in front of the sum because both numbers are negative.

$$-10+(-10)=-20$$

19. 13 yards gained (positive) and
17 yards lost (negative)

$$13+(-17)=-4 \text{ yd}$$

21. The overdrawn amount is negative ($-\$52.50$) and the deposit is positive.

$$-\$52.50+\$50=-\$2.50$$

23. Lost (negative); Won (positive)
Jeff: $-20+75+(-55)=0$ pts
Terry: $42+(-15)+20=47$ pts

25. $7.8+(-14.6)$
The signs are different, so subtract the absolute values.

$$|7.8|=7.8;\quad |-14.6|=14.6$$
$$14.6-7.8=6.8$$

The negative number, -14.6, has the larger absolute value, so the answer is negative.

$$7.8+(-14.6)=-6.8$$

27. $-\frac{1}{2}+\frac{3}{4}$
The signs are different, so subtract the absolute values.

$$\left|-\frac{1}{2}\right|=\frac{1}{2};\quad \left|\frac{3}{4}\right|=\frac{3}{4}$$
$$\frac{3}{4}-\frac{1}{2}=\frac{3}{4}-\frac{2}{4}=\frac{1}{4}$$

The positive number, $\frac{3}{4}$, has the larger absolute value, so the answer is positive.

$$-\frac{1}{2}+\frac{3}{4}=\frac{1}{4}$$

29. $-\frac{7}{10}+\frac{2}{5}$
The signs are different, so subtract the absolute values.

$$\left|-\frac{7}{10}\right|=\frac{7}{10};\quad \left|\frac{2}{5}\right|=\frac{2}{5}$$
$$\frac{7}{10}-\frac{2}{5}=\frac{7}{10}-\frac{4}{10}=\frac{3}{10}$$

The negative number, $-\frac{7}{10}$, has the larger absolute value, so the answer is negative.

$$-\frac{7}{10}+\frac{2}{5}=-\frac{3}{10}$$

31. $-\frac{7}{3}+\left(-\frac{5}{9}\right)$
Add the absolute values.

$$\left|-\frac{7}{3}\right|=\frac{7}{3};\quad \left|-\frac{5}{9}\right|=\frac{5}{9}$$
$$\frac{7}{3}+\frac{5}{9}=\frac{21}{9}+\frac{5}{9}=\frac{26}{9}$$

Write a negative sign in front of the sum because both numbers are negative.

$$-\frac{7}{3}+\left(-\frac{5}{9}\right)=-\frac{26}{9}$$

33. The additive inverse of 3 is -3.

35. The additive inverse of -9 is $-(-9)$ or 9.

37. The additive inverse of $\frac{1}{2}$ is $-\frac{1}{2}$.

39. The additive inverse of -6.2 is $-(-6.2)$ or 6.2.

41. $19-5=19+(-5)$
$=14$

43. $10-12=10+(-12)$
$=-2$

45. $7-19=7+(-19)$
$=-12$

47. $-15-10=-15+(-10)=-25$

49. $-9-14=-9+(-14)$
$=-23$

51. $-3-(-8)=-3+(+8)$
$=5$

53. $6-(-14)=6+(+14)$
$=20$

55. $1-(-10)=1+(+10)$
$=11$

57. $-30-30=-30+(-30)$
$=-60$

59. $-16-(-16)=-16+(+16)$
$=0$

61. $-\frac{7}{10}-\frac{4}{5}=-\frac{7}{10}+\left(-\frac{4}{5}\right)$
$=-\frac{7}{10}+\left(-\frac{8}{10}\right)$
$=-\frac{15}{10}=-\frac{3}{2}$

63. $\frac{1}{2}-\frac{9}{10}=\frac{5}{10}-\frac{9}{10}$
$=\frac{5}{10}+\left(-\frac{9}{10}\right)$
$=-\frac{4}{10}=-\frac{2}{5}$

65. $-8.3-(-9)=-8.3+(+9)$
$=0.7$

67. **(a)** Change 6 to its opposite, -6. Correct answer is $-6+(-6)=-12$.

(b) Do *not* change 9 to its opposite. Correct answer is $-9+(-5)=-14$.

69. **(a)** The 30°F column and the 10 mph wind row intersect at 21°F. The difference between the actual temperature and the wind chill temperature is $30-21=9$ degrees.

(b) The 15°F column and the 15 mph wind row intersect at 0°F. The difference between the actual temperature and the wind chill temperature is $15-0=15$ degrees.

71. **(a)** The 5°F column and the 25 mph wind row intersect at -17°F. The difference between the actual temperature and the wind chill temperature is $5-(-17)=5+17=22$ degrees.

(b) The -10°F column and the 35 mph wind row intersect at -41°F. The difference between the actual temperature and the wind chill temperature is $-10-(-41)=-10+41=31$ degrees.

73. $-2+(-11)-(-3)=-13-(-3)$
$=-13+3$
$=-10$

75. $4-(-13)+(-5)=4+(+13)+(-5)$
$=17+(-5)$
$=12$

77. $-12-(-3)-(-2)=-12+3-(-2)$
$=-9-(-2)$
$=-9+2$
$=-7$

79. $4-(-4)-3=4+(+4)-3$
$=8-3$
$=5$

81. $\frac{1}{2}-\frac{2}{3}+\left(-\frac{5}{6}\right)=\frac{1}{2}+\left(-\frac{2}{3}\right)+\left(-\frac{5}{6}\right)$
$=\frac{3}{6}+\left(-\frac{4}{6}\right)+\left(-\frac{5}{6}\right)$
$=-\frac{1}{6}+\left(-\frac{5}{6}\right)$
$=-\frac{6}{6}=-1$

83. $-5.7-(-9.4)-8.1=-5.7+(+9.4)-8.1$
$=3.7+(-8.1)$
$=-4.4$

85. $-2+(-11)+|-2|=-13+|-2|$
$=-13+2$
$=-11$

87. $-3-(-2+4)+(-5)=-3-(+2)+(-5)$
$=-3+(-2)+(-5)$
$=-5+(-5)$
$=-10$

89. $37 − $689 + $908 − $60
= $37 + (−$689) + $908 + (−$60)
= −$652 + $908 + (−$60)
= $256 + (−$60)
= $196

The balance is $196.

91. $-\$89.62 - \$110.70 - \$99.68 + 3(\$100)$
$= -\$89.62 + (-\$110.70) + (-\$99.68) + \$300$
$= -\$200.32 + (-\$99.68) + \$300$
$= -\$300 + \$300$
$= \$0$
The balance is \$0.

93. **(a)** $-5 + 3 = \underline{-2}$ $\quad 3 + (-5) = \underline{-2}$

(b) $-2 + (-6) = \underline{-8}$ $\quad -6 + (-2) = \underline{-8}$

(c) $17 + (-7) = \underline{10}$ $\quad -7 + 17 = \underline{10}$

Addition is commutative, so changing the order of the addends does *not* change the sum.

94. **(a)** $-3 - 5 = \underline{-8}$ $\quad 5 - (-3) = 5 + 3 = \underline{8}$

(b) $-4 - (-6) = -4 + 6 = \underline{2}$

$-6 - (-4) = -6 + 4 = \underline{-2}$

(c) $3 - 10 = \underline{-7}$ $\quad 10 - 3 = \underline{7}$

Changing the order of the numbers in subtraction *does change* the result.

95. **(a)** The answers have the same absolute value but opposite signs.

(b) When the order of the numbers in a subtraction problem is switched, change the sign on the answer to its opposite.

96. **(a)** $-18 + 0 = \underline{-18}$ $\quad 20 + 0 = \underline{20}$
$0 + (-5) = \underline{-5}$ $\quad 0 + 4 = \underline{4}$

Adding 0 to any number leaves the number unchanged.

(b) $2 - 0 = \underline{2}$ $\quad 0 - 10 = \underline{-10}$
$-3 - 0 = \underline{-3}$ $\quad 0 - (-7) = 0 + 7 = \underline{7}$

Subtracting 0 from a number leaves the number unchanged, but subtracting a *number from 0* changes the sign of the number.

9.3 Multiplying and Dividing Signed Numbers

9.3 Margin Exercises

1. **(a)** $5 \cdot (-4) = -20$
The product of two numbers with *different* signs is *negative.*

(b) $-9(15) = -135$

(c) $12(-1) = -12$

(d) $-6(6) = -36$

(e) $\left(-\frac{7}{8}\right)\left(\frac{4}{3}\right) = \left(-\frac{7}{\underset{2}{\cancel{8}}}\right)\left(\frac{\overset{1}{\cancel{4}}}{3}\right) = -\frac{7}{6}$

2. **(a)** $(-5)(-5) = 25$
The product of two numbers with the *same* sign is *positive.*

(b) $(-14)(-1) = 14$

(c) $-7(-8) = 56$

(d) $3(12) = 36$

(e) $\left(-\frac{2}{3}\right)\left(-\frac{6}{5}\right) = \left(-\frac{2}{\underset{1}{\cancel{3}}}\right)\left(-\frac{\overset{2}{\cancel{6}}}{5}\right) = \frac{4}{5}$

3. **(a)** $\frac{-20}{4} = -5$
The numbers have *different* signs, so the quotient is *negative.*

(b) $\frac{-50}{-5} = 10$
The numbers have the *same* sign, so the quotient is *positive.*

(c) $\frac{44}{2} = 22$

(d) $\frac{6}{-6} = -1$

(e) $\frac{-15}{-1} = 15$

(f) $\frac{-\frac{3}{5}}{\frac{9}{10}} = -\frac{3}{5} \div \frac{9}{10} = -\frac{\overset{1}{\cancel{3}}}{\underset{1}{\cancel{5}}} \cdot \frac{\overset{2}{\cancel{10}}}{\underset{3}{\cancel{9}}} = -\frac{2}{3}$

(g) $\frac{-35}{0}$ is undefined.

9.3 Section Exercises

1. $-5 \cdot 7 = -35$
The numbers have *different* signs, so the product is *negative.*

3. $(-5)(9) = -45$
The numbers have *different* signs, so the product is *negative.*

5. $3(-6) = -18$
The numbers have *different* signs, so the product is *negative.*

7. $10 \cdot (-5) = -50$
The numbers have *different* signs, so the product is *negative.*

9. $(-1)(40) = -40$
The numbers have *different* signs, so the product is *negative.*

11. $-8(-4) = 32$
The numbers have the *same* sign, so the product is *positive.*

13. $11(7) = 77$
The numbers have the *same* sign, so the product is *positive.*

15. $-19 \cdot (-7) = 133$
The numbers have the *same* sign, so the product is *positive.*

17. $-13(-1) = 13$
The numbers have the *same* sign, so the product is *positive.*

19. $0 \cdot (-25) = 0$

21. $-\frac{1}{2} \cdot (-8) = -\frac{1}{\cancel{2}_1} \cdot \left(-\frac{\cancel{8}^4}{1}\right) = \frac{4}{1} = 4$

23. $-10\left(\frac{2}{5}\right) = -\frac{\cancel{10}^2}{1} \cdot \frac{2}{\cancel{5}_1} = -2 \cdot 2 = -4$

25. $\left(\frac{3}{5}\right)\left(-\frac{1}{6}\right) = \left(\frac{\cancel{3}^1}{5}\right)\left(-\frac{1}{\cancel{6}_2}\right) = -\frac{1}{10}$

27. $-\frac{7}{5}\left(-\frac{10}{3}\right) = -\frac{7}{\cancel{5}_1}\left(-\frac{\cancel{10}^2}{3}\right) = \frac{14}{3}$

29. $-\frac{7}{15} \cdot \frac{25}{14} = -\frac{\cancel{7}^1}{\cancel{15}_3} \cdot \frac{\cancel{25}^5}{\cancel{14}_2} = -\frac{5}{6}$

31. $-\frac{5}{2} \cdot \left(-\frac{7}{10}\right) = -\frac{\cancel{5}^1}{2}\left(-\frac{7}{\cancel{10}_2}\right) = \frac{7}{4}$

33. $9(-4.7) = -42.3$

35. $(-0.5)(-12) = 6$

37. $(-6.2)(5.1) = -31.62$

39. $-1.25(-3.6) = 4.5$

41. $(-8.23)(-1) = 8.23$

43. $0(-58.6) = 0$

45. $\frac{-14}{7} = -2$

The numbers have *different* signs, so the quotient is *negative.*

47. $\frac{30}{-6} = -5$

The numbers have *different* signs, so the quotient is *negative.*

49. $\frac{-28}{0}$ is undefined.

51. $\frac{14}{-1} = -14$

The numbers have *different* signs, so the quotient is *negative.*

53. $\frac{-20}{-2} = 10$

The numbers have the *same* sign, so the quotient is *positive.*

55. $\frac{-48}{-12} = 4$

The numbers have the *same* sign, so the quotient is *positive.*

57. $\frac{-18}{18} = -1$

59. $\frac{-573}{-3} = 191$

61. $\frac{0}{-9} = 0$

63. $\frac{-30}{-30} = 1$

65. $\frac{-\frac{5}{7}}{-\frac{15}{14}} = -\frac{\cancel{5}^1}{\cancel{7}_1} \cdot \left(-\frac{\cancel{14}^2}{\cancel{15}_3}\right) = \frac{2}{3}$

67. $-\frac{2}{3} \div (-2) = -\frac{\cancel{2}^1}{3} \cdot \left(-\frac{1}{\cancel{2}_1}\right) = \frac{1}{3}$

69. $5 \div \left(-\frac{5}{8}\right) = \frac{\cancel{5}^1}{1} \cdot \left(-\frac{8}{\cancel{5}_1}\right) = -8$

71. $-\frac{7}{5} \div \frac{3}{10} = -\frac{7}{\cancel{5}_1} \cdot \frac{\cancel{10}^2}{3} = -\frac{14}{3}$

73. $\frac{-18.92}{-4} = 4.73$

The numbers have the *same* sign, so the quotient is *positive.*

```
    4.7 3
4 ) 1 8.9 2
    1 6
      2 9
      2 8
        1 2
        1 2
          0
```

75. $\dfrac{-7.05}{1.5} = -4.7$

The numbers have *different* signs, so the quotient is *negative*.

$$\begin{array}{r} 4.\ 7 \\ 1.5_\wedge \overline{)\,7.\ 0_\wedge\ 5} \\ 6\ 0 \\ \hline 1\ 0\ 5 \\ 1\ 0\ 5 \\ \hline 0 \end{array}$$

77. $\dfrac{45.58}{-8.6} = -5.3$

The numbers have *different* signs, so the quotient is *negative*.

$$\begin{array}{r} 5.\ 3 \\ 8.6_\wedge \overline{)\,4\ 5.\ 5_\wedge\ 8} \\ 4\ 3\ 0 \\ \hline 2\ 5\ 8 \\ 2\ 5\ 8 \\ \hline 0 \end{array}$$

79. $(-4)(-6)\frac{1}{2} = 24 \cdot \frac{1}{2} = 12$

81. $(-0.6)(-0.2)(-3) = (0.12)(-3) = -0.36$

83. $\left(-\frac{1}{2}\right)\left(\frac{2}{5}\right)\left(\frac{7}{8}\right) = -\frac{1}{5}\left(\frac{7}{8}\right) = -\frac{7}{40}$

85. $-36 \div (-2) \div (-3) \div (-3) \div (-1)$
$= 18 \div (-3) \div (-3) \div (-1)$
$= -6 \div (-3) \div (-1)$
$= 2 \div (-1) = -2$

87. $|-8| \div (-4) \cdot |-5|$
$= 8 \div (-4) \cdot (5)$
$= -2 \cdot 5 = -10$

89. $12(-9950) = -119{,}400$
The total loss for the year was $-\$119{,}400$.

91. $\dfrac{-36{,}198}{18} = -2011$

It dives -2011 ft in each step.

93. $13(182) = 2366$
Wei Chan will pay $2366 for 13 credits.

95. $\dfrac{24{,}000}{1000} = 24$

The temperature will change $24(-3) = -72$ degrees in 24,000 ft.

The temperature will be $50 + (-72) = -22$ degrees.

97. **(a)** Examples will vary. Some possibilities:
$6(-1) = -6;\ 2(-1) = -2;\ 15(-1) = -15$

(b) Examples will vary. Some possibilities:
$-6(-1) = 6;\ -2(-1) = 2;\ -15(-1) = 15$

The result of multiplying any nonzero number times -1 is the number with the opposite sign.

98. **(a)** Examples will vary. Some possibilities:
$\frac{-6}{-1} = 6;\ \frac{-2}{-1} = 2;\ \frac{-15}{-1} = 15$

(b) Examples will vary. Some possibilities:
$\frac{6}{-1} = -6;\ \frac{2}{-1} = -2;\ \frac{15}{-1} = -15$

(c) Examples will vary. Some possibilities:
$\frac{-6}{-6} = 1;\ \frac{-2}{-2} = 1;\ \frac{-15}{-15} = 1$

When dividing by -1, the sign of the number changes to its opposite. A negative number divided by itself gives a result of 1.

9.4 Order of Operations

9.4 Margin Exercises

1. **(a)** $-9 + (-15) + (-3)$ *Add*
$= -24 + (-3)$ *Add*
$= -27$

(b) $-8 - (-2) + (-6)$ *Change subtraction to addition*
$= -8 + (+2) + (-6)$ *Add*
$= -6 + (-6)$ *Add*
$= -12$

(c) $-2 - (-7) - (-4)$ *Change subtraction to addition*
$= -2 + (+7) - (-4)$ *Add*
$= 5 - (-4)$ *Change subt.*
$= 5 + (+4)$ *Add*
$= 9$

(d) $3(-4) \div (-6)$ *Multiply*
$= -12 \div (-6)$ *Divide*
$= 2$

(e) $-18 \div 9(-4)$ *Divide*
$= -2(-4)$ *Multiply*
$= 8$

2. **(a)** $10 + 8 \div 2$ *Divide*
$= 10 + 4$ *Add*
$= 14$

(b) $4 - 6(-2)$ *Multiply*
$= 4 - (-12)$ *Change subt.*
$= 4 + (+12)$ *Add*
$= 16$

(c) $-3 + (-5) \cdot 2 - 1$ *Multiply*
$= -3 + (-10) - 1$ *Add*
$= -13 - 1$ *Change subt.*
$= -13 + (-1)$ *Add*
$= -14$

(d) $-6 \div 2 + 3(-2)$ *Divide*
$= -3 + 3(-2)$ *Multiply*
$= -3 + (-6)$ *Add*
$= -9$

(e) $7 - 6(2) \div (-3)$ *Multiply*
$= 7 - 12 \div (-3)$ *Divide*
$= 7 - (-4)$ *Change subt.*
$= 7 + (+4)$ *Add*
$= 11$

3. **(a)** $2 + 40 \div (-5 + 3)$ *Parentheses*
$= 2 + 40 \div (-2)$ *Divide*
$= 2 + (-20)$ *Add*
$= -18$

(b) $5 - 3(2 + 4)$ *Parentheses*
$= 5 - 3(6)$ *Multiply*
$= 5 - 18$ *Change subt.*
$= 5 + (-18)$ *Add*
$= -13$

(c) $(-24 \div 2) + (15 - 3)$ *Parentheses*
$= -12 + 12$ *Add*
$= 0$

(d) $-3(2 - 8) - 5(4 - 3)$ *Parentheses*
$= -3(-6) - 5(1)$ *Multiply*
$= 18 - 5$ *Change subt.*
$= 18 + (-5)$ *Add*
$= 13$

(e) $3(3) - 10(3) \div 5$ *Multiply*
$= 9 - 10(3) \div 5$ *Multiply*
$= 9 - 30 \div 5$ *Divide*
$= 9 - 6$ *Change subt.*
$= 9 + (-6)$ *Add*
$= 3$

(f) $6 - (2 + 7) \div (-4 + 1)$ *Parentheses*
$= 6 - 9 \div (-3)$ *Divide*
$= 6 - (-3)$ *Change subt.*
$= 6 + (+3)$ *Add*
$= 9$

4. **(a)** $2^3 - 3^2$ *Exponents*
$= 8 - 9$ *Change subt.*
$= 8 + (-9)$ *Add*
$= -1$

(b) $4^2 - 3^2(5 - 2)$ *Parentheses*
$= 4^2 - 3^2(3)$ *Exponents*
$= 16 - 9(3)$ *Multiply*
$= 16 - 27$ *Change subt.*
$= 16 + (-27)$ *Add*
$= -11$

(c) $-18 \div (-3)(2^3)$ *Exponent*
$= -18 \div (-3)(8)$ *Divide*
$= 6 \cdot 8$ *Multiply*
$= 48$

(d) $(-3)^3 + (3 - 8)^2$ *Parentheses*
$= (-3)^3 + (-5)^2$ *Exponents*
$= -27 + 25$ *Add*
$= -2$

(e) $\frac{3}{8} + \left(-\frac{1}{2}\right)^2 \div \frac{1}{4}$ *Exponent*
$= \frac{3}{8} + \left(\frac{1}{4}\right) \div \frac{1}{4}$ *Divide*
$= \frac{3}{8} + 1$ *Add*
$= \frac{3}{8} + \frac{8}{8}$
$= \frac{11}{8}$

5. **(a)** $\frac{-3(2^3)}{-10 - 6 + 8}$

First do the work in the numerator.

$-3(2^3)$ *Exponent*
$= -3 \cdot 8$ *Multiply*
$= -24$ *Numerator*

Now do the work in the denominator.

$-10 - 6 + 8$ *Change subt.*
$= -10 + (-6) + 8$ *Add*
$= -16 + 8$ *Add*
$= -8$

The last step is the division.

$$\frac{-24}{-8} = 3$$

(b) $\frac{(-10)(-5)}{-6 \div 3(5)}$

Numerator:
$(-10)(-5) = 50$

Denominator:
$-6 \div 3(5)$ *Divide*
$= -2(5)$ *Multiply*
$= -10$

The last step is the division.

$$\frac{50}{-10} = -5$$

(c) $\dfrac{6+18\div(-2)}{(1-10)\div 3}$

Numerator:

$$\begin{aligned}&6+18\div(-2) && \textit{Divide}\\ &=6+(-9) && \textit{Add}\\ &=-3\end{aligned}$$

Denominator:

$$\begin{aligned}&(1-10)\div 3 && \textit{Parentheses}\\ &=(-9)\div 3 && \textit{Divide}\\ &=-3\end{aligned}$$

The last step is the division.

$$\frac{-3}{-3}=1$$

(d) $\dfrac{6^2-3^2(4)}{5+(3-7)^2}$

Numerator:

$$\begin{aligned}&6^2-3^2(4) && \textit{Exponents}\\ &=36-9(4) && \textit{Multiply}\\ &=36-36 && \textit{Subtract}\\ &=0\end{aligned}$$

Denominator:

$$\begin{aligned}&5+(3-7)^2 && \textit{Parentheses}\\ &=5+(-4)^2 && \textit{Exponent}\\ &=5+16 && \textit{Add}\\ &=21\end{aligned}$$

The last step is the division.

$$\frac{0}{21}=0$$

9.4 Section Exercises

1.
$$\begin{aligned}&6+3(-4) && \textit{Multiply}\\ &=6+(-12) && \textit{Add}\\ &=-6\end{aligned}$$

3.
$$\begin{aligned}&-1+15+2(-7) && \textit{Multiply}\\ &=-1+15+(-14) && \textit{Add}\\ &=14+(-14) && \textit{Add}\\ &=0\end{aligned}$$

5.
$$\begin{aligned}&6^2+4^2 && \textit{Exponents}\\ &=36+16 && \textit{Add}\\ &=52\end{aligned}$$

7.
$$\begin{aligned}&10-7^2 && \textit{Exponent}\\ &=10-49 && \textit{Change subtraction to addition}\\ &=10+(-49) && \textit{Add}\\ &=-39\end{aligned}$$

9.
$$\begin{aligned}&(-2)^5+2 && \textit{Exponent}\\ &=-32+2 && \textit{Add}\\ &=-30\end{aligned}$$

11.
$$\begin{aligned}&4^2+3^2+(-8) && \textit{Exponents}\\ &=16+9+(-8) && \textit{Add}\\ &=25+(-8) && \textit{Add}\\ &=17\end{aligned}$$

13.
$$\begin{aligned}&2-(-5)+3^2 && \textit{Exponent}\\ &=2-(-5)+9 && \textit{Change subt.}\\ &=2+(+5)+9 && \textit{Add}\\ &=7+9 && \textit{Add}\\ &=16\end{aligned}$$

15.
$$\begin{aligned}&(-4)^2+(-3)^2+5 && \textit{Exponents}\\ &=16+9+5 && \textit{Add}\\ &=25+5 && \textit{Add}\\ &=30\end{aligned}$$

17.
$$\begin{aligned}&3+5(6-2) && \textit{Parentheses}\\ &=3+5(4) && \textit{Multiply}\\ &=3+20 && \textit{Add}\\ &=23\end{aligned}$$

19.
$$\begin{aligned}&-7+6(8-14) && \textit{Parentheses}\\ &=-7+6(-6) && \textit{Multiply}\\ &=-7+(-36) && \textit{Add}\\ &=-43\end{aligned}$$

21.
$$\begin{aligned}&-6+(-5)(9-14) && \textit{Parentheses}\\ &=-6+(-5)(-5) && \textit{Multiply}\\ &=-6+25 && \textit{Add}\\ &=19\end{aligned}$$

23.
$$\begin{aligned}&(-5)(7-13)\div(-10) && \textit{Parentheses}\\ &=(-5)(-6)\div(-10) && \textit{Multiply}\\ &=30\div(-10) && \textit{Divide}\\ &=-3\end{aligned}$$

25.
$$\begin{aligned}&9\div(-3)^2+(-1) && \textit{Exponent}\\ &=9\div 9+(-1) && \textit{Divide}\\ &=1+(-1) && \textit{Add}\\ &=0\end{aligned}$$

27.
$$\begin{aligned}&2-(-5)(-3)^2 && \textit{Exponent}\\ &=2-(-5)(9) && \textit{Multiply}\\ &=2-(-45) && \textit{Change subt.}\\ &=2+(+45) && \textit{Add}\\ &=47\end{aligned}$$

29.
$$\begin{aligned}&(-2)(-7)+3(9) && \textit{Multiply}\\ &=14+27 && \textit{Add}\\ &=41\end{aligned}$$

31. $30 \div (-5) - 36 \div (-9)$ *Divide*
$= -6 - 36 \div (-9)$ *Divide*
$= -6 - (-4)$ *Change subt.*
$= -6 + 4$ *Add*
$= -2$

33. $2(5) - 3(4) + 5(3)$ *Multiply*
$= 10 - 12 + 15$ *Change subt.*
$= 10 + (-12) + 15$ *Add*
$= -2 + 15$ *Add*
$= 13$

35. $4(3^2) + 7(3 + 9) - (-6)$ *Parentheses*
$= 4 \cdot 3^2 + 7(12) - (-6)$ *Exponent*
$= 4 \cdot 9 + 7 \cdot 12 - (-6)$ *Multiply*
$= 36 + 84 - (-6)$ *Add; change subt.*
$= 120 + (+6)$ *Add*
$= 126$

37. $\dfrac{-1 + 5^2 - (-3)}{-6 - 9 + 12}$

Numerator:
$(-1) + 5^2 - (-3)$
$= -1 + 25 - (-3)$
$= 24 - (-3)$
$= 24 + (+3)$
$= 27$

Denominator:
$-6 - 9 + 12$
$= -6 + (-9) + 12$
$= -15 + 12$
$= -3$

The last step is the division.

$$\frac{27}{-3} = -9$$

39. $\dfrac{-2(4^2) - 4(6 - 2)}{-4(8 - 13) \div (-5)}$

Numerator:
$-2(4^2) - 4(6 - 2)$
$= -2(4^2) - 4(4)$
$= -2(16) - 4(4)$
$= -32 - 16$
$= -32 + (-16)$
$= -48$

Denominator:
$-4(8 - 13) \div (-5)$
$= -4(-5) \div (-5)$
$= 20 \div (-5)$
$= -4$

The last step is the division.

$$\frac{-48}{-4} = 12$$

41. $\dfrac{2^3(-2 - 5) + 4(-1)}{4 + 5(-6 \cdot 2) + (5 \cdot 11)}$

Numerator:
$2^3(-2 - 5) + 4(-1)$
$= 2^3(-7) + 4(-1)$
$= 8(-7) + 4(-1)$
$= -56 + (-4)$
$= -60$

Denominator:
$4 + 5(-6 \cdot 2) + (5 \cdot 11)$
$= 4 + 5(-12) + 55$
$= 4 + (-60) + 55$
$= -56 + 55$
$= -1$

The last step is the division.

$$\frac{-60}{-1} = 60$$

43. $(-4)^2(7 - 9)^2 \div 2^3$
$= (-4)^2(-2)^2 \div 2^3$
$= 16(4) \div 8$
$= 64 \div 8$
$= 8$

45. $(-0.3)^2 + (-0.5)^2 + 0.9$
$= 0.09 + 0.25 + 0.9$
$= 0.34 + 0.9$
$= 1.24$

47. $(-0.75)(3.6 - 5)^2$
$= (-0.75)(-1.4)^2$
$= (-0.75)(1.96)$
$= -1.47$

49. $(0.5)^2(-8) - (0.31)$
$= 0.25 \cdot (-8) - (0.31)$
$= -2 + (-0.31)$
$= -2.31$

51. $\dfrac{2}{3} \div \left(-\dfrac{5}{6}\right) - \dfrac{1}{2}$
$= \dfrac{2}{\cancel{3}_1} \cdot \left(-\dfrac{\cancel{6}^2}{5}\right) - \dfrac{1}{2}$
$= -\dfrac{4}{5} + \left(-\dfrac{1}{2}\right)$
$= -\dfrac{8}{10} + \left(-\dfrac{5}{10}\right)$
$= -\dfrac{13}{10}$

53. $\left(-\frac{1}{2}\right)^2 - \left(\frac{3}{4} - \frac{7}{4}\right)$
$= \left(-\frac{1}{2}\right)^2 - \left(-\frac{4}{4}\right)$
$= \frac{1}{4} + \left(+\frac{4}{4}\right)$
$= \frac{5}{4}$

55. $\frac{3}{5}\left(-\frac{7}{6}\right) - \left(\frac{1}{6} - \frac{5}{3}\right)$
$= \frac{\overset{1}{\cancel{3}}}{5} \cdot \left(-\frac{7}{\underset{2}{\cancel{6}}}\right) - \left(\frac{1}{6} - \frac{10}{6}\right)$
$= -\frac{7}{10} - \left(-\frac{9}{6}\right)$
$= -\frac{7}{10} + \left(+\frac{3}{2}\right)$
$= -\frac{7}{10} + \frac{15}{10}$
$= \frac{8}{10}$
$= \frac{4}{5}$

57. $5^2(9 - 11)(-3)(-2)^3$
$= 5^2(-2)(-3)(-2)^3$
$= 25(-2)(-3)(-8)$
$= -50(-3)(-8)$
$= 150 \cdot (-8)$
$= -1200$

59. $1.6(-0.8) \div (-0.32) \div 2^2$
$= 1.6(-0.8) \div (-0.32) \div 4$
$= -1.28 \div (-0.32) \div 4$
$= 4 \div 4$
$= 1$

61. **(a)** $(-2)^2 = \underline{4}$ $\quad (-2)^6 = \underline{64}$
$(-2)^3 = \underline{-8}$ $\quad (-2)^7 = \underline{-128}$
$(-2)^4 = \underline{16}$ $\quad (-2)^8 = \underline{256}$
$(-2)^5 = \underline{-32}$ $\quad (-2)^9 = \underline{-512}$

There is a pattern. When a negative number is raised to an even power, the answer is positive; when raised to an odd power, the answer is negative.

(b) $(-2)^{10}$ is positive since 10 is an even number; $(-2)^{15}$ is negative since 15 is an odd number; $(-2)^{24}$ is positive since 24 is an even number.

63. $\dfrac{-9 + 18 \div (-3)(-6)}{5 - 4(12) \div 3(2)}$

Numerator:
$-9 + 18 \div (-3)(-6)$
$= -9 + (-6)(-6)$
$= -9 + 36$
$= 27$

Denominator:
$5 - 4(12) \div 3(2)$
$= 5 - 48 \div 3(2)$
$= 5 - 16(2)$
$= 5 - 32$
$= 5 + (-32)$
$= -27$

Divide: $\dfrac{27}{-27} = -1$

65. $-7\left(6 - \frac{5}{8} \cdot 24 + 3 \cdot \frac{8}{3}\right)$
$= -7(6 - 15 + 8)$
$= -7(-9 + 8)$
$= -7(-1)$
$= 7$

67. $|-12| \div 4 + 2(3^2) \div 6$
$= |-12| \div 4 + 2(9) \div 6$
$= 12 \div 4 + 2(9) \div 6$
$= 3 + 2(9) \div 6$
$= 3 + 18 \div 6$
$= 3 + 3$
$= 6$

Summary Exercises on Operations with Signed Numbers

1. $2 - 8 = 2 + (-8) = -6$

3. $-14 - (-7) = -14 + (+7) = -7$

5. $-9(-7) = 63$

7. $(1)(-56) = -56$

9. $5 - (-7) = 5 + (+7) = 12$

11. $-18 + 5 = -13$

13. $-40 - 40 = -40 + (-40) = -80$

15. $8(-6) = -48$

17. $\left(-\frac{5}{6}\right)\left(\frac{1}{5}\right) = \left(-\frac{\overset{1}{\cancel{5}}}{6}\right)\left(\frac{1}{\underset{1}{\cancel{5}}}\right) = -\frac{1}{6}$

19. $0 - 14.6 = 0 + (-14.6) = -14.6$

21. $-4(-6)(2) = 24(2) = 48$

23. $-60 \div 10 \div (-3) = -6 \div (-3) = 2$

25. $64(0) \div (-8) = 0 \div (-8) = 0$

27. $-9 + 8 + (-2) = -1 + (-2) = -3$

29. $8 + 6 + (-8) = 14 + (-8) = 6$

31. $$\begin{aligned} -7 + 28 + (-56) + 3 &= 21 + (-56) + 3 \\ &= -35 + 3 \\ &= -32 \end{aligned}$$

33. $$\begin{aligned} -6(-8) \div (-5 - 7) &= -6(-8) \div (-12) \\ &= 48 \div (-12) \\ &= -4 \end{aligned}$$

35. $$\begin{aligned} -80 \div 4(-5) &= -20(-5) \\ &= 100 \end{aligned}$$

37. $-7 \cdot |7| \cdot |-7| = -7 \cdot 7 \cdot 7 = -49 \cdot 7 = -343$

39. $$\begin{aligned} -2(-3)(7) \div (-7) &= 6(7) \div (-7) \\ &= 42 \div (-7) \\ &= -6 \end{aligned}$$

41. $$\begin{aligned} 0 - |-7 + 2| &= 0 - |-5| \\ &= 0 - 5 \\ &= 0 + (-5) \\ &= -5 \end{aligned}$$

43. $$\begin{aligned} &12 \div 4 + 2(-2)^2 \div (-4) \\ &= 12 \div 4 + 2(4) \div (-4) \\ &= 3 + 2(4) \div (-4) \\ &= 3 + 8 \div (-4) \\ &= 3 + (-2) \\ &= 1 \end{aligned}$$

45. $$\begin{aligned} &\frac{5 - |2 - 4 \cdot 4| + (-5)^2 \div 5^2}{-9 \div 3(2 - 2) - (-8)} \\ &= \frac{5 - |2 - 16| + (-5)^2 \div 5^2}{-9 \div 3(0) - (-8)} \\ &= \frac{5 - |-14| + (-5)^2 \div 5^2}{-9 \div 3(0) - (-8)} \\ &= \frac{5 - 14 + 25 \div 25}{-3(0) + 8} \\ &= \frac{5 - 14 + 1}{0 + 8} \\ &= \frac{-9 + 1}{8} \\ &= \frac{-8}{8} \\ &= -1 \end{aligned}$$

47. $-\$4176 \div 174 = -\$24$
The average loss on each item was \$24.

49. $37(730) = 27{,}010$
The plane ascended 27,010 ft.

51. $-65 + 24 + (-49) = -41 + (-49) = -90$
Her final depth was -90 ft.

53. January had the greatest loss at $-\$700$.
April had the greatest profit at \$900.

55. The total expenses were

$$\begin{aligned} &\$3100 + \$2000 + \$1800 + \$1400 + \$1600 + \$1200 \\ &= \$11{,}100, \end{aligned}$$

which can be represented as $-\$11{,}100$.

Her average monthly amount of expenses was $-\$11{,}100 \div 6 = -\$1850$.

57. Similar: If the signs match, the result is positive. If the signs are different, the result is negative.
Different: Multiplication is commutative, division is not. You can multiply by 0, but dividing by 0 is not allowed.

9.5 Evaluating Expressions and Formulas

9.5 Margin Exercises

1. $5x - 3y$

(a) $x = 5, y = 2$
Replace x with 5. Replace y with 2.

$$\begin{aligned} 5x - 3y &= 5(5) - 3(2) \\ &= 25 - 6 \\ &= 19 \end{aligned}$$

(b) $x = -3, y = 4$
Replace x with -3. Replace y with 4.

$$\begin{aligned} 5x - 3y &= 5(-3) - 3(4) \\ &= -15 - 12 \\ &= -27 \end{aligned}$$

(c) $x = 0, y = 6$
Replace x with 0. Replace y with 6.

$$\begin{aligned} 5x - 3y &= 5(0) - 3(6) \\ &= 0 - 18 \\ &= -18 \end{aligned}$$

2. $\dfrac{3k + r}{2s}$

(a) $k = 1, r = 1, s = 2$
Replace k with 1, r with 1, and s with 2.

$$\begin{aligned} \frac{3k + r}{2s} &= \frac{3(1) + (1)}{2(2)} \\ &= \frac{3 + 1}{4} \\ &= \frac{4}{4} = 1 \end{aligned}$$

(b) $k = 8, r = -2, s = -4$
Replace k with 8, r with -2, and s with -4.

$$\begin{aligned}\frac{3k+r}{2s} &= \frac{3(8)+(-2)}{2(-4)}\\ &= \frac{24+(-2)}{-8}\\ &= \frac{22}{-8} = -\frac{11}{4}\end{aligned}$$

(c) $k = -3, r = 1, s = -2$
Replace k with -3, r with 1, and s with -2.

$$\begin{aligned}\frac{3k+r}{2s} &= \frac{3(-3)+(1)}{2(-2)}\\ &= \frac{-9+1}{-4}\\ &= \frac{-8}{-4} = 2\end{aligned}$$

3. **(a)** $-x - 6y; x = -1, y = -4$

$$\begin{aligned}-x-6y &= -1(-1)-6(-4)\\ &= 1+24\\ &= 25\end{aligned}$$

(b) $-4a - b; a = 4, b = -2$

$$\begin{aligned}-4a-b &= -4(4)-(-2)\\ &= -16+2\\ &= -14\end{aligned}$$

(c) $-w - 3x - y; w = 10, x = -5, y = -6$

$$\begin{aligned}-w-3x-y &= -10-3(-5)-(-6)\\ &= -10+15+6\\ &= 5+6\\ &= 11\end{aligned}$$

4. **(a)** $A = \frac{1}{2}bh; b = 6$ yd, $h = 12$ yd

$$\begin{aligned}A &= \frac{1}{2}(6\text{ yd})(12\text{ yd})\\ &= \frac{1}{\cancel{2}_1}(\cancel{6}^3\text{ yd})(12\text{ yd})\\ &= 36\text{ yd}^2\end{aligned}$$

(b) $P = 2l + 2w; l = 10$ cm, $w = 8$ cm

$$\begin{aligned}P &= 2(10\text{ cm})+2(8\text{ cm})\\ &= 20\text{ cm}+16\text{ cm}\\ &= 36\text{ cm}\end{aligned}$$

(c) $C = 2\pi r; \pi \approx 3.14, r = 6$ ft

$$\begin{aligned}C &\approx 2(3.14)(6\text{ ft})\\ &= 37.68\text{ ft}\end{aligned}$$

9.5 Section Exercises

1. $r = 2, s = 6$
Replace r with 2. Replace s with 6.

$$\begin{aligned}2r+4s &= 2(2)+4(6)\\ &= 4+24\\ &= 28\end{aligned}$$

3. $r = 1, s = -3$
Replace r with 1. Replace s with -3.

$$\begin{aligned}2r+4s &= 2(1)+4(-3)\\ &= 2+(-12)\\ &= -10\end{aligned}$$

5. $r = -4, s = 4$
Replace r with -4. Replace s with 4.

$$\begin{aligned}2r+4s &= 2(-4)+4(4)\\ &= -8+16\\ &= 8\end{aligned}$$

7. $r = -1, s = -7$
Replace r with -1. Replace s with -7.

$$\begin{aligned}2r+4s &= 2(-1)+4(-7)\\ &= -2+(-28)\\ &= -30\end{aligned}$$

9. $r = 0, s = -2$
Replace r with 0. Replace s with -2.

$$\begin{aligned}2r+4s &= 2(0)+4(-2)\\ &= 0+(-8)\\ &= -8\end{aligned}$$

11. $8x - y; x = 1, y = 8$
Replace x with 1. Replace y with 8.

$$\begin{aligned}8x-y &= 8(1)-8\\ &= 8-8\\ &= 0\end{aligned}$$

13. $6k + 2s; k = 1, s = -2$
Replace k with 1. Replace s with -2.

$$\begin{aligned}6k+2s &= 6(1)+2(-2)\\ &= 6+(-4)\\ &= 2\end{aligned}$$

15. $\frac{-m+5n}{2s+2}; m = 4, n = -8, s = 0$
Replace m with 4, n with -8, and s with 0.

$$\begin{aligned}\frac{-m+5n}{2s+2} &= \frac{-4+5(-8)}{2(0)+2}\\ &= \frac{-4+(-40)}{0+2}\\ &= \frac{-44}{2} = -22\end{aligned}$$

17. $-m-3n; m=\frac{1}{2}, n=\frac{3}{8}$
Replace m with $\frac{1}{2}$. Replace n with $\frac{3}{8}$.

$$\begin{aligned}&-m-3n\\&=-\tfrac{1}{2}-3\left(\tfrac{3}{8}\right)\\&=-\tfrac{1}{2}-\tfrac{9}{8}\\&=-\tfrac{4}{8}+\left(-\tfrac{9}{8}\right)\\&=-\tfrac{13}{8}\end{aligned}$$

19. $-c-5b; c=-8, b=-4$
Replace c with -8. Replace b with -4.

$$\begin{aligned}-c-5b&=-(-8)-5(-4)\\&=8-(-20)\\&=8+(+20)\\&=28\end{aligned}$$

21. $-4x-y; x=5, y=-15$
Replace x with 5. Replace y with -15.

$$\begin{aligned}-4x-y&=-4(5)-(-15)\\&=-20-(-15)\\&=-20+(+15)\\&=-5\end{aligned}$$

23. $-k-m-8n; k=6, m=-9, n=0$
Replace k with 6, m with -9, and n with 0.

$$\begin{aligned}-k-m-8n&=-6-(-9)-8(0)\\&=-6+(+9)-0\\&=3\end{aligned}$$

25. $\dfrac{-3s-t-4}{-s+6+t}; s=-1, t=-13$
Replace s with -1. Replace t with -13.

$$\begin{aligned}\frac{-3s-t-4}{-s+6+t}&=\frac{-3(-1)-(-13)-4}{-(-1)+6+(-13)}\\&=\frac{3+(+13)-4}{1+6+(-13)}\\&=\frac{16-4}{7+(-13)}\\&=\frac{12}{-6}=-2\end{aligned}$$

27. $P=4s; s=7.5$

$$\begin{aligned}P&=4(7.5)\\&=30\end{aligned}$$

29. $P=2l+2w; l=9, w=5$

$$\begin{aligned}P&=2(9)+2(5)\\&=18+10\\&=28\end{aligned}$$

31. $A=\pi r^2; \pi\approx 3.14, r=5$

$$\begin{aligned}A&\approx 3.14\cdot 5\cdot 5\\&=78.5\end{aligned}$$

33. $A=\frac{1}{2}bh; b=15, h=3$

$$\begin{aligned}A&=\tfrac{1}{2}(15)(3)\\&=\tfrac{15}{2}(3)\\&=\tfrac{45}{2}\text{ or }22\tfrac{1}{2}\end{aligned}$$

35. $V=\frac{1}{3}Bh; B=30, h=60$

$$\begin{aligned}V&=\tfrac{1}{3}\cdot 30\cdot 60\\&=10\cdot 60\\&=600\end{aligned}$$

37. $d=rt; r=53, t=6$

$$\begin{aligned}d&=53\cdot 6\\&=318\end{aligned}$$

39. $C=2\pi r; \pi\approx 3.14, r=4$

$$\begin{aligned}C&\approx 2\cdot(3.14)\cdot 4\\&=6.28\cdot 4\\&=25.12\end{aligned}$$

41. **(a)** $P=3s; s=11$
$P=3(11\text{ in.})=33\text{ in.}$
The perimeter is 33 in.

(b) $P=3s; s=3$
$P=3(3\text{ ft})=9\text{ ft}$
The perimeter is 9 ft.

43. **(a)** $\dfrac{p}{t}; p=332, t=4$

$$\frac{p}{t}=\frac{332}{4}=83$$

The average score is 83.

(b) $\dfrac{p}{t}; p=637, t=7$

$$\frac{p}{t}=\frac{637}{7}=91$$

The average score is 91.

45. The error was made when replacing x with -3; should be $-(-3)$, not -3.

$-x-4y$	Replace x with (-3) and y with (-1).
$-(-3)-4(-1)$	Opposite of (-3) is $+3$.
$(+3)-(-4)$	Change subtraction to addition.
$\underbrace{3+(+4)}_{7}$	Add.

47. $F=\dfrac{9C}{5}+32$

$$\begin{aligned}&=\frac{9\cdot(-40)}{5}+32\quad (C=-40)\\&=-72+32\\&=-40\end{aligned}$$

This formula is used to convert a Celsius temperature to Fahrenheit.

48. $C = \frac{5(F-32)}{9}$

$= \frac{5(-4-32)}{9} \quad (F = -4)$

$= \frac{5(-36)}{9}$

$= -20$

This formula is used to convert a Fahrenheit temperature to Celsius.

49. $V = \frac{4\pi r^3}{3}$

$\approx \frac{4(3.14)(3)^3}{3} \quad (\pi \approx 3.14, r = 3)$

$= \frac{4(3.14)(27)}{3}$

$= \frac{12.56(27)}{3}$

$= \frac{339.12}{3} = 113.04$

This formula is used to find the volume of a sphere.

50. $c^2 = a^2 + b^2$

$= 3^2 + 4^2 \quad (a = 3, b = 4)$

$= 9 + 16$

$= 25$

Since $c^2 = 25$ and $5^2 = 25$, $c = 5$.
The Pythagorean Theorem is used to find the hypotenuse or a leg in a right triangle.

51. $A = \frac{1}{2}h(b+B)$

$= \frac{1}{2} \cdot 7(4+12) \quad (h = 7, b = 4, B = 12)$

$= \frac{1}{2} \cdot 7(16)$

$= \frac{7}{2}(16)$

$= 56$

This formula is used to find the area of a trapezoid.

52. $V = \frac{\pi r^2 h}{3}$

$\approx \frac{3.14 \cdot \overset{2}{\cancel{6}} \cdot 6 \cdot 10}{\underset{1}{\cancel{3}}} \quad (\pi \approx 3.14, r = 6, h = 10)$

$= 376.8$

This formula is used to find the volume of a cone.

9.6 Solving Equations

9.6 Margin Exercises

1. **(a)** $p + 1 = 8; 7$
Replace p with 7.

$7 + 1 = 8$
$8 = 8$ *True*

Because the statement is *true*, 7 *is* a solution of the equation $p + 1 = 8$.

(b) $30 = 5r; 6$
Replace r with 6.

$30 = 5(6)$
$30 = 30$ *True*

6 is a solution.

(c) $3k - 2 = 4; 3$
Replace k with 3.

$3(3) - 2 = 4$
$7 = 4$ *False*

Because the statement is *false*, 3 *is not* a solution of the equation $3k - 2 = 4$.

(d) $23 = 4y + 3; 5$
Replace y with 5.

$23 = 4(5) + 3$
$23 = 23$ *True*

23 is a solution.

2. **(a)** $n - 5 = 8$

$n - 5 + 5 = 8 + 5$ Add 5 to both sides.

$n = 13$

Check: $13 - 5 = 8$
$8 = 8$ *True*

(b) $5 = r - 10$

$5 + 10 = r - 10 + 10$

$15 = r$

Check: $5 = 15 - 10$
$5 = 5$ *True*

(c) $3 = z + 1$

$3 - 1 = z + 1 - 1$ Subtract 1 from both sides.

$2 = z$

Check: $3 = 2 + 1$
$3 = 3$ *True*

(d) $k + 9 = 0$

$k + 9 - 9 = 0 - 9$

$k = -9$

Check: $-9 + 9 = 0$

$0 = 0$ *True*

(e) $-2 = y + 9$

$-2 - 9 = y + 9 - 9$

$-11 = y$

Check: $-2 = -11 + 9$

$-2 = -2$ *True*

(f) $x - 2 = -6$

$x - 2 + 2 = -6 + 2$

$x = -4$

Check: $-4 - 2 = -6$

$-6 = -6$ *True*

3. **(a)** $2y = 14$

$\frac{\overset{1}{\cancel{2}} \cdot y}{\underset{1}{\cancel{2}}} = \frac{14}{2}$ Divide both sides by 2.

$y = 7$

Check: $2 \cdot 7 = 14$

$14 = 14$ *True*

(b) $42 = 7p$

$\frac{\overset{6}{\cancel{42}}}{7} = \frac{7 \cdot p}{7}$

$6 = p$

Check: $42 = 7 \cdot 6$

$42 = 42$ *True*

(c) $-8a = 32$

$\frac{-\overset{1}{\cancel{8}}a}{-\underset{1}{\cancel{8}}} = \frac{32}{-8}$

$a = -4$

Check: $-8(-4) = 32$

$32 = 32$ *True*

(d) $-3r = -15$

$\frac{-\overset{1}{\cancel{3}}r}{-\underset{1}{\cancel{3}}} = \frac{-15}{-3}$

$r = 5$

Check: $-3(5) = -15$

$-15 = -15$ *True*

(e) $-60 = -6k$

$\frac{-60}{-6} = \frac{-\overset{1}{\cancel{6}}k}{-\underset{1}{\cancel{6}}}$

$10 = k$

Check: $-60 = -6(10)$

$-60 = -60$ *True*

(f) $10x = 0$

$\frac{10x}{10} = \frac{0}{10}$

$x = 0$

Check: $10 \cdot 0 = 0$

$0 = 0$ *True*

4. **(a)** $\frac{a}{4} = 2$

$\frac{1}{4}a = 2$ Rewrite $a/4$.

$\left(\frac{4}{1}\right) \cdot \left(\frac{1}{4}\right)a = 2 \cdot 4$ Multiply both sides by $4/1$.

$a = 8$

Check: $\frac{8}{4} = 2$

$2 = 2$ *True*

(b) $\frac{y}{7} = -3$

$\frac{1}{7}y = -3$

$\left(\frac{7}{1}\right) \cdot \left(\frac{1}{7}\right)y = -3 \cdot 7$

$y = -21$

Check: $\frac{-21}{7} = -3$

$-3 = -3$ *True*

(c) $-8 = \frac{k}{6}$

$-8 = \frac{1}{6}k$

$6 \cdot (-8) = \left(\frac{6}{1}\right) \cdot \left(\frac{1}{6}\right)k$

$-48 = k$

Check: $-8 = \frac{-48}{6}$

$-8 = -8$ *True*

(d) $8 = -\frac{4}{5}z$

$\left(-\frac{5}{4}\right) \cdot \left(\frac{8}{1}\right) = \left(-\frac{5}{4}\right) \cdot \left(-\frac{4}{5}\right)z$

$-10 = z$

Check: $8 = \left(-\frac{4}{5}\right) \cdot (-10)$

$8 = 8$ *True*

(e) $-\frac{5}{8}p = -10$

$$\left(-\frac{8}{5}\right) \cdot \left(-\frac{5}{8}\right)p = (-10) \cdot \left(-\frac{8}{5}\right)$$

$$p = 16$$

Check: $\left(-\frac{5}{8}\right) \cdot (16) = -10$

$-10 = -10$ *True*

9.6 Section Exercises

1. $x + 7 = 11;\ 4$
Replace x with 4.

$$4 + 7 = 11$$
$$11 = 11 \quad \textit{True}$$

Because the statement is *true*, 4 *is* a solution of the equation $x + 7 = 11$.

3. $4y = 28;\ 7$
Replace y with 7.

$$4(7) = 28$$
$$28 = 28 \quad \textit{True}$$

7 *is* a solution.

5. $2z - 1 = -15;\ -8$
Replace z with -8.

$$2(-8) - 1 = -15$$
$$-16 - 1 = -15$$
$$-17 = -15 \quad \textit{False}$$

Because the statement is *false*, -8 *is not* a solution of the equation $2z - 1 = -15$.

7. $p + 5 = 9$
Subtract 5 from both sides.

$$p + 5 - 5 = 9 - 5$$
$$p = 4$$

Check: $4 + 5 = 9$
$9 = 9$ *True*

The solution is 4.

9. $k + 15 = 0$
Subtract 15 from both sides.

$$k + 15 - 15 = 0 - 15$$
$$k = -15$$

Check: $-15 + 15 = 0$
$0 = 0$ *True*

The solution is -15.

11. $z - 5 = 3$
Add 5 to both sides.

$$z - 5 + 5 = 3 + 5$$
$$z = 8$$

Check: $8 - 5 = 3$
$3 = 3$ *True*

The solution is 8.

13. $8 = r - 2$
Add 2 to both sides.

$$8 + 2 = r - 2 + 2$$
$$10 = r$$

Check: $8 = 10 - 2$
$8 = 8$ *True*

The solution is 10.

15. $-5 = n + 3$
Subtract 3 from both sides.

$$-5 + (-3) = n + 3 - 3$$
$$-8 = n$$

Check: $-5 = -8 + 3$
$-5 = -5$ *True*

The solution is -8.

17. $7 = r + 13$
Subtract 13 from both sides.

$$7 - 13 = r + 13 - 13$$
$$-6 = r$$

Check: $7 = -6 + 13$
$7 = 7$ *True*

The solution is -6.

19. $-4 + k = 14$
Add 4 to both sides.

$$-4 + 4 + k = 14 + 4$$
$$k = 18$$

Check: $-4 + 18 = 14$
$14 = 14$ *True*

The solution is 18.

21. $-12 + x = -1$
Add 12 to both sides.

$$-12 + 12 + x = -1 + 12$$
$$x = 11$$

Check: $-12 + 11 = -1$
$-1 = -1$ *True*

The solution is 11.

23. $-5 = -2 + r$
Add 2 to both sides.

$$-5 + 2 = -2 + 2 + r$$
$$-3 = r$$

Check: $-5 = -2 + (-3)$
$-5 = -5$ *True*

The solution is -3.

25. $d + \frac{2}{3} = 3$
Subtract $\frac{2}{3}$ from both sides.

$$d + \frac{2}{3} - \frac{2}{3} = 3 - \frac{2}{3}$$
$$d = \frac{7}{3} \text{ or } 2\frac{1}{3}$$

Check: $\frac{7}{3} + \frac{2}{3} = 3$
$\frac{9}{3} = 3$
$3 = 3$ *True*

The solution is $\frac{7}{3}$ or $2\frac{1}{3}$.

27. $z - \frac{7}{8} = 10$
Add $\frac{7}{8}$ to both sides.

$$z - \frac{7}{8} + \frac{7}{8} = 10 + \frac{7}{8}$$
$$z = \frac{80}{8} + \frac{7}{8}$$
$$= \frac{87}{8} \text{ or } 10\frac{7}{8}$$

Check: $\frac{87}{8} - \frac{7}{8} = 10$
$\frac{80}{8} = 10$
$10 = 10$ *True*

The solution is $\frac{87}{8}$ or $10\frac{7}{8}$.

29. $\frac{1}{2} = k - 2$
Add 2 to both sides.

$$\frac{1}{2} + 2 = k - 2 + 2$$
$$\frac{5}{2} = k$$

Check: $\frac{1}{2} = \frac{5}{2} - 2$
$\frac{1}{2} = \frac{1}{2}$ *True*

The solution is $\frac{5}{2}$ or $2\frac{1}{2}$.

31. $m - \frac{7}{5} = \frac{11}{4}$
Add $\frac{7}{5}$ to both sides.

$$m - \frac{7}{5} + \frac{7}{5} = \frac{11}{4} + \frac{7}{5}$$
$$m = \frac{55}{20} + \frac{28}{20}$$
$$= \frac{83}{20} \text{ or } 4\frac{3}{20}$$

Check: $\frac{83}{20} - \frac{7}{5} = \frac{11}{4}$
$\frac{83}{20} - \frac{28}{20} = \frac{55}{20}$
$\frac{55}{20} = \frac{55}{20}$ *True*

The solution is $\frac{83}{20}$ or $4\frac{3}{20}$.

33. $x - 0.8 = 5.07$
Add 0.8 to both sides.

$$x - 0.8 + 0.8 = 5.07 + 0.8$$
$$x = 5.87$$

Check: $5.87 - 0.8 = 5.07$
$5.07 = 5.07$ *True*

The solution is 5.87.

35. $3.25 = 4.76 + r$
Subtract 4.76 from both sides.

$$3.25 - 4.76 = 4.76 - 4.76 + r$$
$$-1.51 = r$$

Check: $3.25 = 4.76 - 1.51$
$3.25 = 3.25$ *True*

The solution is -1.51.

37. $6z = 12$
Divide both sides by 6.

$$\frac{\overset{1}{\cancel{6}} \cdot z}{\underset{1}{\cancel{6}}} = \frac{12}{6}$$
$$z = 2$$

Check: $6(2) = 12$
$12 = 12$ *True*

The solution is 2.

39. $48 = 12r$
Divide both sides by 12.

$$\frac{48}{12} = \frac{\overset{1}{\cancel{12}} \cdot r}{\underset{1}{\cancel{12}}}$$
$$4 = r$$

Check: $48 = 12(4)$
$48 = 48$ *True*

The solution is 4.

41. $3y = 0$
Divide both sides by 3.

$$\frac{\overset{1}{\cancel{3}} \cdot y}{\underset{1}{\cancel{3}}} = \frac{0}{3}$$
$$y = 0$$

Check: $3(0) = 0$
$0 = 0$ *True*

The solution is 0.

43. $-6k = 36$
Divide both sides by -6.

$$\frac{-6k}{-6} = \frac{36}{-6}$$
$$k = -6$$

Check: $-6(-6) = 36$
$36 = 36$ *True*

The solution is -6.

45. $-36 = -4p$
Divide both sides by -4.

$$\frac{-36}{-4} = \frac{\overset{1}{\cancel{-4}} \cdot p}{\underset{1}{\cancel{-4}}}$$
$$9 = p$$

Check: $-36 = -4(9)$
$-36 = -36$ *True*

The solution is 9.

47. $-1.2m = 8.4$
Divide both sides by -1.2.

$$\frac{\overset{1}{\cancel{-1.2}}m}{\underset{1}{\cancel{-1.2}}} = \frac{8.4}{-1.2}$$
$$m = -7$$

Check: $-1.2(-7) = 8.4$
$8.4 = 8.4$ *True*

The solution is -7.

49. $-8.4p = -9.24$
Divide both sides by -8.4.

$$\frac{-8.4p}{-8.4} = \frac{-9.24}{-8.4}$$
$$p = 1.1$$

Check: $-8.4(1.1) = -9.24$
$-9.24 = -9.24$ *True*

The solution is 1.1.

51. $\frac{k}{2} = 17$ or $\frac{1}{2}k = 17$
Multiply both sides by 2.

$$\frac{\overset{1}{\cancel{2}}}{1} \cdot \frac{1}{\underset{1}{\cancel{2}}}k = 17 \cdot 2$$
$$k = 34$$

Check: $\frac{34}{2} = 17$
$17 = 17$ *True*

The solution is 34.

53. $11 = \frac{a}{6}$ or $11 = \frac{1}{6}a$
Multiply both sides by 6.

$$6 \cdot 11 = \frac{\overset{1}{\cancel{6}}}{1} \cdot \frac{1}{\underset{1}{\cancel{6}}}a$$
$$66 = a$$

Check: $11 = \frac{66}{6}$
$11 = 11$ *True*

The solution is 66.

55. $\frac{r}{3} = -12$ or $\frac{1}{3}r = -12$
Multiply both sides by 3.

$$\frac{\overset{1}{\cancel{3}}}{1} \cdot \frac{1}{\underset{1}{\cancel{3}}}r = -12 \cdot 3$$
$$r = -36$$

Check: $\frac{-36}{3} = -12$
$-12 = -12$ *True*

The solution is -36.

57. $-\frac{2}{5}p = 8$
Multiply both sides by $-\frac{5}{2}$.

$$-\frac{\overset{1}{\cancel{5}}}{\underset{1}{\cancel{2}}} \cdot \left(-\frac{\overset{1}{\cancel{2}}}{\underset{1}{\cancel{5}}}p\right) = -\frac{5}{\underset{1}{\cancel{2}}} \cdot \overset{4}{\cancel{8}}$$
$$p = -20$$

Check: $-\frac{2}{5}(-20) = 8$
$8 = 8$ *True*

The solution is -20.

59. $-\frac{3}{4}m = -3$
Multiply both sides by $-\frac{4}{3}$.

$$-\frac{\overset{1}{\cancel{4}}}{\underset{1}{\cancel{3}}} \cdot \left(-\frac{\overset{1}{\cancel{3}}}{\underset{1}{\cancel{4}}}m\right) = -\cancel{3} \cdot \left(-\frac{4}{\cancel{3}}\right)$$
$$m = 4$$

Check: $-\frac{3}{4}(4) = -3$
$-3 = -3$ *True*

The solution is 4.

61. $6 = \frac{3}{8}x$

Multiply both sides by $\frac{8}{3}$.

$$\frac{8}{3} \cdot 6 = \frac{8}{3} \cdot \frac{3}{8}x$$
$$16 = x$$

Check: $6 = \frac{3}{8}(16)$

$6 = 6$ *True*

The solution is 16.

63. $\frac{y}{2.6} = 0.5$ or $\frac{1}{2.6}y = 0.5$

Multiply both sides by 2.6.

$$2.6 \cdot \frac{y}{2.6} = 0.5(2.6)$$
$$y = 1.3$$

Check: $\frac{1.3}{2.6} = 0.5$

$0.5 = 0.5$ *True*

The solution is 1.3.

65. $\frac{z}{-3.8} = 1.3$ or $\frac{1}{-3.8}z = 1.3$

Multiply both sides by -3.8.

$$\frac{-3.8}{1} \cdot \frac{1}{-3.8}z = (-3.8)(1.3)$$
$$z = -4.94$$

Check: $\frac{-4.94}{-3.8} = 1.3$

$1.3 = 1.3$ *True*

The solution is -4.94.

67. You may add or subtract the same number on both sides of an equation. Many different equations could have -3 as the solution. One possibility is:

$$x + 5 = 2$$
$$x + 5 - 5 = 2 - 5$$
$$x = -3$$

69.
$$x - 17 = 5 - 3$$
$$x - 17 = 2$$
$$x - 17 + 17 = 2 + 17$$
$$x = 19$$

The solution is 19.

71.
$$3 = x + 9 - 15$$
$$3 = x - 6$$
$$3 + 6 = x - 6 + 6$$
$$9 = x$$

The solution is 9.

73. $\frac{7}{2}x = \frac{4}{3}$

$$\frac{2}{7} \cdot \frac{7}{2}x = \frac{2}{7} \cdot \frac{4}{3}$$
$$x = \frac{8}{21}$$

The solution is $\frac{8}{21}$.

75.
$$\frac{1}{2} - \frac{3}{4} = \frac{a}{5}$$
$$\frac{2}{4} - \frac{3}{4} = \frac{a}{5}$$
$$-\frac{1}{4} = \frac{a}{5}$$
$$\frac{5}{1} \cdot \left(-\frac{1}{4}\right) = \frac{a}{5} \cdot \frac{5}{1}$$
$$-\frac{5}{4} = a$$

The solution is $-\frac{5}{4}$.

77.
$$m - 2 + 18 = |-3 - 4| + 5$$
$$m + 16 = |-7| + 5$$
$$m + 16 = 7 + 5$$
$$m + 16 = 12$$
$$m + 16 - 16 = 12 - 16$$
$$m = -4$$

The solution is -4.

9.7 Solving Equations with Several Steps

9.7 Margin Exercises

1. (a)
$$2r + 7 = 13$$
$$2r + 7 - 7 = 13 - 7$$
$$2r = 6$$
$$\frac{2r}{2} = \frac{6}{2}$$
$$r = 3$$

Check: Replace r with 3.

$2(3) + 7 = 13$

$13 = 13$ *True*

The solution is 3.

(b)
$$20 = 6y - 4$$
$$20 + 4 = 6y - 4 + 4$$
$$24 = 6y$$
$$\frac{24}{6} = \frac{6y}{6}$$
$$4 = y$$

Check: Replace y with 4.

$$20 = 6(4) - 4$$
$$20 = 20 \quad \textit{True}$$

The solution is 4.

(c)
$$7m + 9 = 9$$
$$7m + 9 - 9 = 9 - 9$$
$$7m = 0$$
$$\frac{7m}{7} = \frac{0}{7}$$
$$m = 0$$

Check: Replace m with 0.

$$7(0) + 9 = 9$$
$$9 = 9 \quad \textit{True}$$

The solution is 0.

(d)
$$-2 = 4p + 10$$
$$-2 - 10 = 4p + 10 - 10$$
$$-12 = 4p$$
$$\frac{-12}{4} = \frac{\overset{1}{\cancel{4}}p}{\underset{1}{\cancel{4}}}$$
$$-3 = p$$

Check: Replace p with -3.

$$-2 = 4(-3) + 10$$
$$-2 = -12 + 10$$
$$-2 = -2 \quad \textit{True}$$

The solution is -3.

(e)
$$-10z - 9 = 11$$
$$-10z - 9 + 9 = 11 + 9$$
$$-10z = 20$$
$$\frac{-\overset{1}{\cancel{10}}z}{-\underset{1}{\cancel{10}}} = \frac{20}{-10}$$
$$z = -2$$

Check: Replace z with -2.

$$-10(-2) - 9 = 11$$
$$20 - 9 = 11$$
$$11 = 11 \quad \textit{True}$$

The solution is -2.

2. **(a)** $3(2 + 6) = 3 \cdot 2 + 3 \cdot 6$
$= 6 + 18$
$= 24$

(b) $8(k - 3) = 8 \cdot k - 8 \cdot 3$
$= 8k - 24$

(c) $-6(r + 5) = -6 \cdot r + (-6) \cdot 5$
$= -6r + (-30)$
$= -6r - 30$

(d) $-9(s - 8) = -9 \cdot s - (-9) \cdot 8$
$= -9s - (-72)$
$= -9s + 72$

3. **(a)** $5y + 11y = (5 + 11)y = 16y$

(b) $10a - 28a = (10 - 28)a = -18a$

(c) $3x + 3x - 9x = (3 + 3 - 9)x = -3x$

(d) $k + k = 1k + 1k = (1 + 1)k = 2k$

(e) $6b - b - 7b = (6 - 1 - 7)b = -2b$

4. **(a)**
$$3y - 1 = 2y + 7$$
$$3y - 2y - 1 = 2y - 2y + 7$$
$$y - 1 = 7$$
$$y - 1 + 1 = 7 + 1$$
$$y = 8$$

Check: Replace y with 8.

$$3(8) - 1 = 2(8) + 7$$
$$24 - 1 = 16 + 7$$
$$23 = 23 \quad \textit{True}$$

The solution is 8.

(b)
$$5a + 7 = 3a - 9$$
$$5a - 3a + 7 = 3a - 3a - 9$$
$$2a + 7 = -9$$
$$2a + 7 - 7 = -9 - 7$$
$$2a = -16$$
$$\frac{2a}{2} = \frac{-16}{2}$$
$$a = -8$$

Check: Replace a with -8.

$$5(-8) + 7 = 3(-8) - 9$$
$$-40 + 7 = -24 + (-9)$$
$$-33 = -33 \quad \textit{True}$$

The solution is -8.

(c)
$$3p - 2 = p - 6$$
$$3p - p - 2 = p - p - 6$$
$$2p - 2 = -6$$
$$2p - 2 + 2 = -6 + 2$$
$$2p = -4$$
$$\frac{2p}{2} = \frac{-4}{2}$$
$$p = -2$$

Check: Replace p with -2.

$$3(-2) - 2 = -2 - 6$$
$$-6 - 2 = -8$$
$$-8 = -8 \quad \textit{True}$$

The solution is -2.

5. **(a)**

$$\begin{aligned}-12 &= 4(y-1)\\ -12 &= 4y-4\\ -12+4 &= 4y-4+4\\ -8 &= 4y\\ \frac{-8}{4} &= \frac{4y}{4}\\ -2 &= y\end{aligned}$$

Check: Replace y with -2.

$$\begin{aligned}-12 &= 4(-2-1)\\ -12 &= 4(-3)\\ -12 &= -12 \quad \textit{True}\end{aligned}$$

The solution is -2.

(b)

$$\begin{aligned}5(m+4) &= 20\\ 5m+20 &= 20\\ 5m+20-20 &= 20-20\\ 5m &= 0\\ \frac{5m}{5} &= \frac{0}{5}\\ m &= 0\end{aligned}$$

Check: Replace m with 0.

$$\begin{aligned}5(0+4) &= 20\\ 20 &= 20 \quad \textit{True}\end{aligned}$$

The solution is 0.

(c)

$$\begin{aligned}6(t-2) &= 18\\ 6t-12 &= 18\\ 6t-12+12 &= 18+12\\ 6t &= 30\\ \frac{6t}{6} &= \frac{30}{6}\\ t &= 5\end{aligned}$$

Check: Replace t with 5.

$$\begin{aligned}6(5-2) &= 18\\ 6(3) &= 18\\ 18 &= 18 \quad \textit{True}\end{aligned}$$

The solution is 5.

9.7 Section Exercises

1.

$$\begin{aligned}7p+5 &= 12\\ 7p+5-5 &= 12-5\\ 7p &= 7\\ \frac{7p}{7} &= \frac{7}{7}\\ p &= 1\end{aligned}$$

Check: Replace p with 1.

$$\begin{aligned}7(1)+5 &= 12\\ 7+5 &= 12\\ 12 &= 12 \quad \textit{True}\end{aligned}$$

The solution is 1.

3.

$$\begin{aligned}2 &= 8y-6\\ 2+6 &= 8y-6+6\\ 8 &= 8y\\ \frac{8}{8} &= \frac{8y}{8}\\ 1 &= y\end{aligned}$$

Check: Replace y with 1.

$$\begin{aligned}2 &= 8y-6\\ 2 &= 8(1)-6\\ 2 &= 2 \quad \textit{True}\end{aligned}$$

The solution is 1.

5.

$$\begin{aligned}-3m+1 &= 1\\ -3m+1-1 &= 1-1\\ -3m &= 0\\ \frac{-3m}{-3} &= \frac{0}{-3}\\ m &= 0\end{aligned}$$

Check: Replace m with 0.

$$\begin{aligned}-3(0)+1 &= 1\\ 1 &= 1 \quad \textit{True}\end{aligned}$$

The solution is 0.

7.

$$\begin{aligned}28 &= -9a+10\\ 28-10 &= -9a+10-10\\ 18 &= -9a\\ \frac{18}{-9} &= \frac{-9a}{-9}\\ -2 &= a\end{aligned}$$

Check: Replace a with -2.

$$\begin{aligned}28 &= -9(-2)+10\\ 28 &= 18+10\\ 28 &= 28 \quad \textit{True}\end{aligned}$$

The solution is -2.

9.

$$\begin{aligned}-5x-4 &= 16\\ -5x-4+4 &= 16+4\\ -5x &= 20\\ \frac{-5x}{-5} &= \frac{20}{-5}\\ x &= -4\end{aligned}$$

Check: Replace x with -4.

$$\begin{aligned}-5(-4)-4 &= 16\\ 20-4 &= 16\\ 16 &= 16 \quad \textit{True}\end{aligned}$$

The solution is -4.

11.
$$-\frac{1}{2}z + 2 = -1$$
$$-\frac{1}{2}z + 2 - 2 = -1 - 2$$
$$-\frac{1}{2}z = -3$$
$$-\frac{2}{1}\cdot\left(-\frac{1}{2}z\right) = -3\cdot\left(-\frac{2}{1}\right)$$
$$z = 6$$

Check: Replace z with 6.
$$-\frac{1}{2}(6) + 2 = -1$$
$$-3 + 2 = -1$$
$$-1 = -1 \quad \textit{True}$$

The solution is 6.

13.
$$-0.7 = 5b - 5.2$$
$$-0.7 + 5.2 = 5b - 5.2 + 5.2$$
$$4.5 = 5b$$
$$\frac{4.5}{5} = \frac{5b}{5}$$
$$0.9 = b$$

Check: Replace b with 0.9.
$$-0.7 = 5(0.9) - 5.2$$
$$-0.7 = 4.5 - 5.2$$
$$-0.7 = -0.7 \quad \textit{True}$$

The solution is 0.9.

15. $6(x+4) = 6\cdot x + 6\cdot 4$
$= 6x + 24$

17. $7(p-8) = 7\cdot p - 7\cdot 8$
$= 7p - 56$

19. $-3(m+6) = -3\cdot m + (-3)\cdot 6$
$= -3m - 18$

21. $-2(y-3) = -2\cdot y - (-2)\cdot 3$
$= -2y - (-6)$
$= -2y + 6$

23. $-8(c+8) = -8\cdot c + (-8)\cdot 8$
$= -8c - 64$

25. $-10(w-9) = -10\cdot w - (-10)\cdot 9$
$= -10w + 90$

27. $11r + 6r = (11+6)r = 17r$

29. $8z - 7z = (8-7)z = 1z$ or z

31. $y - 3y = (1-3)y = -2y$

33. $-4t + t - 4t = (-4+1-4)t = -7t$

35. $7p - 9p + 2p = (7-9+2)p = 0p = 0$

37. $\frac{5}{2}b - \frac{11}{2}b = \left(\frac{5}{2} - \frac{11}{2}\right)b = -\frac{6}{2}b = -3b$

39.
$$4k + 6k = 50$$
$$(4+6)k = 50$$
$$10k = 50$$
$$\frac{10k}{10} = \frac{50}{10}$$
$$k = 5$$

Check: Replace k with 5.
$$4(5) + 6(5) = 50$$
$$20 + 30 = 50$$
$$50 = 50 \quad \textit{True}$$

The solution is 5.

41.
$$54 = 10m - m$$
$$54 = (10-1)m$$
$$54 = 9m$$
$$\frac{54}{9} = \frac{9m}{9}$$
$$6 = m$$

Check: Replace m with 6.
$$54 = 10(6) - 6$$
$$54 = 60 - 6$$
$$54 = 54 \quad \textit{True}$$

The solution is 6.

43.
$$2b - 6b = 24$$
$$(2-6)b = 24$$
$$-4b = 24$$
$$\frac{-4b}{-4} = \frac{24}{-4}$$
$$b = -6$$

Check: Replace b with -6.
$$2(-6) - 6(-6) = 24$$
$$-12 - (-36) = 24$$
$$-12 + 36 = 24$$
$$24 = 24 \quad \textit{True}$$

The solution is -6.

45.
$$-12 = 6y - 18y$$
$$-12 = (6-18)y$$
$$-12 = -12y$$
$$\frac{-12}{-12} = \frac{-12y}{-12}$$
$$1 = y$$

Check: Replace y with 1.
$$-12 = 6(1) - 18(1)$$
$$-12 = 6 - 18$$
$$-12 = -12 \quad \textit{True}$$

The solution is 1.

47.

$$\begin{aligned} 6p-2&=4p+6\\ 6p-2+2&=4p+6+2\\ 6p&=4p+8\\ 6p-4p&=4p-4p+8\\ 2p&=8\\ \frac{2p}{2}&=\frac{8}{2}\\ p&=4 \end{aligned}$$

Check: Replace p with 4.

$$\begin{aligned} 6(4)-2&=4(4)+6\\ 24-2&=16+6\\ 22&=22 \quad \textit{True} \end{aligned}$$

The solution is 4.

49.

$$\begin{aligned} 9+7z&=9z+13\\ 9+7z-9z&=9z+13-9z\\ 9-2z&=13\\ 9-2z-9&=13-9\\ -2z&=4\\ \frac{-2z}{-2}&=\frac{4}{-2}\\ z&=-2 \end{aligned}$$

Check: Replace z with -2.

$$\begin{aligned} 9+7(-2)&=9(-2)+13\\ 9+(-14)&=-18+13\\ -5&=-5 \quad \textit{True} \end{aligned}$$

The solution is -2.

51.

$$\begin{aligned} -2y+6&=6y-10\\ -2y+6+10&=6y-10+10\\ -2y+16&=6y\\ -2y+2y+16&=6y+2y\\ 16&=8y\\ \frac{16}{8}&=\frac{8y}{8}\\ 2&=y \end{aligned}$$

Check: Replace y with 2.

$$\begin{aligned} -2(2)+6&=6(2)-10\\ (-4)+6&=12-10\\ 2&=2 \quad \textit{True} \end{aligned}$$

The solution is 2.

53.

$$\begin{aligned} b+3.05&=2\\ b+3.05-3.05&=2-3.05\\ b&=-1.05 \end{aligned}$$

Check: Replace b with -1.05.

$$\begin{aligned} -1.05+3.05&=2\\ 2&=2 \quad \textit{True} \end{aligned}$$

The solution is -1.05.

55.

$$\begin{aligned} 2.5r+9&=-1\\ 2.5r+9-9&=-1-9\\ 2.5r&=-10\\ \frac{2.5r}{2.5}&=\frac{-10}{2.5}\\ r&=-4 \end{aligned}$$

Check: Replace r with -4.

$$\begin{aligned} 2.5(-4)+9&=-1\\ -10+9&=-1\\ -1&=-1 \quad \textit{True} \end{aligned}$$

The solution is -4.

57.

$$\begin{aligned} -10&=2(y+4)\\ -10&=2y+8\\ -10-8&=2y+8-8\\ -18&=2y\\ \frac{-18}{2}&=\frac{2y}{2}\\ -9&=y \end{aligned}$$

Check: Replace y with -9.

$$\begin{aligned} -10&=2(-9+4)\\ -10&=2(-5)\\ -10&=-10 \quad \textit{True} \end{aligned}$$

The solution is -9.

59.

$$\begin{aligned} -4(t+2)&=12\\ -4t+(-8)&=12\\ -4t-8+8&=12+8\\ -4t&=20\\ \frac{-4t}{-4}&=\frac{20}{-4}\\ t&=-5 \end{aligned}$$

Check: Replace t with -5.

$$\begin{aligned} -4(-5+2)&=12\\ -4(-3)&=12\\ 12&=12 \quad \textit{True} \end{aligned}$$

The solution is -5.

61.

$$\begin{aligned} 6(x-5)&=-30\\ 6x-30&=-30\\ 6x-30+30&=-30+30\\ 6x&=0\\ \frac{6x}{6}&=\frac{0}{6}\\ x&=0 \end{aligned}$$

Check: Replace x with 0.

$$\begin{aligned} 6(0-5)&=-30\\ 6(-5)&=-30\\ -30&=-30 \quad \textit{True} \end{aligned}$$

The solution is 0.

63.

$$-2t - 10 = 3t + 5$$

$$-2t - 3t - 10 = 3t - 3t + 5$$ Subtract $3t$ from both sides (addition property).

$$-5t - 10 = 5$$

$$-5t - 10 + 10 = 5 + 10$$ Add 10 to both sides (addition property).

$$-5t = 15$$

$$\frac{\cancel{-5} \cdot t}{\cancel{-5}} = \frac{15}{-5}$$ Divide both sides by -5 (multiplication property).

$$t = -3$$

65.

$$\begin{aligned} 30 - 40 &= -2x + 7x - 4x \\ -10 &= (-2 + 7 - 4)x \\ -10 &= 1x \\ -10 &= x \end{aligned}$$

The solution is -10.

67.

$$\begin{aligned} 0 &= -2(y - 2) \\ 0 &= -2y - (-2) \cdot 2 \\ 0 &= -2y + 4 \\ 0 + 2y &= -2y + 2y + 4 \\ 2y &= 4 \\ \frac{2y}{2} &= \frac{4}{2} \\ y &= 2 \end{aligned}$$

The solution is 2.

69.

$$\begin{aligned} \frac{y}{2} - 2 &= \frac{y}{4} + 3 \\ \frac{y}{2} - 2 + 2 &= \frac{y}{4} + 3 + 2 \\ \frac{y}{2} &= \frac{y}{4} + 5 \\ \frac{1}{2}y &= \frac{1}{4}y + 5 \\ \frac{1}{2}y - \frac{1}{4}y &= \frac{1}{4}y - \frac{1}{4}y + 5 \\ \frac{1}{4}y &= 5 \\ \frac{4}{1} \cdot \frac{1}{4}y &= \frac{4}{1} \cdot 5 \\ y &= 20 \end{aligned}$$

The solution is 20.

71.

$$\begin{aligned} -3(w - 2) &= |0 - 13| + 4w \\ -3 \cdot w - (-3) \cdot 2 &= |-13| + 4w \\ -3w + 6 &= 13 + 4w \\ -3w + 6 - 6 &= 13 + 4w - 6 \\ -3w &= 7 + 4w \\ -3w - 4w &= 7 + 4w - 4w \\ -7w &= 7 \\ \frac{-7w}{-7} &= \frac{7}{-7} \\ w &= -1 \end{aligned}$$

The solution is -1.

73.

$$\begin{aligned} 2(a + 0.3) &= 1.2(a - 4) \\ 2 \cdot a + 2(0.3) &= 1.2 \cdot a - 1.2 \cdot 4 \\ 2a + 0.6 &= 1.2a - 4.8 \\ 2a + 0.6 - 0.6 &= 1.2a - 4.8 - 0.6 \\ 2a &= 1.2a - 5.4 \\ 2a - 1.2a &= 1.2a - 5.4 - 1.2a \\ 0.8a &= -5.4 \\ \frac{0.8a}{0.8} &= \frac{-5.4}{0.8} \\ a &= -6.75 \end{aligned}$$

The solution is -6.75.

9.8 Using Equations to Solve Application Problems

9.8 Margin Exercises

1. **(a)** 15 less than a number

$$x - 15$$

(b) 12 more than a number

$$x + 12 \text{ or } 12 + x$$

(c) A number increased by 13

$$x + 13 \text{ or } 13 + x$$

(d) A number minus 8

$$x - 8$$

(e) -10 plus a number

$$-10 + x \text{ or } x + (-10)$$

(f) A number subtracted from 6

$$6 - x$$

2. **(a)** Double a number

$$2x$$

(b) The product of -8 and a number

$$-8x$$

(c) The quotient of 15 and a number

$$\frac{15}{x}$$

(d) One-half of a number

$$\frac{1}{2}x \text{ or } \frac{x}{2}$$

3. **(a)** Let x represent the unknown number.

3 times a number	added to	4	is	19.
↓	↓	↓	↓	↓
$3x$	$+$	4	$=$	19

$$\begin{aligned} 3x + 4 &= 19 \\ 3x + 4 - 4 &= 19 - 4 \\ 3x &= 15 \\ \frac{3x}{3} &= \frac{15}{3} \\ x &= 5 \end{aligned}$$

The number is 5.

Check: 3 times 5 [= 15] is added to 4 [= 19]. *True*

(b) Let x represent the unknown number.

-6 times a number	added to	5	is	-13.
↓	↓	↓	↓	↓
$-6x$	$+$	5	$=$	-13

$$\begin{aligned} -6x + 5 &= -13 \\ -6x + 5 - 5 &= -13 - 5 \\ -6x &= -18 \\ \frac{-6x}{-6} &= \frac{-18}{-6} \\ x &= 3 \end{aligned}$$

The number is 3.

Check: -6 times 3 [= -18] is added to 5 [= -13]. *True*

(c) Let x represent the unknown number.

$$\begin{aligned} 65 - 2x &= -21 \\ 65 - 2x - 65 &= -21 - 65 \\ -2x &= -86 \\ \frac{-2x}{-2} &= \frac{-86}{-2} \\ x &= 43 \end{aligned}$$

The number is 43.

Check: Twice 43 [= 86] is subtracted from 65 [= -21]. *True*

4. *Step 1*
Asks for amount of LuAnn's donation.

Step 2
Let d be LuAnn's donation.

Step 3
$2d + 10 = 22$
(or, $10 + 2d = 22$)

Step 4

$$\begin{aligned} 2d + 10 - 10 &= 22 - 10 \\ 2d &= 12 \\ \frac{2d}{2} &= \frac{12}{2} \\ d &= 6 \end{aligned}$$

Step 5
LuAnn donated \$6.

Step 6
Susan donated \$10 more than twice \$6, which is $\$10 + 2 \cdot \$6 = \$10 + \$12 = \$22$. That matches the amount given in the problem for Susan.

5. **(a)** *Step 2*
Let x represent the amount made by the daughter and $x + 12$ the amount made by Keonda.

Step 3
$x + (x + 12) = 182$

Step 4

$$\begin{aligned} 2x + 12 &= 182 \\ 2x + 12 - 12 &= 182 - 12 \\ 2x &= 170 \\ \frac{2x}{2} &= \frac{170}{2} \\ x &= 85 \end{aligned}$$

Step 5
The daughter made \$85.
Keonda made $\$85 + \$12 = \$97$.

Step 6
Check: \$97 is \$12 more than \$85, and the sum of \$85 and \$97 is \$182.

(b) *Step 2* Let x represent the length of the shorter piece and $x + 3$ the length of the longer piece.

Step 3
$x + (x + 3) = 21$

Step 4

$$\begin{aligned} 2x + 3 &= 21 \\ 2x + 3 - 3 &= 21 - 3 \\ 2x &= 18 \\ \frac{2x}{2} &= \frac{18}{2} \\ x &= 9 \end{aligned}$$

Step 5
The length of the shorter piece is 9 yd. The length of the longer piece is 9 yd + 3 yd = 12 yd.

Step 6
Check: 12 yards is 3 yards longer than 9 yards and the sum of 9 yards and 12 yards is 21 yards.

6. *Step 2*
Let x represents the width and $x + 3$ the length.

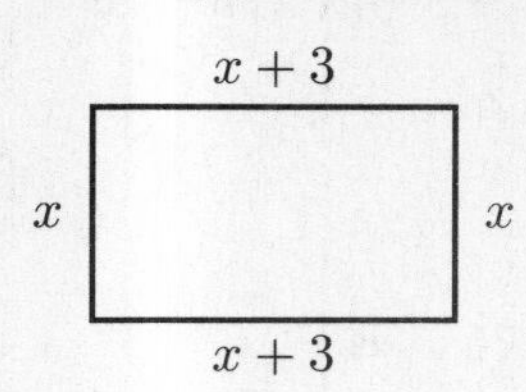

Step 3
Use the formula for the perimeter of a rectangle.

$$P = 2 \cdot l + 2 \cdot w$$
$$22 = 2(x + 3) + 2 \cdot x$$

Step 4

$$22 = 2x + 6 + 2x$$
$$22 = 4x + 6$$
$$22 - 6 = 4x + 6 - 6$$
$$16 = 4x$$
$$\frac{16}{4} = \frac{4x}{4}$$
$$4 = x$$

Step 5
The width is 4 yd and the length is $4 + 3 = 7$ yd.

Step 6
Check: 7 yd is 3 more than 4 yd.

$$P = 2 \cdot 7 \text{ yd} + 2 \cdot 4 \text{ yd}$$
$$P = 14 \text{ yd} + 8 \text{ yd}$$
$$P = 22 \text{ yd}$$

9.8 Section Exercises

1. 14 plus a number

$$14 + x \quad \text{or} \quad x + 14$$

3. -5 added to a number

$$-5 + x \quad \text{or} \quad x + (-5)$$

5. 20 minus a number

$$20 - x$$

7. 9 less than a number

$$x - 9$$

9. Subtract 4 from a number

$$x - 4$$

11. Six times a number

$$6x$$

13. Double a number

$$2x$$

15. A number divided by 2

$$\frac{x}{2} \quad \text{or} \quad \frac{1}{2}x$$

17. Twice a number added to 8

$$8 + 2x \quad \text{or} \quad 2x + 8$$

19. 10 fewer than seven times a number

$$7x - 10$$

21. The sum of twice a number and the number

$$2x + x \quad \text{or} \quad x + 2x$$

23. **(a)** A variable is a letter that represents an unknown quantity. Examples: x, w, p

(b) An expression is a combination of operations on variables and numbers. Examples: $6x, w - 5, 2p + 3x$

(c) An equation has an $=$ sign and shows that two expressions are equal. Examples: $2y = 14$; $x + 5 = 2x$; $8p - 10 = 54$

25. Let n represent the unknown number.

four times a number	decreased by 2	result is	26
↓	↓	↓	↓
$4n$	-2	$=$	26

$$4n - 2 = 26$$
$$4n - 2 + 2 = 26 + 2$$
$$4n = 28$$
$$\frac{4n}{4} = \frac{28}{4}$$
$$n = 7$$

The number is 7.

Check: Four times 7 [= 28] is decreased by 2 [= 26]. *True*

27. Let n represent the unknown number.

Twice a number	added to	the number	is	-15.
↓	↓	↓	↓	↓
$2n$	$+$	n	$=$	-15

$$2n + n = -15$$
$$3n = -15$$
$$\frac{3n}{3} = \frac{-15}{3}$$
$$n = -5$$

The number is -5.

Check: Twice -5 [$= -10$] is added to -5 [$= -15$]. *True*

29. Let n represent the unknown number.

Product of a number and 5	increased by	12	the result is	7 times the number
↓	↓	↓	↓	↓
$5n$	$+$	12	$=$	$7n$

$$
\begin{aligned}
5n + 12 &= 7n \\
5n + 12 - 5n &= 7n - 5n \\
12 &= 2n \\
\frac{12}{2} &= \frac{2n}{2} \\
6 &= n
\end{aligned}
$$

The number is 6.

Check: The product of 6 and 5 [= 30] is increased by 12 [= 42] is seven times 6 [= 42]. *True*

31. Let n represent the unknown number.

30	subtract	3 times a number	is	2	plus	the number
↓	↓	↓	↓	↓	↓	↓
30	$-$	$3n$	$=$	2	$+$	n

$$
\begin{aligned}
30 - 3n &= 2 + n \\
30 - 3n + 3n &= 2 + n + 3n \\
30 &= 2 + 4n \\
30 - 2 &= 2 + 4n - 2 \\
28 &= 4n \\
\frac{28}{4} &= \frac{4n}{4} \\
7 &= n
\end{aligned}
$$

The number is 7.

Check: Three times 7 [= 21] is subtracted from 30 [= 9] is 2 plus 7 [= 9]. *True*

33. Let n represent the unknown number.

half a number	is added	to twice the number	answer is	50
↓	↓	↓	↓	↓
$\frac{1}{2}n$	$+$	$2n$	$=$	50

$$
\begin{aligned}
\frac{1}{2}n + 2n &= 50 \\
\left(\frac{1}{2} + 2\right)n &= 50 \\
\left(\frac{1}{2} + \frac{4}{2}\right)n &= 50 \\
\frac{5}{2}n &= 50 \\
\frac{2}{5} \cdot \frac{5}{2}n &= \frac{2}{5} \cdot 50 \\
n &= 20
\end{aligned}
$$

The number is 20.

Check: If half of 20 [= 10] is added to twice 20 [= 40], the answer is 50. *True*

35. *Step 2*
Let x be Tamu's age.

Step 3

Four times Tamu's age	less	75	is	x
↓	↓	↓	↓	↓
$4x$	$-$	75	$=$	x

Step 4

$$
\begin{aligned}
4x - 75 &= x \\
4x - 75 + 75 &= x + 75 \\
4x &= x + 75 \\
4x - x &= x + 75 - x \\
3x &= 75 \\
\frac{3x}{3} &= \frac{75}{3} \\
x &= 25
\end{aligned}
$$

Step 5
Tamu is 25 years old.

Step 6
Check: When 75 is subtracted from four times 25 [= 100], the result is 25. *True*

37. *Step 2*
Let x be the amount Brenda spent.

Step 3

Twice what Brenda spent	less	\$3	is	the amount Consuelo spent
↓	↓	↓	↓	↓
$2x$	$-$	3	$=$	81

Step 4

$$
\begin{aligned}
2x - 3 &= 81 \\
2x - 3 + 3 &= 81 + 3 \\
2x &= 84 \\
\frac{2x}{2} &= \frac{84}{2} \\
x &= 42
\end{aligned}
$$

Step 5
Brenda spent \$42.

Step 6
Check: Consuelo spent \$3 less than twice \$42 [= \$84], which is \$81. *True*

39. *Step 2*
Let x be my age; $x + 9$ is my sister's age.

Step 3

The sum of the ages	is	51
↓	↓	↓
$x + (x + 9)$	$=$	51

Step 4

$$x + x + 9 = 51$$
$$2x + 9 = 51$$
$$2x + 9 - 9 = 51 - 9$$
$$2x = 42$$
$$\frac{2x}{2} = \frac{42}{2}$$
$$x = 21$$
$$x + 9 = 30$$

Step 5
I am 21; my sister is 30.

Step 6
Check: 30 is 9 more than 21 and the sum of 30 and 21 is 51. *True*

41. *Step 2*
Let x be husband's earnings; $x + 1500$ is Lien's earnings.

Step 3

Together (add)	they earned	37,500.
↓	↓	↓
$x + (x + 1500)$	$=$	37,500

Step 4

$$x + x + 1500 = 37{,}500$$
$$2x + 1500 = 37{,}500$$
$$2x + 1500 - 1500 = 37{,}500 - 1500$$
$$2x = 36{,}000$$
$$\frac{2x}{2} = \frac{36{,}000}{2}$$
$$x = 18{,}000$$
$$x + 1500 = 19{,}500$$

Step 5
Lien earned $19,500 and her husband earned $18,000.

Step 6
Check: $19,500 is $1500 more than $18,000 and the sum of $18,000 and $19,500 is $37,500. *True*

43. *Step 2*
Let x be the cost of the printer; $5x$ is the cost of the computer.

Step 3

The total cost (add)	is	$1320.
↓	↓	↓
$x + 5x$	$=$	1320

Step 4

$$x + 5x = 1320$$
$$6x = 1320$$
$$\frac{6x}{6} = \frac{1320}{6}$$
$$x = 220$$
$$5x = 1100$$

Step 5
The computer cost $1100 and the printer cost $220.

Step 6
Check: Five times $220 is $1100 and the sum of $220 and $1100 is $1320. *True*

45. *Step 2*
Let x be the length of the shorter piece, and $x + 10$ the length of the longer piece.

Step 3

The length of the shorter piece	and	the length of the longer piece	are	78 cm.
↓	↓	↓	↓	↓
x	$+$	$x + 10$	$=$	78

Step 4

$$x + x + 10 = 78$$
$$2x + 10 = 78$$
$$2x + 10 - 10 = 78 - 10$$
$$2x = 68$$
$$\frac{2x}{2} = \frac{68}{2}$$
$$x = 34$$

Step 5
The length of the shorter piece is 34 cm and the length of the longer piece is $34 + 10 = 44$ cm.

Step 6
Check: 44 cm is 10 cm longer than 34 cm and the sum of 34 cm and 44 cm is 78 cm. *True*

47. *Step 2*
Let x be the length of the longer piece, and $x - 7$ the length of the shorter piece.

Step 3

The length of the longer piece	and	the length of the shorter piece	are	31 ft.
↓	↓	↓	↓	↓
x	$+$	$x - 7$	$=$	31

Step 4

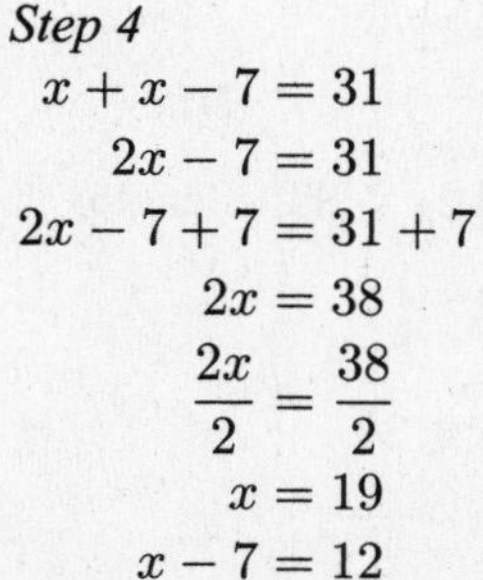

$$x + x - 7 = 31$$
$$2x - 7 = 31$$
$$2x - 7 + 7 = 31 + 7$$
$$2x = 38$$
$$\frac{2x}{2} = \frac{38}{2}$$
$$x = 19$$
$$x - 7 = 12$$

Step 5
The pieces are 19 ft and 12 ft.

Step 6
Check: 12 ft is 7 ft shorter than 19 ft and the sum of 12 ft and 19 ft is 31 ft. *True*

49. Let x be the length of the rectangle.

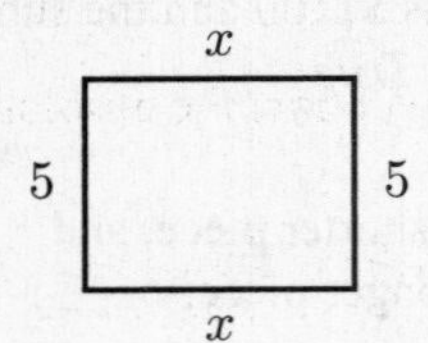

$$\begin{aligned} P &= 2\cdot l + 2\cdot w \\ 48 &= 2x + 2\cdot 5 \\ 48 - 10 &= 2x + 10 - 10 \\ 38 &= 2x \\ \frac{38}{2} &= \frac{2x}{2} \\ 19 &= x \end{aligned}$$

The length is 19 yd.

Check: $48 = 2\cdot 19 + 2\cdot 5$
$48 = 38 + 10$
$48 = 48$ *True*

51. Let w be the width of the rectangle, and $2w$ the length.

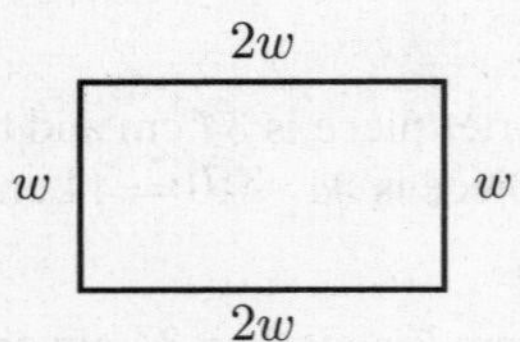

$$\begin{aligned} P &= 2\cdot l + 2\cdot w \\ 36 &= 2\cdot 2w + 2\cdot w \\ 36 &= 4w + 2w \\ 36 &= 6w \\ \frac{36}{6} &= \frac{6w}{6} \\ 6 &= w \\ 12 &= 2w \end{aligned}$$

The length is 12 ft and the width is 6 ft.

Check: $36 = 2(12) + 2(6)$
$36 = 24 + 12$
$36 = 36$ *True*

53. Let x be the width, and $2x + 3$ the length.

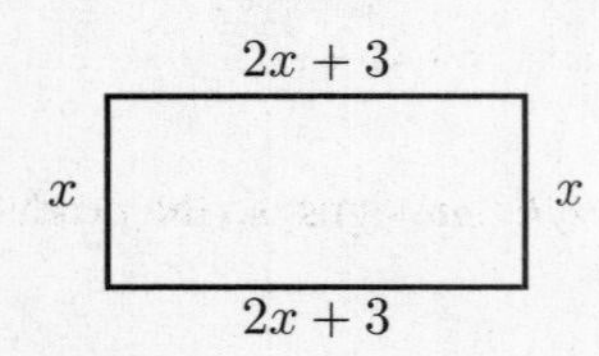

$$\begin{aligned} P &= 2\cdot l + 2\cdot w \\ 36 &= 2\cdot(2x+3) + 2\cdot x \\ 36 &= 4x + 6 + 2x \\ 36 &= 6x + 6 \\ 36 - 6 &= 6x + 6 - 6 \\ 30 &= 6x \\ \frac{30}{6} &= \frac{6x}{6} \\ 5 &= x \\ 13 &= 2x + 3 \end{aligned}$$

The width is 5 in. and the length is 13 in.

Check: $36 = 2(13) + 2(5)$
$36 = 26 + 10$
$36 = 36$ *True*

Chapter 9 Review Exercises

1. $2, -3, 4, 1, 0, -5$

−7 −6 −5 −4 −3 −2 −1 0 1 2 3 4

2. $-2, 5, -4, -1, 3, -6$

−6 −5 −4 −3 −2 −1 0 1 2 3 4 5

3. $-1\frac{1}{4}, -\frac{5}{8}, -3\frac{3}{4}, 2\frac{1}{8}, \frac{3}{2}, -2\frac{1}{8}$

−8 −7 −6 −5 −4 −3 −2 −1 0 1 2 3

4. $0, -\frac{3}{4}, \frac{5}{4}, -4\frac{1}{2}, \frac{7}{8}, -7\frac{2}{3}$

−8 −7 −6 −5 −4 −3 −2 −1 0 1 2 3

5. $0 > -2$ because 0 is to the *right* of -2 on a number line.

6. $-5 < 0$ because -5 is to the *left* of 0 on a number line.

7. $-1 > -4$ because -1 is to the *right* of -4 on a number line.

8. $-9 < -6$ because -9 is to the *left* of -6 on a number line.

9. The *distance* from 0 to 8 is 8, so $|8| = 8$.

10. The *distance* from 0 to -19 is 19, so $|-19| = 19$.

11. $-|-7|$
First, $|-7| = 7$. But there is a negative sign outside the absolute value bars. So, -7 is the simplified expression.

12. $-|15| = -(15) = -15$
Because the negative sign is outside the absolute value bars, the answer is negative.

13. $-4+6$
The signs are different, so subtract the absolute values.

$$|-4|=4; \quad |6|=6$$
$$6-4=2$$

The positive number, 6, has the larger absolute value, so the answer is positive.

$$-4+6=2$$

14. $-10+3$
The signs are different, so subtract the absolute values.

$$|-10|=10; \quad |3|=3$$
$$10-3=7$$

The negative number, 10, has the larger absolute value, so the answer is negative.

$$-10+3=-7$$

15. $-11+(-8)$
Add the absolute values.

$$|-11|=11; \quad |-8|=8$$
$$11+8=19$$

Write a negative sign in front of the sum because both numbers are negative.

$$-11+(-8)=-19$$

16. $-9+(-24)$
Add the absolute values.

$$|-9|=9; \quad |-24|=24$$
$$9+24=33$$

Write a negative sign in front of the sum because both numbers are negative.

$$-9+(-24)=-33$$

17. $12+(-11)$
The signs are different, so subtract the absolute values.

$$|12|=12; \quad |-11|=11$$
$$12-11=1$$

The positive number, 12, has the larger absolute value, so the answer is positive.

$$12+(-11)=1$$

18. $1+(-20)$
The signs are different, so subtract the absolute values.

$$|1|=1; \quad |-20|=20$$
$$20-1=19$$

The negative number, -20, has the larger absolute value, so the answer is negative.

$$1+(-20)=-19$$

19. $\frac{9}{10}+\left(-\frac{3}{5}\right)=\frac{9}{10}+\left(-\frac{6}{10}\right)=\frac{3}{10}$

20. $-\frac{7}{8}+\frac{1}{2}=-\frac{7}{8}+\frac{4}{8}=-\frac{3}{8}$

21. $-6.7+1.5$
The signs are different, so subtract the absolute values.

$$|-6.7|=6.7; \quad |1.5|=1.5$$
$$6.7-1.5=5.2$$

The negative number, -6.7, has the larger absolute value, so the answer is negative.

$$-6.7+1.5=-5.2$$

22. $-0.8+(-0.7)$
Add the absolute values.

$$|-0.8|=0.8; \quad |-0.7|=0.7$$
$$0.8+0.7=1.5$$

Write a negative sign in front of the sum because both numbers are negative.

$$-0.8+(-0.7)=-1.5$$

23. The additive inverse of 6 is -6.

24. The additive inverse of -14 is $-(-14)$ or 14.

25. The additive inverse of $-\frac{5}{8}$ is $-\left(-\frac{5}{8}\right)$ or $\frac{5}{8}$.

26. The additive inverse of 3.75 is -3.75.

27. $4-10=4+(-10)=-6$

28. $7-15=7+(-15)=-8$

29. $-6-1=-6+(-1)=-7$

30. $-12-5=-12+(-5)=-17$

31. $8-(-3)=8+(+3)=11$

32. $2-(-9)=2+(+9)=11$

33. $-1-(-14)=-1+(+14)=13$

34. $-10-(-4)=-10+(+4)=-6$

35. $-40-40=-40+(-40)=-80$

36. $-15-(-15)=-15+(+15)=0$

37. $\frac{1}{3}-\frac{5}{6}=\frac{2}{6}+\left(-\frac{5}{6}\right)=-\frac{3}{6}=-\frac{1}{2}$

38. $2.8-(-6.2)=2.8+(+6.2)=9$

39. $-4(6)=-24$
The numbers have *different* signs, so the product is *negative.*

40. $5\cdot(-4)=-20$
The numbers have *different* signs, so the product is *negative.*

41. $-3(-5) = 15$
The numbers have the *same* sign, so the product is *positive.*

42. $-8(-8) = 64$
The numbers have the *same* sign, so the product is *positive.*

43. $\frac{80}{-10} = -8$
The numbers have *different* signs, so the quotient is *negative.*

44. $\frac{-9}{3} = -3$
The numbers have *different* signs, so the quotient is *negative.*

45. $\frac{-25}{-5} = 5$
The numbers have *same* sign, so the quotient is *positive.*

46. $\frac{-120}{-6} = 20$
The numbers have *same* sign, so the quotient is *positive.*

47. $(-37)(0) = 0$

48. $(-1)(81) = -81$

49. $\frac{0}{-10} = 0$

50. $\frac{-20}{0}$ is undefined.

51. $\left(\frac{2}{3}\right)\cdot\left(-\frac{6}{7}\right) = \frac{2}{\cancel{3}_1}\cdot\left(-\frac{\cancel{6}^2}{7}\right) = -\frac{4}{7}$

52. $-\frac{4}{5} \div \left(-\frac{2}{15}\right) = -\frac{4}{5}\cdot\left(-\frac{15}{2}\right) = \frac{\cancel{4}^2}{\cancel{5}_1}\cdot\frac{\cancel{15}^3}{\cancel{2}_1} = 6$

53. $(-0.5)(-2.8) = 1.4$

54. $\frac{-5.28}{0.8} = -6.6$
Move the decimal place to the right in both the divisor and the dividend.

$$\begin{array}{r} 6.6 \\ 8\overline{)52.8} \\ \underline{48} \\ 4\,8 \\ \underline{4\,8} \\ 0 \end{array}$$

The numbers have *different* signs, so the quotient is *negative.*

55.
$$\begin{aligned} &2 - 11(-5) \\ &= 2 - (-55) \\ &= 2 + (+55) \\ &= 57 \end{aligned}$$

56.
$$\begin{aligned} &(-4)(-8) - 9 \\ &= 32 - 9 \\ &= 23 \end{aligned}$$

57.
$$\begin{aligned} &48 \div (-2)^3 - (-5) \\ &= 48 \div (-8) - (-5) \\ &= -6 - (-5) \\ &= -6 + (+5) \\ &= -1 \end{aligned}$$

58.
$$\begin{aligned} &-36 \div (-3)^2 - (-2) \\ &= -36 \div 9 - (-2) \\ &= -4 - (-2) \\ &= -4 + (+2) \\ &= -2 \end{aligned}$$

59.
$$\begin{aligned} &5(4) - 7(6) + 3(-4) \\ &= 20 - 42 + (-12) \\ &= 20 + (-42) + (-12) \\ &= -22 + (-12) \\ &= -34 \end{aligned}$$

60.
$$\begin{aligned} &2(8) - 4(9) + 2(-6) \\ &= 16 - 36 + (-12) \\ &= 16 + (-36) + (-12) \\ &= -20 + (-12) \\ &= -32 \end{aligned}$$

61.
$$\begin{aligned} &3^3(-4) - 2(5 - 9) \\ &= 27(-4) - 2\cdot(-4) \\ &= -108 - (-8) \\ &= -108 + (+8) \\ &= -100 \end{aligned}$$

62.
$$\begin{aligned} &6(-4)^2 - 3(7 - 14) \\ &= 6(-4)^2 - 3\cdot(-7) \\ &= 6(16) - 3\cdot(-7) \\ &= 96 - (-21) \\ &= 96 + (+21) \\ &= 117 \end{aligned}$$

63. $\frac{3 - (5^2 - 4^2)}{14 + 24 \div (-3)}$

Numerator:
$$\begin{aligned} 3 - (5^2 - 4^2) &= 3 - (25 - 16) \\ &= 3 - 9 \\ &= 3 + (-9) \\ &= -6 \end{aligned}$$

Denominator:
$$\begin{aligned} 14 + 24 \div (-3) &= 14 + (-8) \\ &= 6 \end{aligned}$$

Divide: $\frac{-6}{6} = -1$

64. $(-0.8)^2(0.2) - (-1.2)$
$= (0.64)(0.2) - (-1.2)$
$= 0.128 + (+1.2)$
$= 1.328$

65. $\left(-\frac{1}{3}\right)^2 + \frac{1}{4}\left(-\frac{4}{9}\right)$
$= \frac{1}{9} + \left(\frac{1}{\not{4}_1}\right)\left(-\frac{\not{4}^1}{9}\right)$
$= \frac{1}{9} + \left(-\frac{1}{9}\right)$
$= 0$

66. $\frac{12 \div (2-5) + 12(-1)}{2^3 - (-4)^2}$

Numerator:
$12 \div (2-5) + 12(-1) = 12 \div (-3) + 12(-1)$
$= -4 + 12(-1)$
$= -4 + (-12)$
$= -16$

Denominator:
$2^3 - (-4)^2 = 8 - 16$
$= 8 + (-16)$
$= -8$

Divide: $\frac{-16}{-8} = 2$

67. $3k + 5m; k = 4; m = 3$
Replace k with 4. Replace m with 3.

$3k + 5m = 3(4) + 5(3)$
$= 12 + 15$
$= 27$

68. $3k + 5m; k = -6; m = 2$
Replace k with -6. Replace m with 2.

$3k + 5m = 3(-6) + 5(2)$
$= -18 + 10$
$= -8$

69. $2p - q; p = -5; q = -10$
Replace p with -5. Replace q with -10.

$2p - q = 2(-5) - (-10)$
$= -10 + (+10)$
$= 0$

70. $2p - q; p = 6; q = -7$
Replace p with 6. Replace q with -7.

$2p - q = 2(6) - (-7)$
$= 12 + (+7)$
$= 19$

71. $\frac{5a - 7y}{2 + m}; a = 1, y = 4, m = -3$

Replace a with 1, y with 4, and m with -3.

$\frac{5a - 7y}{2 + m} = \frac{5(1) - 7(4)}{2 + (-3)}$
$= \frac{5 - 28}{-1}$
$= \frac{5 + (-28)}{-1}$
$= \frac{-23}{-1}$
$= 23$

72. $\frac{5a - 7y}{2 + m}; a = 2, y = -2, m = -26$

Replace a with 2, y with -2, and m with -26.

$\frac{5a - 7y}{2 + m} = \frac{5(2) - 7(-2)}{2 + (-26)}$
$= \frac{10 - (-14)}{-24}$
$= \frac{10 + (+14)}{-24}$
$= \frac{24}{-24}$
$= -1$

73. $P = a + b + c; a = 9, b = 12, c = 14$

$P = 9 + 12 + 14$
$= 21 + 14$
$= 35$

74. $A = \frac{1}{2}bh; b = 6, h = 9$

$A = \frac{1}{2}(6)(9)$
$= 3(9)$
$= 27$

75. $y + 3 = 0$
Subtract 3 from both sides.

$y + 3 - 3 = 0 - 3$
$y = -3$

Check: $-3 + 3 = 0$
$0 = 0$ *True*

The solution is -3.

76. $a - 8 = 8$
$a - 8 + 8 = 8 + 8$
$a = 16$

Check: $16 - 8 = 8$
$8 = 8$ *True*

The solution is 16.

77.
$$\begin{aligned} -5 &= z-6 \\ -5+6 &= z-6+6 \\ 1 &= z \end{aligned}$$

Check: $-5 = 1-6$
$-5 = -5$ *True*

The solution is 1.

78.
$$\begin{aligned} -8 &= -9+r \\ -8+9 &= -9+9+r \\ 1 &= r \end{aligned}$$

Check: $-8 = -9+1$
$-8 = -8$ *True*

The solution is 1.

79.
$$\begin{aligned} -\frac{3}{4}+x &= -2 \\ -\frac{3}{4}+x+\frac{3}{4} &= -2+\frac{3}{4} \\ x &= -\frac{8}{4}+\frac{3}{4} \\ &= -\frac{5}{4} \text{ or } -1\frac{1}{4} \end{aligned}$$

Check: $-\frac{3}{4}+\left(-\frac{5}{4}\right) = -2$
$-\frac{8}{4} = -2$
$-2 = -2$ *True*

The solution is $-\frac{5}{4}$ or $-1\frac{1}{4}$.

80.
$$\begin{aligned} 12.92+k &= 4.87 \\ 12.92-12.92+k &= 4.87+(-12.92) \\ k &= -8.05 \end{aligned}$$

Check: $12.92+(-8.05) = 4.87$
$4.87 = 4.87$ *True*

The solution is -8.05.

81.
$$\begin{aligned} -8r &= 56 \\ \frac{\overset{1}{-\cancel{8}r}}{\underset{1}{-\cancel{8}}} &= \frac{56}{-8} \\ r &= -7 \end{aligned}$$

Check: $(-8)(-7) = 56$
$56 = 56$ *True*

The solution is -7.

82.
$$\begin{aligned} 3p &= 24 \\ \frac{3p}{3} &= \frac{24}{3} \\ p &= 8 \end{aligned}$$

Check: $3(8) = 24$
$24 = 24$ *True*

The solution is 8.

83.
$$\begin{aligned} \frac{z}{4} &= 5 \\ \frac{4}{1}\cdot\frac{z}{4} &= 5\cdot 4 \\ z &= 20 \end{aligned}$$

Check: $\frac{20}{4} = 5$
$5 = 5$ *True*

The solution is 20.

84.
$$\begin{aligned} \frac{a}{5} &= -11 \\ \frac{5}{1}\cdot\frac{a}{5} &= -11\cdot 5 \\ a &= -55 \end{aligned}$$

Check: $\frac{-55}{5} = -11$
$-11 = -11$ *True*

The solution is -55.

85.
$$\begin{aligned} 20 &= 3y-7 \\ 20+7 &= 3y-7+7 \\ 27 &= 3y \\ \frac{27}{3} &= \frac{3y}{3} \\ 9 &= y \end{aligned}$$

Check: $20 = 3(9)-7$
$20 = 27-7$
$20 = 20$ *True*

The solution is 9.

86.
$$\begin{aligned} -5 &= 2b+3 \\ -5-3 &= 2b+3-3 \\ -8 &= 2b \\ \frac{-8}{2} &= \frac{2b}{2} \\ -4 &= b \end{aligned}$$

Check: $-5 = 2(-4)+3$
$-5 = -8+3$
$-5 = -5$ *True*

The solution is -4.

87.
$$\begin{aligned} 6(r-5) &= 6\cdot r - 6\cdot 5 \\ &= 6r-30 \end{aligned}$$

88.
$$\begin{aligned} 11(p+7) &= 11\cdot p + 11\cdot 7 \\ &= 11p+77 \end{aligned}$$

89.
$$\begin{aligned} -9(z-3) &= -9\cdot z - (-9)(3) \\ &= -9z-(-27) \\ &= -9z+27 \end{aligned}$$

90.
$$\begin{aligned} -8(x+4) &= -8\cdot x + (-8)\cdot 4 \\ &= -8x-32 \end{aligned}$$

91. $3r + 8r = (3 + 8)r$
$= 11r$

92. $10z - 15z = (10 - 15)z$
$= -5z$

93. $3p - 12p + p = (3 - 12 + 1)p$
$= -8p$

94. $-6x - x + 9x = (-6 - 1 + 9)x$
$= 2x$

95.
$$-4z + 2z = 18$$
$$(-4 + 2)z = 18$$
$$-2z = 18$$
$$\frac{-2z}{-2} = \frac{18}{-2}$$
$$z = -9$$

Check: $-4(-9) + 2(-9) = 18$
$36 + (-18) = 18$
$18 = 18$ *True*

The solution is -9.

96.
$$-35 = 9k - 2k$$
$$-35 = (9 - 2)k$$
$$-35 = 7k$$
$$\frac{-35}{7} = \frac{7k}{7}$$
$$-5 = k$$

Check: $-35 = 9(-5) - 2(-5)$
$-35 = -45 - (-10)$
$-35 = -45 + (+10)$
$-35 = -35$ *True*

The solution is -5.

97.
$$4y - 3 = 7y + 12$$
$$4y - 3 - 4y = 7y + 12 - 4y$$
$$-3 = 3y + 12$$
$$-3 + (-12) = 3y + 12 + (-12)$$
$$-15 = 3y$$
$$\frac{-15}{3} = \frac{3y}{3}$$
$$-5 = y$$

Check: $4(-5) + (-3) = 7(-5) + 12$
$-20 + (-3) = -35 + 12$
$-23 = -23$ *True*

The solution is -5.

98.
$$b + 6 = 3b - 8$$
$$b + 6 - b = 3b - 8 - b$$
$$6 = 2b - 8$$
$$6 + 8 = 2b - 8 + 8$$
$$14 = 2b$$
$$\frac{14}{2} = \frac{2b}{2}$$
$$7 = b$$

Check: $7 + 6 = 3 \cdot 7 - 8$
$13 = 21 - 8$
$13 = 13$ *True*

The solution is 7.

99.
$$-14 = 2(a - 3)$$
$$-14 = 2 \cdot a - 2 \cdot 3$$
$$-14 = 2a - 6$$
$$-14 + 6 = 2a - 6 + 6$$
$$-8 = 2a$$
$$\frac{-8}{2} = \frac{2a}{2}$$
$$-4 = a$$

Check: $-14 = 2(-4 - 3)$
$-14 = 2(-7)$
$-14 = -14$ *True*

The solution is -4.

100.
$$42 = 7(t + 6)$$
$$42 = 7t + 7 \cdot 6$$
$$42 = 7t + 42$$
$$42 - 42 = 7t + 42 - 42$$
$$0 = 7t$$
$$0 = t$$

Check: $42 = 7(0 + 6)$
$42 = 7(6)$
$42 = 42$ *True*

The solution is 0.

101. 18 plus a number

$18 + x$ or $x + 18$

102. Half a number

$\frac{1}{2}x$ or $\frac{x}{2}$

103. -5 times a number

$-5x$

104. A number subtracted from 20

$20 - x$

105. Let x represent the number.

four times a number	the sum of	and 6	is	-14
↓	↓	↓	↓	↓
$4x$	$+$	6	$=$	-14

$$\begin{aligned} 4x + 6 &= -14 \\ 4x + 6 - 6 &= -14 - 6 \\ 4x &= -20 \\ \frac{4x}{4} &= \frac{-20}{4} \\ x &= -5 \end{aligned}$$

The number is -5.

Check: The sum of four times -5 [$= -20$] and 6 is -14. *True*

106. Let x represent the number.

five times a number	subtract	the number	is	100
↓	↓	↓	↓	↓
$5x$	$-$	x	$=$	100

$$\begin{aligned} 5x - x &= 100 \\ 4x &= 100 \\ \frac{4x}{4} &= \frac{100}{4} \\ x &= 25 \end{aligned}$$

The number is 25.

Check: If 25 is subtracted from five times 25 [$= 125$], the result is 100. *True*

107. *Step 2*
Let x be the amount received by one student, and $3x$ the amount received by the other student. The sum of the amounts is \$30,000.

Step 3
$x + 3x = 30{,}000$

Step 4
$$\begin{aligned} 4x &= 30{,}000 \\ \frac{4x}{4} &= \frac{30{,}000}{4} \\ x &= 7500 \\ 3x &= 22{,}500 \end{aligned}$$

Step 5
The students will receive \$7500 and \$22,500.

Step 6
Check: \$22,500 is three times \$7500 and the sum of \$22,500 and \$7500 is \$30,000. *True*

108. *Step 2*
Let w represent the width, and $w + 4$ the length.

$w + 4$

w w

$w + 4$

Step 3
$$\begin{aligned} P &= 2 \cdot l + 2 \cdot w \\ 48 &= 2 \cdot (w + 4) + 2 \cdot w \end{aligned}$$

Step 4
$$\begin{aligned} 48 &= 2w + 8 + 2w \\ 48 &= 4w + 8 \\ 40 &= 4w \\ \frac{40}{4} &= \frac{4w}{4} \\ 10 &= w \\ 14 &= w + 4 \end{aligned}$$

Step 5
The width is 10 in. and the length is 14 in.

Step 6
Check: $48 = 2(14) + 2(10)$
$48 = 28 + 20$
$48 = 48$ *True*

109. **[9.2]** $-6 - (-9) = -6 + (+9) = 3$

110. **[9.3]** $-8 \cdot (-5) = 40$

111. **[9.2]** $-12 + 11 = -1$

112. **[9.3]** $\frac{-70}{10} = -7$

113. **[9.3]** $-4(4) = -16$

114. **[9.2]** $5 - 14 = 5 + (-14) = -9$

115. **[9.3]** $\frac{-42}{-7} = 6$

116. **[9.2]** $16 + (-11) = 5$

117. **[9.2]** $-10 - 10 = -10 + (-10) = -20$

118. **[9.3]** $\frac{-5}{0}$ is undefined

119. **[9.2]** $-\frac{2}{3} + \frac{1}{9} = -\frac{6}{9} + \frac{1}{9} = -\frac{5}{9}$

120. **[9.3]** $0.7(-0.5) = -0.35$

121. **[9.4]** $|-6| + 2 - 3(-8) - 5^2$
$= |-6| + 2 - 3(-8) - 25$
$= 6 + 2 - 3(-8) - 25$
$= 6 + 2 - (-24) - 25$

$$\begin{aligned}&= 6 + 2 + 24 - 25\\&= 8 + 24 - 25\\&= 32 - 25\\&= 7\end{aligned}$$

122. **[9.4]** $9 \div |-3| + 6(-5) + 2^3$
$$\begin{aligned}&= 9 \div |-3| + 6(-5) + 8\\&= 9 \div 3 + 6(-5) + 8\\&= 3 + 6(-5) + 8\\&= 3 + (-30) + 8\\&= -27 + 8\\&= -19\end{aligned}$$

123. **[9.6]**
$$\begin{aligned}-45 &= -5y\\\frac{-45}{-5} &= \frac{-5y}{-5}\\9 &= y\end{aligned}$$

The solution is 9.

124. **[9.6]**
$$\begin{aligned}b - 8 &= -12\\b - 8 + 8 &= -12 + 8\\b &= -4\end{aligned}$$

The solution is −4.

125. **[9.7]**
$$\begin{aligned}6z - 3 &= 3z + 9\\6z - 3 + 3 &= 3z + 9 + 3\\6z &= 3z + 12\\6z - 3z &= 3z + 12 - 3z\\3z &= 12\\\frac{3z}{3} &= \frac{12}{3}\\z &= 4\end{aligned}$$

The solution is 4.

126. **[9.6]**
$$\begin{aligned}-5 &= r + 5\\-5 + (-5) &= r + 5 + (-5)\\-10 &= r\end{aligned}$$

The solution is −10.

127. **[9.6]**
$$\begin{aligned}-3x &= 33\\\frac{-3x}{-3} &= \frac{33}{-3}\\x &= -11\end{aligned}$$

The solution is −11.

128. **[9.7]**
$$\begin{aligned}2z - 7z &= -15\\-5z &= -15\\\frac{-5z}{-5} &= \frac{-15}{-5}\\z &= 3\end{aligned}$$

The solution is 3.

129. **[9.7]**
$$\begin{aligned}3(k - 6) &= 6 - 12\\3k - 18 &= -6\\3k - 18 + 18 &= -6 + 18\\3k &= 12\\\frac{3k}{3} &= \frac{12}{3}\\k &= 4\end{aligned}$$

The solution is 4.

130. **[9.7]**
$$\begin{aligned}6(t + 3) &= -2 + 20\\6t + 18 &= 18\\6t + 18 - 18 &= 18 - 18\\6t &= 0\\t &= 0\end{aligned}$$

The solution is 0.

131. **[9.7]**
$$\begin{aligned}-10 &= \frac{a}{5} - 2\\-10 &= \frac{1}{5}a - 2\\-10 + 2 &= \frac{1}{5}a - 2 + 2\\-8 &= \frac{1}{5}a\\5 \cdot (-8) &= \left(\frac{5}{1}\right) \cdot \left(\frac{1}{5}\right)a\\-40 &= a\end{aligned}$$

The solution is −40.

132. **[9.7]**
$$\begin{aligned}4 + 8p &= 4p + 16\\4 + 8p - 4p &= 4p + 16 - 4p\\4 + 4p &= 16\\4 + 4p - 4 &= 16 - 4\\4p &= 12\\\frac{4p}{4} &= \frac{12}{4}\\p &= 3\end{aligned}$$

The solution is 3.

133. **[9.8]** *Step 2*
Let x represent the recommended amount for a newborn infant, and $2x + 3$ the recommended amount for an adult female.

Step 3
$2x + 3 = 15$

Step 4
$$\begin{aligned}2x + 3 - 3 &= 15 - 3\\2x &= 12\\\frac{2x}{2} &= \frac{12}{2}\\x &= 6\end{aligned}$$

Step 5
An infant should receive 6 mg.

Step 6
Check: 3 mg more than twice 6 mg [= 12] is 15 mg. *True*

134. [9.8] *Step 2*
Let x represent the number of senators, and $5x - 65$ the number of representatives.
There are a total (add) of 535 members.

Step 3
$x + (5x - 65) = 535$

Step 4
$$6x - 65 = 535$$
$$6x - 65 + 65 = 535 + 65$$
$$6x = 600$$
$$\frac{6x}{6} = \frac{600}{6}$$
$$x = 100$$
$$5x - 65 = 435$$

Step 5
There are 100 senators and 435 representatives.

Step 6
Check: 65 less than 5 times 100 [= 500] is 435, and the sum of 100 and 435 is 535. *True*

135. [9.8] *Step 2*
Let x represent a zebra's running speed and $x + 25$ a cheetah's running speed.

Step 3
$x + x + 25 = 111$

Step 4
$$2x + 25 = 111$$
$$2x + 25 - 25 = 111 - 25$$
$$2x = 86$$
$$\frac{2x}{2} = \frac{86}{2}$$
$$x = 43$$

Step 5
A zebra runs 43 miles per hour, and a cheetah sprints $43 + 25 =$ 68 miles per hour.

Step 6
Check: 68 mph is 25 mph faster than 43 mph and the sum of 68 mph and 43 mph is 111 mph. *True*

136. [9.8] *Step 2*
Let $w =$ the width, and $5w =$ the length.

Step 3
$$P = 2 \cdot l + 2 \cdot w$$
$$1320 = 2 \cdot 5w + 2 \cdot w$$

Step 4
$$1320 = 10w + 2w$$
$$1320 = 12w$$
$$\frac{1320}{12} = \frac{12w}{12}$$
$$110 = w$$
$$550 = 5w$$

Step 5
The length is 550 yd and the width is 110 yd.

Step 6
Check: $1320 = 2(550) + 2(110)$
$1320 = 1100 + 220$
$1320 = 1320$ *True*

Chapter 9 Test

1. $-4, -1, 1\frac{1}{2}, 3, 0$

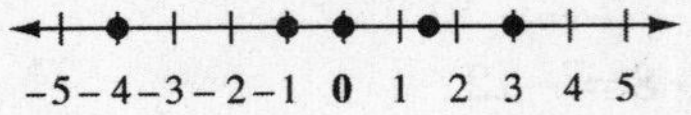

   (number line from −5 to 5 with points graphed)

2. $-3 < 0$ because -3 is to the *left* of 0 on a number line.
   $-4 > -8$ because -4 is to the *right* of -8 on a number line.

3. $|-7| = 7$ and $|15| = 15$ because the distances from -7 to 0 and 15 to 0 are 7 and 15, respectively.

4. $-8 + 7 = -1$

5. $-11 + (-2) = -13$

6. $6.7 + (-1.4) = 5.3$

7. $8 - 15 = 8 + (-15) = -7$

8. $4 - (-12) = 4 + (+12) = 16$

9. $-\frac{1}{2} - \left(-\frac{3}{4}\right) = -\frac{1}{2} + \left(+\frac{3}{4}\right) = -\frac{2}{4} + \frac{3}{4} = \frac{1}{4}$

10. $8(-4) = -32$

11. $-7(-12) = 84$

12. $(-16)(0) = 0$

13. $\frac{-100}{4} = -25$

14. $\frac{-24}{-3} = 8$

15. $-\frac{1}{4} \div \frac{5}{12} = -\frac{1}{\cancel{4}_1} \cdot \frac{\cancel{12}^3}{5} = -\frac{3}{5}$

16. $-5+3(-2)-(-12)$
$=-5+(-6)-(-12)$
$=-11-(-12)$
$=-11+(+12)$
$=1$

17. $2-(6-8)-(-5)^2$
$=2-(-2)-(-5)^2$
$=2-(-2)-(25)$
$=2+(+2)-25$
$=4+(-25)$
$=-21$

18. $8k-3m;\ k=-4,\ m=2$
Replace k with -4. Replace m with 2.

$$\begin{aligned} 8k-3m &= 8(-4)-3(2) \\ &= -32-6 \\ &= -32+(-6) \\ &= -38 \end{aligned}$$

19. $8k-3m;\ k=3,\ m=-1$
Replace k with 3. Replace m with -1.

$$\begin{aligned} 8k-3m &= 8(3)-3(-1) \\ &= 24-(-3) \\ &= 24+(+3) \\ &= 27 \end{aligned}$$

20. When evaluating, you are given specific values to replace each variable. When solving an equation, you are not given the value of the variable. You must find a value that "works"; that is, when your solution is substituted for the variable, the two sides of the equation are equal.

21. $A=\frac{1}{2}bh;\ b=20, h=11$

$$\begin{aligned} A &= \frac{1}{2}(20)(11) \\ &= 10(11) \\ &= 110 \end{aligned}$$

22.

$$\begin{aligned} x-9 &= -4 \\ x-9+9 &= -4+9 \\ x &= 5 \end{aligned}$$

The solution is 5.

23.

$$\begin{aligned} 30 &= -1+r \\ 30+1 &= -1+r+1 \\ 31 &= r \end{aligned}$$

The solution is 31.

24.

$$\begin{aligned} 3t-8t &= -25 \\ (3-8)t &= -25 \\ -5t &= -25 \\ \frac{-5t}{-5} &= \frac{-25}{-5} \\ t &= 5 \end{aligned}$$

The solution is 5.

25.

$$\begin{aligned} \frac{p}{5} &= -3 \\ \frac{1}{5}p &= -3 \\ \left(\frac{5}{1}\right)\cdot\left(\frac{1}{5}\right)p &= -3\cdot 5 \\ p &= -15 \end{aligned}$$

The solution is -15.

26.

$$\begin{aligned} -15 &= 3(a-2) \\ -15 &= 3\cdot a-3\cdot 2 \\ -15 &= 3a-6 \\ -15+6 &= 3a-6+6 \\ -9 &= 3a \\ \frac{-9}{3} &= \frac{3a}{3} \\ -3 &= a \end{aligned}$$

The solution is -3.

27.

$$\begin{aligned} 3m-5 &= 7m-13 \\ 3m-7m-5 &= 7m-7m-13 \\ (3-7)m-5 &= -13 \\ -4m-5 &= -13 \\ -4m-5+5 &= -13+5 \\ -4m &= -8 \\ \frac{-4m}{-4} &= \frac{-8}{-4} \\ m &= 2 \end{aligned}$$

The solution is 2.

28. *Step 2*
Let x represent the shorter piece and $x+4$ the longer piece.

Step 3
$x+x+4=118$

Step 4

$$\begin{aligned} 2x+4 &= 118 \\ 2x+4-4 &= 118-4 \\ 2x &= 114 \\ \frac{2x}{2} &= \frac{114}{2} \\ x &= 57 \end{aligned}$$

Step 5
The length of the shorter piece is 57 cm, and the length of the longer piece is $57+4=$ 61 cm.

Step 6
Check: 61 cm is 4 cm longer than 57 cm and the sum of 57 cm and 61 cm is 118 cm. *True*

29. *Step 2*
Let w represent the width and $4w$ the length.

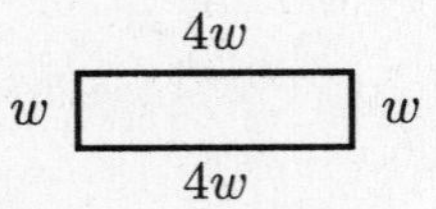

$P = 2 \cdot l + 2 \cdot w$

Step 3
$420 = 2(4w) + 2w$

Step 4
$$\begin{aligned} 420 &= 8w + 2w \\ 420 &= 10w \\ \frac{420}{10} &= \frac{10w}{10} \\ 42 &= w \end{aligned}$$

Step 5
The width is 42 ft. The length is $4(42 \text{ ft}) = 168$ ft.

Step 6
Check:
$$\begin{aligned} 420 &= 2(4w) + 2w \\ 420 &= 2(4 \cdot 42) + 2(42) \\ 420 &= 2(168) + 84 \\ 420 &= 336 + 84 \\ 420 &= 420 \quad \textit{True} \end{aligned}$$

30. *Step 2*
Let h represent Marcella's time and $h - 3$ Tim's time.

Hours Marcella spent	+	hours Tim spent	=	19 hours
↓	↓	↓	↓	↓
h	+	$h - 3$	=	19

Step 3
$h + h - 3 = 19$

Step 4
$$\begin{aligned} 2h - 3 &= 19 \\ 2h - 3 + 3 &= 19 + 3 \\ 2h &= 22 \\ \frac{2h}{2} &= \frac{22}{2} \\ h &= 11 \end{aligned}$$

Step 5
Marcella spent 11 hours redecorating their living room and Tim spent $11 - 3 = 8$ hours.

Step 6
Check: Tim's time is 3 hours less than Marcella's time and the sum of their times is 19 hours. *True*

Cumulative Review Exercises (Chapters 1–9)

1. *Estimate:*
$$\begin{array}{r} 9 \\ 1 \\ +\,40 \\ \hline 50 \end{array}$$
Exact:
$$\begin{array}{r} 8.700 \\ 0.902 \\ +\,41.000 \\ \hline 50.602 \end{array}$$

2. *Estimate:*
$$\begin{array}{r} 6 \\ \times\,50 \\ \hline 300 \end{array}$$
Exact:
$$\begin{array}{r} 6.27 \\ \times\ \ 49.2 \\ \hline 1\,254 \\ 56\,43 \\ 250\,8 \\ \hline 308.484 \end{array}$$

3. *Estimate:*
$$\begin{array}{r} 5\,0\,0 \\ 80\overline{)4\,0,\,0\,0\,0} \end{array}$$
Exact:
$$\begin{array}{r} 5\,0\,3 \\ 78\overline{)3\,9,\,2\,3\,4} \\ 3\,9\,0 \\ \hline 2\,3 \\ 0 \\ \hline 2\,3\,4 \\ 2\,3\,4 \\ \hline 0 \end{array}$$

4. *Estimate:*
$$\begin{array}{r} 4 \\ -\,3 \\ \hline 1 \end{array}$$
Exact:
$$\begin{array}{rcccc} 3\frac{3}{5} & = & \frac{18}{5} & = & \frac{72}{20} \\ -\,2\frac{3}{4} & = & \frac{11}{4} & = & \frac{55}{20} \\ \hline & & & & \frac{17}{20} \end{array}$$

5. *Exact:*
$$5\frac{5}{6} \cdot \frac{9}{10} = \frac{\overset{7}{\cancel{35}}}{\underset{2}{\cancel{6}}} \cdot \frac{\overset{3}{\cancel{9}}}{\underset{2}{\cancel{10}}} = \frac{21}{4} = 5\frac{1}{4}$$

Estimate: $6 \cdot 1 = 6$

6. *Exact:*
$$4\frac{1}{6} \div 1\frac{2}{3} = \frac{25}{6} \div \frac{5}{3} = \frac{\overset{5}{\cancel{25}}}{\underset{2}{\cancel{6}}} \cdot \frac{\overset{1}{\cancel{3}}}{\underset{1}{\cancel{5}}} = \frac{5}{2} = 2\frac{1}{2}$$

Estimate: $4 \div 2 = 2$

7. $17 - 8.094$

$$\begin{array}{r} 17.000 \\ -\ 8.094 \\ \hline 8.906 \end{array}$$

8. $(1309)(408)$

$$\begin{array}{r} 1309 \\ \times\ \ 408 \\ \hline 10\ 472 \\ 523\ 60 \\ \hline 534,072 \end{array}$$

9. $4.06 \div 0.072$

$$\begin{array}{r} 5\,6.\,3\,8 \\ 0.072_\wedge \overline{)4.\,0\,6\,0_\wedge\,0\,0} \\ 3\,6\,0 \\ \hline 4\,6\,0 \\ 4\,3\,2 \\ \hline 2\,8\,0 \\ 2\,1\,6 \\ \hline 6\,4\,0 \\ 5\,7\,6 \\ \hline 6\,4 \end{array}$$

56.38 rounded to the nearest tenth is 56.4.

10. $-12 + 7 = -5$

11. $-5(-8) = 40$

12. $-3 - (-7) = -3 + (+7) = 4$

13. $3.2 + (-4.5) = -1.3$

14. $\frac{30}{-6} = -5$

15. $\frac{1}{4} - \frac{3}{4} = \frac{1}{4} + \left(-\frac{3}{4}\right) = -\frac{2}{4} = -\frac{1}{2}$

16. $45 \div \sqrt{25} - 2(3) + (10 \div 5)$

$= 45 \div \sqrt{25} - 2(3) + 2$
$= 45 \div 5 - 2(3) + 2$
$= 9 - 2(3) + 2$
$= 9 - 6 + 2$
$= 3 + 2$
$= 5$

17. $-6 - (4 - 5) + (-3)^2$

$= -6 - (-1) + (-3)^2$
$= -6 - (-1) + 9$
$= -6 + 1 + 9$
$= -5 + 9$
$= 4$

18. $\frac{3}{10} \_\_ \frac{4}{15} \quad LCD = 30$

$\frac{3}{10} = \frac{9}{30}$ and $\frac{4}{15} = \frac{8}{30}$

$\frac{9}{30} > \frac{8}{30}$, so $\frac{3}{10}\ \boxed{>}\ \frac{4}{15}$.

19. $0.7072 < 0.72$ because $0.7072 < 0.7200$.

20. $-5 < -2$ because -5 is to the *left* of -2 on a number line.

21. $8\% = 0.08 = \frac{8}{100} = \frac{8 \div 4}{100 \div 4} = \frac{2}{25}$

22. $4\frac{1}{2} = 4.5 = 450\%$

23. **(a)** IA to MI

$\frac{32}{16} = \frac{32 \div 16}{16 \div 16} = \frac{2}{1}$

(b) CA to OH

$\frac{5}{15} = \frac{5 \div 5}{15 \div 5} = \frac{1}{3}$

(c) GA to IN

$\frac{20}{20} = \frac{20 \div 20}{20 \div 20} = \frac{1}{1}$

(d) FL to OK

$\frac{48}{52} = \frac{48 \div 4}{52 \div 4} = \frac{12}{13}$

24. $\frac{x}{12} = \frac{1.5}{45}$

$45 \cdot x = 12 \cdot 1.5$

$\frac{45 \cdot x}{45} = \frac{18}{45}$

$x = \frac{18}{45} = \frac{18 \div 9}{45 \div 9} = \frac{2}{5} = 0.4$

25. $\frac{350}{x} = \frac{3}{2}$

$3 \cdot x = 2 \cdot 350$

$\frac{3x}{3} = \frac{700}{3}$

$x \approx 233.33$

26. $\frac{38}{190} = \frac{9}{x}$ **OR** $\frac{1}{5} = \frac{9}{x}$

$x = 5 \cdot 9 = 45$

27. whole is 3000; percent is 0.5; part is unknown. Use the percent proportion.

$\frac{x}{3000} = \frac{0.5}{100}$

$100 \cdot x = 3000 \cdot 0.5$

$\frac{100 \cdot x}{100} = \frac{1500}{100}$

$x = 15$

0.5% of 3000 students is 15 students.

28. part is 6.8; whole is 12.5; percent is unknown. Use the percent proportion.

$$\frac{6.8}{12.5} = \frac{x}{100}$$
$$12.5 \cdot x = 6.8 \cdot 100$$
$$\frac{12.5 \cdot x}{12.5} = \frac{680}{12.5}$$
$$x = 54.4$$

54.4% of 12.5 miles is 6.8 miles.

29. part is 90; percent is 180; whole is unknown. Use the percent proportion.

$$\frac{90}{x} = \frac{180}{100} \quad \textbf{OR} \quad \frac{90}{x} = \frac{9}{5}$$
$$9 \cdot x = 90 \cdot 5$$
$$\frac{9 \cdot x}{9} = \frac{450}{9}$$
$$x = 50$$

90 cars is 180% of 50 cars.

30. part is 5.8; whole is 145; percent is unknown. Use the percent proportion.

$$\frac{5.8}{145} = \frac{x}{100}$$
$$145 \cdot x = 5.8 \cdot 100$$
$$\frac{145 \cdot x}{145} = \frac{580}{145}$$
$$x = 4$$

$5.80 is 4% of $145.

31. $3\frac{1}{2}$ gal to quarts

$$\frac{3\frac{1}{2}\ \cancel{\text{gallons}}}{1} \cdot \frac{4 \text{ quarts}}{1\ \cancel{\text{gallon}}} = \frac{7}{\underset{1}{\cancel{2}}} \cdot \frac{\overset{2}{\cancel{4}}}{1} \text{ quarts}$$
$$= 14 \text{ quarts}$$

32. 72 hr to days

$$\frac{72\ \cancel{\text{hours}}}{1} \cdot \frac{1 \text{ day}}{24\ \cancel{\text{hours}}} = \frac{72 \cdot 1 \text{ day}}{24} = 3 \text{ days}$$

33. 3.7 L to milliliters

$$\frac{3.7\ \cancel{\text{L}}}{1} \cdot \frac{1000 \text{ mL}}{1\ \cancel{\text{L}}} = 3.7 \cdot 1000 \text{ mL}$$
$$= 3700 \text{ mL}$$

34. 40 cm to meters

$$\frac{40\ \cancel{\text{cm}}}{1} \cdot \frac{1 \text{ m}}{100\ \cancel{\text{cm}}} = \frac{40 \text{ m}}{100} = 0.4 \text{ m}$$

35. The building is 15 m high.

36. Rita took 15 mL of cough syrup.

37. Bruce walked 2 km to work.

38. The robin weighs 100 g.

39. Normal body temperature is 37°C (= 98.6°F). A comfortable room temperature is 20°C (= 68°F).

40. The shape is a square.

$$P = 4 \cdot s$$
$$= 4 \cdot 2\tfrac{1}{4} \text{ ft}$$
$$= \frac{\overset{1}{\cancel{4}}}{1} \cdot \frac{9}{\underset{1}{\cancel{4}}} \text{ ft} = 9 \text{ ft}$$

$$A = s \cdot s$$
$$= 2\tfrac{1}{4} \text{ ft} \cdot 2\tfrac{1}{4} \text{ ft}$$
$$= \frac{9}{4} \text{ ft} \cdot \frac{9}{4} \text{ ft}$$
$$= \frac{81}{16} \text{ ft}^2$$
$$= 5\tfrac{1}{16} \text{ ft}^2 \text{ or } 5.0625 \text{ ft}^2$$

41. The shape is a circle.

$$C = \pi \cdot d$$
$$\approx 3.14 \cdot 9 \text{ mm}$$
$$= 28.26 \text{ mm}$$

$$A = \pi r^2$$
$$\approx 3.14 \cdot 4.5 \text{ mm} \cdot 4.5 \text{ mm}$$
$$= 63.585 \text{ mm}^2$$

42. The shape is a right triangle.

$$P = a + b + c$$
$$= 24 \text{ mi} + 7 \text{ mi} + 25 \text{ mi}$$
$$= 56 \text{ mi}$$

$$A = \frac{1}{2}bh$$
$$= \frac{1}{2} \cdot 24 \text{ mi} \cdot 7 \text{ mi}$$
$$= 12 \text{ mi} \cdot 7 \text{ mi}$$
$$= 84 \text{ mi}^2$$

43. The shape is a trapezoid.

$$P = 2.4 \text{ cm} + 1.15 \text{ cm} + 1.4 \text{ cm} + 1.5 \text{ cm}$$
$$= 6.45 \text{ cm}$$

$$A = 0.5 \cdot h(b + B)$$
$$= 0.5 \cdot 1.15 \text{ cm}(1.4 \text{ cm} + 2.4 \text{ cm})$$
$$= 0.5 \cdot 1.15 \text{ cm}(3.8 \text{ cm})$$
$$= 2.185 \text{ cm}^2$$

44. The shape is a parallelogram.

$$P = 2 \cdot l + 2 \cdot w$$
$$= 2 \cdot 10 \text{ ft} + 2 \cdot 8 \text{ ft}$$
$$= 20 \text{ ft} + 16 \text{ ft}$$
$$= 36 \text{ ft}$$

$$\begin{aligned} A &= bh \\ &= 10 \text{ ft} \cdot 7 \text{ ft} \\ &= 70 \text{ ft}^2 \end{aligned}$$

45. $$\begin{aligned} P &= 48 \text{ m} + 46 \text{ m} + 22 \text{ m} + 22 \text{ m} \\ &\quad + 26 \text{ m} + 24 \text{ m} \\ &= 188 \text{ m} \end{aligned}$$

To find the area, draw a horizontal line to break the figure into two parts.
A rectangle with length 48 m and width 24 m:

$$\begin{aligned} A &= l \cdot w \\ &= 48 \text{ m} \cdot 24 \text{ m} \\ &= 1152 \text{ m}^2 \end{aligned}$$

A square with side 22 m:

$$\begin{aligned} A &= s^2 \\ &= 22 \text{ m} \cdot 22 \text{ m} \\ &= 484 \text{ m}^2 \end{aligned}$$

Total area:
$1152 \text{ m}^2 + 484 \text{ m}^2 = 1636 \text{ m}^2$

46. Hypotenuse: 20 yd, leg: 15 yd

$$\begin{aligned} y &= \sqrt{(\text{hypotenuse})^2 - (\text{leg})^2} \\ &= \sqrt{(20)^2 - (15)^2} \\ &= \sqrt{400 - 225} \\ &= \sqrt{175} \\ &\approx 13.2 \text{ yd} \end{aligned}$$

47. Set up a ratio of corresponding sides.

$\dfrac{9 \text{ in.}}{18 \text{ in.}}$, which reduces to $\dfrac{1}{2}$

Now write a proportion to find x.

$$\begin{aligned} \frac{7}{x} &= \frac{1}{2} \\ 1 \cdot x &= 7 \cdot 2 \\ x &= 14 \text{ in.} \end{aligned}$$

48. $$\begin{aligned} -20 &= 6 + y \\ -20 + (-6) &= 6 + y + (-6) \\ -26 &= y \end{aligned}$$

The solution is -26.

49. $$\begin{aligned} -2t - 6t &= 40 \\ -2t + (-6t) &= 40 \\ -8t &= 40 \\ \frac{-8t}{-8} &= \frac{40}{-8} \\ t &= -5 \end{aligned}$$

The solution is -5.

50. $$\begin{aligned} 3x + 5 &= 5x - 11 \\ 3x + 5 + 11 &= 5x - 11 + 11 \\ 3x + 16 &= 5x \\ 3x + 16 - 3x &= 5x - 3x \\ 16 &= 2x \\ \frac{16}{2} &= \frac{2x}{2} \\ 8 &= x \end{aligned}$$

The solution is 8.

51. $$\begin{aligned} 6(p + 3) &= -6 \\ 6(p) + 6(3) &= -6 \\ 6p + 18 &= -6 \\ 6p + 18 - 18 &= -6 - 18 \\ 6p &= -24 \\ \frac{6p}{6} &= \frac{-24}{6} \\ p &= -4 \end{aligned}$$

The solution is -4.

52. Let x represent the number.

40	added to	4 times a number	is	0.
↓	↓	↓	↓	↓
40	$+$	$4x$	$=$	0

$$\begin{aligned} 40 + 4x &= 0 \\ 40 - 40 + 4x &= 0 - 40 \\ 4x &= -40 \\ \frac{4x}{4} &= \frac{-40}{4} \\ x &= -10 \end{aligned}$$

The number is -10.

Check: If 40 is added to four times $-10[= -40]$, the result is zero. *True*

53. *Step 2*
Let m represent the money Reggie will receive and $m + 300$ the money Donald will receive.

Step 3
$m + m + 300 = 1000$

Step 4
$$\begin{aligned} 2m + 300 &= 1000 \\ 2m + 300 - 300 &= 1000 - 300 \\ 2m &= 700 \\ \frac{2m}{2} &= \frac{700}{2} \\ m &= 350 \end{aligned}$$

Step 5
Reggie will receive \$350 and Donald will receive \$350 + \$300 = \$650.

Step 6
Check: \$650 is \$300 more than \$350 and the sum of \$650 and \$350 is \$1000. *True*

54. *Step 2*
Make a drawing.

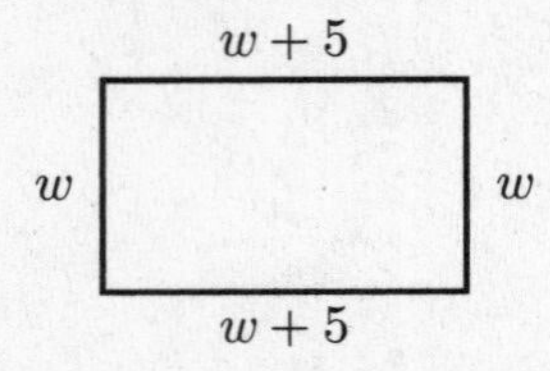

Let w represent the width and $w+5$ the length.
$P = 2\cdot l + 2\cdot w$

Step 3
$82 = 2\cdot(w+5) + 2\cdot w$

Step 4
$$\begin{aligned} 82 &= 4w + 10 \\ 82 - 10 &= 4w + 10 - 10 \\ 72 &= 4w \\ \frac{72}{4} &= \frac{4w}{4} \\ 18 &= w \end{aligned}$$

Step 5
The width is 18 cm and the length is $18 + 5 = 23$ cm.

Step 6
Check:
$$\begin{aligned} 82 &= 2\cdot 23 + 2\cdot 18 \\ 82 &= 46 + 36 \\ 82 &= 82 \quad \textit{True} \end{aligned}$$

55. Find the cost for two CDs.
$\$14.98 + \$14.98 = \$29.96$

$$\begin{aligned} \text{Sales Tax} &= \$29.96\cdot 6\tfrac{1}{2}\% \\ &= \$29.96(0.065) \\ &\approx \$1.95 \end{aligned}$$

$\text{Total cost} = \$29.96 + \$1.95 = \$31.91$

56. Multiply $3\frac{1}{3}$ cups times $2\frac{1}{2}$.

$$3\frac{1}{3}\cdot 2\frac{1}{2} = \frac{\overset{5}{\cancel{10}}}{3}\cdot\frac{5}{\underset{1}{\cancel{2}}} = \frac{25}{3} = 8\frac{1}{3}$$

He needs $8\frac{1}{3}$ cups of tomato sauce.

57. Part is 2480; whole is 2000; percent is unknown. Use the percent proportion.

$$\frac{2480}{2000} = \frac{x}{100} \quad \textbf{OR} \quad \frac{31}{25} = \frac{x}{100}$$
$$\begin{aligned} 25\cdot x &= 31\cdot 100 \\ \frac{25\cdot x}{25} &= \frac{3100}{25} \\ x &= 124 \end{aligned}$$

2480 pounds is 124% of 2000 pounds.

58. First, convert grams to kilograms.
$720 \text{ g} = 0.72 \text{ kg}$

Next, multiply the weight of the chicken by the cost per kilogram.

$$\begin{array}{r} \$5.97 \\ \times\ 0.72 \\ \hline 1194 \\ 4\ 179 \\ \hline \$4.2984 \end{array}$$

She paid \$4.30 (rounded) for the chicken.

59. Find the volume of the crate.

$$\begin{aligned} V &= l\cdot w\cdot h \\ &= 2.4 \text{ m}\cdot 1.2 \text{ m}\cdot 1.2 \text{ m} \\ &= 3.456 \text{ m}^3 \end{aligned}$$

The crate's volume is
$4 - 3.456 \text{ km} = 0.544 \text{ m}^3$ less than 4 m^3.

60. Use the formula to find the perimeter of a rectangle.

$$\begin{aligned} P &= 2\cdot\text{length} + 2\cdot\text{width} \\ 35 &= 2l + 2(6\tfrac{1}{2}) \\ 35 &= 2l + 13 \\ 35 - 13 &= 2l + 13 - 13 \\ 22 &= 2l \\ 11 &= l \end{aligned}$$

The length of the plot is 11 feet.

61. Write a proportion.

$$\begin{aligned} \frac{25}{14} &= \frac{x}{30} \\ 14\cdot x &= 25\cdot 30 \\ 14x &= 750 \\ \frac{14x}{14} &= \frac{750}{14} \\ x &\approx 53.571 \end{aligned}$$

It will take Rich approximately 54 minutes to read 30 pages.

62. Jackie: $\dfrac{364 \text{ mi}}{14.5 \text{ gal}} \approx 25.1 \text{ mpg}$

Maya: $\dfrac{406 \text{ mi}}{16.3 \text{ gal}} \approx 24.9 \text{ mpg}$

Naomi: $\dfrac{300 \text{ mi}}{11.9 \text{ gal}} \approx 25.2 \text{ mpg} (*)$

Naomi's car had the highest number of miles per gallon with 25.2 miles per gallon (rounded).

CHAPTER 10 STATISTICS

10.1 Circle Graphs

10.1 Margin Exercises

1. **(a)** The circle graph shows that the greatest number of hours is spent sleeping.

(b)
$$\begin{array}{rl} 6 & \text{hours working} \\ -4 & \text{hours studying} \\ \hline 2 & \text{hours} \end{array}$$

Two more hours are spent working than studying.

(c)
$$\begin{array}{rl} 4 & \text{hours studying} \\ 6 & \text{hours working} \\ +3 & \text{hours attending class} \\ \hline 13 & \text{hours} \end{array}$$

Thirteen hours are spent studying, working, and attending classes.

2. **(a)** Hours spent driving to whole day:

$$\frac{2 \text{ hours}}{24 \text{ hours}} = \frac{2 \div 2}{24 \div 2} = \frac{1}{12}$$

(b) Hours spent sleeping to whole day:

$$\frac{7 \text{ hours}}{24 \text{ hours}} = \frac{7}{24}$$

(c) Hours spent attending class and studying to whole day:

$$\begin{array}{rl} 3 & \text{hours attending class} \\ +4 & \text{hours studying} \\ \hline 7 & \text{hours} \end{array}$$

$$\frac{7 \text{ hours}}{24 \text{ hours}} = \frac{7}{24}$$

(d) Hours spent driving and working to whole day:

$$\begin{array}{rl} 2 & \text{hours driving} \\ +6 & \text{hours working} \\ \hline 8 & \text{hours} \end{array}$$

$$\frac{8 \text{ hours}}{24 \text{ hours}} = \frac{8 \div 8}{24 \div 8} = \frac{1}{3}$$

3. **(a)** Hours spent in class to hours spent studying:

$$\frac{3 \text{ hours}}{4 \text{ hours}} = \frac{3}{4}$$

(b) Hours spent working to hours spent sleeping:

$$\frac{6 \text{ hours}}{7 \text{ hours}} = \frac{6}{7}$$

(c) Hours spent driving to hours spent working:

$$\frac{2 \text{ hours}}{6 \text{ hours}} = \frac{2 \div 2}{6 \div 2} = \frac{1}{3}$$

(d) Hours spent in class to hours spent for "other:"

$$\frac{3 \text{ hours}}{2 \text{ hours}} = \frac{3}{2}$$

4. Use the percent equation:

$$\text{part} = \text{percent} \cdot \text{whole}$$

(a) whole = 36 billion; percent = 5.1% = 0.051; find the part.

$$x = 0.051 \cdot 36 \text{ billion} = 1.836 \text{ billion}$$

The amount spent on corn chips is $1.836 billion or $1,836,000,000.

(b) whole = 36 billion; percent = 14.4% = 0.144; find the part.

$$x = 0.144 \cdot 36 \text{ billion} = 5.184 \text{ billion}$$

The amount spent on miscellaneous is $5.184 billion or $5,184,000,000.

(c) whole = 36 billion; percent = 5.3% = 0.053; find the part.

$$x = 0.053 \cdot 36 \text{ billion} = 1.908 \text{ billion}$$

The amount spent on cakes/pies is $1.908 billion or $1,908,000,000.

(d) whole = 36 billion; percent = 23.8% = 0.238; find the part.

$$x = 0.238 \cdot 36 \text{ billion} = 8.568 \text{ billion}$$

The amount spent on cookies/crackers is $8.568 billion or $8,568,000,000.

5. **(a)** Ages 10–11:

$$(360°)(25\%) = (360°)(0.25) = 90°$$

(b) Ages 12–13:

$$(360°)(25\%) = (360°)(0.25) = 90°$$

(c) Ages 14–15:

$$(360°)(15\%) = (360°)(0.15) = 54°$$

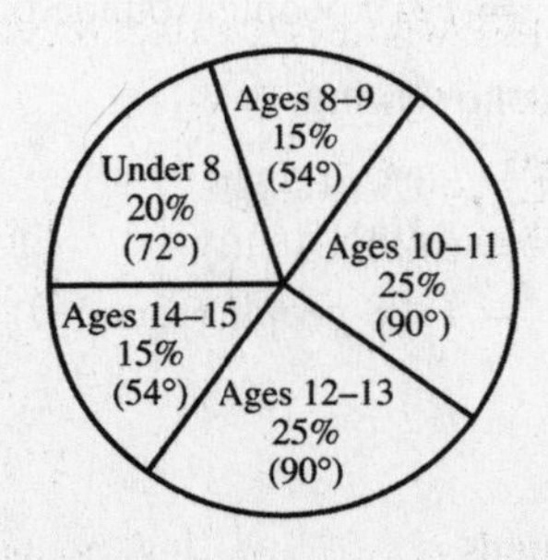

10.1 Section Exercises

1. The total cost of adding the art studio is $12,100 + $9800 + $2000 + $900 + $1800 + $3000 + $2400 = $32,000

3. $\frac{\text{cost of materials}}{\text{total remodeling cost}} = \frac{\$9800}{\$32,000} = \frac{9800 \div 200}{32,000 \div 200} = \frac{49}{160}$

5. $\frac{\text{cost of carpentry}}{\text{cost of window coverings}} = \frac{\$12,100}{\$900} = \frac{121}{9}$

7. The smallest number and the smallest sector represent the reason "Don't know."

9. Total people $= 1740 + 1200 + 1140 + 180 + 1020 + 720 = 6000$

"Quicker" to total:

$$\frac{720}{6000} = \frac{720 \div 240}{6000 \div 240} = \frac{3}{25}$$

11. "Less work/No clean up" to "Atmosphere":

$$\frac{1020}{1200} = \frac{1020 \div 60}{1200 \div 60} = \frac{17}{20}$$

13. "Wanted food they couldn't cook at home" to "Less work/No clean up":

$$\frac{1740}{1020} = \frac{1740 \div 60}{1020 \div 60} = \frac{29}{17}$$

15. To determine how many people prefer onions for their hot dog topping, use the percent equation.

$$\begin{aligned} \text{part} &= \text{percent} \cdot \text{whole} \\ x &= (0.05)(3200) \\ &= 160 \text{ people} \end{aligned}$$

17. Sauerkraut = 3% of 3200

$$\begin{aligned} x &= (0.03)(3200) \\ &= 96 \text{ people} \end{aligned}$$

19. Mustard = 30% of 3200

$$\begin{aligned} x &= (0.30)(3200) \\ &= 960 \text{ people} \end{aligned}$$

21. Consume none or very few milk products:

$$\begin{aligned} x &= 22\% \text{ of } 5540 \\ &= (0.22)(5540) \\ &= 1219 \text{ people (rounded)} \end{aligned}$$

23. Consume one serving:

$$\begin{aligned} x &= 33\% \text{ of } 5540 \\ &= (0.33)(5540) \\ &= 1828 \text{ people (rounded)} \end{aligned}$$

25. Don't know:

$$\begin{aligned} x &= 5\% \text{ of } 5540 \\ &= (0.05)(5540) \\ &= 277 \text{ people (rounded)} \end{aligned}$$

27. First find the percent of the total that is to be represented by each item. Next, multiply the percent by 360° to find the size of each sector. Finally, use a protractor to draw each sector.

29. 25% of total is rent.

$$\begin{aligned} \text{Degrees of a circle} &= 25\% \text{ of } 360° \\ &= (0.25)(360) \\ &= 90° \end{aligned}$$

31. $\text{Percent for clothing} = \frac{\$546}{\$5460} = 0.10 = 10\%$

$$\begin{aligned} \text{Degrees of a circle} &= 10\% \text{ of } 360° \\ &= (0.10)(360°) \\ &= 36° \end{aligned}$$

33. 15% of total is tuition.

$$\begin{aligned} \text{Degrees of a circle} &= 15\% \text{ of } 360° \\ &= (0.15)(360°) \\ &= 54° \end{aligned}$$

35. $\text{Percent for entertainment} = \frac{54°}{360°} = 0.15 = 15\%$

37. **(a)** Total sales = $12,500 + $40,000 + $60,000 + $50,000 + $37,500 = $200,000

(b) Adventure classes = $12,500

$\text{percent of total} = \frac{12,500}{200,000} = 0.0625 = 6.25\%$

$\text{number of degrees} = (0.0625)(360°) = 22.5°$

Grocery and provision sales = $40,000

$\text{percent of total} = \frac{40,000}{200,000} = 0.2 = 20\%$

$\text{number of degrees} = (0.2)(360°) = 72°$

Equipment rentals = $60,000

$\text{percent of total} = \frac{60,000}{200,000} = 0.3 = 30\%$

$\text{number of degrees} = (0.3)(360°) = 108°$

Rafting tours = $50,000

$\text{percent of total} = \frac{50,000}{200,000} = 0.25 = 25\%$

$\text{number of degrees} = (0.25)(360°) = 90°$

Equipment sales = $37,500

$\text{percent of total} = \frac{37,500}{200,000} = 0.1875 = 18.75\%$

$\text{number of degrees} = (0.1875)(360°) = 67.5°$

(c)

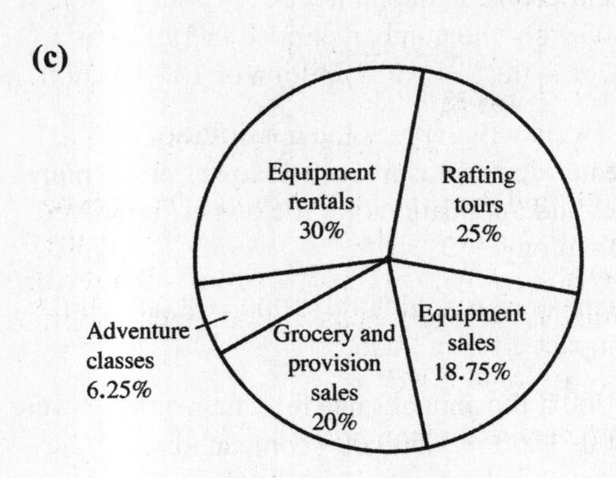

39. **(a)** Percent for Side $= \frac{2464}{4488} \approx 0.55 = 55\%$

Degrees for Side $= (0.55)(360°) = 198°$

Percent for Not sure $= \frac{220}{4488} \approx 0.05 = 5\%$

Degrees for Not sure $= (0.05)(360°) = 18°$

Percent for Stomach $= \frac{536}{4488} \approx 0.12 = 12\%$

Degrees for Stomach $= (0.12)(360°) \approx 43°$

Percent for Varies $= \frac{520}{4488} \approx 0.12 = 12\%$

Degrees for Varies $= (0.12)(360°) \approx 43°$

Percent for Back $= \frac{748}{4488} \approx 0.17 = 17\%$

Degrees for Back $= (0.17)(360°) \approx 61°$

Sleeping Position	*Number of Americans*	*Percent of Total*	*Number of Degrees*
Side	2464	55% (≈)	198°
Not sure	220	5% (≈)	18°
Stomach	536	12% (≈)	43° (≈)
Varies	520	12% (≈)	43° (≈)
Back	748	17% (≈)	61° (≈)

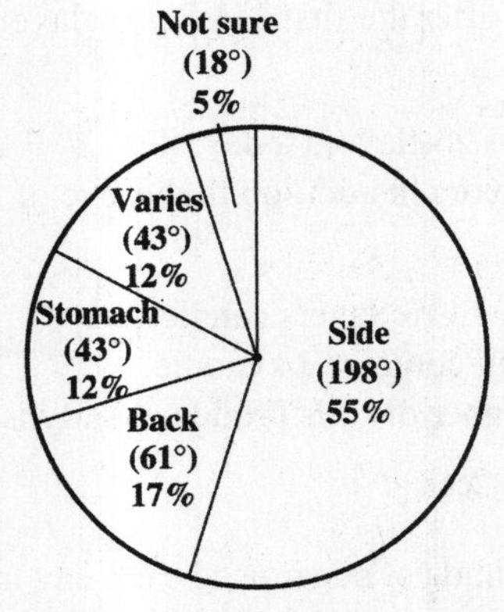

(b) Total percent =

$55\% + 5\% + 12\% + 12\% + 17\% = 101\%$

No. The total is 101% due to rounding.

(c) Total degrees =

$198° + 18° + 43° + 43° + 61° = 363°$

No. The total is 363° due to rounding.

10.2 Bar Graphs and Line Graphs

10.2 Margin Exercises

1. **(a)** The bar for 2002 rises to 25, showing that the number of college graduates was

$$25 \times 1000 = 25{,}000.$$

(b) 2003: $20 \times 1000 = 20{,}000$ graduates

(c) 2005: $30 \times 1000 = 30{,}000$ graduates

(d) 2006: $35 \times 1000 = 35{,}000$ graduates

2. **(a)** 1st quarter, 2005:

$$4 \times 1000 = 4000$$

1st quarter, 2006:

$$3 \times 1000 = 3000$$

(b) 3rd quarter, 2005:

$$7 \times 1000 = 7000$$

3rd quarter, 2006:

$$8 \times 1000 = 8000$$

(c) 4th quarter, 2005:

$$5 \times 1000 = 5000$$

4th quarter, 2006:

$$4 \times 1000 = 4000$$

(d) The tallest bar rose to 8, showing 8000 installations for the 3rd quarter of 2006.

3. **(a)** June

The dot is at 5.5 on the vertical axis.

$$5.5 \times 10{,}000 = 55{,}000$$

(b) May: $3 \times 10{,}000 = 30{,}000$

(c) April: $4 \times 10{,}000 = 40{,}000$

(d) July: $6 \times 10{,}000 = 60{,}000$

4. **(a)** The number of minivans sold:

2002: $30 \times 1000 = 30{,}000$
2004: $40 \times 1000 = 40{,}000$
2005: $20 \times 1000 = 20{,}000$
2006: $15 \times 1000 = 15{,}000$

(b) The number of SUVs sold:

2002: $10 \times 1000 = 10{,}000$
2003: $20 \times 1000 = 20{,}000$
2004: $30 \times 1000 = 30{,}000$
2005: $50 \times 1000 = 50{,}000$

(c) The number of SUVs sold was greater than the number of minivans sold when the red line was above the blue line. The first full year in which this happened was 2005.

10.2 Section Exercises

1. The country which had the highest percent of income spent on food was India, which spent 51.4% of household income.

3. The countries in which less than 15% of household income is spent, on average, for food are USA, Britain, and Australia.

5. What is 33.2% of 365 days?

$$\begin{aligned} x &= (0.332)(365 \text{ days}) \\ &= 121 \text{ days (rounded)} \end{aligned}$$

7. May had the greatest number of unemployed workers. The total was 10,000 unemployed workers.

9. Unemployed workers in February of 2006 = 7000
Unemployed workers in February of 2005 = 5500

$$7000 - 5500 = 1500$$

There were 1500 more workers unemployed in February of 2006.

11. The number of unemployed workers increased from 5500 in February of 2005 to 9500 in April of 2006. The increase was $9500 - 5500 = 4000$ workers.

13. 150,000 gallons of supreme unleaded gasoline were sold in 2002.

15. The greatest difference occurred in 2002. The difference was $400{,}000 - 150{,}000 = 250{,}000$ gallons.

17. 700,000 gallons of supreme unleaded gasoline were sold in 2006. 150,000 gallons of supreme unleaded gasoline were sold in 2002.

$$700{,}000 - 150{,}000 = 550{,}000$$

There was an increase of 550,000 gallons of supreme unleaded gasoline sales.

19. The number of PCs shipped in 1990 was 24.1 million or 24,100,000.

21. The increase in the number of PCs shipped in 2005 from the number shipped in 1995 was $197.4 - 62.3 = 135.1$ million or 135,100,000.

23. Answers will vary. Some possibilities are: greater demand as a result of lower price; more uses and applications for the owner; improved technology.

25. Chain store A sold $3000 \cdot 1000 = 3{,}000{,}000$ compact discs in 2006.

27. In 2004, the annual sales for Chain store A were $1500 \cdot 1000 = 1{,}500{,}000$ compact discs.

29. Chain store B sold $3500 \cdot 1000 = 3{,}500{,}000$ compact discs in 2005.

31. Probably store B with greater sales. Predicted sales might be 4,500,000 CDs to 5,000,000 CDs in 2007.

33. A single bar or a single line must be used for each set of data. To show multiple sets of data, multiple sets of bars or lines must be used.

35. The total sales in 2006 were $40 \cdot \$1000 = \$40,000$.

37. The total sales in 2004 were $25 \cdot \$1000 = \$25,000$.

39. The profit in 2005 was $5 \cdot \$1000 = \$5000$.

41. The answers will vary. Some possibilities are that the decrease in sales may have resulted from poor service or greater competition. The increase in sales may have been a result of more advertising or better service.

43. The five-flavor roll was invented in 1935. The Pep-O-Mint flavor was invented in 1912. $1935 - 1912 = 23$, so the five-flavor roll was invented 23 years after the first LifeSaver flavor was invented.

44. There are a total of 25 flavors today. $25 - 5 = 20$, so there are 20 flavors in addition to the five of the five-flavor roll.

45. Each day 3,000,000 LifeSavers candies are produced. Each roll contains 14 candies. Therefore, the number of rolls produced daily is $\frac{3{,}000{,}000}{14} \approx 214{,}286$.

46. The number of pounds of sugar used each day in the making of LifeSavers equals 250,000. According to Exercise 45, 214,286 rolls are produced each day. So, the amount of sugar in one roll of LifeSavers equals roughly $\frac{250{,}000}{214{,}286} \approx 1.17$ lb (rounded).

47. The weight of one roll of LifeSavers is much less than 1.17 lb. The answer is "correct" using the information given, but some of the data given must not be accurate.

48. Answers will vary. Possible answers are: Misprints or typographical errors; careless reporting of data; math errors.

10.3 Frequency Distributions and Histograms

10.3 Margin Exercises

1. **(a)** The least number of sales calls made in a week is 20 calls.

(b) The most common number of sales calls made in a week is 30 calls (frequency = 9).

(c) The number of weeks in which 35 calls were made is 4 weeks.

(d) The number of weeks in which 45 calls were made is 5 weeks.

2. **(a)** Less than 50 calls were made in $9 + 13 + 13 = 35$ weeks.

(b) In $5 + 4 + 6 = 15$ weeks, 50 or more calls were made.

3. **(a)** Less than 60 calls were made in $9 + 13 + 13 + 5 = 40$ weeks.

(b) In $4 + 6 = 10$ weeks, 60 or more calls were made.

10.3 Section Exercises

1. The age group of 61–65 years has the greatest number of members, which is 16,000.

3. To find the number of members 35 years of age and under, add the number of members in the 16–20, 21–25, 26–30, and 31–35 groups.

$1000 + 2000 + 4000 + 6000 = 13{,}000$ members

5. To find the number of members ages 41 to 60, add the number of members in the 41–45, 46–50, 51–55, and 56–60 groups.

$10{,}000 + 15{,}000 + 11{,}000 + 14{,}000 = 50{,}000$ members

7. The greatest number of employees is in the $4100 to $5000 group. There are 16 employees in that salary group.

9. There are 11 employees who earn $3100 to $4000.

11. The number of employees earning $5000 or less is $6 + 10 + 6 + 11 + 16 = 49$ employees.

13. Class intervals are the result of combining data into groupings. Class frequency is the number of data items that fit in each class interval. These are used to group data and to have multiple responses (frequency) in a class interval—this makes the data easier to interpret.

15. 140–149: 6 tally marks and a class frequency of 6

17. 160–169: 1 tally mark and a class frequency of 1

19. 180–189: 2 tally marks and a class frequency of 2

21. 0–24: 4 tally marks and a class frequency of 4

23. 50–74: 6 tally marks and a class frequency of 6

25. 100–124: 5 tally marks and a class frequency of 5

27. 150–174: 1 tally mark and a class frequency of 1

29. 1–5: 14 tally marks and a class frequency of 14

31. 11–15: 16 tally marks and a class frequency of 16

33. 21–25: 11 tally marks and a class frequency of 11

35. 31–35: 5 tally marks and a class frequency of 5

10.4 Mean, Median, and Mode

10.4 Margin Exercises

1.
$$\text{mean} = \frac{\text{sum of all values}}{\text{number of values}} = \frac{96 + 98 + 84 + 88 + 82 + 92}{6} = \frac{540}{6} = 90$$

2. **(a)** The sum of the 12 values is $50.28 + $85.16 + $110.50 + $78 + $120.70 + $58.64 + $73.80 + $86.24 + $67.85 + $96.56 + $138.65 + $48.90 = $1015.28.

$$\text{mean} = \frac{\$1015.28}{12} \approx \$84.61 \text{ (rounded)}$$

(b) The sum of the 8 values is $749,820 + $765,480 + $643,744 + $824,222 + $485,886 + $668,178 + $702,294 + $525,800 = $5,365,424.

$$\text{mean} = \frac{\$5{,}365{,}424}{8} = \$670{,}678$$

3.

Value	Frequency	Product
\$2	4	\$8
\$4	6	\$24
\$6	5	\$30
\$8	6	\$48
\$10	12	\$120
\$12	5	\$60
\$14	8	\$112
\$16	4	\$64
Totals	50	\$466

$$\text{mean} = \frac{\$466}{50} = \$9.32$$

The mean daily amount spent for lottery tickets was \$9.32.

4.

Course	Credits	Grade	Credits • Grade
Mathematics	3	A (= 4)	3 • 4 = 12
P.E.	1	C (= 2)	1 • 2 = 2
English	3	C (= 2)	3 • 2 = 6
Keyboarding	2	B (= 3)	2 • 3 = 6
Recreation	4	B (= 3)	4 • 3 = 12
Totals	13		38

$$\text{GPA} = \frac{\text{sum of Credits} \cdot \text{Grade}}{\text{total number of Credits}} = \frac{38}{13} \approx 2.92$$

5. Arrange the numbers in numerical order from smallest to largest (they already are).

10, 14, 15, 17, 18, 19, 20

The list has 7 numbers. The middle number is the 4th number, so the median is 17 lb.

6. Arrange the numbers in numerical order from smallest to largest.

74, 87, 96, 108, 125, 136

$$\frac{96\text{ m} + 108\text{ m}}{2} = \frac{204\text{ m}}{2} = 102\text{ m.}$$

7. **(a)** 19, <u>18</u>, 22, 20, <u>18</u>

The only number that occurs more than once is 18. The mode is 18 years.

(b) <u>312</u>, <u>219</u>, 782, <u>312</u>, <u>219</u>, 426

Because both 219 and 312 occur two times, each is a mode. This list is *bimodal.*

(c) \$1706, \$1289, \$1653, \$1892, \$1301, \$1782

No number occurs more than once. This list has *no mode.*

10.4 Section Exercises

1.
$$\text{Mean} = \frac{\text{sum of all values}}{\text{number of values}} = \frac{6 + 22 + 15 + 2 + 8 + 13}{6} = \frac{66}{6} = 11\text{ yr}$$

The mean (average) age of the shopping centers was 11 years.

3.
$$\text{Mean} = \frac{3.1 + 1.5 + 2.8 + 0.8 + 4.1}{5} = \frac{12.3}{5} \approx 2.5\text{ in. of rain (rounded)}$$

The mean (average) inches of rain per month was 2.5.

5. Mean
$$= \frac{\$38{,}500 + \$39{,}720 + \$42{,}183 + \$21{,}982 + \$43{,}250}{5} = \frac{\$185{,}635}{5} = \$37{,}127$$

The mean (average) annual salary was \$37,127.

7.
$$\text{mean} = \frac{\text{sum of all values}}{\text{number of values}}$$
$$= \frac{(75.52 + 36.15 + 58.24 + 21.86 + 47.68 + 106.57 + 82.72 + 52.14 + 28.60 + 72.92)}{10}$$
$$= \frac{582.40}{10} = 58.24$$

The mean (average) shoe sales amount was \$58.24.

9.

Customers Each Hour	Frequency	Product
8	2	(8 • 2) = 16
11	12	(11 • 12) = 132
15	5	(15 • 5) = 75
26	1	(26 • 1) = 26
Totals	20	249

$$\text{weighted mean} = \frac{\text{sum of products}}{\text{total number of customers}} = \frac{249}{20} \approx 12.5\text{ customers (rounded)}$$

11.

Fish per Boat	Frequency	Product
12	4	$(12 \cdot 4) = 48$
13	2	$(13 \cdot 2) = 26$
15	5	$(15 \cdot 5) = 75$
19	3	$(19 \cdot 3) = 57$
22	1	$(22 \cdot 1) = 22$
23	5	$(23 \cdot 5) = 115$
Totals	20	343

$$\text{weighted mean} = \frac{\text{sum of products}}{\text{total number of fish}}$$
$$= \frac{343}{20} = 17.15 \approx 17.2 \text{ fish (rounded)}$$

13.

Policy Amount ($)	Number of Policies Sold	Product ($)
10,000	6	60,000
20,000	24	480,000
25,000	12	300,000
30,000	8	240,000
50,000	5	250,000
100,000	3	300,000
250,000	2	500,000
Totals	60	2,130,000

$$\text{weighted mean} = \frac{\text{sum of products}}{\text{total number of policies}}$$
$$= \frac{2{,}130{,}000}{60} = 35{,}500$$

The mean (average) amount for the policies sold was $35,500.

15. Arrange the numbers in numerical order from smallest to largest.

$$100, 114, 125, 135, 150, 172$$

The list has 6 numbers. The middle numbers are the 3rd and 4th numbers, so the median is

$$\frac{125 + 135}{2} = \frac{260}{2} = 130 \text{ books.}$$

17. Arrange the numbers in numerical order from smallest to largest.

$$298, 346, 412, 501, 515, 521, 528, 621$$

The list has 8 numbers. The middle numbers are the 4th and 5th numbers, so the median is

$$\frac{501 + 515}{2} = \frac{1016}{2} = 508 \text{ calories.}$$

19. <u>21%</u>, 18%, <u>21%</u>, 28%, 22%, <u>21%</u>, 25%

The mode is the value that occurs most often. The mode is 21%.

21. <u>74</u>, <u>68</u>, <u>68</u>, <u>68</u>, 75, 75, <u>74</u>, <u>74</u>, 70

Because both 68 and 74 occur three times, each is a mode. This list is *bimodal.*

23. When the data contains a few very low or a few very high values, the mean will give a poor indication of the average. Consider using the median or mode instead.

25. The median is a better measure of central tendency when the list contains one or more extreme values. Find the mean and the median of the following home values.
$82,000; $64,000; $91,000; $115,000; $982,000
mean home value = $266,800;
median home value = $91,000

27.

Credits	Grade	Credits • Grade
4	B (= 3)	$4 \cdot 3 = 12$
2	A (= 4)	$2 \cdot 4 = 8$
5	C (= 2)	$5 \cdot 2 = 10$
1	F (= 0)	$1 \cdot 0 = 0$
3	B (= 3)	$3 \cdot 3 = 9$
15		39

$$\text{GPA} = \frac{\text{sum of Credits} \cdot \text{Grade}}{\text{total number of credits}}$$
$$= \frac{39}{15} = 2.60$$

29.

Credits	Grade	Credits • Grade
2	A (= 4)	$2 \cdot 4 = 8$
3	C (= 2)	$3 \cdot 2 = 6$
4	A (= 4)	$4 \cdot 4 = 16$
1	C (= 2)	$1 \cdot 2 = 2$
4	B (= 3)	$4 \cdot 3 = 12$
14		44

$$\text{GPA} = \frac{\text{sum of Credits} \cdot \text{Grade}}{\text{total number of credits}}$$
$$= \frac{44}{14} \approx 3.14 \text{ (rounded)}$$

31. Samuels total:
$= 39 + 15 + 40 + 22 + 13 + 22 + 17 + 8$
$= 176$ calls

Stricker total:
$= 21 + 22 + 20 + 23 + 19 + 24 + 25 + 22$
$= 176$ calls

32. Samuels mean $= \frac{176}{8} = 22$

Stricker mean $= \frac{176}{8} = 22$

33. Arrange the data in numerical order.

Samuels: 8, 13, 15, 17, <u>22</u>, <u>22</u>, 39, 40

$$\text{Median} = \frac{17 + 22}{2} = \frac{39}{2} = 19.5$$

Stricker: 19, 20, 21, <u>22</u>, <u>22</u>, 23, 24, 25

$$\text{Median} = \frac{22 + 22}{2} = \frac{44}{2} = 22$$

34. The mode is the value that occurs most often.

Samuels mode = 22

Stricker mode = 22

35. The mean and mode are identical for both sales representatives and the medians are close.

36. The number of weekly sales calls made by Samuels varies greatly from week to week while the number of weekly sales calls made by Stricker remains fairly constant.

37. Samuels range = 40 − 8 = 32

38. Stricker range = 25 − 19 = 6

39. No, not with any certainty. What else might you want to know? There probably are additional questions that need to be answered before a determination could be made, such as the dollar amount of sales, number of repeat customers, and so on.

40. Answers will vary. Some possible answers are: He works hard one week, then takes it easy the next week; the characteristics of the sales territories vary greatly; illness or personal problems may be affecting performance.

Chapter 10 Review Exercises

1. The largest expense was lodging at $560.

2. Total cost was
$280 + $300 + $560 + $400 + $160 = $1700.

$$\frac{\text{food}}{\text{total cost}} = \frac{\$400}{\$1700} = \frac{400}{1700} = \frac{4}{17}$$

3. $$\frac{\text{gasoline}}{\text{total cost}} = \frac{\$300}{\$1700} = \frac{300}{1700} = \frac{3}{17}$$

4. $$\frac{\text{sightseeing}}{\text{total cost}} = \frac{\$280}{\$1700} = \frac{280 \div 20}{1700 \div 20} = \frac{14}{85}$$

5. $$\frac{\text{gasoline}}{\text{other}} = \frac{\$300}{\$160} = \frac{300 \div 20}{160 \div 20} = \frac{15}{8}$$

6. $$\frac{\text{lodging}}{\text{food}} = \frac{\$560}{\$400} = \frac{560 \div 80}{400 \div 80} = \frac{7}{5}$$

7. Casual dress
82% of 4800 companies

$$x = 0.82 \cdot 4800$$
$$= 3936 \text{ companies}$$

8. Free food/Beverages
36% of 4800 companies

$$x = 0.36 \cdot 4800$$
$$= 1728 \text{ companies}$$

9. Fitness centers
15% of 4800 companies

$$x = 0.15 \cdot 4800$$
$$= 720 \text{ companies}$$

10. Flexible hours
60.5% of 4800 companies

$$x = 0.605 \cdot 4800$$
$$= 2904 \text{ companies}$$

11. On-site child care and Fitness centers
Answers will vary. Perhaps employers feel that they are not needed or would not be used. Or, it may be that they would be too expensive for the benefit derived.

12. Flexible hours and Casual dress
Answers will vary. Perhaps employees request them and appreciate them. Or, it may be that neither of them cost the employer anything to offer.

13. In 2006, the greatest amount of water in the lake occurred in March when there were 8,000,000 acre-feet of water.

14. In 2005, the least amount of water in the lake occurred in June when there were 2,000,000 acre-feet of water.

15. In June of 2006, there were 5,000,000 acre-feet of water in the lake.

16. In May of 2005, there were 4,000,000 acre-feet of water in the lake.

17. From March, 2005 to June, 2005 the water went from 7 to 2 million acre-feet, a decrease of 5,000,000 acre-feet of water.

18. From April, 2006 to June, 2006 the water went from 7 to 5 million acre-feet, a decrease of 2,000,000 acre-feet of water.

19. In 2003, Center A sold $50,000,000 worth of floor covering.

20. In 2005, Center A sold $20,000,000 worth of floor covering.

21. In 2004, Center B sold $20,000,000 worth of floor covering.

22. In 2006, Center B sold $40,000,000 worth of floor covering.

23. The floor-covering sales decreased for 2 years and then moved up slightly. Answers will vary. Perhaps there is less new home construction, remodeling, and home improvement in the area near center A, or better product selection and service have reversed the decline in sales.

24. The floor-covering sales are increasing. Answers will vary. Perhaps new construction and home remodeling have increased in the area near center B, or greater advertising has attracted more customers.

25. Mean

$$= \frac{18 + 12 + 15 + 24 + 9 + 42 + 54 + 87 + 21 + 3}{10}$$

$$= \frac{285}{10} = 28.5 \text{ digital cameras}$$

26. Mean

$$= \frac{31 + 9 + 8 + 22 + 46 + 51 + 48 + 42 + 53 + 42}{10}$$

$$= \frac{352}{10} = 35.2 \text{ complaints}$$

27.

Dollar Value	Frequency	Product
\$42	3	\$126
\$47	7	\$329
\$53	2	\$106
\$55	3	\$165
\$59	5	\$295
	20	\$1021

$$\text{Weighted Mean} = \frac{\$1021}{20} = \$51.05$$

28.

Total Points	Frequency	Product
243	1	243
247	3	741
251	5	1255
255	7	1785
263	4	1052
271	2	542
279	2	558
	24	6176

$$\text{Weighted Mean} = \frac{6176}{24}$$

$$\approx 257.3 \text{ points (rounded)}$$

29. Arrange the numbers in numerical order from smallest to largest.

13, 28, 35, 37, 39, 43, 54, 68, 75

The list has 9 numbers. The middle number is the 5th number, so the median is 39 forms.

30. Arrange the numbers in numerical order from smallest to largest.

\$151, \$525, \$542, \$559, \$565, \$576, \$578, \$590

The list has 8 numbers. The middle numbers are the 4th and 5th numbers, so the median is

$$\frac{\$559 + \$565}{2} = \$562.$$

31. $\underline{\$79}$, \$56, \$110, $\underline{\$79}$, \$72, \$86, $\underline{\$79}$

The value \$79 occurs three times, which is more often than any other value. Therefore, \$79 is the mode.

32. $\underline{18}$, 25, 63, $\underline{32}$, 28, 37, $\underline{32}$, 26, $\underline{18}$

Because both 18 and 32 occur two times, each is a mode. This list is *bimodal.*

33. **[10.1]** Degrees for Plumbing and electrical changes

$$= 10\% \text{ of } 360°$$
$$= (0.10)(360°) = 36°$$

34. **[10.1]** Percent for Work stations (total = \$22,400)

$$= \frac{\$7840}{\$22{,}400} = 0.35 = 35\%$$

35. **[10.1]** Percent for Small appliances

$$= \frac{\$4480}{\$22{,}400} = 0.20 = 20\%$$

36. **[10.1]** Percent for Interior decoration

$$= \frac{\$5600}{\$22{,}400} = 0.25 = 25\%$$

37. **[10.1]** Degrees for Supplies

$$= 10\% \text{ of } 360°$$
$$= (0.10)(360°) = 36°$$

38. **[10.1]**

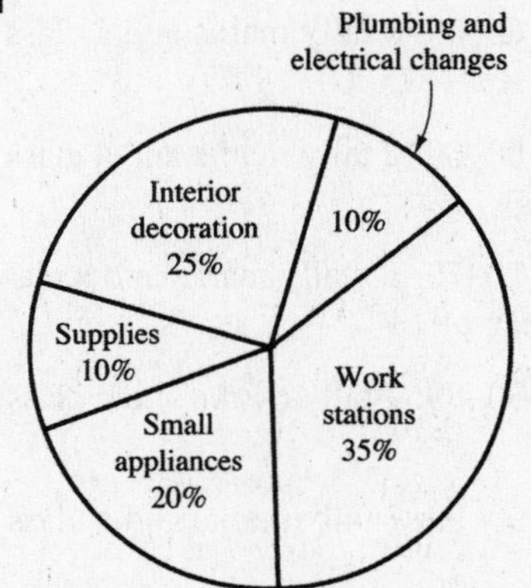

39. **[10.4]** Mean $= \frac{48 + 72 + 52 + 148 + 180}{5}$

$$= \frac{500}{5} = 100 \text{ volunteers}$$

40. **[10.4]**

Mean

$$= \frac{122 + 135 + 146 + 159 + 128 + 147 + 168 + 139 + 158}{9}$$

$$= \frac{1302}{9} \approx 144.7 \text{ vaccinations (rounded)}$$

41. **[10.4]** $\underline{48}, 43, 46, 47, \underline{48}, \underline{48}, 43$

The value 48 occurs three times, which is more often than any other value. Therefore, 48 applicants is the mode.

42. **[10.4]** $26, \underline{31}, \underline{31}, 37, \underline{43}, 51, \underline{31}, \underline{43}, \underline{43}$

Because both 31 and 43 occur three times, each is a mode. This list is *bimodal*.

43. **[10.4]** Arrange the numbers in numerical order from smallest to largest.

$$1.0, 2.9, 3.2, 4.7, 5.3, 7.1, 8.2, 9.4$$

The list has 8 numbers. The middle numbers are the 4th and 5th numbers, so the median is

$$\frac{4.7 + 5.3}{2} = 5.0 \text{ hours.}$$

44. **[10.4]** Arrange the numbers in numerical order from smallest to largest.

$$2, 3, 9, 12, 17, 19, 23, 27, 35, 39, 46, 51$$

The list has 12 numbers. The middle numbers are the 6th and 7th numbers, so the median is

$$\frac{19 + 23}{2} = 21 \text{ e-mails.}$$

45. **[10.3]** 30–39: 4 tally marks and a class frequency of 4

46. **[10.3]** 40–49: 1 tally mark and a class frequency of 1

47. **[10.3]** 50–59: 6 tally marks and a class frequency of 6

48. **[10.3]** 60–69: 2 tally marks and a class frequency of 2

49. **[10.3]** 70–79: 13 tally marks and a class frequency of 13

50. **[10.3]** 80–89: 7 tally marks and a class frequency of 7

51. **[10.3]** 90–99: 7 tally marks and a class frequency of 7

52. **[10.2]**

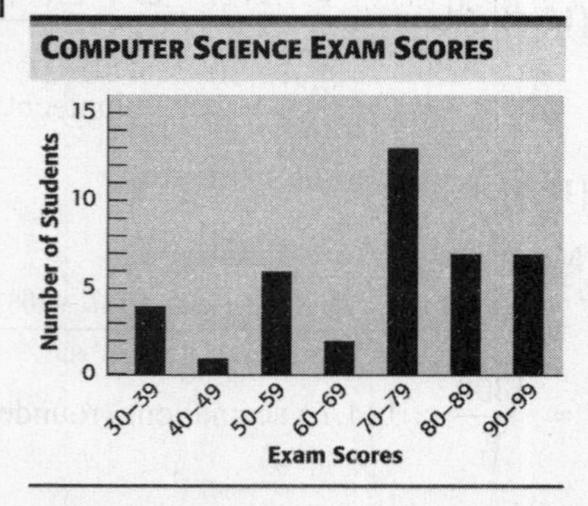

53. **[10.4]**

Test Score	Frequency	Product
46	4	184
54	10	540
62	8	496
70	12	840
78	10	780
	44	2840

$$\text{Weighted Mean} = \frac{2840}{44}$$
$$\approx 64.5 \text{ (rounded)}$$

54. **[10.4]**

Units Sold	Frequency	Product
104	6	624
112	14	1 568
115	21	2 415
119	13	1 547
123	22	2 706
127	6	762
132	9	1 188
	91	10,810

$$\text{Weighted Mean} = \frac{10{,}810}{91}$$
$$\approx 118.8 \text{ units (rounded)}$$

Chapter 10 Test

1. Newspaper is 23% of \$2,800,000.

$$x = 0.23 \cdot \$2{,}800{,}000$$
$$= \$644{,}000$$

2. Outdoor advertising is 19% of \$2,800,000.

$$x = 0.19 \cdot \$2{,}800{,}000$$
$$= \$532{,}000$$

3. Television is 29% of \$2,800,000.

$$x = 0.29 \cdot \$2{,}800{,}000$$
$$= \$812{,}000$$

4. Radio is 15% of \$2,800,000.

$$x = 0.15 \cdot \$2{,}800{,}000$$
$$= \$420{,}000$$

5. Miscellaneous is 3% of \$2,800,000.

$$x = 0.03 \cdot \$2{,}800{,}000$$
$$= \$84{,}000$$

6. Internet is 11% of \$2,800,000.

$$x = 0.11 \cdot \$2{,}800{,}000$$
$$= \$308{,}000$$

7. Degrees for Salaries

$$= 30\% \text{ of } 360°$$
$$= 0.30 \times 360°$$
$$= 108°$$

8. Degrees for Delivery expense
$= 10\%$ of $360°$
$= 0.1 \times 360°$
$= 36°$

9. Degrees for Advertising
$= 20\%$ of $360°$
$= 0.2 \times 360°$
$= 72°$

10. Degrees for Rent
$= 30\%$ of $360°$
$= 0.3 \times 360°$
$= 108°$

11. Percent for Other
$= \frac{36°}{360°} = 10\%$

12.

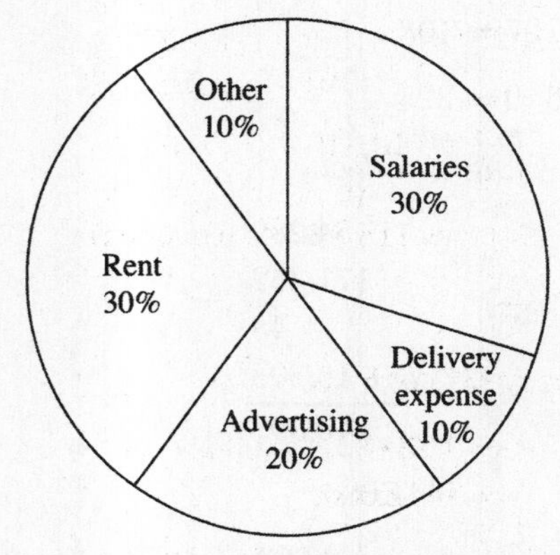

	Profit	*Tally*	*Number of Weeks*
13.	\$120–\$129	|||	3
14.	\$130–\$139	||	2
15.	\$140–\$149	||||	4
16.	\$150–\$159	|||	3
17.	\$160–\$169	|||	3
18.	\$170–\$179	~~||||~~	5

19.

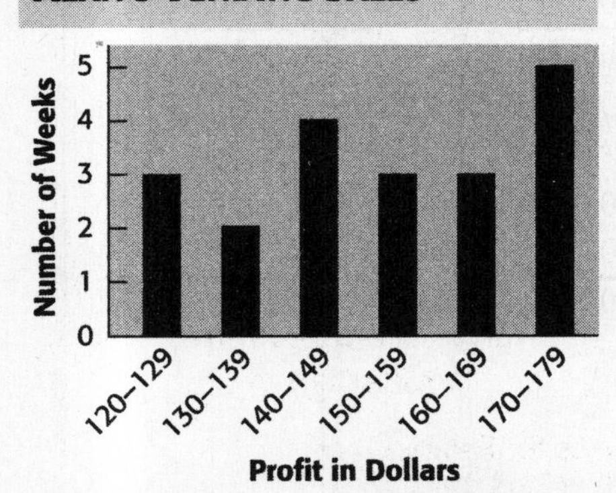

20. Mean
$$= \frac{52 + 61 + 68 + 69 + 73 + 75 + 79 + 84 + 91 + 98}{10}$$
$$= \frac{750}{10} = 75 \text{ mi}$$

21. Mean
$$= \frac{11 + 14 + 12 + 14 + 20 + 16 + 17 + 18}{8}$$
$$= \frac{122}{8} \approx 15.3 \text{ lb (rounded)}$$

22. Mean
$$= \frac{(458 + 432 + 496 + 491 + 500 + 508 + 512 + 396 + 492 + 504)}{10}$$
$$= \frac{4789}{10} = 478.9 \text{ miles per hour}$$

23. The weighted mean must be used because different classes are worth different numbers of credits.

Credits	Grade	Credits • Grade
3	A (= 4)	3 • 4 = 12
2	C (= 2)	2 • 2 = 4
4	B (= 3)	4 • 3 = 12
9		28

$$\text{GPA} = \frac{\text{sum of Credits} \cdot \text{Grade}}{\text{total number of Credits}} = \frac{28}{9} \approx 3.11$$

24. Arrange the values in order, from smallest to largest. When there is an odd number of values in a list, the median is the middle value.

$$8, 17, \underline{23}, 32, 64$$

The median is 23.

25.

Cost	Frequency	Product
12	5	60
20	6	120
22	8	176
28	4	112
38	6	228
48	2	96
	31	792

$$\text{Weighted Mean} = \frac{792}{31}$$
$$\approx 25.5 \text{ (rounded)}$$

26.

Value	Frequency	Product
150	15	2250
160	17	2720
170	21	3570
180	28	5040
190	19	3610
200	7	1400
	107	18,590

$$\text{Weighted Mean} = \frac{18{,}590}{107} \approx 173.7 \text{ (rounded)}$$

27. Arrange the numbers in numerical order from smallest to largest.

$$16, 28, 28, 31, 32, 35, 37, 41$$

The list has 8 numbers. The middle numbers are the 4th and 5th numbers, so the median is

$$\frac{31+32}{2} = 31.5 \text{ degrees.}$$

28. Arrange the numbers in numerical order from smallest to largest.

$$6.2, 7.6, 9.1, 9.5, 10.0, 11.4, 12.5, 22.8, 31.7$$

The list has 9 numbers. The middle number is the 5th number, so the median is 10.0 meters.

29. $72, 46, \underline{52}, 37, 28, 18, \underline{52}, 61$

The value 52 occurs two times, which is more often than any other value. Therefore, 52 milliliters is the mode.

30. $96, \underline{104}, \underline{103}, \underline{104}, \underline{103}, \underline{104}, 91, 74, \underline{103}$

Because both 103 and 104 degrees occur three times, each is a mode. This list is *bimodal.*

Cumulative Review Exercises (Chapters 1–10)

1. \$65.236 to the nearest cent
Draw a cut-off line: \$65.23|6
The first digit cut is 6, which is 5 or more, so round up. **\$65.24**

2. \$781.499 to the nearest dollar
Draw a cut-off line: \$781.|499
The first digit cut is 4, which is less than 5, so the part you keep stays the same. **\$781**

3. 93,367 to the nearest ten: 93,3$\underline{6}$7
Next digit is 5 or more.

Tens place changes $(6+1=7)$. All digits to the right of the underlined place change to 0. **93,370**

4. 854,279 to the nearest ten thousand: 8$\underline{5}$4,279
Next digit is less than 5.

Ten thousands place stays the same. All digits to the right of the underlined place change to 0.
850,000

5.
$$\begin{aligned} &4 + 10 \div 2 + 7(2) \\ &= 4 + 5 + 7(2) \\ &= 4 + 5 + 14 \\ &= 9 + 14 \\ &= 23 \end{aligned}$$

6.
$$\begin{aligned} &\sqrt{81} - 4(2) + 9 \\ &= 9 - 4(2) + 9 \\ &= 9 - 8 + 9 \\ &= 1 + 9 \\ &= 10 \end{aligned}$$

7. $2^2 \cdot 3^3 = 4 \cdot 27 = 108$

8. $6^2 \cdot 3^2 = 36 \cdot 9 = 324$

9. *Estimate:*

$$\begin{array}{r} 50{,}000 \\ 200{,}000 \\ 50 \\ +\ 20{,}000 \\ \hline 270{,}050 \end{array}$$

Exact:

$$\begin{array}{r} 54{,}289 \\ 171{,}352 \\ 48 \\ +\ 18{,}208 \\ \hline 243{,}897 \end{array}$$

10. *Estimate:*

$$\begin{array}{r} 3 \\ 800 \\ 40 \\ 50 \\ +\ 8 \\ \hline 901 \end{array}$$

Exact:

$$\begin{array}{r} 2.607 \\ 796.200 \\ 37.960 \\ 53.720 \\ +\ 8.060 \\ \hline 898.547 \end{array}$$

11. *Estimate:*

$$\begin{array}{r} 400{,}000 \\ -\ 200{,}000 \\ \hline 200{,}000 \end{array}$$

Exact:

$$\begin{array}{r} \scriptstyle 4\ 12\ 9\ 16 \\ 44\not{5},\not{3}\not{0}\not{6} \\ -\ 234{,}867 \\ \hline 210{,}439 \end{array}$$

12. *Estimate:*

$$\begin{array}{r} 900 \\ -\ 60 \\ \hline 840 \end{array}$$

Exact:

$$\begin{array}{r} \scriptstyle 4\ 15\ 11\ 10 \\ 87\not{5}.\not{6}\not{2}\not{0} \\ -\ 63.757 \\ \hline 811.863 \end{array}$$

13. *Estimate:*

$$\begin{array}{r} 7000 \\ \times \quad 600 \\ \hline 4{,}200{,}000 \end{array}$$

Exact:

$$\begin{array}{r} 7064 \\ \times \quad 635 \\ \hline 35\,320 \\ 211\,92 \\ 4\,238\,4 \\ \hline 4{,}485{,}640 \end{array}$$

14. *Estimate:*

$$\begin{array}{r} 60 \\ \times\ 3 \\ \hline 180 \end{array}$$

Exact:

$$\begin{array}{r} 62.75 \\ \times \quad 2.644 \\ \hline 25100 \\ 2\,5100 \\ 37\,650 \\ 125\,50 \\ \hline 165.91100 \end{array}$$

15. *Estimate:*

$$\begin{array}{r} 5\,0\,0 \\ 20\overline{)1\,0{,}\,0\,0\,0} \end{array}$$

Exact:

$$\begin{array}{r} 6\,4\,2 \\ 18\overline{)1\,1{,}\,5\,5\,6} \\ \underline{1\,0\,8} \\ 7\,5 \\ \underline{7\,2} \\ 3\,6 \\ \underline{3\,6} \\ 0 \end{array}$$

16. *Estimate:*

$$\begin{array}{r} 1\,5 \\ 4\overline{)6\,0} \\ \underline{4} \\ 2\,0 \\ \underline{2\,0} \\ 0 \end{array}$$

Exact:

$$\begin{array}{r} 1\,4.\,7\,2 \\ 4.25_{\wedge}\overline{)6\,2.\,5\,6_{\wedge}\,0\,0} \\ \underline{4\,2\,5} \\ 2\,0\,0\,6 \\ \underline{1\,7\,0\,0} \\ 3\,0\,6\,0 \\ \underline{2\,9\,7\,5} \\ 8\,5\,0 \\ \underline{8\,5\,0} \\ 0 \end{array}$$

17. $\frac{7}{8}+\frac{3}{4}=\frac{7}{8}+\frac{6}{8}=\frac{7+6}{8}=\frac{13}{8}=1\frac{5}{8}$

18. $\frac{3}{4}+\frac{5}{8}+\frac{1}{2}=\frac{6}{8}+\frac{5}{8}+\frac{4}{8}=\frac{15}{8}=1\frac{7}{8}$

19.

$$\begin{array}{rcl} 4\frac{3}{5} & = & 4\frac{9}{15} \\ +\,5\frac{2}{3} & = & 5\frac{10}{15} \\ \hline & & 9\frac{19}{15} = 10\frac{4}{15} \end{array}$$

20. $\frac{5}{6}-\frac{1}{3}=\frac{5}{6}-\frac{2}{6}=\frac{5-2}{6}=\frac{3}{6}=\frac{1}{2}$

21.

$$\begin{array}{rcccl} 6\frac{2}{3} & = & 6\frac{8}{12} & = & 5\frac{20}{12} \\ -\,4\frac{3}{4} & = & 4\frac{9}{12} & = & 4\frac{9}{12} \\ \hline & & & & 1\frac{11}{12} \end{array}$$

22.

$$\begin{array}{rcl} 8 & = & 7\frac{5}{5} \\ -\,5\frac{2}{5} & = & 5\frac{2}{5} \\ \hline & & 2\frac{3}{5} \end{array}$$

23. $\frac{3}{5}\cdot\frac{5}{8}=\frac{3}{\cancel{5}_1}\cdot\frac{\cancel{5}^1}{8}=\frac{3}{8}$

24. $\left(9\frac{3}{5}\right)\cdot\left(4\frac{5}{8}\right)=\frac{\cancel{48}^6}{5}\cdot\frac{37}{\cancel{8}_1}=\frac{6\cdot 37}{5\cdot 1}=\frac{222}{5}=44\frac{2}{5}$

25. $22\left(\frac{2}{5}\right)=\frac{22}{1}\cdot\frac{2}{5}=\frac{44}{5}=8\frac{4}{5}$

26. $\frac{5}{6}\div\frac{5}{8}=\frac{\cancel{5}^1}{\cancel{6}_3}\cdot\frac{\cancel{8}^4}{\cancel{5}_1}=\frac{4}{3}=1\frac{1}{3}$

27. $18\div\frac{3}{4}=18\cdot\frac{4}{3}=\frac{\cancel{18}^6}{1}\cdot\frac{4}{\cancel{3}_1}=24$

28. $3\frac{1}{3}\div 8\frac{3}{4}=\frac{10}{3}\div\frac{35}{4}=\frac{\cancel{10}^2}{3}\cdot\frac{4}{\cancel{35}_7}=\frac{2\cdot 4}{3\cdot 7}=\frac{8}{21}$

29. $\frac{2}{3}\left(\frac{7}{8}-\frac{3}{4}\right)=\frac{2}{3}\left(\frac{7}{8}-\frac{6}{8}\right)=\frac{\cancel{2}^1}{3}\cdot\frac{1}{\cancel{8}_4}=\frac{1}{12}$

30. $\left(\frac{5}{6}-\frac{1}{3}\right)+\left(\frac{1}{2}\right)^2\cdot\frac{3}{4}$

$=\left(\frac{5}{6}-\frac{2}{6}\right)+\left(\frac{1}{2}\right)^2\cdot\frac{3}{4}$

$=\frac{3}{6}+\left(\frac{1}{2}\right)^2\cdot\frac{3}{4}$

$=\frac{3}{6}+\frac{1}{4}\cdot\frac{3}{4}$

$=\frac{3}{6}+\frac{3}{16}$

$=\frac{24}{48}+\frac{9}{48}$

$=\frac{33}{48}=\frac{11}{16}$

31. $\frac{3}{8}$

$$\begin{array}{r} 0.375 \\ 8\overline{)3.000} \\ \underline{2\,4} \\ 60 \\ \underline{56} \\ 40 \\ \underline{40} \\ 0 \end{array}$$

$\frac{3}{8} = 0.375$

32. $\frac{4}{5} = \frac{4 \cdot 2}{5 \cdot 2} = \frac{8}{10} = 0.8$

33. $\frac{1}{6}$

$$\begin{array}{r} 0.1666 \\ 6\overline{)1.0000} \\ \underline{6} \\ 40 \\ \underline{36} \\ 40 \\ \underline{36} \\ 40 \\ \underline{36} \\ 4 \end{array}$$

$\frac{1}{6} \approx 0.167$ (rounded)

34. $\frac{17}{20} = \frac{17 \cdot 5}{20 \cdot 5} = \frac{85}{100} = 0.85$

35.

$$\begin{aligned} 0.218 &= 0.2180 \\ 0.22 &= 0.2200 \\ 0.199 &= 0.1990 \\ 0.207 &= 0.2070 \\ 0.2215 &= 0.2215 \end{aligned}$$

From smallest to largest:
0.199, 0.207, 0.218, 0.22, 0.2215

36.

$$\begin{aligned} 0.6319 &= 0.6319 \\ \frac{5}{8} &= 0.6250 \\ 0.608 &= 0.6080 \\ \frac{13}{20} &= 0.6500 \\ 0.58 &= 0.5800 \end{aligned}$$

From smallest to largest:
$0.58, 0.608, \frac{5}{8}, 0.6319, \frac{13}{20}$

37. $5\frac{1}{2}$ in. to 44 in.

$$\frac{5\frac{1}{2}\text{ in.}}{44\text{ in.}} = \frac{5\frac{1}{2} \div 5\frac{1}{2}}{44 \div 5\frac{1}{2}} = \frac{1}{\frac{44}{1} \cdot \frac{2}{11}} = \frac{1}{4 \cdot 2} = \frac{1}{8}$$

38. 3 hr to 45 min

$3\text{ hr} = 3(60\text{ min}) = 180\text{ min}$

$$\frac{180\text{ min}}{45\text{ min}} = \frac{180 \div 45}{45 \div 45} = \frac{4}{1}$$

39.

$$\begin{aligned} \frac{6}{15} &= \frac{18}{45} \\ 6 \cdot 45 &= 18 \cdot 15 \\ 270 &= 270 \quad \textit{True} \end{aligned}$$

40.

$$\begin{aligned} \frac{52}{180} &= \frac{36}{120} \\ 52 \cdot 120 &= 180 \cdot 36 \\ 6240 &= 6480 \quad \textit{False} \end{aligned}$$

41.

$$\begin{aligned} \frac{1}{5} &= \frac{x}{30} \\ 5 \cdot x &= 1 \cdot 30 \\ \frac{5 \cdot x}{5} &= \frac{30}{5} \\ x &= 6 \end{aligned}$$

42.

$$\begin{aligned} \frac{15}{x} &= \frac{390}{156} \\ x \cdot 390 &= 15 \cdot 156 \\ x \cdot 390 &= 2340 \\ \frac{x \cdot 390}{390} &= \frac{2340}{390} \\ x &= 6 \end{aligned}$$

43.

$$\begin{aligned} \frac{200}{135} &= \frac{24}{x} \\ 200 \cdot x &= 24 \cdot 135 \\ 200x &= 3240 \\ \frac{200x}{200} &= \frac{3240}{200} \\ x &= 16.2 \end{aligned}$$

44.

$$\begin{aligned} \frac{x}{208} &= \frac{6.5}{26} \\ 26 \cdot x &= 6.5 \cdot 208 \\ 26x &= 1352 \\ \frac{26x}{26} &= \frac{1352}{26} \\ x &= 52 \end{aligned}$$

45. $65\% = \frac{65}{100} = 0.65$

46. $0.035 = \frac{35}{1000} = \frac{3.5}{100} = 3.5\%$

47. $380\% = \frac{380}{100} = 3.80$ or 3.8

48. $7.75\% = \frac{7.75}{100} = \frac{775}{10{,}000} = 0.0775$

49. $4\% = \frac{4}{100} = \frac{4 \div 4}{100 \div 4} = \frac{1}{25}$

50. $87\frac{1}{2}\% = \frac{87\frac{1}{2}}{100} = \frac{\frac{175}{2}}{100} = \frac{175}{2} \div \frac{100}{1}$
$= \frac{175}{2} \cdot \frac{1}{100} = \frac{175}{200} = \frac{175 \div 25}{200 \div 25} = \frac{7}{8}$

51. $\frac{7}{20} = \frac{7 \cdot 5}{20 \cdot 5} = \frac{35}{100} = 35\%$

52. $5\frac{3}{4} = 5\frac{75}{100} = 5.75 = 575\%$

53. $x = 0.65 \cdot 780$
$= 507$

65% of $780 is $507.

54. $x = 0.054 \cdot 6000$
$= 324$

5.4% of 6000 homes is 324 homes.

55. Use the percent proportion.
part is 238; percent is 8.5; whole is unknown

$$\frac{238}{x} = \frac{8.5}{100}$$
$$8.5 \cdot x = 238 \cdot 100$$
$$\frac{8.5x}{8.5} = \frac{23{,}800}{8.5}$$
$$x = 2800$$

$8\frac{1}{2}\%$ of 2800 people is 238 people.

56. Use the percent proportion.
part is 72; percent is 18; whole is unknown

$$\frac{72}{x} = \frac{18}{100}$$
$$x \cdot 18 = 72 \cdot 100$$
$$\frac{x \cdot 18}{18} = \frac{7200}{18}$$
$$x = 400$$

72 is 18% of 400 DVDs.

57. Use the percent proportion.
part is 468; whole is 1040; percent is unknown

$$\frac{468}{1040} = \frac{x}{100}$$
$$1040 \cdot x = 468 \cdot 100$$
$$\frac{1040 \cdot x}{1040} = \frac{46{,}800}{1040}$$
$$x = 45$$

45% of $1040 is $468.

58. Use the percent proportion.
part is 13; whole is 52; percent is unknown

$$\frac{13}{52} = \frac{x}{100}$$
$$52 \cdot x = 13 \cdot 100$$
$$\frac{52 \cdot x}{52} = \frac{1300}{52}$$
$$x = 25$$

13 weeks is 25% of 52 weeks.

59. $\underline{9}$ ft = 3 yd

$$\frac{3 \text{ yd}}{1} \cdot \frac{3 \text{ ft}}{1 \text{ yd}} = 9 \text{ ft}$$

60. 28 qt = $\underline{7}$ gal

$$\frac{28 \text{ qt}}{1} \cdot \frac{1 \text{ gal}}{4 \text{ qt}} = 7 \text{ gal}$$

61. 5 days = $\underline{120}$ hr

$$\frac{5 \text{ days}}{1} \cdot \frac{24 \text{ hr}}{1 \text{ day}} = 120 \text{ hr}$$

62. $\underline{12{,}000}$ lb = 6 T

$$\frac{6 \text{ T}}{1} \cdot \frac{2000 \text{ lb}}{1 \text{ T}} = 12{,}000 \text{ lb}$$

63. 10 km to m

Count 3 places to the *right* on the metric conversion line.

10 km = 10,000 m

64. 3815 mm to m

Count 3 places to the *left* on the metric conversion line.

3815 mm to 3.815 m

65. 8.3 g to mg

$$\frac{8.3 \not{g}}{1} \cdot \frac{1000 \text{ mg}}{1 \not{g}} = 8300 \text{ mg}$$

66. 230 g to kg

$$\frac{230 \not{g}}{1} \cdot \frac{1 \text{ kg}}{1000 \not{g}} = \frac{230}{1000} \text{ kg} = 0.23 \text{ kg}$$

67. 72 mL to L

Count 3 places to the *left* on the metric conversion line.

72 mL = 0.072 L

68. 0.28 L to mL

$$\frac{0.28 \not{L}}{1} \cdot \frac{1000 \text{ mL}}{1 \not{L}} = 280 \text{ mL}$$

69. The fuel tank on the chain saw has a capacity of 750 <u>mL</u> of fuel.

70. A nickel weighs 5 g.

71. The distance of the run this Saturday is 10 <u>km</u>.

72. The heaviest player on the team weighs 108 <u>kg</u>.

73. a rectangle 8.45 m by 5.6 m

$$\begin{aligned} A &= l \cdot w \\ &= 8.45 \text{ m} \cdot 5.6 \text{ m} \\ &= 47.32 \text{ m}^2 \\ &\approx 47.3 \text{ m}^2 \end{aligned}$$

74. a trapezoid with bases 6.2 cm and 8.4 cm and height 5.3 cm

$$\begin{aligned} A &= \frac{1}{2} \cdot h \cdot (b + B) \\ &= 0.5 \cdot 5.3 \text{ cm} \cdot (6.2 \text{ cm} + 8.4 \text{ cm}) \\ &= 0.5 \cdot 5.3 \text{ cm} \cdot (14.6 \text{ cm}) \\ &= 38.69 \text{ cm}^2 \\ &\approx 38.7 \text{ cm}^2 \end{aligned}$$

75. a triangle with base 4.25 ft and height 4.5 ft

$$\begin{aligned} A &= \frac{1}{2} \cdot b \cdot h \\ &= 0.5 \cdot 4.25 \text{ ft} \cdot 4.5 \text{ ft} \\ &= 9.5625 \text{ ft}^2 \\ &\approx 9.6 \text{ ft}^2 \end{aligned}$$

76. a circle with diameter of 13 cm

$$\text{radius} = \frac{13 \text{ cm}}{2} = 6.5 \text{ cm}$$

$$\begin{aligned} A &= \pi \cdot r^2 \\ &\approx 3.14 \cdot 6.5 \text{ cm} \cdot 6.5 \text{ cm} \\ &\approx 132.7 \text{ cm}^2 \end{aligned}$$

77. a cylinder with radius 4.8 cm and height 7.6 cm

$$\begin{aligned} V &= \pi \cdot r^2 \cdot h \\ &\approx 3.14 \cdot 4.8 \text{ cm} \cdot 4.8 \text{ cm} \cdot 7.6 \text{ cm} \\ &\approx 549.8 \text{ cm}^3 \end{aligned}$$

78. a rectangular solid with length 9.5 m, width 3 m, and height 7 m

$$\begin{aligned} V &= l \cdot w \cdot h \\ &= 9.5 \text{ m} \cdot 3 \text{ m} \cdot 7 \text{ m} \\ &= 199.5 \text{ m}^3 \end{aligned}$$

79.

$$\begin{aligned} \text{leg} &= \sqrt{(\text{hypotenuse})^2 - (\text{leg})^2} \\ &= \sqrt{(10)^2 - (6)^2} \\ &= \sqrt{100 - 36} \\ &= \sqrt{64} \\ &= 8 \end{aligned}$$

The unknown length is 8 m.

80.

$$\begin{aligned} \text{hypotenuse} &= \sqrt{(\text{leg})^2 + (\text{leg})^2} \\ &= \sqrt{(30)^2 + (16)^2} \\ &= \sqrt{900 + 256} \\ &= \sqrt{1156} = 34 \end{aligned}$$

The hypotenuse is 34 cm.

81. $-12 + (-10) = -22$

82. $-5.7 - (-12.6) = -5.7 + (+12.6) = 6.9$

83. $9(-5) = -45$

84. $(-14.6)(-5.7) = 83.22$

85. $\dfrac{-42}{-7} = 6$

86. $\dfrac{-34.04}{14.8} = -2.3$

87.

$$\begin{aligned} 3x - 5 &= 16 \\ 3x - 5 + 5 &= 16 + 5 \\ 3x &= 21 \\ \frac{3x}{3} &= \frac{21}{3} \\ x &= 7 \end{aligned}$$

The solution is 7.

88.

$$\begin{aligned} -12 &= 3(x + 2) \\ -12 &= 3x + 6 \\ -12 + (-6) &= 3x + 6 + (-6) \\ -18 &= 3x \\ \frac{-18}{3} &= \frac{3x}{3} \\ -6 &= x \end{aligned}$$

The solution is -6.

89.

$$\begin{aligned} 15x - 11x &= 12 \\ (15 - 11)x &= 12 \\ 4x &= 12 \\ \frac{4x}{4} &= \frac{12}{4} \\ x &= 3 \end{aligned}$$

The solution is 3.

90.
$$\begin{aligned} 3.4x + 6 &= 1.4x - 8 \\ 3.4x + 6 - 6 &= 1.4x - 8 - 6 \\ 3.4x &= 1.4x - 14 \\ 3.4x - 1.4x &= 1.4x - 1.4x - 14 \\ 2x &= -14 \\ x &= -7 \end{aligned}$$

The solution is -7.

91. Mean

$$= \frac{16 + 37 + 27 + 31 + 19 + 25 + 15 + 38 + 43 + 19}{10}$$

$$= \frac{270}{10} = 27 \text{ cable hookups}$$

Arrange the numbers in numerical order from smallest to largest.

$$15, 16, \underline{19}, \underline{19}, 25, 27, 31, 37, 38, 43$$

The list has 10 numbers. The middle numbers are the 5th and 6th numbers, so the median is

$$\frac{25 + 27}{2} = 26 \text{ cable hookups.}$$

The number 19 occurs two times, which is more often than any other number. Therefore, 19 is the mode.

92. Mean

$$= \frac{10.3 + 4.3 + 1.65 + 2.85 + 5.3 + 5.7 + 2.3 + 4.35 + 2.85}{9}$$

$$= \frac{39.6}{9} = 4.4 \text{ acres}$$

Arrange the numbers in numerical order from smallest to largest.

$$1.65, 2.3, \underline{2.85}, \underline{2.85}, 4.3, 4.35, 5.3, 5.7, 10.3$$

The list has 9 numbers. The middle number is the 5th number, so the median is 4.3 acres.

The number 2.85 occurs two times, which is more often than any other number. Therefore, 2.85 is the mode.

93. Use the percent proportion.
part is 1267; whole is 3620; percent is unknown

$$\begin{aligned} \frac{1267}{3620} &= \frac{x}{100} \\ 3620 \cdot x &= 1267 \cdot 100 \\ 3620 \cdot x &= 126{,}700 \\ \frac{3620 \cdot x}{3620} &= \frac{126{,}700}{3620} \\ x &= 35 \end{aligned}$$

35% were baked.

94. Use the percent proportion.
part is 78.68; percent is 7; whole is unknown

$$\begin{aligned} \frac{78.68}{x} &= \frac{7}{100} \\ 7 \cdot x &= 78.68 \cdot 100 \\ \frac{7x}{7} &= \frac{7868}{7} \\ x &= 1124 \end{aligned}$$

The cost of the item was \$1124.

95. Divide the number of gallons of the additive by the number of gallons needed for each tank.

$$280\frac{1}{2} \div 2\frac{3}{4} = \frac{561}{2} \div \frac{11}{4} = \frac{\overset{51}{\cancel{561}}}{\underset{1}{\cancel{2}}} \cdot \frac{\overset{2}{\cancel{4}}}{\underset{1}{\cancel{11}}}$$

$$= \frac{51 \cdot 2}{1 \cdot 1} = 102$$

102 storage tanks can receive the additive.

96. Write a proportion.

$$\begin{aligned} \frac{34}{50} &= \frac{x}{1200} \\ 50 \cdot x &= 34 \cdot 1200 \\ \frac{50 \cdot x}{50} &= \frac{40{,}800}{50} \\ x &= 816 \end{aligned}$$

One would expect 816 rooms to be for nonsmokers.

97. 12 nasal strips for \$6.50 – \$2 (coupon) = \$4.50

$$\frac{\$4.50}{12} \approx \$0.38 \text{ per strip}$$

24 nasal strips for \$7.50

$$\frac{\$7.50}{24} \approx \$0.31 \text{ per strip}$$

30 nasal strips for \$8.95 – \$1 (coupon) = \$7.95

$$\frac{\$7.95}{30} \approx \$0.27 \text{ per strip}$$

38 nasal strips for \$9.95

$$\frac{\$9.95}{38} \approx \$0.26 \text{ per strip } (*)$$

The best buy is 38 nasal strips for \$9.95.

98. First find the total area.

$$\begin{aligned} A &= l \cdot w \\ &= 45 \text{ yd} \cdot 32 \text{ yd} \\ &= 1440 \text{ yd}^2 \end{aligned}$$

Now find the area of the atrium.

$$\begin{aligned} A &= \pi \cdot r^2 \\ &\approx 3.14 \cdot 5 \text{ yd} \cdot 5 \text{ yd} \\ &= 78.5 \text{ yd}^2 \end{aligned}$$

Subtract the area of the atrium from the total area.

$$1440 \text{ yd}^2 - 78.5 \text{ yd}^2 = 1361.5 \text{ yd}^2$$

Multiply the cost of the carpeting by the number of yards needed.

$$\frac{\$43.50}{\text{yd}^2} \cdot \frac{1361.5 \text{ yd}^2}{1} = \$59{,}225.25$$

The cost is $59,225.25.

99. Change milliliters to liters.

$$\frac{125 \text{ mL}}{1} \cdot \frac{1 \text{ L}}{1000 \text{ mL}} = \frac{125}{1000} \text{ L} = 0.125 \text{ L}$$

Multiply the amount of muriatic acid needed for each spa by the number of spas to be serviced.

$$\begin{array}{r} 0.125 \\ \times\ 140 \\ \hline 5\,000 \\ 12\,5 \\ \hline 17.500 \end{array}$$

To service 140 spas, 17.5 L of muriatic acid will be needed.

100. $I = p \cdot r \cdot t \quad (8.5\% = 0.085)$

$$\begin{aligned} &= \$9400(0.085)\left(\tfrac{9}{12}\right) \\ &= \$599.25 \end{aligned}$$

Total amount due is
$9400 + $599.25 = $9999.25.

APPENDIX B INDUCTIVE AND DEDUCTIVE REASONING

Margin Exercises

1. $2, 8, 14, 20, \ldots$

Find the difference between each pair of successive numbers.

$$8 - 2 = 6$$
$$14 - 8 = 6$$
$$20 - 14 = 6$$

Note that each number is 6 greater than the previous number. So the next number is $20 + 6 = 26$.

2. $6, 11, 7, 12, 8, 13, \ldots$

Find the difference between each pair of successive numbers.

$$11 - 6 = 5$$
$$7 - 11 = -4$$
$$12 - 7 = 5$$
$$8 - 12 = -4$$
$$13 - 8 = 5$$

Note that to obtain the next number, either 5 or -4 is added. To obtain the next number, -4 should be added. So the next number is $13 + (-4) = 9$.

3. $2, 6, 18, 54, \ldots$

To see the pattern, use division.

$$6 \div 2 = 3$$
$$18 \div 6 = 3$$
$$54 \div 18 = 3$$

To obtain the next number, multiply the previous number by 3. So the next number is $(54)(3) = 162$.

4. The next figure is obtained by rotating the previous figure clockwise the same amount of rotation ($\frac{1}{4}$ turn) as the second figure was from the first.

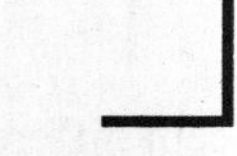

5. All cars have four wheels.
All Fords are cars.

$\therefore$ All Fords have four wheels.

The statement "All cars have four wheels" is shown by a large circle that represents all items that have 4 wheels with a small circle inside that represents cars.

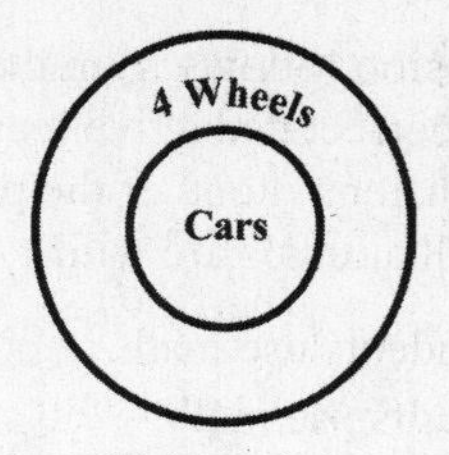

The statement "All Fords are cars" is represented by adding a third circle representing Fords inside the circle representing cars.

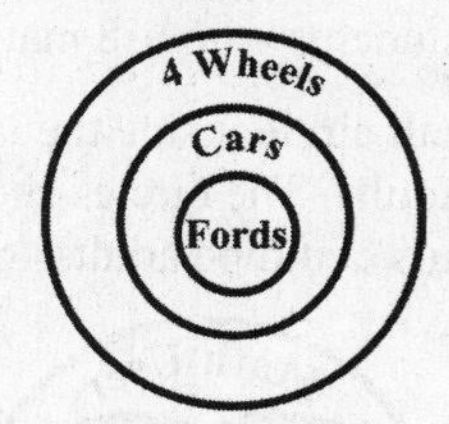

Since the circle representing Fords is completely inside the circle representing items with 4 wheels, it follows that:

All Fords have four wheels.

The conclusion follows from the premises.

6. (a) All animals are wild.
All cats are animals.

$\therefore$ All cats are wild.

"All animals are wild" is represented by a large circle representing wild creatures with a smaller circle inside representing animals.

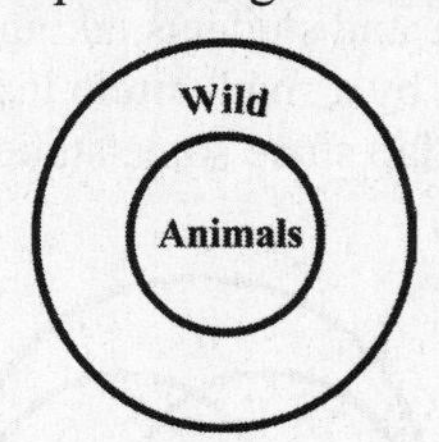

"All cats are animals" is represented by a small circle representing cats, inside the circle representing animals.

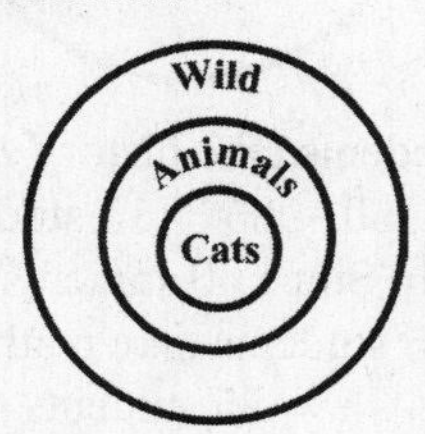

Since the circle representing cats is completely inside the circle representing wild creatures, it follows that:

All cats are wild.

continued

The conclusion follows from the premises. (Note that correct deductive reasoning may lead to false conclusions if one of the premises is false, in this case, all animals are wild.)

(b) All students use math.
All adults use math.
$\therefore$ All adults are students.

A larger circle is used to represent people who use math. A small circle inside the larger circle represents students who use math.

Another small circle inside the larger circle represents adults. The circles should overlap since some students could be adults.

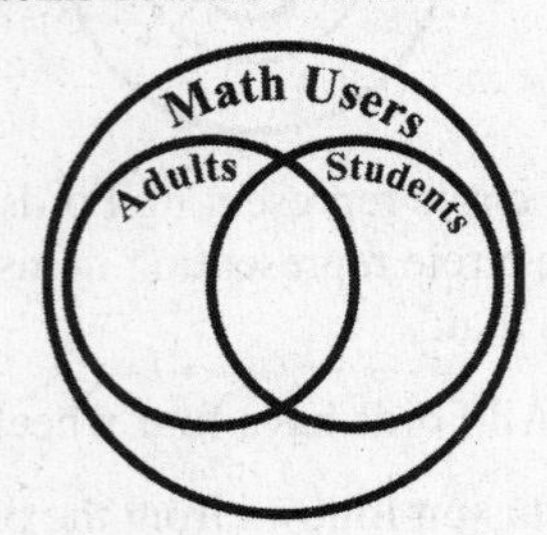

From the diagram, we can see that some adults are not students. Thus, the conclusion does *not* follow from the premises.

7. All 100 students in the class are represented by a large circle.
Students taking history are represented by a small circle inside and students taking math are also represented by a small circle inside. The small circles overlap since some students take both math and history.

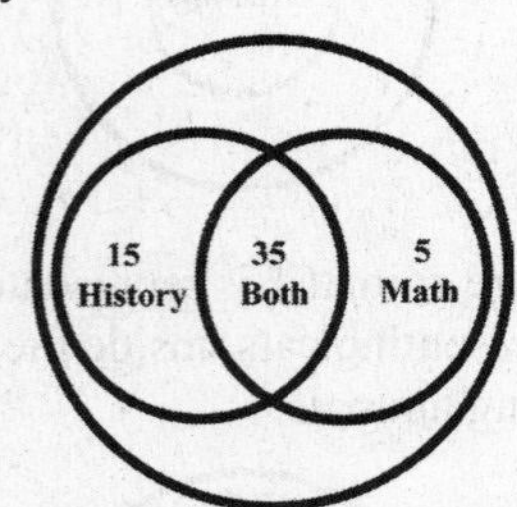

Since 35 students take history and math and 50 take history, $50 - 35 = 15$ students take history, but not math. Since 40 students take math, $40 - 35 = 5$ students take math, but not history. So $15 + 35 + 5 = 55$ students take history, math or both subjects. Therefore, $100 - 55 = 45$ students take neither math nor history.

8. A Chevy, BMW, Cadillac, and Ford are parked side by side.

1. The Ford is on the right side (fact a), so write "Ford" at the right of a line.

Ford

2. The Chevy is between the Ford and the Cadillac (fact c). So write "Chevy" between Ford and Cadillac.

Cadillac Chevy Ford

3. The BMW is next to the Cadillac (fact b), so the BMW must be on the other side of the Cadillac.

BMW Cadillac Chevy Ford

Therefore, the BMW is parked on the left end.

Appendix B Exercises

1. $2, 9, 16, 23, 30, \ldots$

Inspect the sequence and note that 7 is added to a term to obtain the next term. So the term immediately following 30 is $30 + 7 = 37$.

3. $0, 10, 8, 18, 16, \ldots$

Inspect the sequence and note that 10 is added, then 2 is subtracted, then 10 is added, then 2 is subtracted, and so on. So the term immediately following 16 is $16 + 10 = 26$.

5. $1, 2, 4, 8, \ldots$

Inspect the sequence and note that 2 is multiplied times a term to obtain the next term. So, the term immediately following 8 is $(8)(2) = 16$.

7. $1, 3, 9, 27, 81, \ldots$

Inspect the sequence and note that 3 is multiplied times a term to obtain the next term. So, the term immediately following 81 is $(81)(3) = 243$.

9. $1, 4, 9, 16, 25, \ldots$

Inspect the sequence and note that the pattern is add 3, add 5, add 7, etc., or 1^1, 2^2, 3^2, etc. So, the term immediately following 25 is $6^2 = 36$.

11. The first three shapes are unique. The fourth shape is the same as the first shape except that it is reversed, and reversing the position of the second shape gives the fifth shape. So, the next shape will be the reverse of the third shape.

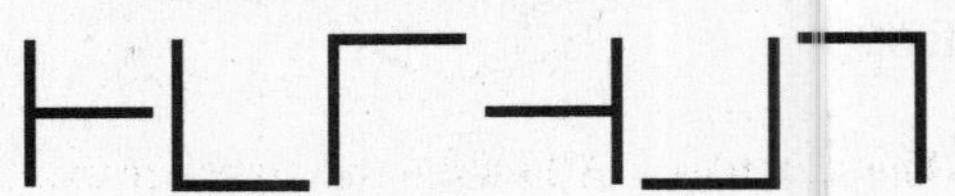

13. The first three shapes are the same except different images. Since the third shape is the reverse of the first shape, the fourth shape should be a reverse of the second shape.

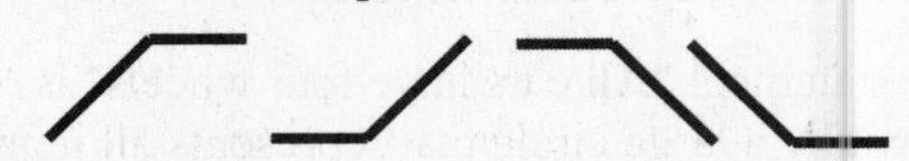

15. All animals are wild.
All lions are animals.
∴ All lions are wild.

The statement "All animals are wild" is shown by a large circle that represents all creatures that are wild with a smaller circle inside that represents animals.

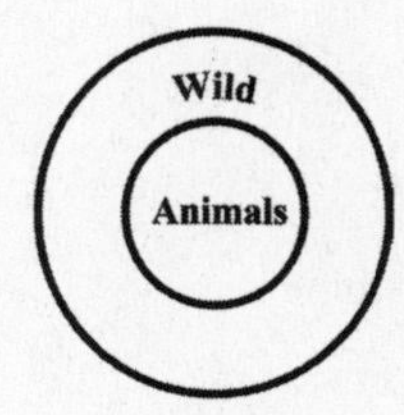

The statement "All lions are animals" is represented by adding a third circle representing lions inside the circle representing animals.

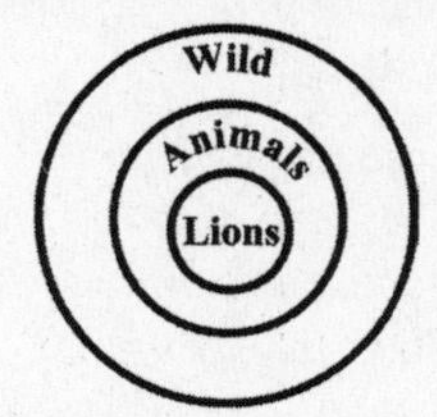

Since the circle representing lions is completely inside the circle representing creatures that are wild, it follows that:

All lions are wild.

The conclusion follows from the premises.

17. All teachers are serious.
All mathematicians are serious.
∴ All mathematicians are teachers.

The statement "All teachers are serious" is represented by a large circle representing serious and a smaller circle inside the large circle representing teachers.

The statement "All mathematicians are serious" is represented by another circle that represents mathematicians inside the larger circle that represents serious. (The circle for teachers would overlap the one for mathematicians.)

Since the circle representing mathematicians is not completely inside the circle representing teachers, the conclusion does *not* follow from the premises.

19. Two intersecting circles represent the days of television watching by the husband and wife. Since they watched 18 days together, place an 18 in the area shared by the two smaller circles. Since the wife watched a total of 25 days, place a $25 - 18 = 7$ in the other region of the circle labeled Wife.

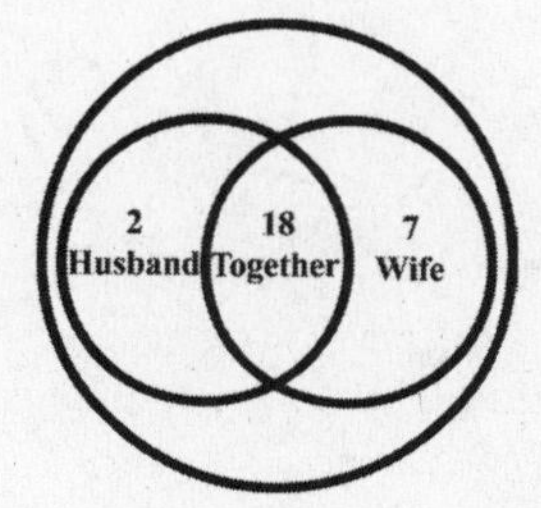

Since the husband watched a total of 20 days, place a $20 - 18 = 2$ in the other region labeled Husband. The 2, 18, and 7 give a total of 27 days. This results in $30 - 27 = 3$ days for the region outside the intersecting circles. This represents 3 days of neither one watching television.

21. Tom, Dick, Mary, and Joan

One is a secretary, one is a computer operator, one is a receptionist, and one is a mail clerk. Find the one who is a computer operator.

1. Tom and Joan eat dinner with the computer operator (fact a), so neither Tom nor Joan is the computer operator.
2. Mary works on the same floor as the computer operator and the mail clerk (fact c), so Mary is not the computer operator.

We know that if Tom, Joan, and Mary are not the computer operator, then Dick must be the computer operator. Fact b is not needed.